Animal Intelligence

From Individual to Social Cognition

Zhanna Reznikova

Novosibirsk State University; Institute for Animal Systematics and Ecology, Siberian Branch RAS

CAMBRIDGE UNIVERSITY PRESS
Cambridge, New York, Melbourne, Madrid, Cape Town, Singapore, São Paulo

Cambridge University Press
The Edinburgh Building, Cambridge CB2 8RU, UK

Published in the United States of America by Cambridge University Press, New York

www.cambridge.org
Information on this title: www.cambridge.org/9780521825047

First published 2007

Printed in the United Kingdom at the University Press, Cambridge

A catalogue record for this publication is available from the British Library

ISBN 978-0-521-82504-7 hardback
ISBN 978-0-521-53202-0 paperback

Animal Intelligence

From ants to whales, the lives of animals are filled with challenges that demand minute-by-minute decisions: to fight or flee, dominate or obey, take off, share, eat, spit out or court. Learning develops adaptive tuning to a changeable environment, while intelligence helps animals use their learned experiences in new situations. Using examples from field to laboratory, *Animal Intelligence* pools resources from ethology, behavioural ecology and comparative psychology to help the reader enter the world of wild intelligence through the analysis of adventures of ideas and methods, rather than through theoretical modelling. It reminds us in the gentlest way that there is a world of intellectual biodiversity out there, giving a multi-faceted panorama of animal intelligence using the ant as its touchstone. Written in an accessible and charming style, and with undergraduates in mind, this book should be read by anyone with an interest in the world of animal behaviour.

ZHANNA REZNIKOVA is Head of the Department of Comparative Psychology at the Novosibirsk State University, and Head of the Laboratory of Community Ethology at the Institute for Animal Systematics and Ecology, Siberia. She is a researcher and professor in the fields of ethology, behavioural ecology, and the behaviour and ecology of social insects.

Contents

Foreword: An ant's eye view of animal intelligence

Synthetic and critical reviews of intelligence in non-human animals have typically focused on large-brained mammals, especially primates (apes, monkeys, prosimians) and cetaceans (dolphins and whales) (e.g. Tomasello and Call, 1997; de Waal and Tyack, 2003). At first glance, this may seem entirely reasonable, given the phylogenetic closeness of the other primate species to *Homo sapiens*, and the extraordinary and exotic abilities of our aquatic counterparts. But is this cosy arrangement based on comfortable links to Curious George and Flipper really sufficient?

Now comes an alternative viewpoint that is far more inclusive in drawing upon all the vertebrate classes; moreover it takes as its touchstone the world seen through the eyes of ants. Small and underfoot these arthropods may be, but again and again, the complexity of their behaviour fascinates and astonishes us. (Of course, this is not the first time that a student of the Formicidae has taken such a starting point and developed a masterly treatment on a grand scale. Wilson's (1976) *Sociobiology* comes to mind as another fruitful example . . .) What is new here is Reznikova's demonstration that this complexity is not just the invariant and inevitable manifestation of hard-wired automatons, but is instead a rich array of flexibility and local adaptability, e.g. small foraging squads led by individualised leaders.

However, by no means is this compendium a paean to ants, but rather it repeatedly reminds us (in the gentlest ways) that there is a wonderful world of intellectual biodiversity out there. In many cases, species long known from classic natural history or even folklore have been re-studied with modern methods to exacting standards. Take the common beaver, an engineering rodent of impressive plasticity, whose aversion to the sound of running water can lead ultimately to the transformation of a landscape. There is no need to rely on large-brained mammals such as bulldozing African elephants to find examples of niche construction in action (Odling-Smee *et al.*, 2003). Similarly, the resurgence of research on the Corvidae (crows, ravens, jays, jackdaws) is another emerging case of intellectual convergences between bird and mammals, in a range of cognitive domains from numeracy to theory of mind (Emery and Clayton, 2004).

Reznikova brings far more to the table than an encyclopaedic (with more than 1500 references in the bibliography) knowledge of animal cognition. Although a self-professed experimentalist, she is methodologically catholic and refreshingly all-embracing and even-handed. There is no dogma here. The book encompasses studies in settings from field to laboratory, and various points in between, whether these be zoological garden, farmyard, or companion animals at home. Disciplinarily, she roams over the behavioural literature from ethology to comparative/developmental/cognitive psychology, as underpinned by neurophysiology. Aspects of learning from spinal reflex to cultural transmission are included.

A further bonus is that this is the first such book published in English to take a non-North American/western European perspective. As a Russian scientist working in Siberia, Reznikova has a somewhat detached position from which to view the trans-Atlantic squabbles (quibbles?) that sometimes generate more heat than light. Further, she alerts us to neglected (in the West) work published in Russian, such as the pioneering studies of chimpanzees done by Firsov. As such, she adds to the rediscovery in the West of such early work as the classic study of ape home-rearing done by Ladygina Koths (2002). Of course, there are bound to be gaps, e.g. the work of Matsuzawa's (2001) group at Kyoto is under-represented, and the problem-solving studies of Döhl (1968) and Lethmate (1977) are omitted altogether. Yet no other volume available today gives such comprehensive coverage.

The author's writing style is gratifyingly accessible, even charming, and is sometimes enlivened with understated humour. She grounds each phenomenon in its historical context, but always seems to know when enough background is enough, just as she does with the selective insertion of telling stories as illustrations, e.g. the linguistic tantrums thrown by Alex the grey parrot, when he has had enough testing for the day. What this amounts to is a wide-ranging, but artfully spare crash course on animal intelligence for the non-specialist. This is a book for the non-psychologist who wants a down-to-earth, readable treatment of the subject, mercifully free of jargon. This is accomplished by clear writing and apt metaphors, e.g. the ants' counting system being likened to that of Roman numerals.

Any doubts about whether or not the author is on the track are dispelled by more recent findings that have appeared since she finished the manuscript and before I sat down to compose this foreword. The first convincing experimental treatments of teaching in animals have since come not from apes or dolphins, but from ants (Franks and Richardson, 2006) and meerkats (Thornton and McAuliffe, 2006). A recent review in *Science* (Pennisi, 2006) saying that 'a new generation of experiments reveals that group-living animals have a surprising degree of intelligence' (p. 1734) gives equal treatment to carnivores, corvids and insects. By the time you read this, doubtless much more will have been added. If only Darwin, Romanes, Pavlov, Thorndike, etc. were alive today to see what they have wrought!

W. C. McGrew
Leverhulme Centre for Human Evolutionary Studies,
University of Cambridge

References

Döhl, J. (1968) Über die Fähigkeit einer Schimpansin, Umwege mit selständigen Zwischenzielen zu überblicken. *Zeitschift für Tierpsychologie*, **25**, 89–103.

Emery, N. J. and Clayton, N. S. (2004) The mentality of crows: convergent evolution of intelligence in corvids and apes. *Science*, **306**, 1903–1907.

Franks, N. R. and Richardson, T. (2006) Teaching in tandem-running ants. *Nature*, **439**, 153.

Ladygina Koths, N. N. (2002) *Infant Chimpanzee and Human Child*. Oxford, UK: Oxford University Press.

Lethmate, J. (1977) *Problemloseverhalten von Orang-Utans* (Pongo pygmaeus). Berlin: Paul Parey.

Matsuzawa, T. (ed.) (2001) *Primate Origins of Human Cognition and Behavior*. Tokyo: Springer Verlag.

Odling-Smee, F. J., Laland, K. N. and Feldman, M. W. (2003) *Niche Construction: The Neglected Process in Evolution*. Princeton, NJ: Princeton University Press.

Pennisi, E. (2006) Social animals prove their smarts. *Science*, **312**, 1734–1738.

Thornton, A. and McAuliffe, K. (2006) Teaching in wild meerkats. *Science*, **313**, 227–229.

Tomasello, M. and Call, J. (1997) *Primate Cognition*. New York: Oxford University Press.

Waal, F. B. M. de and Tyack, P. L. (eds.) (2003) *Animal Social Complexity: Intelligence, Culture, and Individualized Societies*, Cambridge, MA: Harvard University Press.

Wilson, E. O. (1976) *Sociobiology: The New Synthesis*. Cambridge, MA: The Belknap Press of Harvard University Press.

Preface

The lives of animals, from ants to whales, are full of challenges that demand minute-to-minute decisions from them: to fight or flee, to court or ignore, to dominate or obey, to take off or share with, to eat or spit out. Learning serves for adaptive tuning of species-specific stereotypes to the changeable environment, and intelligence helps animals to use the experience gained in new situations. Studying animal intelligence seems to be a daunting task due to numerous different species to investigate which possess hardly comparable intellectual skills. Although animals and researchers have achieved remarkable results in displaying cognitive skills from one end and measuring them from the other end, and many excellent books and papers on animal intelligence have been published, we are still far from a complete understanding of the specificity of thinking both in human and non-human beings.

In this book I am trying to present a unified approach to studying a quantity of displays of animal intelligence and thus help the readers to enter the world of wild intelligence through the analysis of adventures of ideas and methods, rather than through theoretical modelling. I think that such an approach reflects the real situation in the field, because now, more than 125 years since the very first scientific books about animal intelligence were published, it has became apparent that the rise of the scientific study of animal intelligence may be portrayed in terms of progressive changes in experimental methods. Measuring the most pronounced manifestations of intelligence more and more elegantly, experimenters have demonstrated animals' abilities to catch regularities, to communicate in complex and flexible ways and to navigate a social landscape.

In this book many possible forms of learning and cognition, from simple discriminations to categorisation, social learning and theory of mind, are analysed from the experimenter's standpoint. Pooling resources of ethology, behavioural ecology and comparative psychology, I develop here the idea of 'species geniuses', whether they are pig-headed or bird-brained: members of some species display very fast and advanced learning within specific domains, closely connected with their ecological traits and evolutionary history. Based on the ideas of sociobiology, this book raises an intriguing problem of whether there is room for intelligence within the framework of social specialisation in animal communities. It is also worth noting that, being a student of social insects and advocating their intelligence, I am trying to bridge the gap between studies of intelligence in vertebrates and invertebrates. It seems a common rule that investigators of vertebrate animals refer to results obtained by the explorers of invertebrates much less than vice versa. Besides, people do not expect intellectual feats from creatures with more than four legs. Here I hope to attract attention to the cognitive ability of insects and other invertebrates.

In general, if this book increases the interest of the new generation of ethologists, ecologists, evolutionists, zoologists, physiologists and psychologists in interdisciplinary problems in the field of animal intelligence, and establishes links between laboratory experimentalists and field researchers, it will have done its job.

Acknowledgements

I thank many friends and colleagues for their help and encouragement.

While writing this book, I discussed general aspects of animal intelligence as well as my own contributions in this field, with the following colleagues: Patrick Bateson, Donald Broom, Nicola Clayton, Renee Feneron, Raghavendra Gadagkar, Bennett Galef, Linda Gottfredson, Jürgen Heinze, Bert Hölldobler, Ludwig Huber, Tanya Humle, Kevin Laland, William McGrew, Linda Marchant, Ulrich Maschwitz, Randolf Menzel, Irene Pepperberg, Carel van Schaik, Peter Slater, Jürgen Tautz and Frans B. M. de Waal.

I am very grateful to William McGrew for writing the foreword.

I am truly grateful to the colleagues who kindly provided me with photographs of their experiments which helped me to make my reasoning about animal intelligence more vivid: Shigeyuki Aoki, Raimund Apfelbach, Michael Beran, Andrew Bray, Anton Chernenko, Marietta Dindo, Elena Dorosheva, Ann Göth, Linda Hollén, Ludwig Huber, Timothy Judd, William McGrew, Tatyana Novgorodova, Sofia Panteleeva, Duane Rumbaugh, Emily Sue Savage-Rumbaugh, Adam Seward, Paul Sherman, Harunobu Shibao, Anna Smirnova, Caroline Tutin, Marina Vančatová, Frans de Waal, Alex Weir, Ivan Yakovlev and Zoya Zorina.

A special word of thanks is due to Daniil Ryabko who provided detailed comments on many drafts of the manuscript and contributed enormously to improve logical structure and technical aspects of the book.

It was supportive for writing the manuscript that Donald Michie visited me and my co-author Boris Ryabko in Novosibirsk after reading our paper about ant language in *Complexity* (1996), and then published a paper about his impressions in the *Independent on Sunday*.

I also wish to thank the people at Cambridge University Press who helped me to prepare the manuscript and to bring it to publication; among them are Martin Griffiths, Shana Coats, Tracey Sanderson, Clare Georgy, Anna Hodson and Dawn Preston.

While writing the book, I have been supported by the Russian Fund for Basic Research (02-04-48386 and 05-04-48104).

Part I

Development of ideas and methods in studying animal intelligence

'Good day, Maestro Antonio,' said Geppetto, 'What are you doing on the floor?'
'I am teaching the ants their ABCs.'

Carlo Collodi, *Adventures of Pinocchio*

Animal intelligence has been experimentally studied for not much longer than a century and controversial ideas about how animals learn and to what limits they understand still exist. This part is about dramatic adventures of ideas. This is not just a historical background which is usually considered by readers as a 'second preface' and the boring duty of the author to give due to predecessors and key players, before descending to particulars of modern discussible issues. This part also has no pretensions of being historically complete as a review. The reader can find historical analyses of the development of ideas and researches in animal behaviour, comparative psychology and comparative cognition in many books and journal reviews such as: Evans (1976), Griffin (1978, 1984), Macphail (1993), Dewsbury (1985), Kline (1989), Ristau (1991), Kimble *et al.* (1991, 1996, 1998), Balda *et al.* (1998), Slater (1999), Kimble and Wertheimer (2000), Shettleworth (2001) and Bekoff *et al.* (2002), as well as in other reviews which are cited in the relevant places. This part represents a biased author's sample of ideas and methods in studying animal intelligence. Even from this brief review we will see that the rise of the scientific study of animal intelligence may be portrayed as progressive changes in experimental methods.

Chapter 1

Evolution of views on animal intelligence

A tendency to study human and animal intelligence in parallel has been in evidence from the very beginning of regular observations and investigations. It was not an easy task to locate the seat of intelligence. The famous Greek physician Hippocrates (b. 460 BC) was the first thinker to prefer to avoid mystical interpretations of mental disease and to stick close to the empirical evidence. For example, in a treatise entitled 'On the sacred disease' (meaning epilepsy), he dismissed the usual demonic-possession theory and suggested that it was a hereditary disease of the brain. Hippocrites and also Plato recognised the significance of the brain. Later, around 280 BC, Erasistratus of Chios dissected the brain and differentiated the various parts.

Galen, a physician to Roman emperors and an early author of works on anatomy and physiology, who acquired great experience from practising for some time as physician to the gladiators, gave public lectures and anatomical demonstrations in Rome. In 177 AD he gave his famous lecture 'On the brain'. Galen described the frontal lobes as being the seat of thought (and the soul). He expertly dissected and accurately observed many kinds of animals (mostly dogs, monkeys and pigs, because dissection of humans was forbidden) and applied, sometimes mistakenly, what he saw to the human body. In addition to giving a great deal of fairly decent, concrete advice, Galen theorised that all life is based on pneuma or spirit, and he believed that the brain generates and transmits special vital spirit through the hollow nerves to the muscles, allowing movement and sensation (for details see Young, 1970).

After the time of Galen, there was only very slow progress in the development of the concept of intelligence until René Descartes, the French mathematician, philosopher, and physiologist, proposed his model of the nervous system and thus established the first systematic account of the mind–body relationship. The first of Descartes' works, *De homine*, was completed in Holland about 1633, on the eve of the condemnation of Galileo. When Descartes learned of Galileo's fate at the hands of the Inquisition, he immediately suppressed his own treatise. As a result, the world's first extended essay on physiological psychology was published only a long time after its author's death.

Perhaps one of the earliest to write about animal behaviour in a modern scientific fashion was the British zoologist John Ray. In 1676 he published a scientific text on the study of 'instinctive behaviour' in birds. He was astonished by the fact that birds, removed from their nests when young, would still build species-typical nests when adult. He did not try to explain this, but noted the fact that very complex behaviour could develop without learning or practice. Almost 100 years later, the French naturalist Charles Georges Leroy published a book on intelligence and adaptations in animals. Leroy criticised those philosophers who spent their time indoors, thinking about the world, rather than observing animals in their natural environments. Only by doing this, he argued, would it be possible to fully appreciate the adaptive capacity and flexibility in the behaviour of animals.

The development of objective methods of analysis of animal behaviour is attributed to psychologists studying animal mind in the nineteenth century, based on ideas of evolutionism and Charles Darwin's representation of the common processes that govern natural selection in humans and other species and the psychological continuity between humans and other animals.

The theory of evolution went back to the ideas of Buffon's and Lamarck's theory of evolution by means of the inheritance of acquired characteristics, and to the poem *Zoonomia* of Erasmus Darwin (1794–96) who elaborated similar ideas. Carl Linnaeus devised a system for classifying the diversity of known plants and animals, and Georges Buffon pursued a massive compilation of facts about animals, resulting in an encyclopaedia of 36 volumes. This stimulated energetic pursuit of discovery, an activity that continues up to the present time.

Evolutionism in the nineteenth century was primarily concerned with the interpretation of the geological, palaeontological and biological evidence. In the mid 1840s two crucial works – Charles Lyell's *Principles of Geology* and Robert Chambers's *Vestiges of the Natural History of Creation* – had become the Bible of many explorers and had instilled into naturalists an appreciation of how long-term change could be effected through the operation of slow, ongoing processes.

The evolutionism of Herbert Spencer was significant for the development of concepts of mind and brain. Spencer's first serious intellectual endeavours were devoted to the study of phrenology, and it was from phrenology that he drew the conception of society as an organism in which interdependent, specialised structures serve diverse functions (see Young, 1970). To Spencer is certainly due the immense credit of having been the first to see in evolution an absolutely universal principle. Spencer's theory of evolution actually preceded Charles Darwin's, when he wrote *The Development Hypothesis* in 1852, which is an advocacy against creationism and in favour of a progressive (evolutionary) view. While in *The Development Hypothesis*, Spencer simply turned the creationists' arguments against themselves, he also published 'A theory of population, deduced from the general law of animal fertility' in 1852. In this essay, an early formulation of the mechanism of natural selection can be found. Unlike Darwin, Spencer was never much of an observer, and his independent formulation of a theory of evolution developed from his speculations in social theory and psychology. His theory was not taken into serious consideration, largely because of a lack of an effective theoretical system for natural selection. Nevertheless, it was Spencer and not Darwin who first popularised the term 'evolution', and few people outside the field realise that the oft-used phrase 'survival of the fittest' was actually coined by Spencer. Although Spencer was wrong about the mechanism of evolution, modern views support his main theme: the adaptations of living things to their surroundings are evoked by problems posed by their environments.

Darwin published his theory of natural selection in *On the Origin of Species by Means of Natural Selection, or the Preservation of Favoured Races in the Struggle for Life*, often abbreviated to *The Origin of Species*, in 1859. This book revolutionised biology, and is one of the most revolutionary works ever published. Darwin had not read Spencer's *Principles of Psychology* (1855) when he wrote this but added complimentary references into later editions. Darwin's ideas were based on an enormous amount of natural observations. He had studied Lamarck's theory of evolution when he was a medical student, but neither this nor his grandfather's (Erasmus Darwin) theory had shaken his belief in the fixity of species by the time he became a professional naturalist in 1830 when he secured a position as ship's naturalist aboard HMS *Beagle*. Upon returning to England, he published his observations in *Zoology of the Voyage of the* Beagle (1840). Darwin, reflecting on his observations, came up with a theory that had a non-Lamarckian basis for the variation that leads to adaptation.

Darwin assumed that organisms naturally vary in almost every attribute that they display, such as morphology, physiology and behaviour. Such variation is heritable; on average offspring tend to resemble their parents more than other individuals in the population. Organisms have a huge capacity for increase in numbers; they produce far more offspring than give rise to breeding individuals, but the number of individuals in a population tends to remain relatively constant over time. Thus there must be competition for scarce resources such as food, mates and places to live. Darwin called this a 'struggle for existence'. As a result of this competition, some variants will leave more offspring than others. These offspring will inherit the characteristics of their parents and so evolutionary change will take place by natural selection. As a consequence of natural selection organisms will slowly acquire changes in traits, behavioural traits included, and become adapted to their environment. The blind force of natural selection drives the evolution of such changes. The key to Darwin's argument is that there are heritable variations among individuals of a single species, and that such differences lead to heritable changes from generation to generation, which ultimately lead to the origin of an entirely new species from existing species.

Upon returning to England, Darwin began to develop his ideas on evolution in several 'sketchbooks' and developed the kernel of his theory, that of natural selection, by the year 1838. He held on to these ideas for nearly 20 years before publishing them, and only moved to publish because of the fear of being scooped. Alfred Russel Wallace, English naturalist, evolutionist, geographer, anthropologist and social critic, had sent Darwin a manuscript to read and was asking his advice on the content of the manuscript before he presented the ideas on natural selection to the scientific community; these ideas were very similar to Darwin's own. By that time Wallace had travelled in the Amazon Basin and Malay Archipelago for about ten years and his collecting efforts had produced the astonishing total of 125 660 specimens, including more than a thousand species new to science. After publishing three books, Wallace in 1855 wrote and published the essay 'On the law which has regulated the introduction of new species' introducing the concept of evolution through natural selection basing on his observations on huge species diversity. In 1858, while on the island of Ternate, Wallace drafted his ideas on 'the survival of the fittest'. He posted the letter by mail-boat from the island with the request that if Darwin thought the ideas worthy he should send the letter on to Charles Lyell. Through actions of his friends, the geologist Lyell and the botanist Joseph Dalton Hooker it was made possible for Darwin's long efforts to be acknowledged jointly with those of Wallace. The paper that is made available here is the result of that forced union. Darwin published in 1858, communicating with Wallace on a joint paper to the Royal Society's meeting in which they described the role of natural selection in evolution (Darwin and Wallace, 1858). Darwin (1859) then published his famous book *On the Origin of Species*.

Darwin's *On the Origin of Species* contained a chapter devoted to an attempt to explain how instincts could evolve by natural selection. He considered this issue one of the most formidable objections to his theory. In another book, *The Descent of Man and Selection in Relation to Sex* (1871) Darwin introduced the idea that the difference in mind between humans and 'higher' animals, great as it is, certainly is one of a degree and not of a kind. His last book, published in 1872, *The Expression of the Emotions in Man and Animals*, was probably the first work on comparative ethology.

From the early 1870s experimental ethology (called psychology that time) had started to develop, and it was Douglas Spalding who initiated this, being a long way ahead of his time in experimental approaches. He was the first to study the phenomenon that we now call *imprinting*; he began the study of anti-predator reactions; he experimented with both visual and auditory releasers and set out the logic of comparing the behaviour of species born in different developmental stages. Perhaps Spalding would have been thought of as the founder of ethology if only he had lived longer.

Spalding published series of papers (one of them in *Nature*, 1872) in which the antecedent conditions of behaviour in young animals were systematically manipulated and variations in response observed. For example, he hatched chicks from eggs by using the heat from a steaming kettle, in order to examine the development of the visual and acoustic senses without the influence of a mother hen. In sum, he described a number of experiments on young chicks and ducklings, carefully observed for the first few hours after birth. His conclusions were that these young birds not only showed 'intuitive' powers of walking, scratching and pecking, but also possessed intuitive knowledge of various kinds. He asserted that they were afraid of bees and of the cry of a hawk, and that they intuitively knew the meaning of a hen's call note and danger signal when heard for the first time. Spalding also kept young house martins in narrow cages thus depriving them of the possibility of flight and then, after maturation, compared them with control birds. It turned out that birds did not need practice to develop their vital behavioural patterns.

In his numerous experiments, Spalding tried to demonstrate the unwinding of a strategy of life as the development of an organism. 'When, as by a miracle, the lovely butterfly bursts from the chrysalis full-winged and perfect', he wrote in *Macmillan's Magazine* in 1873 – 'it has, for the most part, nothing to learn, because its little life flows from its organisation like melody from a music box.'

Although systematic and experimental, Spalding's investigations were based merely on naturalistic observations. One of the first to introduce apparatus and quantification into the study of animal behaviour was John Lubbock. He was introduced to natural history particularly through the efforts of Charles Darwin, a friend of the family who came to visit the family's home in County Down when Lubbock was only seven years old.

Lubbock published a dozen of books, several of which report his observations of animal behaviour. Among these, the book *Ants, Bees, and Wasps: A Record of Observations on the Habits of the Social Hymenoptera*, first published in 1882 and reprinted many times, is possibly the most remarkable. I would like to note that reading this book, in its Russian translation, was the event that influenced the author of this book, when a schoolgirl. Indeed, this book moved me from a decision to become a musician to take up entomology and ethology. I used to construct my first ant maze completely in Lubbock's manner, with the use of small parts from my parent's vacuum cleaner (it stopped working after that).

Lubbock's book included a chapter entitled 'General intelligence, and power of finding their way'. In this chapter Lubbock took a number of critical steps away from natural history towards the modern animal laboratory. His first innovation was to provide precise, detailed, quantitative descriptions of the conditions on an observation, not much different from those one would find in the methods section of a modern journal article. One of his innovations was to include actual data in the body of his text and to use these data to compute simple summary statistics.

For example, in observing ants learning to take an experimentally contrived route between food and the nest, Lubbock employed experimental techniques of importance for future research. Following his ants with a pencil as they pursued their way, Lubbock made and, in his text reproduced, detailed tracings of the ants' paths. This is certainly one of the first attempts to make an analogue record of behaviour for later coding. To observe the progress made by his ants in learning to follow a new path from food to nest, Lubbock designed a number of simple pieces of apparatus that constrained the ants' movements. These pieces of apparatus were, in effect, the first animal mazes.

Apart from being a powerful stimulus to the development of experimental investigations, Darwin's ideas of succession in animal and human thinking gave new arguments for anthropomorphic and anecdotal approaches to animal behaviour. Basically, anthropomorphism, that is, attributing to animals the same qualities as humans, goes back to the Roman author Pliny the Elder, who considered, for example, the elephant as the nearest to humans in intelligence (and surpassing humans in chastity and prudence), to present-day examples (Cartwright, 2002). At the turn of the nineteenth century the conflict between the anthropomorphic and the objective study of psychology was reaching a boiling point.

For example, Romanes regarded animals as possessing the intelligence to solve a problem by reasoning. In his book *Animal Intelligence* (1881) he suggested the following criterion of mind in animals:

> Does the organism learn to make new adjustments, or to modify old ones, in accordance with the results of its own individual experience? If it does so, the fact cannot be due merely to reflex action in the sense above described, for it is impossible that heredity can have provided in advance for innovations upon, or alterations of, its machinery during the lifetime of a particular individual.

This was quite a reasonable criterion but the claim was based on anecdotal evidences rather than on careful experimentation. After the publication in 1883 of Romanes' book *Mental Evolution in Animals*, the use of the anecdotal method was seriously discredited and retrieval of manifestations of intelligence in animals met strong criticism.

One of the early lessons about the importance of controlled procedures stems from work that was done with a horse nicknamed 'Clever Hans'. Hans's owner von Osten devoted 14 years to the detailed education of this horse following to syllabus of a grammar school. He claimed that the horse could count and was capable of mental arithmetic operations such as subtraction, multiplication and division. Thorough examinations demonstrated that the trainer unwittingly gave the horse barely perceptible cues for the right answer (Pfungst, 1908). When Hans reached the correct number of hoof strikes the trainer's expression changed slightly. The horse possessed such a great level of keenness of observation that he could catch sight of the pulsation of veins on his owner's head. Hans responded to these cues and stopped striking his hoof. If deprived of the possibility of observing his trainer, Hans did not give the correct answer. Although these results were based on an honest delusion, it is worthwhile to note that von Osten first demonstrated the phenomenon which will later be called 'autoshaping' (see Chapter 6), and that he may be first to have suggested a variant of intermediate language (see Chapter 30) with which to develop a direct dialogue with animals.

I think it is appropriate to mention here that, in my observation, frequency of references to an anecdote about Clever Hans in scientific and educational texts correlates with the development of our knowledge about animal intelligence. Of course, fashion changes in science and another set of examples comes to take place of former ones. It would be interesting to note that there are no mentions of Clever Hans in fundamental textbooks on animal behaviour published in 1970s and 1980s (for example, Hinde, 1970; Dewsbury, 1978; Manning, 1979; McFarland, 1985) but the textbooks published during the last decade contain detailed descriptions of this story (Allen and Bekoff, 1997; Shettleworth, 1998; Pearce, 2000; Cook, 2001; Griffin, 2001). I have also noted that frequency of mentioning of the Clever Hans phenomenon in lectures and even in posters at International Ethological Conferences has increased from 1997 to the present.

To my mind, this is caused not only by fashion based on imitation as well as the authors' wishes to underline their data verification, although these factors are at work, but also by attainment of a new level of knowledge concerning animal intelligence. As we will see further in this book, recent experimental studies show animals as doing more and cleverer things: they use artificial intermediary languages for direct dialogue with experimenters, demonstrate their abilities for rule extracting and representation, counting, using mirror images, creating cognitive maps, shaping cultural traditions and manipulating other creatures with Machiavellian tricks. There is a natural aspiration of modern investigators of animal intelligence to underline the fact that although experiments and observations have resulted in the discovery of amazingly clever behaviour in animals, they are completely differ from the Clever Hans phenomenon. Perhaps that is why another citation from those times – Lloyd Morgan's law – is also found more frequently nowadays.

The predominance of anecdotal evidence of animal intelligence led one of the pioneers of comparative psychology, Lloyd Morgan, to suggest that no animal behaviour should be explained in terms of higher mental processes if it could be explained by simpler processes: 'In no case may we interpret an action as the outcome of the exercise of a higher psychical faculty, if it can be interpreted as the outcome of the exercise of one which stands lower in the psychological scale' (Morgan, 1900). This became known as *Lloyd Morgan's law*, or the *Law of Parsimony*.

Being a successor of Romanes, Morgan drew heavily on Romanes' work and was his literary executor. In 1896 Morgan published the important volume *Habit and Instinct* which included, among others, chapters such as 'The relation of consciousness to instinctive behaviour', 'Intelligence and the acquisition of habits', 'Imitation' and 'The emotions in relation to instinct'. A large part of the book was based on Morgan's own experiments and observations. Morgan repeated Spalding's experiments on the early development of behaviour, with a considerable variety of species, and, while confirming many of his observations and conclusions, showed that some of them were erroneous. He decided that newly hatched chicks had no perception of the qualities of objects. They pick up stones as well as grain, bits of red worsted as well as worms, gaudy-coloured inedible caterpillars as well as those that are edible. They do not recognise water till they have felt it, and they do not know that water is drinkable till contact with the beak sets up the nervous and muscular reactions of drinking. By a series of careful experiments Morgan showed that young chicks have no fear of bees as bees, but merely fear them because they are large and unusual. They are equally suspicious of a large fly or beetle, and, though eating small worms greedily, are afraid of a large one. And when the chicks are a few days old, and are no longer afraid of large flies, they show no fear even of wasps, when presented with them for the first time. Experimenting with fear of predators, Morgan came to the conclusion that a similar principle of alarm develops in different species: any sight, or sound or smell very different from what they have been accustomed to alarms them, and they learn what is really dangerous either through the actions of their parents or by their own personal experience.

In fact, this statement was not entirely adequate, and the problem is *still* under discussion. But this was the first time that the problem of the development of complex behavioural patterns had been systematically attacked on the basis of experimental method. Since that time, this problem has been re-examined by many researchers. The first coherent theory of instinctive behaviour within the framework of ethology was developed in the early 1940s (see Chapter 2).

Morgan tried to divide innate from learned behaviour and thus to construct the idea of intelligence. Among other examples of learned behaviour, he described competitions for sheepdogs in the north of England, where dogs have to cope with rather complex tasks, such as driving sheep over a predetermined course, just obeying their masters' six or eight whistle signals, often accompanied by gestures. The dogs behaved like Clever Hans demonstrating a remarkable ability to respond to their masters' cues but in this situation the process was carefully watched. So, Morgan said, the intelligent animal is what he is trained to be – one whose natural powers are under the complete control of his master, with whom the whole plan of action lies.

This idea was corrected by Morgan's own fox terrier Tony. Tony was not trained to lift the latch of the garden gate and thus to let himself out. The way in which the dog learnt the trick was on his own initiative. Tony naturally wanted to go out into the road, and after several attempts he raised the latch with the back of his head, and thus released the gate, which swung open. Once firmly established, the behaviour remained constant throughout the remainder of the dog's life, some five or six years. And the famous psychologist could not succeed, notwithstanding much expenditure of biscuits, in teaching Tony to lift the latch more elegantly with his muzzle instead of the back of his head.

Morgan's dog became the inspiration of infinite numbers of animals doomed to search for specific ways to open doors. It was Morgan's influential lecture at Harvard in 1896 on habit and instinct that prompted Edward Lee Thorndike to elaborate his experimental approach based on the study of animals escaping from puzzle-boxes. Thorndike was experimenting with chicks, and when he moved to Columbia University in 1897, he took along his two best-educated chicks. At Columbia he expanded his sample, and in 1898 when he published his doctoral thesis, 'Animal intelligence: an experimental study of the associative processes in animals', he reported data on chicks, cats and dogs. So 1898 could be considered the year of birth of the puzzle-box method.

The introduction of the puzzle-box method into comparative psychology was the point at which objective experimental methods were introduced into psychology, and prepared the way for its absorption into behaviourism. This was to exert a profound influence on the study of animal intelligence. Like Morgan, Thorndike reacted against anthropomorphic representations of animal reasoning and began his experiments in order to discredit the anecdotal approach for describing the behaviour. Many observers of animals, he argued, have looked for animal intelligence, and never for animal stupidity.

Figuratively speaking, the objective study of intelligent behaviour from the earliest stages has been based on several 'corner boxes'. First of all, there was Thorndike's puzzle-box and, later, Skinner's reinforcement box (the so-called 'Skinner box'). Both were aimed at developing experimental methods for controlling and manipulating the presentation of stimuli, but in different ways. Being placed into the puzzle-box, an animal has to extricate itself, while in the Skinner box its only desire is to stay there as long as possible in order to get more and more rewards. Differing in their presentation of reinforcement, these two ways of studying animals' capacity for solving problems resemble each other in principle: the animal is unaware of how the apparatus works. A combination of these two variants much later resulted in another kind of a box, a so-called 'artificial fruit'. You are not placed into the box; instead, you are presented with a sort of a puzzle-box containing food, so the challenge is for you to extract a reward – but not yourself – from the box. The principle is the same: it is not necessary for a subject to understand machinery, just to find a way of opening the box by trial and error and then to repeat the found route again and again. The artificial fruit has been constructed mainly for studying animals' abilities for imitation (see Chapter 25), and it is worth noting that the first attempts in this direction were based on Thorndike's early experiments with chicks. The first variants of mazes that were applied for studying animal intelligence were also constructed within boxes. The American psychologist W. S. Small described his research which involved an apparatus for the study of intelligence and learning in rats that he had modelled on the sixteenth-century Hampton Court maze near London. While the specific design was taken from a diagram provided in the *Encyclopedia Britannica*, the suggestion for use of the Hampton Court maze was that of Edmund Sanford. Published in two parts in the *American Journal of Psychology* (Small, 1900, 1901) ('An experimental study of the mental processes of the rat'), this paper prompted many investigators to use mazes for studying intelligence in animals. Besides acting with locked boxes (such as 'artificial fruit') and solving problems when they themselves were in boxes (e.g. puzzle-boxes and Skinner boxes), animals demonstrated their intelligence by placing one box on another. This came from an alternative Gestalt approach in which thought was seen as an organisational process by which a problem was reorganised or solved (see Chapter 17). Gestalt was not part of psychology's mainstream in the first half of the twentieth century, in which behaviourism dominated. In the second half of the century cognitive psychology was to come together as a relatively coherent movement.

One of the first presentations of comparative and objective investigations of human and animal mental representation was by Leonard T. Hobhouse. In his book *Mind in Evolution* (1901), Hobhouse articulated a theory of intelligence defined in terms of levels of adaptive behaviour, introduced a series of animal problem-solving tasks that initiated the experimental study of complex animal behaviour, and reported evidence of sudden improvement in the learning curve that appeared to reflect animals' ability to employ perceptual relations in problem-solving. Hobhouse referred to this as 'practical judgement', and later students in animal behaviour have termed it 'insight'. Hobhouse described the development of adaptive ability of animals from the simplest tendencies to the maintenance of 'organic equilibrium, through the relatively stereotyped, inherited, stimulus-specific reflexes and more variable but also inherited patterns of instinctive behaviour, to the individually acquired adaptations of intelligence'. According to his definition, intelligent behaviour was that 'devised by the individual on the basis of its own experience for compassing the ultimate and proximate ends to which it is impelled'. The presence of intelligence was indicated by the modification of action in accordance with the results of experience, showing that in some degree the animal can correlate its own past experiences with its subsequent action. At the lower levels of intelligence, Hobhouse placed behaviour that is the result of habitually acquired correlations which function as general modes of reaction to circumstances. At the higher levels of intelligence, however, behaviour, was, in his view, purposive. Purposive action dealt with the complex and varying circumstances of individual cases; and, unlike habit, which was a general mode of reaction, higher intelligence was selective, choosing appropriate means from among a host of possible actions best suited to achieve the desired aims.

After outlining the broad trend of mental development, Hobhouse then turned to the general question of whether the higher levels of intelligence have been attained by animals other than humans. To address this question, he introduced an ingenious set of problems and described a long series of experiments employing these tasks with dogs, cats, monkeys, an elephant and an otter. The problems involved manipulations with a simple mechanism (e.g. pulling a string, pushing a door, pushing a lever, sliding a lid, lifting a catch) or even more complicated behaviour (e.g. box stacking or rake use with monkeys) to obtain food. In the animals' response to these problems, Hobhouse found evidence of sudden improvement in the learning curve, improvement that he interpreted as the evidence that animals are, indeed, capable of some degree of higher intelligence.

Extending his analysis first to conceptual and then to systematic thought in humans, Hobhouse presented a four-stage theory of the development of intelligent adaptation from *unconscious readjustment* (Stage 1) through *concrete experience and practical judgement* (Stage 2), to *conceptual thinking and will* (Stage 3), and finally *rational system* (Stage 4) that anticipated certain of Jean Piaget's ideas much later. Piaget also accepted ideas from of J. B. Watson and from J. M. Baldwin, with his classic study of infant behavioural development, 'The origin of right-handedness' (Baldwin, 1894). Thus a tendency to study human and animal intelligence hand in hand had been developed.

Among other great persons who rocked the cradle of experimental psychology were Wilhelm Wundt and William James. They established the first two laboratories in the world (in Leipzig and in Harvard, simultaneously in 1875) dedicated to experimental psychology and pioneered the concept of stating mental events in relation to objectively knowable and measurable stimuli and reactions. James used Spencer's *Principles of Psychology* (1855) as the text for his first course in physiological psychology at Harvard (1876/7), although he was very critical of Spencer's detailed formulations. Thorndike was one of the graduate students supervised by James and learned experimental psychology from his two-volume course *The Principles of Psychology* (James, 1890). At that time, students from all over the world journeyed to Leipzig to learn experimental technique and to return to their home institutions imbued with the spirit of scientific psychology.

Wundt considered mind to be an activity, not a substance. The basic mental activity was designated by Wundt as 'apperception'. He discerned the process of excitations from stimulation of the sense organs, through sensory neurons to the lower and higher brain centres, and from these centres to the muscles. Parallel with this process run the events of mental life, known through introspection. Introspection became, for Wundt, the primary tool of experimental psychology. The method that Wundt developed was a sort of experimental introspection: the researcher was to carefully observe some simple events – ones that could be measured as to quality, intensity or duration – and record the responses to variations of those events.

In the next chapters we will consider the early development of two main approaches to the study of intelligence which have expanded in parallel from the beginning.

Chapter 2

The dramatic adventures of behaviourism

Behaviourism was the late nineteenth century's answer to the criticism that psychology was not a true science. Behaviourism is an approach in psychology that argues that the only appropriate subject matter for scientific psychological investigation is directly observable and measurable behaviour. Behaviourists claim that internal mental processes are not appropriate for scientific investigation as they are not directly observable. By the beginning of the twentieth century, psychologists, physiologists and sociologists favoured developing the science of behaviour.

The founding of behaviourism as an avowed movement is credited to John B. Watson, who gave his famous lecture entitled 'Psychology as the behaviorist views it' at Columbia University in New York in 1913. In this 'manifesto' Watson claimed that introspective psychology was unscientific because it did not deal with objective states. He rejected all subjective states such as sensation, imagery and thought unless they could be observed by others. Watson laid the foundations of what was to be labelled as 'behaviourism'. This branch would be a stimulus–response psychology, dealing only with the reactions of muscles and glands to stimulus situations. Its theoretical goal is the prediction and control of behaviour.

The publication of Watson's lecture in the *Psychological Review* marked the formal beginning of behaviourism. In reality, this paper of 1913 was not the revolutionary moment that it is sometimes thought to be. By 1913, the study of human and animal behaviour by means of purely objective methods under conditions of experimental manipulation and control of stimulus conditions had a 40-year history. The very beginning of the twentieth century may be identified as the point of origin of this field.

2.1 Classical behaviourism

Being placed in a title, the word 'Classical' perhaps provokes readers to jump to another, at least, 'neo-classical' piece of scientific history. But for me, these are not badly digested clinkers of antiquity. Methods of classical behaviourism are still alive and well today.

In about 1900, Ivan Pavlov, at that time Director of the physiological department of the Institute of Experimental Medicine in St Petersburg, devised a new method of investigating the physiology of the nervous system in its relations to the psychic reactions of organisms. This method, which later became widely known as the Pavlov salivary reflex method, was first employed not for the study of psychic phenomena but simply as a mean of approach to the physiology of the nervous system. The history of the development of Pavlov's simple, elegant experimental paradigm for studying learning in animals is interesting and dramatic.

Pavlov was born in 1849 at Ryazan, in a small village in Central Russia, where his father was a village priest. He was educated first at the Ryazan Ecclesiastical High School and then at the theological seminary there. Inspired by the progressive ideas which D. I. Pisarev, the most eminent of the Russian literary critics of the 1860s and

I. M. Sechenov, the father of Russian physiology, were spreading, Pavlov abandoned his religious career and left the seminary for the University of St Petersburg. There he studied chemistry and physiology, and received his doctorate in 1879. After earlier studies in Russia, he went to Germany for graduate work under the direction of the cardiovascular physiologist Carl Ludwig and the gastrointestinal physiologist Rudolf Heidenhain.

Pavlov was looking at the digestive processes in dogs, especially the interaction between salivation and the action of the stomach. He realised they were closely linked by reflexes in the autonomic nervous system. Without salivation, the stomach did not receive the message to start digesting. Pavlov wanted to see if external stimuli could affect this process, so he rang a bell at the same time as he gave the experimental dogs their food. After a while, the dogs – which before only salivated when they saw and ate their food – would begin to salivate when the bell rang, even if no food were present. Moreover, around mealtime, as soon as his dogs saw or heard the laboratory assistant who fed them, they began salivating and secreting gastric juices, even before he actually gave them their food.

Pavlov suggested that an animal had formed an association between a previously neutral stimulus and a previously unconditional (unlearned) response. As part of his research on salivation, the physiologist invented an apparatus to enable him to measure precisely the amount of saliva a dog produced. He gave the dog food, in order to make it salivate. Being placed in Pavlov's apparatus the animal had no possibility to act in response to stimuli; the dog's whole business was to drool.

Pavlov carried out many controlled laboratory experiments and established a theory of classical (or Pavlovian) conditioning. The basic idea was to use the salivary secretion as a quantitative measure of the psychical or subjective activity of the animal. It was a new tool for objective study of subjective psychic processes. In essence, Pavlov distilled the complexity of learned behaviour down to one elemental component, a *conditioned reflex*: a change in behaviour produced by the association of two stimuli in time (for instance, the ringing bell and food). Throughout the rest of 1900s, behavioural properties of Pavlovian learning were quantified and compared across the animal kingdom. Pavlov considered the conditioned reflex as a cue to the mechanism of the most highly developed forms of reaction in animals and humans to their environment.

Indeed, Pavlov was neither the first nor the last behaviourist tempted by the idea of discovering the fundamental way that all organisms, humans as well as animals, learn. In his model, all non-innate behaviour is a conditioned response. He told of ambition as a sort of conditioned response – 'the reflex to purpose', he called it (Pavlov, 1927, 1928). He even believed that there is an innate 'reflex of selfdom', which leads the weak to be servile to their oppressors. 'I am convinced profoundly, once and for all', Pavlov proclaimed, 'that it is by following this road that human intelligence will triumph over the most important of the problems it has been solving – namely, the knowledge of the laws and mechanisms of human nature.' Pavlov announced principles of the language function in the human as based on long chains of conditional reflexes involving words. The function of language involves not only words, he held, but an elaboration of abstractions and generalisations is not possible in animals lower than the human.

At the 14th International Medical Congress in Madrid (1903), Pavlov read a paper on 'The experimental psychology and psychopathology of animals', and in 1906 he gave a lecture in London. Enlarged reviews of his papers were published in European languages. Influential American behaviourists, such as R. Yerkes (Yerkes and Morgulis, 1909) and J. B. Watson (1916) published articles that ranked Pavlov's new method as one of the most significant in the development of behaviourism.

After the 1920s, in Communist Russia, science had to develop behind the Iron Curtain. Pavlov, the first Nobel Prize winner in the field of theoretical medicine, was too well known for him to be isolated, so the government used his name as a peculiar talisman. Pavlov was one in a thousand among Soviet scientists who was allowed to go abroad. In the 1920s Pavlov visited the United States of America twice, being hosted by his close colleague Walter Cannon, the father of conception of homeostasis and the author of *The Wisdom of the Body* (1932). Cannon wrote a foreword for English edition of Pavlov's *Conditioned Reflexes* (1927). The Soviet Union became a prominent centre for the study of physiology, and the fact that the 15th International Physiological Congress in 1935 was held in Leningrad and Moscow clearly shows that it was acknowledged as such.

The Pavlov Institute of Physiology of the Russian Academy of Sciences was founded in 1925. The Institute is partially located in St Petersburg, but its main campus is located 10 km from St Petersburg, in the rural town of Koltushi. This research campus was founded in the early 1930s. It now bears Pavlov's name and is rightfully recognised as 'the capital of conditional reflexes'. Pavlov erected a monument to a generalised 'Pavlovian Dog' just near the famous 'Tower of Silence' where experiments were carried out in seclusion from the noisy environment. When Pavlov was still alive, Koltushi was visited by such world leaders in science and culture as F. Hill, L. Lapique, D. Barcroft, W. Cannon, H. Hunt, Herbert Wells and Niels Bohr. In 1993, building on the success of the Institute, the Pavlov International Research Center was founded. The Pavlovian Society was established in 1955 by W. Horsley Gantt at the Johns Hopkins University School of Medicine, Baltimore, USA, and the Pavlovian Laboratory at the School of Medicine is actively working.

About the same time that Pavlov was conducting his first experiments, Thorndike formulated the theory of instrumental conditioning at Columbia University, New York. The Thorndike's method was based on observations on animals' (usually cats') trial-and-error learning to escape from puzzle-boxes to obtain food. Being placed in a puzzle-box, the cat would be able to see a feeder outside and this was its incentive to escape. Inside the puzzle-box Thorndike had rigged up a number of devices that, if pulled or pushed, would lead to the door being opened. Initially the cat would scratch and struggle wildly in the box, and a considerable time elapsed before it responded correctly. Having made the response, the cat was allowed a few moments of access to food before being returned to the box for another trial. It turned out that the time, or latency, to escape decreased over trials. For example, in trying to get free, the cat, by pure accident, performed the correct response – perhaps by pushing the lever – and was able to escape. As a consequence of this the animal gained freedom and the food, thus positively reinforcing the lever-pushing behaviour. After a number of trials the cat eventually learns to connect pushing on the lever with escape. Thus, being placed back in the box, it goes straight to the lever, pushes it, and escapes. Therefore, according to Thorndike, the cat has now developed a connection between pushing the lever and positive consequences.

Thorndike can be considered one of the first true connectionists. In his book *The Fundamentals of Learning* (1932) he differentiated between the principles of British associationism and what he had introduced as 'new connectionism'. His most important contributions to psychology are summarised in *Selected Writings from a Connectionist's Psychology* (Thorndike, 1949).

While Thorndike was engaged in research with his animals at Columbia, two young men, Linus Ward Kline and Willard Stanton Small, were working in the Psychology Laboratory at Clark University. At that time, the Clark laboratory was under the direction of Edmund Clark Sanford, one of the great pioneers in the development of experimental laboratory technique (remember that it was Sanford who had suggested the use of the plan of the Hampton Court maze). Like Thorndike, Kline and Small had been inspired by Morgan's *Habit and Instinct* (Morgan, 1896). Under the guidance of Sanford, Kline (1899) constructed several pieces of laboratory apparatus for the study of the behaviour of vorticella, wasps, chicks and white rats. Indeed, one of these pieces, designed with the assistance of Small, approximated a simple Y-maze. This may be considered a 'Gestalt germ' in the heart of behaviouristic movement: that Kline criticised Thorndike's over-reliance on a purely experimental method, and argued for a combination of the naturalistic and experimental approaches. While Kline involved himself in comparative studies, Small focused exclusively on the rat (Small, 1900, 1901). In studying the behaviour of rats in a maze, Small unwittingly introduced a technique into psychology that became so widespread during the heyday of behaviourism that it came for many to symbolise the science itself.

As was mentioned above, it is Watson who can be credited with founding behaviourism as the school of psychology in which behaviour is described in terms of physiological responses to stimuli. The famous American psychologist, while a doctoral student with James R. Angell at the University of Chicago, carried out animal research in maze learning. He considered maze problems a form of instrumental conditioning in which the animal is faced with a sequence of alternatives. Being one of the first experimentalists who observed animal maze learning in details, Watson at the same time conducted a series of ethological studies of seabirds, researching the relationship between sensory input and learning and bird behaviour. He studied all aspects of the birds' behaviour: imprinting, homing, mating and nesting habits, feeding and chick-rearing. These extensive studies, carried out over four years, are some of the earliest examples of what would later be called 'ethology' and his comprehensive records of the birds' behaviour were some of the earliest examples of an 'ethogram' – a comprehensive record of the naturally occurring behaviour of an organism (for details see: Buckley, 1982; Coon, 1994; Wozniak, 1997; Rilling, 2000).

Jacques Loeb, who had come to Chicago from Germany, developed some new approaches in experimental physiology (for details see: Pauly, 1987; Wozniak, 1997). Being already by that time the author of *Comparative Physiology of the Brain and Comparative Psychology* (1900), Loeb, together with Dewey, Angell and Donaldson, led Watson, then a Ph.D. student, to a highly descriptive, objective approach to the analysis of behaviour that he would later call 'behaviourism' (Watson, 1903). Working mainly with insects, Loeb was studying the general mechanisms by which animal behaviour is controlled, and conceived of animal behaviour as a response of the whole organism. For Loeb there was no distinction in principle between reflex movement, tropism and instinct. He considered 'reflex' to be the term used when reaction to an external stimulus involves only a part of the organism, tropisms to be reactions of the organism as a whole, and instincts to be more complicated reactions composed of tropisms and reflex chains. Regardless of whether the animal's behaviour is reflexive, tropistic or instinctive, it consists of a reaction – usually purposive and coordinated – to an external stimulus. Loeb studied animal activity, holding to the idea of consciousness as associative memory or capacity of the animal to learn from experience. In Chicago Loeb studied the learning abilities of brain-damaged dogs and he was full of enthusiasm for the application of physiology and biochemistry to the problems of associative memory. Watson was duly impressed, both with what he read and heard and by what he saw in Loeb's laboratory.

In 1907 Watson was hired by Johns Hopkins University as a full professor of psychology. Very soon he became Director of the Psychological Laboratory and the editor of *Psychological Review*. At the early stage of his career as psychologist, Watson had been particularly influenced by the reflex studies of Ivan M. Sechenov and Vladimir Bekhterev (see Chapter 3). He apparently did not become interested in the conditioning of motor and autonomic responses until late 1914, when he read a French edition of Bekhterev's *Objective Psychology* (see Hilgard and Marquis, 1940).

Of all Watson's papers and books, the most influential was *Psychology from the Standpoint of a Behaviorist* (1919), the first textbook to extend behaviouristic analysis to human psychological function. Watson believed in a stimulus–response method for predicting and controlling behaviour. His methods, however, differed from those of Pavlov in that he did not use the terms 'stimulus' and 'response' in as narrow a sense as the Russian physiologists did. For him, a stimulus could be a general environmental situation or some internal condition of the organism. He even dealt with language and thought, maintaining that speech was a form of behaviour and that thought was a form of illicit (covert) behaviour. He was able to explain thought in this manner by stating that thinking is accompanied by minute manipulations of the tongue and throat muscles. Thus, thought for Watson was nothing more than implicit speech – tiny movements of the larynx that take place during problem-solving (Watson, 1913). It is worth noteing that another article on subvocal speech by Wyczoikowska (1913) was to appear in the same issue of the *Psychological Review*. The theory of thinking as subvocal speech was not original to these authors. Curtis (1900) had already attempted to measure movements of the larynx during thinking.

In 1920 Watson published a paper, co-authored with Rosalie Rayner, known as 'The little Albert study' which is still one of the most frequently cited psychological publications (Watson and Rayner, 1920). Albert B was born to a woman who was a wet-nurse in the Harriet Lane Home for Invalid Children. Although raised in the hospital environment, Albert developed normally and was emotionally healthy and stable. Before the start of the experiment, when Albert was 9 months old, Watson and Rayner ran little Albert through emotional tests. The infant was confronted briefly and for the first time with a white rat, a rabbit, a dog, masks with and without hair, etc. The infant at no time showed any fear. The actual experiment began when the infant was 11 months old. Watson had exposed Albert to a loud sound (made by a bar being banged right behind Albert's head) while being presented with a white rat. A week later, after a series of tests, Albert cried when being presented only with the rat. Five days later, Albert showed generalisation by reacting with fear to a dog, a fur coat, Watson's hair, cotton wool and a Santa Claus mask with a white beard (perhaps today the experiment would be considered unethical). During the whole experiment, Albert was happy to play with wooden toys at any time. This result was for Watson proof that complex behaviour develops by conditioning out of simple unlearned responses and confirmed him in the idea that behaviourism should apply the techniques of animal research or conditioning to humans. Later Watson (1928) published a book *Psychological Care of Infant and Child*. In it he encouraged parents to approach child-rearing as a professional application of behaviourism and advocated strict routines and tight control over the child's environment and behaviour.

Being at Johns Hopkins University Watson began to look beyond academia for opportunities in the application area. He offered a course on 'The psychology of advertising' in which he instructed future managers in the importance of applied psychology (for details see Petty *et al.*, 1983). While working in the advertising industry (1920–35), Watson revealed that marketing goods depends not upon an appeal to reason but upon emotional conditioning and stimulation of desire. He compared a customer with a green frog and thus considered the marketplace as a laboratory for the advertising industry, where a consumer was akin to the experimental subject whose behaviour should be deliberately controlled. Elaborating practical rules for advertisements, Watson stressed three common features: evoking emotion rather than cognition, providing specific instructions for using the product, and employing direct testimonials.

The widely known novel *Brave New World* by Aldous Huxley (1932) was based on a direct allusion to conditioning of human behaviour and Watson's principles of advertisement. A famous English writer, Aldous Leonard Huxley was born into a family that was closely involved with the development of science and art. In particular, Aldous' father was the son of Thomas Henry Huxley, a great biologist who had helped develop Darwin's theory of evolution. In Huxley's dystopia *Brave New World*, universal human happiness has been achieved by means of control of reproduction, genetic engineering and conditioning. Each class, from the super-intelligent Alpha Pluses down to the dwarfed semi-moron Epsilons, is conditioned to love its type of work and its place in society. Each factory-produced group of clones is shaped in the Centre of Hatcheries and Conditioning. Toddlers are trained to associate flowers and books with electric shocks.

It is clear for us why behaviouristic ideas raised a protest in humanistic writers like Huxley. However, as we will see further in this book, these ideas still underlie many modern methods of studying intelligence. Besides, some behaviouristic data contain germs of new ideas and interpretations. For instance, one of the reasons that 'The little Albert study' is so well known is that it is rediscovered every five or ten years by a new group of psychologists (Harris, 1979). In particular, Seligman, who seized control of the Albert story and used it to attack traditional theories of learning, devoted his attention to the fact that the infant had formed fear reactions to certain stimuli much more readily than to other ones; and that he generalised stimuli but his conditioned reactions were hard to extinguish (see description of these properties of associative learning in Chapter 6). Seligman (1971) then proposed his own reformulation, known as 'preparedness theory'. This theory is important for understanding the polyhedral set of animal intelligence. We will consider this aspect of intelligence in Part VII.

2.2 The Skinnerian branch of behaviourism

Pavlov and Watson's behavioural work lead to B. F. Skinner's operant conditioning experiments about ten years later. Burrhus Frederic Skinner is considered one of the most important psychologists of the twentieth century and also the most misunderstood. During his life, Skinner established and promoted a science of behaviour which he named 'experimental analysis of behaviour'. Skinner's research was concerned with the experimental analysis of behaviour, and specifically, the effect of reinforcement on sequential behavioural acts. Whereas Pavlov used 'hard-wired' reflexes as the raw material for conditioning, Skinner used any overt action of the organism, including human beings, that is, from a baby's smile to a student working harder for a grade.

When a student, Skinner was influenced by Pavlov's ideas and considered his method a cornerstone of his career in psychology (for details see: Catania, 1998; Catania and Laties, 1999). As Skinner wrote in his autobiography (1967), Pavlov had given him a glimpse of the experimental method: 'Control the environment and you will see order in behaviour.' When Pavlov gave the Principal Address at the International Congress of Physiology which was held in Harvard Medical School in 1929, Skinner, then a graduate student, attended the Congress as a volunteer (one of his tasks was to operate a slide projector). He acknowledged this event in his Presidential Address to the Pavlovian Society of North America (Skinner, 1966).

At the age of 24, in 1928, Skinner enrolled in the Psychology Department of Harvard University. With his enthusiasm and talent for building new equipment, after a dozen pieces of apparatus and some lucky accidents (described in his paper 'A case history in scientific method', 1956), Skinner invented the first cumulative recorder, a mechanical device that recorded responses of an animal. This box, along with the attached recording equipment, provided a way to collect more objective data about behaviour than had been possible before. The device came to be known as the 'Skinner box'. In a typical Skinner experiment, an animal (usually a rat in his early experiments) was placed into the box containing a bar. If the rat pressed the bar, it was rewarded with a pellet of food. It is important to note that in these experiments nothing forces an animal to press the bar. The first time it possibly does so accidentally. When the rat finds that food arrives, it presses the bar again. Eventually, the animal learns that if it is hungry, it can obtain food by pressing the bar. In further trials the task is made more difficult. The rat only gets rewarded if it presses the bar while a light is flashing. At first the animal is puzzled. Eventually it learns the trick. Then the task is made more difficult again. This time the animal only receives food if it presses the bar a certain number of times. After initial confusion, it learns to do this also. In that way Skinner had proved that, given time, animals could be trained to perform an amazing variety of seemingly complex tasks, provided two basic principles were followed. Firstly, the tasks must be broken down into a number of carefully graduated steps. Secondly, the animals must be repeatedly rewarded.

Skinner changed the name of instrumental conditioning to *operant conditioning*. 'Operant' refers to Skinner's idea that any organism 'operates' on its environment – that is, performs actions that change the environment around it for better or for worse. Operant psychology is based on the idea that an action made by a person or an animal often has consequences that occur naturally in the environment. This principle is called operant conditioning. Reinforcement is something that makes it more likely that a given behaviour will be repeated. The consequences of a given action either reinforce the behaviour or do not.

Skinner claimed that this type of learning was not the result of a stimulus–response association but instead was an association between the operant response and the reinforcement. He considered all learned behaviour to be the result of selective reinforcement of random responses. Mental states (what goes on in our minds) have no effect on our actions. Skinner did not deny the existence of mental states; he simply denied that they explain behaviour. A person does what he does because he has been 'reinforced' for doing that, not because his mind decided so. Skinner noticed a similarity between reinforcement and natural selection: random mutations are 'selected' by the environment; random behaviour is also selected by the environment. A random action can bring reward (from the environment) that will cause reinforcement and therefore will increase the chances that the action is repeated in the future. An action that does not bring reward will not be repeated. The environment determines which behaviour is learned.

Perhaps Skinner's most significant contribution to the theory of conditioning was his work on partial reinforcement. He worked with 'schedules of reinforcement' to study and manipulate behaviour. Experimenters and circus trainers using Skinner's techniques have taught animals to perform any number of unnatural actions. For instance, it is possible to teach chickens to play toy pianos, and dogs to climb ladders, acting like firemen (see details in Chapter 6). *Shaping*, along with *reinforcing*, and *operant*, are the key words in Skinnerian behaviourism. Like Watson, Skinner claimed that speech is a set of habits gradually built up over the years, as is the process of autoshaping, that is, shaping of the individual's behaviour by itself. According to Skinner, no complicated innate or mental mechanisms are needed. All that is necessary is the systematic observation of the events in the external world which prompt the speaker to utter sounds.

Some Skinner's projects were flamboyant and became a part of 'folk behaviourism'. One such legendary project was Skinner's top-secret plan to train pigeons to guide bombs during World War II. As early as 1940, during a train trip, watching birds flying, lifting and wheeling in formation as they flew alongside the train, Skinner had the idea of 'bombing the bombers' based on birds' excellent vision and extraordinary manoeuvrability. He started in the spring of 1941 to explore the feasibility of this idea and bought pigeons in order to train them to peck when they saw a target through a lens. They were harnessed to a movable hoist and movements of their beaks sent electric motors up and steered the hoist to the target. It is interesting that Skinner's installation initially met a negative response from R. C. Tolman, another famous expert in animal behaviour (see Chapter 7). But after Pearl Harbor Tolman changed his mind. The Japanese kamikaze attacks gave him new incentive to promote his work, defining his pigeons as kamikaze 'substitutes'. At the same time in Russia thousands of dogs were trained by Pavlov's method as to act as kamikaze against German tanks. They were previously trained to associate a dummy German tank with food. During a battle, hungry dogs ran down German tanks and detonated them. Ideally, the dog should drop explosive from its back but in reality it flew into a rage together with a hostile tank. Skinner trained his pigeons to recognise an intersection in an aerial photograph of Stalingrad. He proved that the pigeons could remain astonishingly reliable, even under conditions of stress, acceleration, temperature and pressure difference and noise. By means of films and installations, Skinner reported his latest improvements and finally, after an electrical engineer was sent to Minneapolis to evaluate Skinner's work, he was invited to Washington to present his project now labelled 'Bird's-eye

Bomb' (and later 'Project Pigeon'). After a series of votes and expert reports, Project Pigeon was, nevertheless, discontinued, because of another top-secret project unknown to Skinner – an anti-aircraft predictor. However, Skinner's work inaugurated a new era in behaviourism, 'engineering behaviour'. This work was also useful for further development of the behaviouristic movement. In particular, it turned out that pigeons learn more rapidly than rats, allowing more rapid discoveries of the effect of new contingencies. Skinner never again worked with rats.

Another project, this time peaceful in intent but nevertheless giving rise to a scandal, was Skinner's 'baby tender'. When Skinner's second daughter, Deborah, was born in 1944 he constructed what was essentially a large version of a hospital incubator for her, a tall box with a door at its base and a glass window in front. This 'baby tender', as Skinner called it, allowed the baby to spend time in a safe, thermostatically controlled environment while her father worked, and by all accounts she grew up a happy and healthy child. The trouble began when Skinner published an article about the baby tender in the magazine *Ladies' Home Journal* (Skinner, 1945). People jumped to the conclusion that Skinner was raising his daughter in a box for the purpose of conducting psychological experiments on his hapless child. Magazine articles that were published later painted Skinner as an unfeeling, inhumane parent. This of course was not the case. Deborah Skinner had good relations with her father; she is a successful artist and lives with her husband in London.

Skinner's principles of operant conditioning still play an important role in the way we approach learning and behaviour modification today. Programmed teaching materials providing immediate feedback to students' responses are utilised in today's classrooms to effectively teach certain types of material. Skinner's ideas have also been adopted to teach mentally retarded and autistic children, in industry to reduce workplace accidents, in numerous applications in health-related fields, including such matters as bed-wetting and stuttering, and to improve human learning ability.

It is hard to say whether such concrete applications exactly fulfil the dreams of the great behaviourist. Perhaps, Skinner desired to change the world, having little interest in paltry domestic concerns but willing to apply new science to human behaviour. The basic idea was that what a man does is the result of specifiable conditions and that once these conditions have been discovered, we can anticipate and to some extent determine his actions. Trying to make society a much better thing seemed to be number one priority in Skinner's mind.

Ideas about behaviourism, in particular, the use of operant conditioning techniques to control and engineer human behaviour, are reflected in many literary works, mostly in ironical manner. In fact, behaviouristic ideas were firstly represented in the novels and popular books of Skinner himself. In *Walden Two* (Skinner, 1948a) he gave an account of a utopian society where traditional child-rearing techniques were replaced with behavioural engineering. In 1971, Skinner wrote his most controversial book, *Beyond Freedom and Dignity*. In it he dismissed the notion that individual freedom existed. Man's actions were nothing more than a set of behaviours that were shaped by his environment, over which he had no control. These ideas were developed in *Reflections on Behaviorism and Society* (1978). Skinner says if humanity is to survive we must abandon such pre-scientific ideals as freedom and dignity, and set about controlling our environment and ourselves by means of a technology of behaviour, which will be comparable in power and precision to physical and biological technology and which will induce people not to be good but to behave well.

Despite his many misinterpretations, Skinner did not consider people as puppets or robots guided by managers armed with behaviouristic ideas, but was deeply concerned with individuals as persons. He described the processes of shaping a human being with a light vein of humour in the second part of his autobiography, *The Shaping of a Behaviorist* (1979).

By the middle of twentieth century behaviourists had been nearly buried under tons and tons of articles by hundreds of experimentalists as well as by the efforts of many generations of pigeons, rats, cats and rabbits which had unceasingly pressed levers and pulled out ropes. It seemed to be a case of learning more and more about less and less. This became apparent especially when a short list of species had been enlarged apart from the favourite laboratory animals such as rats, pigeons, dogs and cats. This happened at the same time as the development of the theory of instinct by Lorenz (1935, 1950) and Tinbergen (1942, 1951) and the appearance of new data on innate behavioural patterns obtained by Hess (1956) and Eibl-Eibesfeldt (1961a, 1967). Ethology as a new branch of behavioural sciences came of age, and this changed behaviouristic ideology (see details in Chapter 21). In 1961 Skinner's former students Keller and Marian Breland published a paper 'The misbehavior of organisms' (this title parodies the title of Skinner's 1938 book *The Behavior of Organisms: An Experimental Analysis*) (Breland and Breland, 1961). Starting with the concept of instinctive behaviour, the Brelands predicted that if one begins with evolution and instinct as the basic format for the science, a very illuminating viewpoint can be developed which leads naturally to a drastically revised and simplified conceptual framework of startling explanatory power (see details in Chapter 20). This forecast came true, and this path led to cognitive ethology.

Chapter 3

Intelligence under the scalpel: starts and false starts of neuroscience

From the very beginning, investigations of learning processes in animals were closely tied to the search for 'engrams', or material tracks, within the brain. In one of his last, testamentary, works, the paper entitled 'The origins of cognitive thought', Skinner (1989) pointed out that the position of behavioural analysts is sometimes characterised as treating a person as a black box and ignoring its contents.

Skinner wrote:

> Behaviour analysts would study the invention and uses of clocks without asking how clocks are built. But nothing is being ignored. Behaviour analysts leave what is inside the black box to those who have the instruments and methods needed to study it properly. There are two unavoidable gaps in any behavioural account: one between the stimulating action of the environment and the response of the organism, and one between consequences and the resulting change in behaviour. Only brain science can fill those gaps. In doing so it completes the account; it does not give a different account of the same thing. Human behaviour will eventually be explained, because it can only be explained by the cooperative action of ethology, brain science, and behaviour analysis.

In this chapter we will launch a very short trip to the historical origin of neuroscience, concentrating on establishing physiological psychology as a part of behavioural theory.

3.1 A short history of neurophysiology

Neurophysiology developed in parallel to all the other medical and physiological sciences (see Rose and Bynum, 1982; Finger, 1994). We could point at Thomas Willis' anatomical description of the brain in 1664 as the first major step. His book was illustrated by Christopher Wren, the famous English artist and architect. Willis suggested the term 'neurology' in 1681. It was only a few decades after another major event that was William Harvey's explanation of the circulation of the blood in 1628. Most physicians, still using Galen's text, believed that the blood ebbed and flowed like a tide through the whole body.

In fact, the experimental study of the neural mechanisms of learning has a short history. Since the theoretical advance of understanding the brain to be the 'organ of mind', little progress had been made for a long time in understanding that ablation of successive slices was not a method well suited to the discovery of cortical localisation, still less so to the study of intelligence.

It was Descartes' articulation of a mechanism for automatic, differentiated reaction that led to his generally being credited with the founding of reflex theory. Descartes included far more complex behaviour as reflexes than we imagine today. At the same time, he considered 'reflex' a common basis for explanation both human and animal behaviours.

The term 'reflex' in its modern sense was introduced in 1736 by Jean Astruc. Reflexes were defined as cycles of actions of the senses that led to nearly immediate responses by the muscles. By that time, in 1717, Antony van Leeuwenhoek had described the appearance of nerve fibres in cross-section. In 1791 Luigi Galvani published work on the electrical stimulation of frog nerves. Investigating the effects of electrostatic stimuli applied to the muscle fibre of frogs he discovered he could also make the muscle twitch by touching the nerve with various metals without a source of electrostatic charge, and greater reaction was obtained when two dissimilar metals were used. He attributed the effect to 'animal electricity' and thus claimed to demonstrate that electrical energy was generated in the nervous systems of animals. This work initiated research into electrophysiology.

One of the first to advocate the use of experimental techniques in physiology was Johannes Peter Müller, German physiologist and comparative anatomist, one of the great natural philosophers of the nineteenth century and author of the most influential textbook of that times, *Handbuch der Physiologie des Menschen für Vorlesungen*, (1834–1840) (Elements of Physiology). Müller's most important achievement was the discovery that each of the sense organs responds to different kinds of stimuli in its own particular way or, as Müller wrote, with its own specific energy. The phenomena of the external world are perceived, therefore, only by the changes they produce in sensory systems. Müller's (1826) monograph about imaginary apparitions was published in 1826. According to this theory the eye as a sensory system not only reacts to external optical stimuli but can also be excited by internal stimuli generated by the imagination. Thus, persons who report seeing religious visions, ghosts or phantoms may actually be experiencing optical sensations and believe them to be of external origin, even though they do not in fact have an adequate external stimulus. Müller examined many problems in physiology, evolution and comparative anatomy. He studied the passage of impulses from afferent nerves (going to the brain and spinal cord) to efferent nerves (going away from the same centres), further elucidating the concept of reflex action. By careful experiments on live frogs, he confirmed the law named after Charles Bell and François Magendie, according to which the anterior roots of the nerves originating from the spinal cord are motor and the posterior roots are sensory.

One of the most talented Müller's students, Hermann von Helmholtz, in 1849 measured the speed of frog nerve impulses and thus found that the nerve impulse was perfectly measurable. Helmholtz was equally distinguished in physiology, mathematics, and experimental and mathematical physics. His physiological works are principally connected with the eye, the ear and the nervous system. His work on vision is regarded as fundamental to modern visual science.

Working at the same time as Müller, the French physiologist Marie-Jean-Pierre Flourens was the first to demonstrate the general functions of the major portions of the vertebrate brain. Flourens conducted a series of experiments in 1814–22 to determine physiological changes in pigeons after removal of certain portions of their brains (Flourens, 1924). He found that removal of the cerebral hemispheres, at the front of the brain, destroys will, judgement and all the senses of perception; that removal of the cerebellum, at the base of the brain, destroys the animal's muscular coordination and its sense of equilibrium; and that removal of the medulla oblongata, at the back of the brain, results in death. These experiments led him to conclude that the cerebral hemispheres are responsible for higher psychic and intellectual abilities, that the cerebellum regulates all movements, and that the medulla controls vital functions, especially respiration. Flourens was also the first to recognise the role of the semicircular canals of the inner ear in maintaining body equilibrium and coordination. Later work by Pavlov would prove that conditional responses could not be learned by dogs after removal of the cerebral cortex. Thus cerebral cortex was determined to be critical for the formation and storage of conditioned reflexes.

Flourens is generally credited with being the father of experimental brain research. He measured both behaviour and brain events. He contended, in opposition to the earlier claims of phrenology, that behaviours were not localised in the cortex, and argued, as Carl Lashley later would, that behaviours were diffusely localised, and that lesions led to general losses (Lashley's law of mass action). He additionally noted that if the lesions were not too severe, then recovery of function was possible and that restitution of a function was the result of compensation by the remaining intact brain (Lashley's law of equipotentiality; see details in Chapter 9).

The first anatomical proof of the localisation of brain function has also been attributed to Paul Broca, the French surgeon whose study of brain lesions contributed significantly to the understanding of the origins of aphasia, the loss or impairment of the ability to form or articulate words. Much of Broca's research concerned the comparative study of the crania of the races of mankind. He used original techniques and methods to study the form, structure and topography of the brain and sections of prehistoric crania. In 1861 he announced his discovery of the seats of articulate speech in the left frontal region of the brain, since widely known as the convolution of Broca.

Broca's name is usually associated with those of Wernicke and Jackson. The German neurologist Carl Wernicke related nerve diseases to specific areas of the brain. He is best known for his descriptions of the aphasias, disorders interfering with the ability to communicate in speech or writing. In a small book published in 1874, Wernicke tried to relate the various aphasias to impaired mental processes in different regions of the brain; the book included the first accurate description of a sensory aphasia located in the temporal lobe. Wernicke also demonstrated the dominance of one hemisphere in brain functions in these studies. His *Textbook of Brain Disorders* (1881–83) was an attempt to account comprehensively for the cerebral localisation of all neurological disease. John Hughlings Jackson was British neurologist whose studies of epilepsy, speech defects and nervous-system disorders arising from injury to the brain and spinal cord remain among the most useful and highly documented in the field. One of the first to state that abnormal mental states may result from structural brain damage, Jackson discovered epileptic convulsions, now known as Jacksonian epilepsy. Jackson's epilepsy studies initiated the development of modern methods of clinical localisation of brain lesions and the investigation of localised brain functions.

The concept of the neuron as a discrete structural and functional entity had been proposed by Deiters in 1865 and by Waldeyer in 1891 (who introduced the word neuron), on the basis of microscopic dissection and staining of cells in the nervous system.

Two major questions confronted neurologists at the end of the nineteenth and beginning of the twentieth centuries (Allen, 1998):

What was the basic anatomical element of the nervous system (individual cells, or a continuous nerve network)?

How were parts of the nervous system integrated to produce an overall functioning system?

By the early 1900s the first question was resolved ultimately in favour of the neuron theory: individual nerve cells serve as the basic structural and functional units of the nervous system. Central to that debate was the work of the Spanish cytologist Santiago Ramón y Cajal whose study of the embryological development of the nervous system helped to demonstrate that the nervous system arises from many discrete individual cells. The second question, concerning structural and functional organisation of the nervous system, has been an area of great advancement during the twentieth century. As A. R. Damasio (1994) noted, in the small world of brain science, even as early as 1860s, two camps had begun to form. One held that psychological functions such as language or memory could never be traced to a particular region of the brain. If one had to accept, reluctantly, that the brain did produce the mind, it did so as a whole and not as a collection of parts with special functions. The other camp held that, on the contrary, the brain did have specialised parts and those parts generated separate mind functions. The rift between the two camps was not merely indicative of the infancy of brain research; the argument endured for another century and, to a certain extent, is still with us today.

3.2 Two sides of reflex: Pavlov's and Sherrington's branches of reflexology

Pavlov started using the term 'conditional reflex' in 1903, more than a half a century before the term 'neuroscience' appeared. By that time basic concepts concerning the mind/body problem already existed, such as the 'blood–brain barrier' (M. Lewandowsky), the membrane theory for cells (J. Bernstein) and the cytoarchitecture of the anthropoid cerebral cortex (A. W. Campbell).

In developing the theory of conditioning, Pavlov inherited the ideas and investigations of notable Russian physiologists, among whom were Vladimir Bekhterev and Ivan Sechenov.

Sechenov was the first person to appreciate the significance of inhibition and reflex action. During work in Germany and France Sechenov showed that by placing salt crystals in certain parts of a frog's brain he could reversibly inhibit its leg-withdrawal reflex. He returned to St Petersburg and in 1863 published a major classic, *Reflexes of the Brain*. In it he suggested that thinking is based on reflexes within the brain and mental activity has to be considered as a subject of experimental investigations.

Sechenov's description of reflex had begun to diverge from Descartes' notion of simple fixed 'reflections' of stimuli. Firstly, Sechenov suggested that the strength of stimuli and the responses they elicited need not be similar – very weak stimuli might trigger quite intense reactions. Secondly, Sechenov suggested that reflexes are ubiquitous and flexible; for example, he suggested that, as it was his habit to think of politics before going to bed each night, it might happen that if he were to lie down in the daytime the properties of his bedroom might elicit thoughts of politics in him. Sechenov felt that inhibition played a significant role in both of these extensions of Descartes' concept of the reflex. Having demonstrated the existence of centrally mediated inhibition of reflexes, however, Sechenov did not go on to test these later inferences. Sechenov's pioneering work in the field of reflexes led him to the hypothesis that the psyche was actually a function of the brain and that psychical activity can be explained by reflex activity, either innate or learned.

Another famous Russian physiologist and physician, Sergei Botkin, who had worked with Sechenov on some of his studies in Germany, became Professor of Clinical Medicine at the Military Medical Academy in Moscow where he maintained an animal laboratory for the experimental study of physiology. In 1878, Botkin appointed a highly recommended young physiologist from St Petersburg, Ivan Pavlov, to be its director. Pavlov transformed Sechenov's theoretical attempt to discover the reflex mechanisms of psychic activity into an experimentally proven theory of conditional reflexes.

V. M. Bekhterev was interested in the study of conditioning, under the terminology of associative reflexes. After graduating in 1878 he held a position at the psychiatric clinic in St Petersburg. He was then awarded a fellowship to study and conduct research abroad. In 1884–85 he worked with Wilhelm Wundt who is generally acknowledged to be the founder of experimental psychology, and the psychiatrist and neurologist Paul Emil Flechsig in Leipzig. In 1895 he returned to Russia to become professor of psychiatric diseases at the University of Kazan. At Kazan he established the first laboratory for research on the anatomy and physiology of the nervous system. In 1907 he established a psychoneurological institute with his own means. This later had subsidiary institutes and became the State Psychoneurological Academy. Bekhterev served as the institute's first director even after he was forced to resign his medical school appointment under governmental pressure in 1913, and he founded the State Institute for the Study of the Brain, which today bears his name, also in St Petersburg.

Bekhterev was an academic competitor and faculty colleague of Pavlov, with whom he was frequently in open conflicts at meetings, and there was much acrimonious debate between workers from his laboratory and those from Pavlov's. Bekhterev's work and theories were published in the seven-volume *Foundations of Knowledge about the Functions of the Brain*, in Russian in 1903–07, and later in German, French and English. His three-volume *Objective Psychology* was published in Jena, in 1907–12 (Bekhterev, 1926, 1928 (1973); V. M. Bekhterev's *Collective Reflexology* translated and published in 1994). Bekhterev's fundamental books about mind and life moved many scientists in Russia to investigate mental processes. In particular, his books inspired a young student of Moscow University, Nadezhda Ladygina, to ask herself a question: at what level do living creatures start to possess consciousness? This became the central research problem in her life and generated many impressive results obtained on wolves and birds, as well as on a chimpanzee and a child of her own (see Chapters 15 and 16).

It is worthwhile to note that Bekhterev diagnosed Stalin as suffering from 'grave paranoia'. A couple of days after he had visited the Kremlin he suddenly died.

Another branch of reflexology, based on ideas of associations between neurones, was elaborated by the most notable nerve physiologist of the time, Charles Scott Sherrington (the Nobel Laureate of 1932). Sherrington, 8 years younger than Pavlov, was introduced to neurology through histology. He studied neurophysiology with Santiago Ramón y Cajal, who shared with Camillo Golgi the Nobel Prize in 1906 for their work on the structure and function of nerve cells. Sherrington's research focused on individual neurons, on the basis of which all reflexes really act. With this background he studied spinal reflexes as well as the physiology of perception and reactions and arrived to the theory of the synapse. Sherrington concentrated on studying peripheral nervous system, in particular the efferent nerve supply of muscles. He also studied the connection between the brain and the spinal cord by way of the pyramidal tract. In 1897 Sherrington introduced the term 'synapse'. After series of papers, in 1904 he published the book *The Integrative Action of the Nervous System*, which was a landmark in physiological studies.

In studying the salivary reflexes in dogs, Pavlov based his work on Sherrington's investigations of the pyramidal tract. He sought analogies between the conditional reflex and the spinal reflex. According to Sherrington, the spinal reflex is composed of integrated actions of the nervous system involving such complex components as the excitation and inhibition of many nerves, induction (i.e. the increase or decrease of inhibition brought on by previous excitation) and the radiation of nerve impulses to many nerve centres. To these components, Pavlov added cortical and subcortical influences, the mosaic action of the brain, the effect of sleep on the spread of inhibition, and the origin of neurotic disturbances principally through a collision, or conflict, between cortical excitation and inhibition.

With E. Adrian, Sherrington is often referred to as one of the founders of modern neurophysiology, while Pavlov, with his discovery of elemental units of behaviour such as conditional reflexes, became one of the fathers of behaviourism. Pavlov enriched this branch of behavioural sciences by his main idea that physiological methods could be used to study psychological phenomena and that these phenomena must be described and explained in physiological terms if they are to be understood.

Although Pavlov's work contradicted some ideas concerning processes of central inhibition, the two great physiologists maintained scientific contacts: Sherrington visited Pavlov's laboratory in St Petersburg in 1916, and Pavlov visited Sherrington in 1928.

3.3 The puzzle of the memory trace

At the beginning of the twentieth century an animal that acted within a puzzle-box was considered a black box itself. Since then a wide range of techniques have been developed that help us to look inside the black box. An intriguing question that scientists and psychologists alike have been striving to answer for centuries is how and where memory processes occur in the brain. In his Lecture to the Royal Society of 1894, Santiago Ramón y Cajal proposed a theory of memory storage: memory is stored in the growth of new connections of neurons. This prescient idea was almost entirely neglected for half a century as students of learning fought over newer competing ideas (Kandel, 2001).

The German philosopher Hermann Ebbinghaus is fondly remembered as the founder of the scientific study of memory. Ebbinghaus published a book called *Memory: A Contribution to Experimental Psychology* in 1885. Basically, his research focused on the memorisation of nonsense syllables. The fact that it is easier to remember a short sequence of nonsense syllables than a long one seems to be completely trivial. But from the modern physiologists' point of view, this clearly demonstrates that human memory differs from the memory of a computer or, say, a tape recorder, because machines store all obtained information, if they have room for it, up till the moment when they are switched off. This experimental approach for studying the mechanisms of memory was applied to studying animal intelligence and preceded early behaviouristic and neurophysiological researches.

Ideas of early behaviourism moved one of the founders of modern neurophysiology, Karl Lashley, to the search for the 'engram', the material track of learned units. At the age of 16, Lashley studied general zoology and comparative anatomy with Albert M. Reese at the University of West Virginia. Reese appointed him departmental assistant. One of the new assistant's first tasks was to sort out various materials in the basement. Among them he found a Golgi series on the frog brain and proposed to Reese that he should draw all of the connections between the cells in order to know how the frog's brain works. Only later did Lashley realise that functional variables such as spatial and temporal summation, excitatory and inhibitory states, and micromovements of elements influencing synaptic contact need not be represented microscopically. The lesson is that neurons are not inert and static, like soldered wires. They are live metabolising cells with synaptic contacts that vary.

By formal training Lashley was a geneticist, and his first 30 publications, up to 1917, were dedicated to genetic and behavioural problems. Two of them were co-authored with John B. Watson, the founder of behaviourism, with whom Lashley studied physiology at Johns Hopkins University. The problem, Lashley suggested in the 1920s, was the omission of the brain from the Watsonian stimulus–reaction formula. In 1929, Lashley wrote his famous monograph, *Brain Mechanisms and Intelligence*. In his well-known article 'In search of the engram', published in 1950, Lashley summarised his 33 years of research and theory of memory in two principles:

(1) *The equipotentiality principle*: all cortical areas can substitute each other as far as learning is concerned.

(2) *The mass action principle*: the reduction in learning is proportional to the amount of tissue destroyed, and the more complex the learning task, the more disruptive lesions are. Thus, every brain region participates (to some extent) in all brain processes.

In other words, Lashley believed that learning was a distributed process that could not be isolated within any particular area of the brain. Furthermore, it was not the location of the lesion that was important (within reason), but the amount of tissue destroyed that determined the degree of behavioural dissociation. In sum, his theory stated that within functional regions (such as the visual association cortex), all parts of the region were equally effective in carrying out the function normally served by the entire cortical region.

In 1949 Lashley's student, the Canadian physiologist Donald Hebb, presented the most successful theoretical view of the general nature of the engram. His 'theory of cell assemblies' suggested another model of brain functioning based on a simple but powerful intuition: that strengthening and weakening of connections depend on how often they are used. If a connection is never used, it is likely to decay, just like any muscle that is not exercised. Hebb's theory supports the view that changes that occur during learning develop among interconnections of neurons throughout wide areas of the brain. Particular kinds of learning have been proved to involve the development of particular circuits of neurons. The engram does not appear to be localised, but its existence cannot be questioned. Hebb discarded the term 'neural lattice' that was introduced by Lashley in favour of the term 'cell assembly'. Hebb succeeded in presenting a theory of behaviour based as much as possible on the physiology of the nervous system as well as on establishing physiological psychology as part of behaviour theory. A good reason for his integrative approach was that he worked both on emotions and intelligence in chimpanzees and on surging in the human brain. He said that five years of working with chimpanzees gave him more ideas concerning the nature of the human brain than the whole of the rest of his scientific career.

A relative latecomer to academia, Hebb had worked as a schoolteacher, farmer, labourer and novelist before starting to study psychology. When he began as a part-time graduate student at McGill University in Montreal, one of his supervisors, the Russian physiologist Boris Babkin, who had worked with Pavlov at one time, advised him to work experimentally with animals, and that turned Hebb to the work of Köhler and Lashley. Working with Lashley, he received his Ph.D. from Harvard. He then took a fellowship with the famous brain surgeon Wilder Penfield at the Montreal Neurological Institute, where his research revealed that large lesions in the brain often have little effect on a person's perception, thinking or behaviour. Penfield was trying to treat patients with intractable epilepsy. While patients were fully conscious, though locally anaesthetised, he opened their skulls and tried to pinpoint the source of their epilepsy. Penfield and Jasper (1954) probed the conscious brain (which has no pain receptors) with electrodes and asked the patients to describe their sensations as various parts of the brain were stimulated. Some patients 'heard' conversations that had taken place years before, some heard music and some seemed to find themselves with old friends, long deceased. They seemed almost to relive the experiences, rather than simply to remember them. The powers of recall under such circumstances were phenomenal.

Penfield's experimental surgery led him to a dramatic discovery. Stimulation anywhere on the cerebral cortex could bring responses of one kind or another, but he found that only by stimulating the temporal lobes (the lower parts of the brain on each side) could he elicit meaningful, integrated responses such as memory, including sound, movement and colour. These memories were much more distinct than normal memory, and were often about things unremembered under ordinary circumstances. If Penfield stimulated the same area again, the exact same memory popped up – a certain song, the view from a childhood window – each time. It seemed he had found a physical basis for memory (Penfield and Rasmussen, 1950).

In 1942, Hebb worked with Lashley again, this time at the Yerkes Laboratory of Primate Biology. He then returned to McGill as a professor of psychology, and became the department chairperson in 1948. The following year, he published his most famous book, which made McGill a centre for neuropsychology. The basics of his theory can be summarised by defining three of his terms (see Herenhahn and Olson, 2001). Firstly, there is the Hebb synapse. Repeated firing of a neuron causes growth or metabolic changes at the synapse that increase the efficiency of that synapse in the future. This is often called consolidation theory, and is the most accepted explanation for neural learning today. Secondly, there is the Hebb cell assembly. These are groups of neurons so interconnected that, once activity begins, it persists well after the original stimulus is gone. Today, people call these neural nets. And thirdly, there is the phase sequence. Thinking is what happens when complex sequences of these cell assemblies are activated. Hebb first distinguished between immediate memory (short-term memory, or STM) and long-term memory (LTM). We will return to this matter in Part III.

Further development of knowledge about memory processes in the brain has been closely connected with development of methods. Before the middle of the twentieth century, the techniques for investigating the functions of the brain were relatively primitive. Most research on the relationship of the brain with behaviour was conducted by removing sections of brain (*ablation*) or damaging them (*lesion*). The distinction between memory trace and sensory circuits permits several interpretations of localisation studies. In such studies, animals are typically trained on some response, and a brain structure, hypothesised to be involved in the learning of that response, is destroyed. Post-lesion behaviour is compared to pre-lesion behaviour. If the lesion abolishes the learned behaviour, it is tempting to conclude that the destroyed structure must have contained the memory trace of the behaviour.

In order to determine whether learning-induced changes actually develop in a given brain structure, more information than can be garnered from lesion studies must be obtained. After 1940 the widespread use of the microelectrode and its accompanying technology (electronic amplifiers, oscilloscopes, etc.) had a revolutionary impact on neurophysiological theories of behaviour. More advanced methods have been developed since then, including electrical stimulation of and recording from selected brain areas; chemical stimulation; and much more precisely limited lesion techniques (such as stereotaxic surgery). These research techniques have greatly increased our knowledge of brain functioning and its relationship to learning. The nerve impulse was observed in almost all parts of the peripheral and central nervous systems and came to be regarded as the universal currency, whether it was induced by electrical stimulation or sensory stimulation or occurred 'spontaneously'.

Indeed, progress in neuroscience was based on oscillations from moving behaviour theory away from the physiology of the organism for behavioural explanation, and again into the physiology of the organism itself.

The search for the engram continues even today. But what we perceive to be the engram has changed. No longer do researchers assume that a single anatomical structure in the brain changes when learning takes place. One of central questions in modern neuroscience concerns the nature of the changes in a nerve cell underlying learning and memory. Researchers have joined together a series of different techniques to try to reveal how the brain can memorise and learn. Observations and cognitive tests with people who have lost part of their brain due to accidents or strokes provide information on the functions of different regions of the brain. Studies on the effects of chemotherapies reveal information about neurotransmitters. Electroencephalograms and positron emission tomography (PET) scans reveal activity patterns in different parts of the brain while people carry out cognitive tasks. During brain surgery with awake patients, electrode stimulation of different parts of the brain as well as measures of specific neuronal activities help map the impacts of different neural stimuli, and of different cognitive experiences. The introduction of genetic engineering techniques to manipulate genes that code for particular proteins has brought a new level of accuracy to the study of neuronal basis for memory processes.

Chapter 4

Integrative approaches and coherent movement in studying animal intelligence

4.1 Wholes perceive the wholes: the Gestalt approach to perception and learning

During the first decade of the twentieth century, the opposite approach to studying intelligence was developed in parallel with behaviourism, based on Gestalt theory. The Gestalt approach to learning originated in Europe around 1912, with Max Wertheimer's insights on perception and, initially, emphasised perceptual phenomena. 'Gestalt' is the German word for 'configuration', or 'organisation'. It also means any segregated whole.

Gestalt theory focused on the whole perceptual experience and not on its elements. Gestalt theorists asserted that perception ought to be more than the sum of the things perceived, that the whole is more than the sum of the parts. At the same time, they considered the organism as not really being divided into 'mind' and 'body', because it is the whole that reacts to the environment. The organism does not construct a perception by analysing a myriad data but rather perceives the form as a whole and is able to solve most problems not due to decomposition of the problem but to sudden insight which cannot be broken down into atomic processes.

Gestalt theory was established mainly through the researches of Max Wertheimer, Wolfgang Köhler, Kurt Koffka and Kurt Lewin, who were trying to find natural science laws of integral behaviour. The coming to power of National Socialism in Germany substantially interrupted the fruitful scientific development of Gestalt theory in the German-speaking world; Koffka, Wertheimer, Köhler and Lewin emigrated to the United States.

The term Gestalt as the Gestalt psychologists were to use it was first mentioned by the Austrian philosopher, Christian von Ehrenfels, in his paper 'Über Gestalt qualitäten' ('On Gestalt qualities') (1890). This was the first definite claim that thousands of perceptions have characteristics which cannot be derived from the characteristics of their ultimate components, the so-called sensations. One of von Ehrenfels' students was Max Wertheimer.

Wertheimer was born in Prague in 1880. He studied law and philosophy in Germany. In 1910, he became interested in the perceptions he experienced on a train travelling on a vacation. While stopped at the station, he bought a toy stroboscope – a spinning drum with slots to look through and pictures on the inside, sort of a primitive movie machine or sophisticated flip book. In a hotel room he set up the experiment by substituting strips of paper on which he had drawn series of lines for the pictures in the toy. The results were as he expected: by varying the time interval between the exposures of the lines, he found that he could see one line after another, two lines standing side by side, or a line moving from one position to another. At Frankfurt, his former teacher Friedrich Schumann gave him the use of a tachistoscope to study the effect. His first subjects were Schumann's two young assistants, Wolfgang Köhler and Kurt Koffka. They would become Wertheimer's lifelong partners. In 1912 Wertheimer published his seminal paper 'Experimental studies of the perception of movement'. This marked the birth of Gestalt theory.

Gestalt principles are based on *perception* versus *sensation*. While sensation refers to simple, detectable simulation, perception refers to an organisation and interpretation of sensations. The phi phenomenon discussed by Wertheimer is the perception of movement from stationary stimuli which flash on and off at a certain frequency. The perception of several sensory elements in combination is often different from the sum of individual elements themselves. Gestalt psychologists study the perception of stimuli rather than the actual stimuli in relation with observable behaviour. This sets them apart from behaviourists. Gestalt psychologists consider organisms with their perceived environment as the field. They borrowed the term 'field' from physics where it was defined as a 'dynamic interrelated system, any part of which influences every other part'.

The Gestalt psychologists argued that we are built to experience the structured whole as well as the individual sensations. And not only do we have the ability to do so, we have a strong tendency to do so. Perception of the whole does not depend on perception of all of its parts; we recognise the shape of a landscape long before we recognise each tree and rock in the landscape, and we recognise that a tree is a tree before we recognise what kind of tree it is, because recognising the species requires an analysis of its parts. A set of dots outlining the shape of a star is likely to be perceived as a star, not as a set of dots. We tend to complete the figure, make it the way it 'should' be, finish it.

The question of whether non-human animals are able to complete an amputated to a whole image has been intensively studied over the past century and it remains an issue even today. Most studies have used Pavlovian conditioned procedures in order to reveal these abilities in animals. After learning to discriminate between whole figures, subjects respond to pictures represented only by parts of them. Comparative studies have demonstrated that members of some species including mice, pigeons and primates can perceive and discriminate complex stimuli based on either local parts or the global configuration, much like humans. However, comparative data give a complicated picture because for some species specificity of visual perception shades their ability to complete amputated images (see Vallortigara (2004) for a detailed review).

Gestalt theory is well known for its concept of *insight learning* (see details in Chapter 17). The most famous example of insight learning in animals involved a chimpanzee named Sultan. He was presented with many different practical problems. When, for example, he was allowed to play with sticks that could be put together like a fishing rod, he appeared to consider thoughtfully in a very human fashion the situation of the out-of-reach banana – and then rather suddenly jumped up, assembled the sticks, and reached the banana. Sultan was the most intelligent chimp in the chimpanzee colony established by Köhler, which was the first in the world. Köhler was working at a primate research facility maintained by the Prussian Academy of Sciences in the Canary Islands when World War I broke out. Marooned there, Köhler had at his disposal a colony of nine chimpanzees. He constructed a variety of problems for the chimps, each of which involved obtaining food that was not directly accessible. Chimpanzees appeared to demonstrate a holistic understanding of problems, by arriving at solutions in a sudden moment of revelation or insight (see Chapter 17).

Köhler described his findings in *The Mentality of Apes*. It was published in German in 1917 and translated into English in 1925. This is still one of the most often referred-to books on cognitive ethology. It is important to note that Köhler's long-term experimental work was the first systematic study of animal intelligence based on Gestalt principles and may be considered a new phase in the development of comparative psychology.

At the same time as Köhler, Russian zoopsychologist Nadezhda Ladygina Koths bought a young chimpanzee named Ioni with whom she worked closely for 3 years (Fig. 4.1). This work was based at the Moscow Darwinian Museum which had been established by her husband Alexander Koths who was also one of her first professors in Moscow University. This museum was initially established in 1907 as a scientific collection for the first High Courses for Women in Russia. Ladygina Koths was the first curator of the Darwinian Museum. Just married, the young couple enthusiastically gathered the displays for their museum. Based at the Darwinian Museum, Ladygina Koths established a laboratory of experimental psychology. When a girl, she had dreamt of becoming a psychiatrist, and Ioni became her very first 'patient' and at the same time the first experimental chimpanzee in Russia (and one of the first in the world, indeed). Ladygina Koths conducted many experiments in learning and tool-using with Ioni in 1913–16, and compared her observations of an infant chimpanzee in her laboratory with those of her own son, Rudy, in 1925–29. Her first book, *Study of Cognitive Abilities in the Chimpanzee*, was published in 1923, and the second was *Infant Chimpanzee and Human Child*, first published in 1935 and than re-issued in English and German several times (see the version edited by de Waal (2002)).

Fig. 4.1 Nadezhda Ladygina Koths with her young chimpanzee Ioni. (From the archives of the Moscow Darwinian Museum; courtesy of Zoya Zorina.)

In 1916 Robert M. Yerkes published an article in *Science* in which he called for the establishment of a primate research institute for the systematic study of the 'fundamental instincts' and 'social relations' of primates (Yerkes, 1916a). In 1925 he bought a chimpanzee and a bonobo, Panzy and Chim, from a sailor in Boston and they became the first members of his first non-human primate research laboratory in the United States, where Yerkes acted was Director from 1929 until 1941. The Yale Laboratories of Primate Biology were eventually renamed the Yerkes National Primate Research Center which is now located in Atlanta, Georgia.

Despite being a pupil of the eminent behaviourists Watson and Thorndike, Yerkes later accepted a more holistic approach to study animal behaviour, close to Gestalt ideas. He perhaps was predisposed to comparative investigations of intelligence in non-human primates by his work on human intelligence testing. Yerkes revised the Stanford–Binet Intelligence Scales in 1915 to create a widely used point scale for the measurement of human mental ability. He was a principal figure in the development of human multiple-choice testing. During World War I, Yerkes directed a team of 40 psychologists charged with assessing the abilities of army recruits for training, assignment and discharge purposes. The methods they elaborated had a far-reaching effect on civilian life after the war. Yerkes published the fundamental books *The Mental Life of Monkeys and Apes* (1916b), *The Great Apes* (1929) (co-authored with Ada Yerkes) and *Chimpanzees: A Laboratory Colony* (1943).

As early as 1925 Yerkes suggested the idea of teaching chimpanzees sign language as used by deaf people in order to make dialogue between humans and non-humans possible. The first series of experiments was performed 40 years later by Allen and Beatrix Gardner (see Gardner and Gardner, 1989) and led to a very dramatic shift in ideas of animal intelligence. Since then, Gestalt psychologists have adjusted their research efforts to address more and more complex phenomena, and this has included direct dialogue with apes and other animals based on artificial intermediate languages (see Chapter 30).

4.2 A cognitive map of the learning land: from behaviourism to cognitivism via Gestalt theory

The dramatic shift from behaviourism to cognitivism occurred mainly due to the contributions of Gestalt theory. After years of almost exclusively behaviourist research, psychologists and educators became dissatisfied with the limitations of behaviourism. Gestalt psychologists initiated the study of more complex problem-solving and perceptual problems.

The origin of cognitivism can be traced back to the early part of twentieth century when the Gestalt psychologists in Germany, Edward Chase Tolman in the United States, and Jean Piaget in Switzerland had a tremendous influence on moving psychology away from behaviourist theories. Behaviourists argued that mental events were impossible to observe and measure and could not therefore be studied objectively. Consequently behaviourists could not explain the ways learners attempt to make sense of what they learn. Cognitivists proposed that through empirical research and observation inferences could, indeed, be drawn about the internal, cognitive processes that produce responses.

In the 1920s, Piaget began a research programme in Geneva that focused on epistemology, the origins of knowledge. Piaget began his career with the study of snails, and he extended the use of careful behavioural observations and descriptions to his studies on human cognitive development. Central to Piaget is the idea that children are able to solve certain problems only at certain ages and that these problems can be organised into a developmental sequence that defines discrete stages of cognitive development. Piaget proposed several stages of cognitive development in children: sensorimotor (using sensory and motor capabilities), pre-operational (using symbols and responding to objects and events based on how they appear to us), concrete operations (thinking logically) and formal operations (thinking about thinking). He proposed that children grow and develop through each of these stages until they can reason logically (Piaget, 1932, 1937). Piaget elaborated many diagnostic tests of children's capacities that corresponded with the stages of cognitive development. Since the 1980s, these tests have been widely applied to the study of cognitive abilities in different species of mammals and birds. This approach is currently bringing exciting results which are consequently changing ideas of mental skills in animals (see details in Part V).

Piaget, Tolman and the Gestalt psychologists all share the notion that human knowledge is structured and organised. Tolman was initially trained in electrochemistry. He changed the course of his career after reading the works of William James. After his first year as a graduate student, he went to Giessen in Germany to study for his Ph.D., and was introduced to Gestalt psychology through the teachings and readings of Koffka (see Kimble *et al.*, 1991; Hothersall, 1995). In 1918, Tolman went to the University of California at Berkeley, where he began to study maze learning in rats – a research programme that made the department of psychology at Berkeley world famous. To study learning, he conducted several classical rat experiments. He examined the role that reinforcement plays in the way that rats learn their way through complex mazes. These experiments eventually led to the theory of *latent learning* which describes learning that occurs in the absence of an obvious reward (see Chapter 17).

The most important contribution to the learning theory was Tolman's concept of *cognitive map* based on his ideas of *purposive* (or *goal-directed*) *behaviour*. A cognitive map is a spatial schema or representation. In learning a coordinated set of spatial relations, an organism is sometimes said to be developing a cognitive map. The term is most likely to be invoked when the organism orients toward locations that it cannot see or otherwise respond to directly. Tolman and his co-authors used different mazes that were so designed as to test alternative hypothesis: whether an animal had learned the response, say, 'turning right', or had developed a 'cognitive map' of the maze. The main idea of Tolman's approach is that he took a *molar* rather than a 'molecular' view of the study of behaviour. 'Molar' refers to an approach that studies extensive patterns of behaviour that appear directed toward some goal, while 'molecular' refers to approaches that emphasises specific stimulus–response relationships only (see Part IV).

Tolman identified himself as a behaviourist and eschewed the type of introspection that was practised by Wundt and Titchener. However, he was also opposed to the type of behaviourism dominating the field at that time and expressed this in his paper 'A new formula for behaviorism' (1922). Like the behaviourists, Tolman valued the importance of objective research; however, being influenced by Gestalt theory, he included mental phenomena in his perspective of how learning occurs. Tolman's view of learning was more holistic rather than behaviouristic. In his 1932 book *Purposive Behavior in Animals and Men*, Tolman clarified behaviour as largely reaction and effect-oriented. It is purposive in that it is designed and executed to attain or avoid something, exemplified by 'persistence and the tendency to use the shortest route'. Tolman is classified now as a cognitive behaviourist and the originator of the cognitive theory of learning.

4.3 'Forget about schools': the development of an integrative approach to the study of animal intelligence

'Forget about schools' is a quotation from Köhler's Address of the President at the 67th Annual Convention of the American Psychological Association in 1959. After half a century of battles between behaviouristic and Gestalt psychologists, Köhler invited students of animal intelligence to proceed in another direction.

He said:

> I suggest that, in this situation, we forget about schools. The Behaviorist is convinced that his functional concepts are those which we all ought to use. The Gestalt psychologist, who deals with a greater variety of both phenomenal and physical concepts, expects more from work based on such premises. Both parties feel that their procedures are scientifically sound. Why should we fight? Many experiments done by Behaviorists seem to me to be very good experiments. May I now ask the Behaviorists to regard the use of some phenomenal facts, and also of field physics, as perfectly permissible? If we were to agree on these points, we could, I am sure, do excellent work together. It would be an extraordinary experience – and good for psychology.

The 1950s was a prolific decade for the development of an integrative approach to the study of animal intelligence. Cognitive psychologists generally agree that the date of the birth of cognitive psychology should be reckoned as 1956, when a large number of researchers published influential books and articles on attention, memory, language, concept formation and problem-solving. Enthusiasm for the cognitive approach grew rapidly, so that by about 1960, methodology, approach and attitudes had all changed substantially.

At about that time the first International Ethological Conferences were organised, which directed minds along a path that would subsequently lead to cognitive ethology. In 1949 the Symposium on Physiological Mechanisms in Animal Behaviour brought to Cambridge most of the key players in animal behaviour at the time, such as Niko Tinbergen and Gerard Baerends (the Netherlands), Konrad Lorenz (Austria), William Thorpe and J. Z. Young (UK), Erich von Holst and Otto Köhler (Germany) and Paul Weiss and Karl Lashley (USA), an international group that renewed the friendships they had built before World War II (Brockmann, 2001). The first International Ethological Conference was held at Buldern in 1952, a castle in Westphalia and the site of Konrad Lorenz's first institute. The second was held at Oxford in 1953, and the Conferences have been held at regular 2-year intervals since then. At the turn of the millennium, the new paradigm has been developed based on integration of holistic approach and studying the atomic mechanisms of learning and memory. This helps us better understand not only animal but also human learning and to find congruences between us and other species.

Chapter 5

Ethological approaches for studying animal learning

After many years of studying the behaviour of animals placed in boxes and forced to press bars and pull ropes like Little Red Riding Hood, some naturalists had come to understand that in the natural world the organism has to cope with the challenges of the environment and of its own body rather than with artificial tasks fabricated by experimentalists. In the 1930s, this led to the second birth of ethology, as a branch of science devoted to the study of animal behaviour as it occur in natural contexts, often in natural surroundings.

I would like to clarify the above words about the *second* birth of ethology. The study of animal behaviour as a self-contained discipline has a tradition that stretches before the time of Darwin. The term 'ethology' appeared at the same time as another one, 'comparative psychology'. Both terms came out of the polarisation in French biology created by the debates between Cuvier and Geoffroy Saint-Hilaire (see Jaynes, 1969). Flourens founded *comparative psychology* in 1864, and Saint-Hilaire founded *ethology* in 1859. Whereas comparative psychology as a term was eagerly taken up (five texts with these words as the title appeared in the late 1870s), ethology was less successful.

Saint-Hilaire defined ethology in the same sense as Haeckel implemented ecology, i.e. as a discipline concerning relations of organisms with each others and their environment. In his *Logic*, the British philosopher J. S. Mill (1843) called for a science of character, to be called 'ethology'. This went back to a tradition in the seventeenth century when an actor, often a mime, who portrayed human characters, was called an ethologist. In the eighteenth century ethology was defined as a science of ethic. In Mill's sense, ethology is the term for the study of how a person with certain characteristics acquires others under certain circumstances – the study of personality change. In 1870, A. Giard suggested the distinction between the terms 'ethology' concerning habits and relations of animals, and 'comparative psychology' concerning feelings, intelligence, movement and orientation. At last in 1895 the Belgian palaeontologist L. Dollo suggested 'ethology' as a specific branch of science for studying behaviour of organisms. In the early 1900s, the pioneering work of American entomologist W. Wheeler with ant social organisation and behaviour became instrumental in establishing ethology as a significant and serious branch of modern biology. Wheeler advocated the use of the term ethology to mean the investigation of habits, instincts, intelligence and, in general, mode of life in animals.

In 1911 the German biologist Oskar Heinroth, the founder of 'Vergleichende Verhaltensforschung' (comparative ethology) and the scientific mentor of young Konrad Lorenz, the future founder of classical ethology, published his famous paper on the ethology of ducks. Heinroth defined ethology as investigation of innate habits, manners and rituals in animals, combining them as communicative systems. This interpretation surprisingly matches up with Mill's treatment of the term (Durrant, 1986).

In the 1930s, P. Pelseneer insisted on that ethology should be quantitative, comparative and phylogenetic. At that time, the researches of Konrad Lorenz and Niko Tinbergen established *ethology as a distinct branch of biology*, which has since had an effect on such wide-ranging disciplines as genetics, anthropology, and political science in addition to psychology. Ethologists believe that an animal must be studied on its own terms rather than primarily in relation to human beings, with a focus on its normal behaviour and environment. They study animal behaviour from the dual perspective of both *proximate explanations* (which concern the individual lifetime of an animal) and *ultimate explanations* (which concern an animal's phylogenetic past). Proximate explanations answer questions about how a specific behaviour occurs; ultimate explanations answer questions about why behaviour occurs.

The field of ethology is considered to be the study of evolution and functional significance of behaviour. This definition was framed by C. O. Whitman in the 1800s. Whitman studied the display patterns of pigeons and coined the word *instinct* as a specific term to describe such behaviours (initially this word was used in the same sense by Condillac in 1754). Many of these instincts are triggered by various environmental stimuli. A hundred years later, the German biologist Jakob von Uexküll termed such triggers of instinctive stereotyped behaviours *sign stimuli*. Von Uexküll (1909) also used the German term *Umwelt* to describe the unique world that each species, even each individual, inhabits. Not only are the worlds of sight and smell vastly different from creature to creature, but as von Uexküll realised, our senses of time and space differ as well (Gould, 1982a).

The notion of the 'private world' developed by von Uexküll inspired Karl von Frisch, who blazed a trail for serious exploration of the various and distinct sensory worlds of animals. Von Frisch is most famous for his work on honeybees, most notably for discovering how the foragers communicate the location of food sources using the so-called 'dance language'; this work earned him the Nobel Prize in 1973, 50 years after his first publication in the field. It is relatively little known that he started his investigations from the starting point of pioneering work on how bees and fishes perceive the world, and he developed this theme during his lifetime. His results in this field were often at variance with current concepts at that time. For example, von Frisch showed that fish could hear, and were able to see colours, despite the claims to the contrary by the famous psychologist von Hess. The results obtained on bees were also rather controversial. As von Frisch said in his Nobel lecture, many biologists thought that bees and other insects were totally colour-blind. But he was unable to believe it, because the bright colours of flowers can be understood only as an adaptation to colour-sensitive visitors. This was the beginning of experiments on the colour sense of bees.

The ethological approach had a strong Darwinian tradition underlying its development. Much of the work of early ethologists was further developed by the Dutch biologist Niko Tinbergen and the Austrian biologist Konrad Lorenz. They shared a Nobel Prize in 1973 with von Frisch for their 'discoveries in the field of the organisation and occurrence of individual and social behavioural patterns' in the animal world (Ritter, 1988). This was the first Nobel Prize to be awarded to behavioural scientists and was shared by the founders of the field of ethology. This event definitely placed ethology on the solid ground of well-accepted sciences.

Lorenz and Tinbergen were tied by lifelong friendship and allied research interests. They first met at a small symposium in "Instinct" organised by van der Klaauw in Leiden in 1936. Tinbergen became the second pupil of Lorenz (after Alfred Seitz) and did postdoctoral research under him, but they both learned equally as much from the other. Their collaboration was interrupted by the outbreak of World War II, but soon after they were invited to the United States to lecture on their findings in animal behaviour. It is a common view that the European and American schools of the study of animal behaviour have developed separately in isolation from each other. From the 1930s, European founders of ethology had known and visited investigators of behaviour in the United States, such as Schneirla, Yerkes and Thorpe, as well as the evolutionists Ernst Mayr and David Lack. This all helped, as Tinbergen (1973) said, 'in bridging the gap (so much wider than we had realised) between ethology and neurophysiology'.

Lorenz is noted primarily for his work on genetically programmed behaviour in animals. He formulated many of the fundamental ideas in ethology, and developed the first coherent theory of instinct and innate behaviour (Lorenz, 1950, 1965, 1969). Tinbergen is famous for diverse studies of the natural behaviour of many animals including insects, fish and birds (Tinbergen, 1942, 1951, 1963). He was a pioneer in experimental field ethology and developed a field methodology of high precision. In particular, Tinbergen manipulated the environment to explore the rules of behaviour and formulated a general research programme of studying animal behaviour (see Part VII). Tinbergen argued that the scientific study of animal behaviour must begin with careful observation and description. We cannot explain why an animal acts in some fashion until we know the animal's ordinary activities. Ideally, description results in a complete inventory of the behaviour patterns of a species. This is called an *ethogram*. The ethogram can then be used to formulate questions about the causes of behaviour. However, Tinbergen cautioned against moving to causal issues before a complete ethogram had been obtained.

One of the most famous pupils of Lorenz, Irenaus Eibl-Eibesfeldt, elaborated experiments on young mammals (mainly squirrels, polecats and hamsters) and revealed intricate interlacing of innate and learned behaviours during the early lifetime in animals. His research on the behavioural ontogeny of mammals contributed significantly to the so-called *nature–nurture* debate. Eibl-Eibesfeldt demonstrated ways in which phylogenetic adaptations determine mammalian behaviour and how learning contributes to integrating behaviour into functional units. His studies on communicative behaviour were aimed at tracing the process of *ritualization* by which signals evolve as well as at the functional aspect of how signals control social interactions. In 1970 he published *Ethology: The Biology of Behavior*, the first textbook in which the entire scope of the field was laid out. Eibl-Eibesfeldt (1971, 1979, 1989) was one of the founders of *human ethology* as a distinct field of ethology (see also Eibl-Eibesfeldt and Salter, 1998).

It is worth noting that techniques developed by the behaviourists, such as classical conditioning, proved to be very valuable to ethologists. Animals trained to respond to a particular stimulus in the appropriate behavioural context can demonstrate their ability to discriminate between different types of stimuli. For example, von Frisch used classical conditioning to demonstrate colour vision in bees and fishes. At the same time, classical ethology has been based mainly on experiments that were conducted in natural surroundings for species investigated. Early ethologists demonstrated that even 'simple' animals are capable of learning when their ecological situation requires it.

The coherent development of ethology and experimental comparative psychology has resulted in the modern branches of ethology, in particular, cognitive ethology and neuroethology.

Cognitive ethology is the comparative, evolutionary and ecological study of non-human animal minds including thought processes, beliefs, rationality, information processing and consciousness (Bekoff, 1995). The modern era of cognitive ethology and its concentration on the evolution and evolutionary continuity of animal cognition is thought to have begun with the appearance of Donald R. Griffin's (1976) book *The Question of Animal Awareness: Evolutionary Continuity of Mental Experience*. Cognitive ethologists differ from comparative or cognitive psychologists in their approaches and methods but can learn important lessons from one another (Allen and Bekoff, 1997; Bekoff, 2002). Cognitive ethologists try to observe animals under natural conditions solving the kinds of problems for which their intelligence has become adapted by evolution (Fig. 5.1). Surely they should have regard to the importance of controls which often may be problematic in a changeable environment. Cognitive psychologists specialise in highly controlled experimental

Fig. 5.1 The development of cognitive ethology made Pavlov's experiments a doggy thriller. (Cartoon by P. Ryabko.)

procedures and they could broaden their horizons and learn more about the importance of more naturalistic methods as used by ethologists. While it may be easier to study animals in captivity, they must be provided with the complexity of social and other stimuli to which they are exposed in the field. As a rule, cognitive ethologists avoid talk of 'lower' and 'higher' animals and also mind that in many instances sweeping generalisations about the cognitive skills (or lack thereof) of species are based on small data sets from a limited number of individuals.

One of the main goals of cognitive ethology is to integrate the answers to questions traditionally asked in psychology laboratories with the answers to questions about ecology and evolution. This approach was examined in Sara J. Shettleworth's 1998 book *Cognition, Evolution, and Behavior*. Her own study on the comparison of memory, cognitive abilities and brain processes in food-storing birds and animals may be considered one of the most effective examples of experimental investigation in the field of cognitive ethology.

CONCLUDING COMMENTS

To conclude this part, I would like to note that the turn of a century is the traditional time for asking questions about the place of humans in nature. One of central problems concerns the evolutionary routes that have led to human intelligence. We know now that although many animals, from ants to apes, sometimes behave 'like humans', they seem to use a variety of intricately interlocked specific-purpose subroutines to solve their vital problems. It is a challenge to understand the degree of flexibility of intelligent behaviour in non-human animals, and this is the main problem considered in the rest of this book.

Part II

Animals are welcomed to the class: learning classes

To dogs by culture so refined,
The wised man is well inclined;
And e'en your favour he may earn,
Who from his tutor thus can learn.

Johann Wolfgang von Goethe, *Faust*
(trans. Lord Francis Leveson Gower)

We start by making an integrative portrait of an intelligent animal whilst describing basic forms of learning. As we saw in Part I, our knowledge about animal intelligence has not risen gradually to the discovery of top-level cognitive skills from the investigation of simple forms of learning. At the beginning of the twentieth century two scientific schools that approached learning basing on insight (Gestaltism) and on conditioning (behaviourism) had started almost simultaneously on their efforts to describe learning processes quantitatively and objectively. Recently new experimental methods have been found for studying animal cognition. For example, intermediary languages have been elaborated for carrying on a dialogue with several species including apes, dolphins and parrots, allowing one to ask animals directly what they think about the world in general and their trainers in particular. At the same time, both classical and operant conditioning are assumed as basic for many modern methods for studying animals' ability for abstraction, categorisation and extraction of rules. Discrimination learning based on the principles of stimulus discrimination and stimulus generalisation is actively used by neurophysiologists as an effective tool for studying neurological mechanisms of memory.

As we now know, individual adaptive behaviour involves different kinds of learning together with innate behavioural patterns. For our relative comfort, we will consider a labelling system generally acceptable to ethologists and keeping with Thorpe's (1963) classification of learning classes (see also Wallace, 1979):

(1) Habituation
(2) Associative learning
  (a) Classical conditioning (also called Type I, or the conditioned reflex)
  (b) Operant conditioning (also called Type II, or instrumental conditioning)
  (c) Trial and error
  (d) Gestalt perception
  (e) Learning sets
(3) Latent learning
(4) Insight learning
(5) Imitation
(6) Play
(7) Imprinting

While roughly keeping to this classification, we will not adhere closely to it because our knowledge about learning has changed since the late 1970s; besides, every researcher perhaps has some special feelings concerning the order and subordination of different kinds of learning. In particular, so-called *guided learning*, which is not included in this scheme, will be considered in Part VII. Social learning nowadays is considered an integrative class of learning that includes 'imitation' as a subordinate form of learning. This theme will be considered in Part VIII. Besides it seems more natural in the modern context to consider 'Trial and error' as a subset of instrumental or operant conditioning, and 'Gestalt perception' as a subset of 'Insight'. Such forms of learning as 'Insight', 'Latent learning' and 'Learning sets' as well as 'Concept formation', which does not figure in Thorpe's classification of cognitive learning and intelligence, will only be mentioned in this Part but will be analysed in detail in Parts V and VI. In this Part we will concentrate on *habituation* and *associative learning*, and merely sketch other classes of learning, which require more cognitive explanations and will be considered in later parts.

Chapter 6

Habituation and associative learning

6.1 Habituation

Habituation is usually considered the simplest form of learning, although its mechanisms and correlation with memory processes are still not completely clarified.

Habituation may be defined as the relatively persistent waning of a response as a result of repeated or continuous simulation that is not followed by any kind of reinforcement (Thorpe, 1963; Hinde, 1970), or a decline in the responsiveness to a stimulus as a result of its repeated presentation (Pearce, 2000). In its widest sense, habituation is simply learning not to respond to specific stimuli that tend to be without significance. Habituation is probably one of the more widely occurring types of learning responses and this may be found and experimentally demonstrated in great variety of organisms. Indeed, one finds habituation everywhere. One of the first experimental examples came from *Paramecium*: Jennings (1906) found that an infusorium reacts to being touched by contracting. With continued touching the number of stimulations needed to produce this response increases to 20 or 30. Similar patterns may be observed in different species, from *Hydra* that drops its responses for tapping a glass where it lives if nothing happens, to human beings who ceases to respond to noise and other stimuli that are given repeatedly when nothing either pleasant or unpleasant happens.

Timing is critical in habituation. The marine ragworm *Nereis* will withdraw into its burrow in response to a wide range of stimuli, from touches and shadows to bright light and electric shocks. Habituation to a bright flash or light occurred in fewer than 40 trials when the stimuli were given at 30-second intervals; but 80 trials were necessary for habituation to occur if the stimuli were given 5 minutes apart (Clark, 1960).

The process paired with habituation is *sensitisation*. This is an increase in the response to an innocuous stimulus when that stimulus occurs after a significant stimulus. For example, when the siphon of the sea slug *Aplysia* is gently touched, the animal withdraws its gill for a brief period. However, if preceded by an electrical shock to the tail, the same gentle touch to the siphon will elicit a longer period of withdrawal. The sensitisation response to a single shock dies out after about an hour, and returns to baseline after a day. In this case sensitisation underlies short-term memory. However, if the animal is sensitised with multiple shocks given over several days, its subsequent response to a gentle touch on the siphon is much larger and retained longer. This is an example of long-term memory and requires protein synthesis (Kandel, 2001).

Habituation also can be long-lasting. Thus, if you make an unusual sound in the presence of a dog, it will respond – usually by turning its head toward the sound. If the stimulus is given repeatedly and nothing either pleasant or unpleasant happens to the dog, it will soon cease to respond. When fully habituated, the dog will not respond to the stimulus even though weeks or months have elapsed since it was last presented.

It has been demonstrated in some experiments that habituation and sensitisation move in opposite directions over time, i.e. habituation increases whereas sensitisation declines.

There are a number of factors that can increase or decrease the response strength. Also, it has been found for different species that each type of stimulus (light, electricity, shadows or touch) has its own characteristic rate of habituation at any trial frequency.

Habituation as the basis of test procedures is widely used by modern experimenters. These tests often allow researchers to obtain new results in different fields, from molecular mechanisms of memory to remarkable linguistic abilities of human neonates. As an illustration, let us consider an example from a set of experiments supporting the hypothesis that human newborn infants during the first few months of life can discriminate phonetic constraints. Human newborns provide scientists with the same type of challenge that is offered by non-human species: we cannot simply ask them what they perceive, we must use some tricks to obtain the answer. The trick used by developmental psychologists is just a procedure of habituation/dishabituation (Kellman and Spelke, 1983). Infants are habituated to a particular stimulus, and after habituation has occurred, they are shown a somewhat changed stimulus. Dishabituation indicates that the changed stimulus is perceived by babies as a novel one. The ability to discriminate phonemes in babies was firstly studied in the 1970s by behavioural methods, namely, by repeating an acoustic stimulus for several minutes until some behavioural response of the infant habituates, and then examining whether the response recovers when the stimulus is changed. For example, babies always stop sucking a pacifier when they react to a novel stimulus and return to this pastime when habituated. Later high-density recordings of event-related potentials were applied in order to reveal how fast 3-month-old infants can detect phonetic changes and what brain mechanisms are involved (Dehaene-Lambertz and Dehaene, 1994). The behavioural basis was the same. Infants were seated in a carrier affixed to their parents, and their heads were covered with a geodesic sensor net in a very soft way. They faced a loudspeaker placed on top of a TV monitor where a silent video showed attention-grabbing coloured objects. The video was not synchronised with the auditory stimuli, thus preventing any visually evoked potentials. On each trial, a sequence of five syllables was presented. In half the trials, one syllable, designated as the standard, was repeated five times. In the other half (deviant trials), the standard was repeated only four times, followed by one instance of the other syllable, designated as the deviant. Because repeated and deviant trials were randomly mixed, infants could not predict the nature of the fifth stimuli. Thus any significant difference in event-related potentials indicated that the two syllables had been discriminated. The infants thus reacted to a new stimulus against a background of habitual ones. The infant brain recognises a phonetic change in less than 400 ms. Thus, children at 2 to 3 months of age may already possess a supramodal anterior network for novelty detection which can be activated in less than one second. So the method which basically relies on habituation/dishabitation turned out to be useful for studying cognitive development processes in humans. As we will see further in this book, this method has been successfully applied in many cognitive experiments with animals.

The adaptiveness of habituation in animals may readily be seen in social or colonial animals that habituate to each other and therefore save time and energy that would have been spent in needless squabbling. It also plays an important role in habitat selection: once habituated to an area an animal may feel at ease only in the familiar surroundings where novel stimuli are rare. In any case, the context is important for animals in the wild: a stimulus that has been habituated to must appear in its usual context if it is to be ignored. Out of context, it may arouse the original reactions (Wallace, 1979).

6.2 Associative learning

Associative learning can be said to have taken place when there is a change in an animal's behaviour as a result of one event being paired with another (Pearce, 2000). As we have seen in Part I, the methods elaborated by behaviourists were based on classical and operate conditioning which are considered the simplest form of learned behaviour. Although there are several types of learning that neither classical nor operant conditioning theory can explain, the experimental procedures elaborated by behaviourists still play an important role in studying animal intelligence. In particular, these methods help to test the limits of learning abilities and to evaluate the balance between flexibility and conservatism in animal's behaviour in the real world where one event will readily predict another.

Experimental technique based on the first class of associative learning, i.e. classical (Pavlovian) conditioning, allows one to investigate how animals learn that one stimulus signals another and, vice versa, that one stimulus indicates another will not occur. To learn effectively, organisms should differentiate between similar stimuli in one situation and use generalisation in another. These relatively simple capacities could underline cognitive skills such as concept formation and rule extraction.

When, for example, a trainer asks Alex, a grey parrot, to answer how many things he sees and how many red things a set contains, how many wooden and so on (see Part IX for details), the researcher is testing the cognitive skills of the bird. However, 'simple' properties of associative learning such as stimuli discrimination and generalisation are also involved in this test. Of course, these properties are not so simple in reality. Wasserman and Miller (1997) named their review 'What's elementary about associative learning?' In this review they consider associative learning as the foundation for our understanding of other forms of behaviour and cognition. In the present chapter we will analyse laws of associative learning in order to estimate the formation of associations as an integral part of intellectual activity.

Classical conditioning

A procedure for studying learning based on pairing an unconditional stimulus (such as food or pain) with a conditioned (previously neutral) stimulus is connected with the name of the Russian physiologists Pavlov and called also Pavlovian conditioning (see Chapter 2). The main idea of classical conditioning is that an organism learns new behaviours by establishing an association between an involuntary unconditional stimulus and another stimulus it faces in its life. Every animal that possesses the nervous system has a number of innate stimulus–response associations.

The action of the *unconditional stimulus* (US) is based on a simple inborn reflex. This reflex may involve, say, taste receptors, sensory neurons, networks of interneurons in the brain, and motor neurons running to the salivary glands. For example, hungry vertebrates, including humans, will produce saliva when presented with food and do many other things automatically, just because many connections are 'wired' in their nervous system at birth. As was described in Part I, while studying the salivary reflexes, Pavlov noticed that his laboratory dogs had formed an association between a previously neutral stimulus and an unconditioned, unlearned, response. They then learned to respond to a substitute stimulus, the *conditioned stimulus* (CS).

The acquisition of the response includes several stages (see Cartwright, 2002). At the first stage, the animal is presented with the to-be-conditioned stimulus (for example, a bell ring) called a *neutral stimulus* (NS). This is a control procedure, used in order to ensure that this NS does not cause the unconditioned response (salivation) when it is presented alone. The dog will show some reflex responds, that is, an orienting response or barking, but should not salivate when hearing the bell.

At the second stage, the animal is presented with the unconditional stimulus (US), say, meat, and exhibits an unconditional response (UCR), in this case the salivation reflex. At the next stage the dog is presented with the NS (the bell) and the US (meat) simultaneously on a number of occasions. Each time the animal automatically produces the UCR (in this case the salivation response). After a number of parings of the NS and the US, at the final stage the animal is presented with the NS alone, without the food. If the dog now exhibits the salivation response when only the bell is presented it has been conditioned to associate the NS with the receipt of food. Hence the bell (previously the NS) has become a conditional (learned) stimulus (CS), and the salivation response (previously UCR) to this is now called the conditional (learned) response.

In this example the 'positive' US (the food) was used but it is easy to form the association between the NS and any 'negative' stimulus. For example, the dog will withdraw a paw (or simply lift it) hearing the bell (or, say, seeing flashlight) because in previous series of trials the NS combined with an electric shock. Thus the previously neutral stimulus (NS) such as the bell ring became the conditional (CS): the dog associated this stimulus with pain.

These laboratory experiments reflect many situations that can be observed in real life. It is known, for example, that babies quickly learn to associate a doctor's white coat with injections and other unpleasant procedures.

Any stimulus can play a role of a CS, provided that this stimulus itself does not provoke too strong reaction in an organism. A very hungry dog that will normally wince and be afraid of an electric shock is able to associate this with a food reward if an experimenter combines these two stimuli. The dog thus would be paradoxically wagging, licking and salivating when feeling a light penalty. What happens with the hungry dog if we intensify the strength of the electric shock and finally combine the painful insult with giving the food? In Pavlov's experiments this resulted in development of experimental neurosis which, in turn, led to either stomach ulcer or cancer in experimental dogs.

Pavlov's research also revealed that the main factors that influence the strength of a CR are the intensity of the US and the order and timing of the NS and US. The stronger the US, the stronger the CR, and vice versa. For example, if the bell (the CS) was rung quietly the dog would not produce as much salivation as if the bell was rung loudly.

The order and timing of the NS and US are important, that is, *the temporal contiguity of presentation of the NS and US*. It turned out that the most effective order and timing for the learning procedure were when the NS was presented half a second before the US and remained until the US appeared. This format of presentation of the NS and US is known as forward conditioning.

Using combination of conditional stimuli, it is possible to develop different variants of associations between stimuli – just an idea that lead Pavlov to hypothesis that complex behaviour in animals and humans is governed by conditional reflexes.

Combination of conditioning stimuli forms the so-called *stimulus–stimulus learning (S–S learning)*. By this protocol, a sequence of stimuli precedes US. For example, a tone might be followed by a light, which is followed by food. Although the initial element of the sequence is rather distant from the US, it nevertheless elicits a CR. In this case the CR may be weaker than the response observed in the presence of the element that is closer to the US. For example, rats were trained with the sequence light–tone–food (Ross and Holland, 1981). The normal conditioned response to a light that signals food is rearing or approaching the trough; for a tone it is either head jerking or approaching the trough. After a number of sessions of serial conditioning it was found that during the light there was little trough activity but there was a considerable amount of head jerking. The authors suggested that during serial conditioning the presence of the first element causes animals to anticipate the second one and to respond as if it were actually present.

Another set of examples demonstrating that associations can develop between two stimuli come from *sensory preconditioning*. In an experiment by Brogden (1939) a dog received the sequence of a tone followed by a light in a number of trails. The dog did not display any visible sign of learning. The tone was then paired with electric shock and, finally, the light was found to elicit a substantial fear CR when it was presented for testing. This phenomenon was later studied by many physiologists (Seidel, 1959; Rizley and Rescorla, 1972). Parks (1968) found that in those cases when two neutral stimuli were paired too frequently the first stage of the experiments, subjects hardly acquired S–S associations. He suggested that an orienting reaction serves as a mediator during sensory preconditioning. When two stimuli are paired frequently, this leads to extinction of attention. We later will return to a role of attention in learning processes.

The most known method that shows evidence of associations being formed between two stimuli is known as *second-order* and *higher-order conditioning*. In the laboratory second-order conditioning was observed in cases of pairing of conditional stimulus with the second – still unconditional – stimulus. For example, the first CS (bell) is paired, in close temporal contiguity, with light, but without the original US (food) being presented at the same time. After several combinations of the first-order and second-order stimuli, the dog salivates when the second-order stimulus is presented alone; therefore these are now a second-order CS (light) and CR (salivation).

Pavlov (1927) was the first to report this effect and later many results were obtained and more complex displays of conditioning were studied such as paring a second-order CS (CS2) with one that has already been paired with a US (CS1). It turned out that stimulus similarities (such as lights of different colours but not the light and the tone which are from different modalities) are important in determining the outcome of these forms of conditioning (see detailed analysis in Pearce, 2000).

Conditioning, as a form of associative learning, penetrates behavioural displays in huge variety of species, from planaria to primates.

Although methods applied for conditioning were very clear, the results obtained were not as simple as was desired. Many researchers noted that in their experiments conditional reactions were accompanied by 'superfluous' behavioural acts, that is, a dog may look at a source of sounds, turn its head, bark and whine, chew, lick itself, blink, etc. This raises a problem of separating the acts that need to be considered. It could be difficult to sort out behavioural acts when an animal reacts to some stimuli by hidden behavioural patterns. For example, in an experiment in which pigeons are to be mated, a male and a female were housed in neighbouring chambers divided by a falling door. Once per day a light is switched on and the door is lifted up to allow the male to start courtship. After some trials, males start to display courtship behaviour towards the light signal and behave as if the light signal is a real female (McFarland, 1985).

Some characteristics of conditioning are hidden from experimenters. Such are conditional reactions for internal receptive stimuli as well as reactions for timing. The organism is able to associate US with its own internal changes, say, in blood pressure or in a level of glucose in blood.

Indeed, students of animal and human behaviour should be familiar with specific characteristics and rules of classical conditioning as well as difficulties in interpreting the results obtained. It is a rather arduous task to avoid all possible associations that could appear during any laboratory experiments with animals, such as associations with the time of day when an experimenter comes, with details of equipment and so on. But these possibilities should be, at least, taken into account because the hidden conditional connections could impact on other results.

Classical conditioning cannot explain many situations in which associations are formed contrary to the theory's expectations. To explain them, other forms of learning should be considered.

Trial-and-error learning and instrumental conditioning

Pavlov's dogs were restrained and the response being conditioned (salivation) was innate. But the principles of conditioning can also be used to train animals to perform tasks that are not innate. This procedure was suggested by the American physiologist Thorndike (1911). As was described in Section 2.1., in Thorndike's experiments an animal was placed in a setting where it was able to move about and to engage in different activities. Thorndike did not place a hungry animal in a stall; instead, he allowed animals to make trials and errors in order to get out of a puzzle-box. The method employed by Thorndike to study animal intelligence is now referred to either as *instrumental* or as *operant conditioning*. In perfunctory descriptions it is also called *trial-and-error learning* because the animal is free to try various responses before finding the one that is rewarded. We will see from the further analysis that instrumental and operant conditioning differ in some details and that operant conditioning is not truly based on 'trial-and-error' learning.

In the real world animals and humans quickly remember sequences of movements or actions that once led them to a pleasant event even though these actions are senseless themselves. The environment 'selects' successful actions and thus forms some behavioural traits. For example, in Bloom *et al.* (1985) one can find a photo of a large dog 'Joey' who is just managing to keep his balance on a fire hydrant. Nobody knows why Joey does this with every fire hydrant he meets: he springs on the hydrant and keeps his balance on it for some time. This is similar to some irrational rituals in human routine when, say, a schoolboy will knock three times on a table, or a manager will only make the important decision after a yellow bus has passed the window. Having considered characteristics of instrumental conditions we can explain some behaviours observed in experimental and natural situations.

Thorndike was the first psychologist to propose that animals learn on the basis of the outcome of their actions. He formulated his theory of instrumental conditioning at about the same time that Pavlov was conducting his experiments on classical conditioning. Being placed in Thorndike's puzzle-boxes, animals took progressively less time to escape. From these observations, Thorndike claimed that the animal gradually formed a connection between a situation and a response that led to freedom. Certain stimuli and responses become connected or dissociated from each other. To Thorndike, the most prevalent questions within learning theory were:

(1) What happens when the same situation or stimulus acts repeatedly upon an organism – does the mere frequency of an experience cause useful modifications?
(2) What happens when the same connection occurs repeatedly in a mind?
(3) What effect do rewards and punishments have on connections, and how do they exert this effect?

In trying to answer these questions, Thorndike formulated the principles of instrumental conditioning theory. This theory represents the original 'S–R' framework of behaviourism in that it states that learning involves forming connections between stimuli and responses. According to Thorndike, these are neuronal connections within the brain and learning is the process of 'stamping in' and 'stamping out' these stimulus–response connections. Behaviour is due to the association of stimuli with responses that are generated through those connections.

Thorndike's theory states that the following three main conditions are necessary for learning to occur: the law of effect, the law of recency and the law of exercise.

The law of effect states that what happens as an effect or consequence, or as an accompaniment or close sequel to a situation response, works back upon the connection to strengthen or weaken it. Thus, if an event was followed by a reinforcing stimulus, then the connection was strengthened. If, however, an event was followed by a punishing stimulus, then the connection was weakened.

The law of recency states that the most recent response is likely to govern the recurrence of the response.

The law of exercise states that all things being equal, the more often a situation connects with or evokes or leads to or is followed by a certain response, the stronger becomes the tendency for it to do so in the future. It contains two portions: law of use (the strength of a connection increases when the connection is used) and law of disuse (the strength of a connection diminishes when the connection is not used). Thus connections become strengthened with practice.

Considering a concrete example with an animal, for example, a cat escaping from the puzzle-box in terms of S–R associations, the corresponding laws of learning can be seen in these studies. After many trials and errors the cat learns to associate, say, pulling the loop (S) with escape from the box (R). This S–R connection is established because it results in a positive consequence. The connection was established because the S–R pairing occurred on many occasions (the law of exercise) and resulted in a positive consequence (the law of effect).

Thorndike came very close to formulating Hebb's law (see Part I) when he discovered the law of effect: the probability that a stimulus will cause a given response is proportional to the satisfaction that the response has produced in the past.

Through many series of experiments Thorndike came to the conclusion that there is no sufficient reasons for ascribing any power over and above that of repetition and reward to any 'higher powers' or 'forms of thought' or 'transcendent systems' and that all learning involves the formation of connections and these connections are strengthened according the law of effect. Thus, intelligence is attributed to the ability to form connections, and as humans are the most evolved animals they form more connections than others and learning processes are investigated only on the basis of observable behaviour.

This statement makes instrumental conditioning theory and classical conditioning theory similar. However, the key difference between these theories is that classical conditioning states that learning involves associations between unconditioned reflex behaviours. Instrumental and operant conditioning state that learning involves associations between the performance of specific behaviour and the consequences of these actions.

Operant conditioning

Operant conditioning is a form of learning in which voluntary behaviour becomes more or less likely to be repeated, depending on its consequences. It is also known as Skinnerian conditioning. As was described in Section 2.2, in the 1930s Skinner developed and modified Thorndike's instrumental conditioning theory and formulated operant conditioning theory (Skinner, 1938). Skinner's name is connected with the Skinner box in which animals must press a lever in order to obtain a reward. Nevertheless, as we have already seen in Chapter 2, it is not necessary to be packaged in the box to be involved in operant conditioning. Key terms of Skinner's theory are *reinforcement* and *schedule of reinforcement*. Rewarding even the slightest movement in the desired direction and not encouraging, or even punishing, other actions, in other words selecting the animal's actions, a skilled experimenter can train animals to do many things, for example, train pigeons to play ping-pong or play a toy piano. These behaviours are of no use in themselves for the animals but they lead it to successes in laboratory experiments, that is, food rewards. This way of training is based on *operant conditioning*. Let us consider this approach to study animal learning in more details.

The law of reinforcement: an association between behaviour and a consequence

Thorndike's law of effect was the conceptual starting point for Skinner's work on operant conditioning. Skinner changed the name of instrumental conditioning to operant conditioning because it is more descriptive (i.e. in this type of learning, one is 'operating' on, and is influenced by, the environment). But the main idea is the same: operant (instrumental) behaviour is spontaneous as this is not a reaction of an organism to a certain stimulus. In fact, there is no CS at the starting point of these experiments. With time, some aspects of the situation may start to act as conditional stimuli that let the subject know that its actions are relevant and that reinforcement is coming. Skinner also renamed a positive consequence 'reinforcement' and a negative consequence 'punishment'. The apparatus invented by Skinner, which is known as the Skinner box (or 'operant chamber'), with a cumulative recording device, is still in use today (upgraded with computerised control) for studying learning processes in animals and humans.

In order to make this experimental technique more clear, let us focus once more on the differences between classical, operant and instrumental conditioning.

By a protocol of the technique of classical conditioning the animal is relatively passive being optionally locked in a stall. It is interesting to note that the Skinner box is often used for studying classical conditioning. In this case an operant chamber is transformed into a conditioning chamber. For example, a hungry pigeon

Fig. 6.1 A pigeon in a conditioning chamber. (Photograph by L. Huber; courtesy of L. Huber.)

is placed in a conditioning chamber (Fig. 6.1). At intervals of about 1 minute a response key is illuminated for about 5 seconds, and the offset of this stimulus is followed by the delivery of food to a hopper. At first subjects may be unresponsive to the key, but after a few trials they will peck it rapidly whenever it is illuminated. This is not an instrumental conditioning, as the pigeon does not have to peck the key to obtain food. Instead, it is an example of Pavlovian conditioning as the mere pairing of the illuminated key with food is sufficient to engender a CR of key-pecking. It is important that behaviour is not governed by its consequences in this case.

If we compare instrumental and operant conditioning, in both, behaviour is affected by its consequences, but in operant conditioning the basic process is not 'trial-and-error' learning. Instead, operant conditioning forms the association between the behaviour and the consequence. As was already noted in Part I, it is also called response–stimulus or R–S conditioning because it forms an association between the animal's response (behaviour) and the stimulus that follows (consequence).

Unlike in classical conditioning, where a new response is formed basing on the association with the previously neutral stimulus (US), in operant conditioning there is no creation of a new response to a neutral stimulus. Instead, there is an increase or decrease in a response that is already being exhibited. Operant conditioning is based on the law of reinforcement, which states that the probability of a given response being made is increased if the consequences of performing it are pleasant.

Manifestations of the law of reinforcement clearly support Thorndike's law of effect. Let us consider the simplest case of operant conditioning. When a rat is placed in a Skinner box, at first the animal begins to explore the new space, wandering around and sniffing at everything. During this exploration the rat may accidentally press the lever (operant). This accidental occurrence enables the researcher to manipulate the consequences of the rat's accidental lever-pressing behaviour by making the consequences of this either pleasant or unpleasant. If the researcher positively reinforces the lever-pressing behaviour (positive consequences) then it results in an increase in lever-pressing by the rat. In a case of using punishment (an unpleasant consequence) the lever-pressing behaviour decreases and then stops.

As has been noted above, unlike the experiments of Thorndike, who used 'trials' in instrumental conditioning, Skinner's operant conditioning procedure did not use trials. Instead, the researcher had to wait for manifestation of reactions that were already present in animal's behavioural repertoire.

Both instrumental and operant conditioning essentially differ from classical (Pavlovian) conditioning.

Where classical conditioning illustrates $S \rightarrow R$ learning, operant conditioning is often viewed as $R \rightarrow S$ learning since it is the consequence that follows the response that influences whether the response is likely or unlikely to occur again. It is through operant conditioning that voluntary responses are learned.

The three-term model of operant conditioning ($S \rightarrow R \rightarrow S$) incorporates the concept that responses cannot occur without an environmental event (e.g. an antecedent stimulus) preceding it. While the antecedent stimulus in operant conditioning does not elicit or cause the response, as it does in classical conditioning, it can influence it. When the antecedent does influence the likelihood of a response occurring, it is technically called a discriminative stimulus.

Reinforcement and punishment

Reinforcement is the key element in Skinner's S–R theory. A reinforcement is anything that strengthens the desired response. Reinforcement may be positive or negative. A *positive reinforcement* reinforces when it is presented; a *negative reinforcement* reinforces when it is withdrawn. Anything that increases a behavioural pattern, that is, makes it occur more frequently, makes it stronger, or makes it more likely to occur, is reinforcement. Anything that decreases the behavioural pattern – makes it occur less frequently, makes it weaker or makes it less likely to occur – is punishment. For example, in the Skinner box a rat may have to learn to press the lever to either gain food (positive reinforcement), or an electric shock (negative reinforcement), stop food being taken away (negative punishment) or prevent the receipt of an electric shock (positive punishment).

Elements of casuistry may be found in this terminology. Thus, positive reinforcements are something like rewards, however, the definition of a positive reinforcement is more precise than that of reward. Specifically, we can say that positive reinforcement has occurred when three conditions have been met:

(1) A consequence is presented dependent on behaviour.
(2) The behaviour becomes more likely to occur.
(3) The behaviour becomes more likely to occur because and only because the consequence is presented dependent on the behaviour.

Negative reinforcement is often confused with punishment, but they are different, however. Reinforcements always strengthen behaviour; that is what 'reinforced' means. Punishment is used to suppress behaviour. It consists of removing a positive reinforcement or presenting a negative one. It often seems to operate by conditioning negative reinforcements. Negative reinforcement strengthens behaviour because a negative condition is stopped or avoided as a consequence of the behaviour. Punishment, on the other hand, weakens behaviour because a negative condition is introduced or experienced as a consequence of the behaviour.

Here is an example of how a negative reinforcement works. A rat is placed in a cage and immediately receives a mild electrical shock on its feet. The shock is a negative condition for the rat. The rat presses a bar and the shock stops. The rat receives another shock, presses the bar again, and again the shock stops. The rat's behaviour of pressing the bar is strengthened by the consequence of the stopping of the shock.

There are four major techniques in operant conditioning. They result from combining the two major purposes of operant conditioning (increasing or decreasing the probability that a specific behaviour will occur in the future), the types of stimuli used (positive/pleasant or negative/aversive), and the action taken (adding or removing the stimulus).

Schedules of consequences

One of the most important aspects of Skinner's theory of learning is his concept of *schedules of consequences*. According to Pavlov's theory, effective conditioning is possible when time gap between the NS was presented half a second before the US and remained until the US appeared. This format of presentation of the NS and US is known as *forward conditioning*. Skinner showed that there are two main factors influencing the strength of conditioned reflex (CR):

(1) Ratio (proportion) of reinforcement.
(2) The time interval (delay) between response and reinforcement.

There are two basic categories in this format: continuous reinforcement and intermittent reinforcement.

Continuous reinforcement simply means that the behaviour is followed by a consequence each time it occurs. *Intermittent schedules* are based either on the passage of time (*interval schedules*) or the number of correct responses emitted (*ratio schedules*). The consequence can be delivered based on the same amount of passage of time or the same number of correct responses (*fixed*) or it could be based on a slightly different amount of time or number of correct responses that vary around a particular number (*variable*). This results in four classes of intermittent schedules. Experiments have revealed that each specific type of schedule results in different effects of conditioning.

(1) Fixed interval: the first correct response after a definite amount of time has passed is reinforced (i.e. a consequence is delivered). For example, an animal is given a food pellet 2 minutes after it presses the lever throughout the trial. In the context of positive reinforcement, this schedule produces a scalloping effect during learning (a dramatic drop of responding immediately after reinforcement). The animal often only makes responses towards the last few seconds of the interval. This schedule is not so effective; its effectiveness is probably because the animal is able to learn the amount of time that will elapse before the response is reinforced.

(2) Variable interval: after the first reinforcement, a new time period (shorter or longer) is set with the average equalling a specific number over the total sum of trials. For example, the animal is given the food after 6 minutes, then 3 minutes and then 7 minutes after it has pressed the lever, resulting in an average time interval of 5 minutes between receptions of reinforcement. This schedule is effective, probably because it is not possible for the animal to learn the precise amount of time that will elapse before the response is reinforced, so it should carry on emitting the lever-pressing response.

(3) Fixed ratio: reinforcement is given after a specified number of correct responses. This schedule is best for learning a new behaviour. Behaviour is relatively stable between reinforcements, with a slight delay after reinforcement is given.

(4) Variable ratio: reinforcement is given after a set number of correct responses. After reinforcement the number of correct responses necessary for the reinforcement changes. This schedule is best for maintaining behaviour.

In summary, the number of responses per time period increases as the schedule of reinforcement is changed from fixed interval to variable interval and from fixed ratio to variable ratio.

Conditional reinforcement and clicker-training

Skinner referred to two types of reinforcement: *primary reinforcement* and *conditional* (or *secondary*) *reinforcement*. A primary reinforcement is one that is biologically pre-established, such as food, water or sex. A conditional (or secondary) reinforcement is a previously neutral stimulus which, if paired with a primary reinforcement, acquires the same reinforcement properties that are associated with the primary reinforcement. For instance, for a hungry rat placed in the Skinner box, the sound of the food dispenser may serve as the secondary reinforcement.

For practical animal trainers it is important to know that the secondary reinforcement may serve as a 'bridge' between the desirable behaviour and the receiving of a reward. For example, you use special words (such as 'good dog') or a special clicker (a small hand-held device that emits a clicking sound) to train dogs. The latter has given its name to a new technique of training: clicker-training. Let us clarify this with several examples.

If you want to train, say, a bear to cycle, you use a primary reinforcement, i.e. food. Each time the bear exhibits the desired behaviour, for example, sitting on a bike, it receives a bit of food. But it is very important to provide the pupil with the reward at the same moment as the performance of the desired reaction. Thus, if you are a little bit late with the reinforcement and the bear receives the food at the moment it sets down a paw to dismount from the bike, it perhaps 'concludes' that you reward leaving the bike. You have missed the moment. It is not an easy task for a trainer to hit the bear's mouth with the food pellet at exactly the right moment. If you use a secondary reinforcement such as the clicker, it is not necessary to hit the animal's mouth. What you need is to 'negotiate' with the bear during previous trials that it will receive the food when it hears the clicking noise. In this case the sound of the clicker becomes equivalent to the receipt of the primary reinforcement, the food pellet. Therefore the sound of the clicker now serves as a reward for the desired behavioural model. It is much easier to click at the necessary moment than to aim at the bear with the bit of food.

Using such a simple method as the secondary reinforcement, practical animal trainers can solve many problems. They are now able to 'stop the moment' when shaping the desirable patterns, for example, by rewarding the top of the high jump in dolphins with a whistle and thus training them to jump higher and higher. In general, fixed association between the secondary and the primary reinforcements allows you to inform your subject about what behaviour you are interested in.

In reality this association has been spontaneously used for ages in many domains. Even money may be considered as the secondary reinforcement for human beings as an indication of things that could be purchased. This analogy might be nearly correct especially if we forget about economy and concentrate on psychology remembering experiments performed on primates and based on token reinforcers. These are typically small plastic discs that are earned by performing some response, and once earned they can be exchanged for food, drink and even a possibility to escape unpleasant events such as an encounter with a rat. The fact is that the secondary reinforcement works as a conventional sign. In Wolf's early experiments (1936) with chimpanzees he elaborated something like 'token language': plastic discs of different colours had different values (such as one or two bananas) and gave access to different sorts of material welfare (food, drink) as well as spiritual needs (the chance to play with the trainer or to escape the encounter with the rat).

Returning to clicker reinforcement, it is worth noting that trainers consider the clicker as more than simply a conditioned reinforcement, a substitute for food or 'an event marker'. It is also a bridging stimulus, meaning 'Food is coming, but later' (Pryor, 1975).

The elegant technique that is now called clicker-training is an application of behaviour analysis that was initially invented and developed in 1960s by Keller Breland, Marian Breland Bailey and Bob Bailey on the basis of Skinner's theory (see Chapter 20). Karen Pryor has further developed this method and in 1985 published a book on the new concept of training: *Don't Shoot the Dog! The New Art of Teaching and Training* is concerned with altering animal behaviour without being coercive.

The bridge between operant conditioning and thinking

In fact, it was Skinner who put in place the first stone in this bridge. As he customarily made pigeons perform different tasks in order to receive food, he wondered what would happen to these same pigeons if the receipt of this food suddenly became arbitrary, i.e. having no relationship with tasks the birds might perform. The results of his study were surprising, and Skinner published them in a famous short article entitled 'Superstition in the pigeon' (Skinner, 1948b). When the receipt of food became arbitrary, the birds began to do strange things like putting their heads in the corner of the cage, or tucking their heads under their wings. Since these animals had already gone through many behaviouristic experiments, it would perhaps not be too far-fetched to imagine that they were trying to come up with some kind of behaviour that would give food. That is, they were throwing out 'hypotheses' about how to influence food acquisition.

Many practical trainers use animals' capacity to produce 'hypotheses'. For example, a mare had learned that clicks meant carrots; and that she could make clicks happen. And she had also become aware that the operant had something to do with ears, but what? In order to make clicks happen, she started to make different 'proposals' to her trainer, different variants of ear position and movements until one of them would be rewarded.

The next example explains that such behaviour stands beyond trial and error as animals have to derive an algorithm of a task in order to get a reward. When dolphins jump high out of the water they receive a herring for their behaviour and this is somewhat like an agreement between the animals and their trainer about rewarding the high jump. Once the dolphin jumps high but nothing happens. It jumps again and again but it does not make clicks happen. In a rage the dolphin taps on the water with its tail – and immediately hears the click. Naturally, it repeats the rewarded behaviour but nothing happens. After some abortive attempts, the animal drifts to another pattern, say, stands to its full height above water level. This brings a success but again only once. When the common behavioural repertoire is exhausted, the dolphin starts to contrive new and new actions and elaborate novel 'creative' patterns. It has learnt a rule: in order to hear the click do only things that have been never done before (Pryor, 1969).

Such a capability to 'derive a law' perhaps exceeds the bounds of associative learning and requires cognitive skills. Similar experiments on many species have revealed a great deal of ingenuity. For example, in Skinner's experiments pigeons exhausted the usual repertoire of actions; desperate to get rewarded they went to such unusual behaviour as dancing on their wings, expanding and stretching them under their legs.

How to train all creatures effectively

Basing their work on the principles of operant conditioning, Skinner and his followers achieved great success in training great variety of species including humans. Even invertebrates were included. A crayfish learned to pull out a rope and thus ring a bell; a bivalve clapped its shell at the command of the trainer.

All these results were obtained by the method of secondary reinforcement. In order to train trainers themselves, a special 'training game' has been elaborated that allows trainers to make mistakes, and learn from them, without confusing some poor animal or unsuspecting person. In a group, one person is selected as the Animal, and goes out of hearing range. The others choose a Trainer and a behavioural pattern to be shaped. The behaviour must be something easy to do physically, which everyone can see, such as turning in a circle, pouring or drinking water, turning on a light switch, picking up an object, opening or closing a door or window, or marking on a blackboard. The Trainer will use a clicker, hand-clap, or other noise as a secondary reinforcement. Each time the Animal hears the sound he or she must return to the trainer and get an imaginary treat.

One of the funniest stories came from Skinner himself. He once attended a lecture of a famous psychologist who decried Skinner's 'inhuman' methods but did not know him personally. Skinner sat in the front row and listened to the lecture very passionately thus forcing the lecture to concentrate on this so keen 'student'. He then began pose as bored when the lecturer spoke about love but came alive and nodded his approval for every war-like or aggressive gesture, even in the slightest degree. By Skinner's account, that lecturer ended his presentation by brandishing his fists in the air like Hitler.

Pryor successfully applied her method for shaping the behaviour of all members of her family and strongly recommend using clicker-training in day-to-day life. The idea seems fantastic, but armed with this method you can force a teenager to clean a room, an infant to stop yelling or a grandmother to stop grumbling. Parents are learning to shape appropriate behaviour instead of accidentally reinforcing inappropriate behaviour: to reinforce silence not noise, play not tantrums.

It is really possible to achieve consensus both with humans and animals as if you were in speechless dialogue with them. I myself tried out this method in an 'extreme' (and funny) situation when I needed promptly to return a tit to its cage in an empty flat belonging to my friends who had asked me to feed the bird. I was there with my friend who dealt with animal morphology, not behaviour, and thus considered the situation a *causa mortis*. She nevertheless agreed to help me and to fulfil my instructions. Our relationship with the tit was divided into several stages. At first, we prevented the bird from sitting anywhere with the exception of that zone of the flat where the cage was located. If the bird tried to sit within any other space, it was immediately scared away. When it learnt that this was the only place to sit, visits to that zone became more frequent. At the second stage the bird was permitted to sit only in the vicinity of the cage, then on the cage only and finally on the door of the cage. After that it became easy to have the bird in the cage. This took not more than half an hour and Skinner's method of shaping behaviour triumphed over others that day.

There are some principles of effective training based on Skinner's theory as well as experience gained from practice of clicker-training. Some of them were mentioned before, namely the use of reinforcement rather than punishment, interval schedules and optimal timing for giving a sign of approval. Let us take them and some others as protocols for practice.

OPTIMAL TIMING

If something is going wrong with the learning process, the trainer should first consider the timing of reinforcement. As was already noted, the secondary reinforcement is the message that informs a subject about what is desirable in its behaviour. If this message is late, even by a moment, the subject begins to fix this with another kind of behaviour. Adding once more to examples considered before, if you are training a dog to sit but compliment the animal as soon as it is standing up, your dog 'concludes' that the owner thanks it for standing up. If this situation is repeated at least twice, this brings learning to a standstill. Pryor (1985) considers that encouraging students for unfinished tasks and, in general, giving compliments and gifts for incomplete behaviour does not reinforce appropriate behaviour. We seem to encourage attempts to do something good but in reality even if something is being shaped this is only 'instant' behaviours and most likely the behaviour we really want gets no reinforcement.

NATURE OF REINFORCEMENT

As has been explained in detail above, reinforcement can be both positive and negative. A sharp loud word of an interdiction for many animals (and children) frequently is a primary (unconditional) negative stimulus. But if a subject does not react to such influence, another negative reinforcement should be applied to which the subject is sensitive. For example, Lorenz (1952) in his book *King Solomon's Ring* recommended throwing pebbles at a dog in order to rid it of the unpleasant habit of running away from the owner.

REINFORCEMENT RATE

When dealing with animals, it is better to use relatively small food items in order to keep their interest continuously. For example, bears have learnt many tricks for raisins. Nevertheless, like humans, animals often disclaim their obligations if recompense is small. Thus, killer whales in an aquatic show refused small herrings for high jumping. Only large fishes used by trainers restored their motivation for jumping. Pryor (1985) also used the concept of 'snatching a large sum' designating the largeness of reward obtained for 'insight' and in some situations even for nothing. This pedagogical trick sometimes helps to sustain a competitive spirit in animals and humans and allows astonishing results in learning to be achieved.

INTERVAL SCHEDULES

One common mistake is made by those trainers who having begun to shape some behavioural pattern continue to reinforce this behaviour with monotonous frequency during the whole life of a trainee subject. Indeed, there is no need to reinforce thoroughly learnt behaviour in order to sustain reliability. Quite the contrary, it is necessary to stop giving tips and start giving them fortuitously.

For example, if a dolphin receives a fish for every jump it possibly will jump more and more jauntily, just to show off. But if the dolphin taught to jump for a fish receives its reward for the first jump, then for the third, and then at random, most likely it will jump higher and higher, trying to catch its luck. In turn, this allows the trainer to reward the highest jumps selectively and thus to improve desirable behaviour by the use of variable interval of reinforcement. If one stops rewarding completely then a tendency of extinction will become apparent, but it may be sufficient to give a tip from time to time and the prolonged interval may just strengthen behaviour.

Pryor believes that the efficacy of the variable schedule underlies all gambling games and she even considers deep affections in human beings to be amongst these phenomena. It is enough that a rough and egoistic person occasionally presents his or her partner with 'good behaviour' so as to force the partner to yearn for the return of these marvellous moments, the more passionately the more rarely they happen.

In addition to a variable schedule, a specified regime of reinforcement sometimes gives good results. A subject knows that it must work for a definite time or perform some patterns in order to get a reward. For example, when rewarding a dolphin for each sixth jump, it is possible to obtain a stable series of six jumps. There is a weak point in these conditions (a fish for the sixth jump, a wage-packet every Friday): both animals and humans tend to do the minimal work, just to escape being fired.

With the help of variable and specified schedules it is possible to shape extremely long chains of behaviours. For example, a chicken is able to peck a button up to a hundred times in order to obtain one grain. In experiments with the use of token reinforcements chimpanzees had to press a key 125 times in order to receive a single token, and when they collected 50 tokens they were allowed to push them all into a slot in order to receive food. The animals performed a sequence of more than 6000 responses (Kelleher, 1958). There is a joke among psychologists that school studies are the longest non-rewarded regimes in human life.

6.3 Common basis for different forms of associative learning

Both classical and operant forms of conditioning are based on the formation of associations and involve such common properties as stimulus discrimination, generalisation, extinction and spontaneous recovery.

The role of the motivation in conditioning

One of the most important questions in the field of learning theory is what is it that makes a stimulus a reinforcement? What is the reason for animals and humans to be engaged in an activity such as lever-pressing? In the majority of experiments food serves as the reinforcement for a hungry animal, or water for a thirsty animal. For those investigators of animal intelligence who went beyond a set of diversity limited by rats and pigeons, it rapidly turned out that for some species the greatest reward was to leave an animal alone, while for others (such as chimps) this might be, say, the chance to look at a children's railway through the window. For example, for an ant in a maze a worthy tribute is the opportunity to return smoothly to its laboratory nest. Isolated chicks will quickly learn a simple maze in order to rejoin their nest mates (Gilbert *et al.*, 1989). Even such a favourite and usual subject for laboratory experiments as a rat may surprise experimenters by its preference for a relevant award. It turned out that rats very quickly learnt mazes if they were presented with an opportunity to kill a mouse in the final chamber (Myer and White, 1965).

On the basis of Pavlov's concepts of conditioned and unconditioned stimuli, Konorski (1967) suggested that unconditioned stimuli possess two different characteristics – specific and affective. *Specific characteristics* are those that make the US unique: the place where it is delivered, its duration, intensity and so on. *Affective characteristics*, by contrast, are those that the US has in common with other stimuli and reflects its *motivational* quality. Thus food, water and an opportunity to mate have the common *appetitive* characteristics that animals will actively search for them. Conversely, electric shock, illness and loud noise possess the common aversive characteristic that animals will do their best to minimise their contact with them.

These characteristics of the US come to the concept of *motivation* that was suggested in early 1930s by one of the most famous Pavlov's protégés, P. K. Anokhin. He included motivation into a general statement of properties of learning. This notion is helpful for completing characteristics of stimuli that dictate whether or not they will function as reinforcements. Anokhin first suggested a method of sudden substitution of US in experiments on classical conditioning and thus revealed that animals possess something like 'predisposed stimulation' in their central nervous system in a situation when they are waiting for a definite stimulus. This enabled him to develop a *theory of functional systems*, which was based on a concept of motivation (Anokhin, 1961, 1968, 1974). The role of the motivation is to form a goal and to support the goal-directed forms of behaviour. Physiologists consider motivation to be an active drive, which simulates nervous processes during searching for such a decision that is adequate to the animal's needs in given environmental conditions. The motivation is closely correlated with the notion of *dominanta*, a term that was later introduced by another Russian physiologist, A.A. Ukhtomsky (1950). The dominanta means that animal resources are mobilised to reach a given goal. In particular, the nervous resources are correspondingly mobilised, so the animal's attention is aimed at the goal during purposeful behaviour.

As Anokhin supposed, that behavioural programme starts first which answers the purpose of an organism in solving its vital problem. If the reflex is conditioned on the basis of foraging behaviour it should be supported by a strong motivation for satisfying hunger. A hungry dog always displays a complex of foraging behaviour near a food source. But if a hungry dog is thrown into deep water, it would rather manifest another pattern, just trying to reach land instead of eating even an appetising bit of food that is floating near its mouth.

It is important to note that, unlike in Pavlov's theory, intelligence was included into Anokhin's theory of functional systems. The functional system was introduced to explain the animal purposeful behaviour and that included 'intelligence resources' which should be mobilised for seeking the priority goal, recognition of a situation and 'planning' of actions.

Several authors have since developed concepts very close to Anokhin's theory of functional systems. They have proposed that the determinants of conditioned behaviour are organised into functional systems that are concerned with such activities as feeding, mating, defence and parenting (Davey, 1989; Timberlake, 1994). These systems are activated by the appropriate stimuli, and serve to coordinate patterns of behaviour that are both innate and learned. Each system is assumed to control a wide range of actions, the selection of which is determined by the stimuli that are present.

In the 1930s, Hull and Spence introduced motivation as an intervening variable in the form of homeostasis, the tendency to maintain equilibrium by adjusting physiological responses (Hull, 1932, 1952; Spence, 1936, 1960; see also Pearce, 2000 for detailed analysis). An imbalance creates needs, which in turn create drives. A *drive*, according to Hull, is a single central motivational state activated by a certain need. Actions can be seen as attempts to reduce these drives by meeting the associated needs. This is the *drive-reduction theory*: the association of stimulus and response in classical and operant conditioning only results in learning if accompanied by drive reduction.

Premack (1959, 1962, 1965) suggested a solution to the problem of deciding whether a stimulus could be a reinforcement. His idea was also based on differences in preferences of different kinds of activities. The *Premack principle*, often called 'grandma's rule', states that a high-frequency activity can be used to reinforce low-frequency behaviour. Access to the preferred activity is contingent on completing the low-frequency behaviour. Premack proposed that reinforcements were not stimuli, but opportunities to engage in behaviour. Thus the activity of eating, not the stimulus of food, should be regarded as the reinforcement when an animal has been trained to press a lever for food (see Pearce, 2000).

In order to determine if one activity will serve as reinforcement for another activity, Premack proposed that the animal should be allowed to engage itself freely in both activities. For example, a rat might be placed into a chamber containing a lever and a pile of food pellets. If it shows a greater wiliness to eat the food than to press the lever, then we can conclude that the opportunity to eat will reinforce lever-pressing, but the opportunity to lever press will not reinforce eating.

This seems to be completely naturalistic and trivial. To demonstrate the relative property of reinforcements, Premack conducted an experiment with rats placed into a running wheel and demonstrated that the opportunity to run could serve as a reinforcement for drinking in rats that were not thirsty. In this experiment, animals were placed in a running wheel for 15 minutes a day. When the rats were not thirsty, they preferred to run rather than to drink. For the test phase of the experiment the wheel was locked and the rats had to lick the drinking tube in order to unlock the wheel and so to gain the opportunity to run for 5 seconds. Thus the animals had to drink in order to earn the opportunity to run.

This and other experiments (see, for example, Allison and Timberlake, 1974) led researchers to a concept of the relative value of reinforcements depending on the environmental circumstances and the internal state of an animal which forms motivation. We have to adopt a point that there are no absolute reinforcements and in each experiment their properties should be considered directly.

Common properties and rules of associative learning

Although in previous sections of this chapter we emphasised the difference between Pavlovian and Skinnerian conditioning, it is still not completely clear whether mechanisms of these forms of learning differ in principle. Many experiments have been effectively performed with the use of a hybrid technique in which properties of classical conditioning are studied on subjects placed in operant chambers. For example, in experiments by Rashotte *et al.* (1977) second-order conditioning was studied in pigeons which first received autoshaping and had to peck keys being illuminated with different colours. Even in pure Skinnerian experiments distinct evidence was obtained that pigeons form conditioned reactions for a key they peck, i.e. they associate the key with food or drink, depending on experimental condition (Moore, 1973). When pecking for food, the pigeon's eyes are closed and the beak open (typical for feeding); when pecking for water, the pigeon's eyes are open and its beak nearly closed (a typical drinking motion), so their behaviour seems to follow the reward given. When students watching video recordings of these experiments were asked what sort of reward each pigeon was waiting for - water or food - they never made a mistake in their answers. Similar, although not so impressive, effects have been observed with rats and monkeys. It seems that operant and classical conditioning cannot be considered to be different forms of learning in principle.

Let us consider general principles common to all forms of associative learning. Some of them serve as elements of useful experimental techniques which help experimenters to discover more and more facts concerning animal intelligence.

Timing of the stimulation

We have already seen when considering actual examples that for both classical and instrumental conditioning the order and timing of stimuli are important. Only a few seconds' delay between a rat pressing a lever in the operant chamber and the delivery of food greatly impedes the rat's ability to learn. Such phenomena have presented thorny problems for animal trainers. In many instances the reward can not be given fast enough, but a method of secondary reinforcement has been developed that we considered in detail above.

Classical conditioning is based on the temporal continuity of presentation of the NS and US. Pavlov also found that learning becomes poorer if the time gap between the presentation of the NS and US is increased beyond the optimal half a second. This format of presentation is referred to as *delayed conditioning*. Based on this experimental technique many important results have been obtained concerning short-term retention in animals, which have thrown light on the processes of memorisation and forgetting.

An early demonstration of animal memory in the laboratory was provided by Hunter (1912) in his experiments with racoons, rats, dogs and children. The main principle of Hunter's experiments is that US is absent just in that moment when a subject performs a reaction. A stimulus is substituted by 'something else' which was called an *idea* by Hunter, and this may also be named 'reminiscence about a stimulus'. In these experiments an animal, say, a racoon, being retained in an observational chamber, was allowed to observe three exits. A light above one exit was then briefly illuminated and some time later the animal was released. If it chose the exit that had been indicated by the light it received reward. By this time the light is no longer illuminated, so the subject reacts according to its 'reminiscence about the light'. The time of the delay is important in these experiments. Hunter found that with sufficient training the subjects were able to tolerate a delay of as much as 25 seconds between the switching off of the light and their release. This technique was termed by Hunter the *delayed reaction test*, which was later transformed by Harlow and co-workers (Harlow *et al.*, 1932) and applied to different species of primates, from lemur to orang-utan. These experiments demonstrated that different animals are able to retain information about past events, and raises many interesting questions about how much information can be retained and for how long. These questions will be considered in Chapter 10.

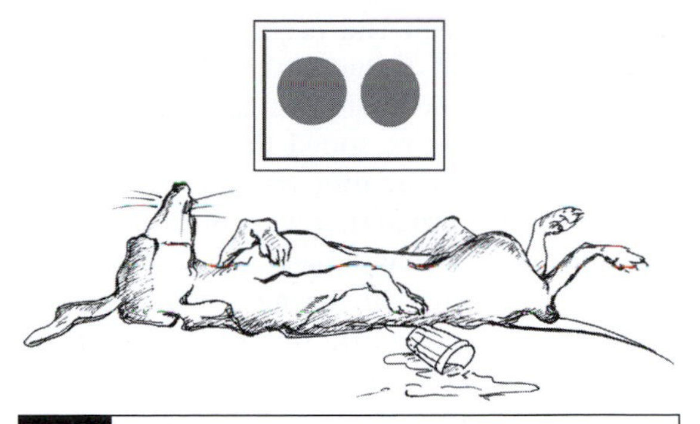

Fig. 6.2 By making a dog discriminate between circles and less and less different ellipses, one can drive it to neurosis. (Cartoon by P. Ryabko.)

Stimulus discrimination

Pavlov discovered that stimulus discrimination was learned in very much the same way as the acquisition of an association between things. In many experiments animals demonstrated their ability to detect the difference between those stimuli that are connected with a reward and those that are not. For example, if a dog is presented with the sound of bells at different decibels on a number of occasions but only receives food when the bell at 5 decibels is sounded, it only exhibits the CR (salivation reflex) with the bell of 5 decibels. Therefore the dog only responds to the stimulus that will result in reinforcement.

Stimulus discrimination learning appears to occur in both classical and instrumental conditioning in a very similar way. If only a very specific response is reinforced eventually the animal will only exhibit this specific response.

This experimental technique was immediately appreciated as an excellent and practically universal tool for determining the sensory abilities of many animals as well as their cognitive and communicative skills. For example, honeybees can be conditioned to seek food on a piece of blue cardboard. By offering other colours to a blue-conditioned bee, Karl von Frisch in the 1920s found that honeybees see four distinct colours: yellow green, blue-green, blue-violet and ultraviolet. Even earlier, in the 1910s, Ladygina Koths had used the same method to reveal colour vision in birds.

Dramatic consequences for animals, which face fatal difficulties in stimulus discrimination during the process of investigation of their sensorial capability, were discovered first in Pavlov's laboratory. For example, experimenters flash a circle on a screen and follow it with food. Then they flash an ellipse on the screen, but leave the bowl empty. The dog quickly learns to distinguish between the two shapes. Then they flash a circle, followed by an ellipse that is a few degrees closer to the circle. The dog still manages to detect the difference. After that, though, the experimenters keep changing the ellipse, so that it gets closer and closer to the shape of a circle. When they reach a certain point, the dog starts whimpering and wriggling, and it bites through the tubes on the collection device. With the use of such experiments it was possible to bring the animal to a serious neurosis (Fig. 6.2).

As has already been noted, the capacity of many animals to give a definite answer to a question about differences between stimuli has been used up to now in many experimental schemes for studying animals' cognitive abilities.

One of the earliest studies of animal cognition based on discrimination learning was carried out by Köhler (1918) who applied the *transposition* test to chickens. Birds were trained with two cards, one of which was darker than the other. Pecks at the dark card (S+) but not at the light card (S−) were rewarded with food. Once the discrimination had been mastered, a transposition test was given in which the subjects had to choose between the original S+ and even more shaded card. If the original discrimination was solved on the basis of relational information, then Köhler reasoned that the new card would be chosen in preference to S+ on the test trials. This prediction was confirmed, even though it meant that subjects rejected just the card they had originally been trained to select.

Köhler interpreted these results in terms of Gestalt principles which were considered in Chapter 4 and will be analysed again in Chapter 17. Here we only mention that the Gestalt explanation says that during training the organism learns a principle related to the two stimuli. The principle might be 'the darker shade of grey is better'. The chicken thus had followed the rule 'choose the darkest card'. When presented with the other stimulus, the animal 'transposes' the solution from the original problem to the new problem and chooses the darker grey even though the medium grey was associated with food originally.

Although a number of more recent theorists have preferred to interpret Köhler's findings in other and maybe more parsimonious ways (as we will see in Chapter 17), there is no doubt that discrimination learning underlies such cognitive abilities as abstraction, categorisation and rule extraction. These domains of animal intelligence will be considered in Part VII. Here we only briefly describe one more class of experiments based on stimulus discrimination and thus add *learning set* to the list of basic classes of learning nominated in this section.

Experiments for studying *learning set formation* were elaborated by Harlow (1949), and involved a succession of discrimination tasks with different stimuli; the focus of interest is whether there is an improvement in the rate at which each discrimination task is solved (see also Harlow and Bromer, 1938). Harlow devised the Wisconsin General Test Apparatus (WGTA) which is still in use for a battery of tests for studying cognitive skills in primates including humans, but the general principle of this test is applicable for many species. In the experiment a subject is presented with two containers of different shapes under one of which a food item is hidden. The subject, say, a monkey, can lift either or both, but is only rewarded in one case. The subject is then successively presented with several more pairs of containers with the same shapes as the first pair, and with the food always under a box of a certain shape. After this set of experiments, a new set starts in which the subject is shown another pair of containers with different shapes from those used in the first set. In several more trials, the food is again always under a box of particular shape. After several such experiences, the monkey begins to make a steady improvement in the percentage of correct choices in the successive trials until, finally, once the animal learns which box the food is under, it chooses the right box every time. In effect, the monkey has learned to learn.

The method of learning set formation is based on more general methodological approach called *matching to sample* (MTS), which in turn came from the concept of stimulus discrimination. This method came from a mental diagnostic elaborated for human children (Baldwin *et al.*, 1898; Binet, 1905). Ladygina Koths (1923) was first to apply this method for studying perception and learning abilities in experiments with her famous young chimpanzee Ioni. The chimpanzee was presented with a sample stimulus such as a simple geometric figure, and a short while later the chimpanzee was presented with two figures, one of which was the same as the sample. In other case the sample stimulus could be of a particular colour, and from two comparison stimuli one would be the same colour as the sample. In order to gain reward Ioni had to choose the stimulus that matched the sample (Fig. 6.3). This method has become popular as a very effective mental test for the comparative study of animal intelligence. Many species including not only mammals but also birds and even insects seem capable of solving discriminations on the basic of relational information (Fig. 6.4). We will consider the concrete results in Chapter 14.

Stimulus generalisation

As an element of animals' ability to judge about things, their aptitude for generalisation is of the same importance as for discrimination.

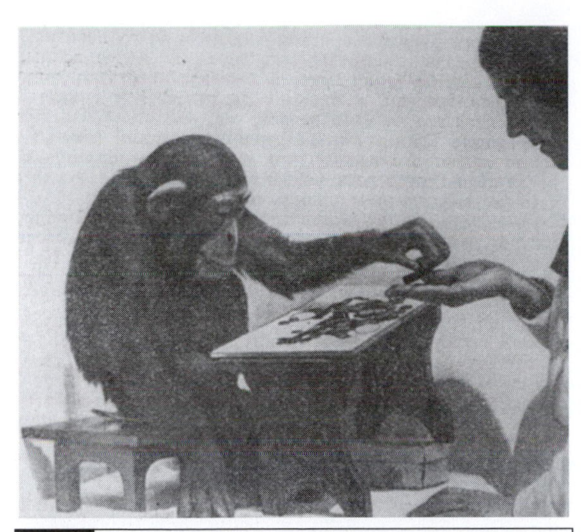

Fig. 6.3 Matching-to-sample experiment by Ladygina Koths: Ioni gives the experimenter a geometrical shape that looks like the given sample. (From the archives of the Museum of Anthropology of Moscow State University.)

Stimulus generalisation underlies the processes of integration of information and determines the procedure of discrimination learning.

In Pavlov's experiments a dog trained to salivate (CR) in response to a 1000-Hz tone (CS) also matched to similar tones, although salivation was not so intensive. The dog had 'generalised' its responses for all stimuli that were similar to an original stimulus. It turned out that the more similar the other stimuli were to the original CS, the less similar the stimuli were to the original CS, the weaker the CR. This is called a *generalisation gradient*.

Generalisation appears to be a characteristic of both classical and operant conditioning. An animal exhibits a very similar response to stimuli that resemble the original stimulus but not identical to it. If a rat is reinforced whenever it presses the red lever it will also attempt to press a number of coloured levers in a Skinner box. Furthermore, the more similar the stimulus is to the original the greater the frequency of responses the animal will emit. Thus if the original lever was deep red then the animal will also press levers of varying shades of red. However, it will press a lever that is nearest to the original shade of red more times per minute than a lever that is least like the original shade of red.

The use of the principle of stimulus generalisation allowed the explanation of some phenomena of associative learning. Thus Spence (1936), by referring to the effects of stimulus generalisation, was able to explain the transposition study of Köhler in a more efficient way. He argued that when animals are presented with a discrimination task between two pairs of stimuli from the same dimension (say, brightness), there will be a measure of generalisation between them. As a consequence, the excitatory tendency to approach S+ (dark) will also be elicited by S− (bright), but to a weaker extent; and the inhibitory tendency to avoid S− will be aroused, albeit slightly, by S+. The strength of approach to either stimulus will then be determined by the interaction between these sources of generalisation. The surprising prediction of Spence's theory is that if the test stimuli are far away from the training stimuli, transposition will break down. The subject will choose the lighter of two stimuli. Gestalt theory predicts that the subject will continue to choose the darker of the stimuli. Experimental data confirmed Spence's theory. Transposition breaks down when the test stimuli differ greatly. Despite this success of Spence's theory, there are a number of problems that it is unable to overcome. This shortcoming is partly connected with the rule that is used to determine to what extent the associative properties of a stimulus will change on any trial. It turned out that these changes did not take place independently of the properties of the other stimuli that were presented (Rescorla and Wagner, 1972).

Detailed analysis of many theories of discrimination learning are to be found in Pearce's *Animal Learning and Cognition* (2000). Despite their difference, they share a common feature. They assume that when animals are presented with a set of stimuli they learn about each one separately. According to *configuration theory* (Pearce, 1987, 1994) if a compound stimulus is presented for conditioning, or discrimination, then a configural representation of the entire pattern of stimulation will be formed.

In brief, configuration theory emphasises the importance of generalisation between different patterns of stimuli. This theory assumes

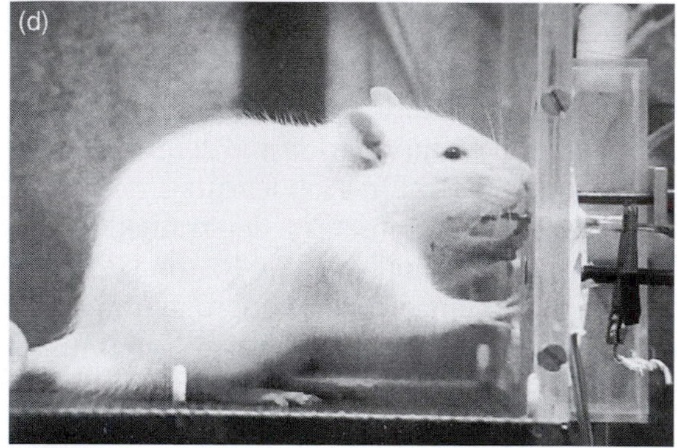

Fig. 6.4 Applying the method of stimulus discrimination for comparative studies in animal intelligence. Different subjects engaged in discrimination tasks: (a) orang-utan (photograph by V. Vrabcova, Prague Zoo; courtesy of M. Vančatová); (b) panda (copyright E. Dungl; Courtesy of L. Huber); (c) bumble-bee (photograph by T. Oganesov; courtesy of T. Oganesov); (d) a rat engaged in discriminating odours in a conditioning chamber (photograph by D. Schmid-Bielenberg; courtesy of R. Apfelbach).

that when two or more stimuli are presented together for conditioning, only a single association will develop. This association is between a unitary, configural representation of all the stimuli and the US. By making the additional assumption that generalisation will occur between configurations, configuration theory can explain most of the findings from discrimination studies based on stimulus generalisation.

The ability of organisms to generalise causes dramatic shifts in nature. For example, many predators quickly learn to withdraw when meeting with brightly coloured and venomous insects. Generalising key features of an image they also avoid similar but harmless insects. The well-known phenomenon of mimicry is based on evolutionary changes of species towards similarity to dangerous prototypes.

Repetition, extinction, recovery and attention

An important characteristic of associative learning is that it improves with *repetition*. Thus, a rat learning a maze reduces its number of errors with each trial until it runs directly to the food box with no wrong turns at all. After this its performance cannot be improved upon. Moreover, if the rat is continually permitted to run the maze after it consistently achieves perfect scores, it remembers the route longer.

In Pavlov's experiments salivation production increased slowly with every pairing of CS with US until it achieved the same level as when only US was presented alone. For example, a dog produced the same amount of saliva when hearing the bell ring as it would have done if presented with the actual meat. As has already been stated, the performance of the reaction can not be improved; however, the longer the stimuli are presented together after the top level of reaction had been achieved (this is called *over-training*), the more solid the reaction becomes regarding to *extinction*. In relation to classical conditioning, this term refers to the disappearance of a given response when the CS is repeatedly presented without the US. In relation to operant conditioning it refers to the elimination of a response by withholding all reinforcements of the response (Cartwright, 2002).

In Pavlov's experiments quantitative characteristics of extinction had been obtained. If a CS (say, a ringing bell) was presented on a number of occasions in the absence of the US (food), production of salivation began to lessen on each presentation of the CS without the US until it simply disappeared. Hence extinction is not a sudden process but occurs slowly, so that the CR slowly becomes progressively weaker until it no longer occurs.

It is interesting that the extinction may not be permanent. Pavlov found that carrying out an extinction procedure did not lead to total loss of the learned association. It turned out that even after a response has been completely extinguished in a series of trials, if the animal is allowed to rest for a few hours and is then given the CS again, the CR may appear through what it called *spontaneous recovery*. Such recovery can be elicited several times before the extinction becomes permanent. To extinguish a CR fully the CS should be presented on a number of occasions. Nevertheless, research has shown that even when the CR has been completely extinguished, compared to naive animals, previously conditioned animals that have had the CR completely extinguished will relearn the response much faster than naive animals learn it. Therefore it seems that what has been learnt once is never entirely forgotten.

Another way to call extinguished reactions into being is the use of a new stimulus together with a known CS. For example, extinguished salivation as a reaction of a dog for a tone may be restored if a blinker light is switched on together with the tone. Similar data were obtained in operant conditioning. Pavlov called this process *disinhibition* and considered extinction a new process of conditioning that inhibits the first conditioning reflex. Thus neutral stimuli, which act at early stages of learning together with conditioning stimuli, often inhibit learning and decrease effectiveness of a process. When a new process, namely extinction, starts new stimuli inhibit it and thus call previously extinguished reactions into being. Pavlov's logic of explanation is quite natural (Pavlov, 1927). He suggested that when facing a novel stimulus an animal immediately activates the *orienting response* (OR) (or, as he called it, an investigatory reflex) and turns its organs of preception to a new source of information. This just suppresses development of the conditioned reaction. This process is called external inhibition. If the novel stimulus appears during a process of extinction, then that extinguished CR is increased. This is called Pavlovian disinhibition. It is important to note that the disinhibition is not caused by competition of two CRs but by excitation of central nervous system.

Many data confirm Pavlov's hypothesis that extinction is a process of learning. During the process of extinction an organism learns that the CS is no longer followed by the UR. Now the CS is connected with the absence of the UR, hence CR is extinguished. It is important that in experiments animals meet the absence of US in the combination with CS under circumstances in which that CS has been rewarded many time previously. This is an unexpected event and being repeated it will form the new CR, namely, 'do not pay attention to this stimulus any more'. In real life an organism meets lots of stimuli that are not connected with any reinforcement and ignore them. Only meeting the absence of the reward suddenly, the individual learns to connect certain stimuli with the absence of reinforcement (Mackintosh, 1976).

This does not mean that it is necessary to reinforce carefully every correct movement of the family dog when teaching it to follow commands. As was explained before, very infrequent awards will do in order to keep the animal's attention active.

In principle, extinction is closely connected with *attention*. The Pavlovian investigatory reflex (orienting response, OR) mentioned above is a consequence of the animal attending to the stimulus, and the vigour of this response may well provide an indication of the attention a stimulus receives.

An experiment by Kaye and Pearce (1984) shows how the simple experience of being presented repeatedly with a stimulus might influence this measure of attention. This experiment was based on Skinner's technique in combination with Pavlov's method. Two groups of rats were placed into a conditioning chamber containing a light bulb and a food dispenser. For the first 12 sessions nothing happened for Group Novel, whereas for Group Familiar the bulb was illuminated for 10 seconds at a time at intervals in each session. Both groups were then given a single pre-test session in which the light was occasionally illuminated for 10 seconds. A typical OR to the light was performed at the outset of training by Group Familiar. But with repeated exposure to the light the frequency with which this response occurred declined progressively across 12 sessions. Such a decline in the responsiveness to a stimulus as a result of its repeated presentations is referred to as *habituation*, a phenomenon discussed in Section 6.1. The strength of the OR was considerably more vigorous in Group Novel than in Group Familiar, which suggests that the groups differed in the amount of attention they paid to the light. All subjects were then conditioned with the light serving as a signal for food. There was no difference in the strength of the CR of activity at the trough during the light for both groups on the pre-test session. As predicted, on test sessions conditioning was more rapid in Group Novel than in Group Familiar, for which extended exposure to the light reduced its conditionability. This effect is known as *latent inhibition* and supports the claim that Group Familiar paid rather little attention to the light as a result of the pre-exposure stage.

Lubow (1973) reviewed experiments showing latent inhibition in goats, dogs, sheep, rats and rabbits, but beyond these, studies of latent inhibition are rare even in mammals. There have been several attempts to demonstrate this phenomenon in pigeons, but these have led to conflicting findings. No success has been achieved in attempts to show latent inhibition in honeybees (Bitterman *et al.*, 1983) or goldfish (Shishimi, 1985). Pearce (2000) suggests that if future research should confirm that changes in attention, as indexed by changes in conditionability, are unique to mammals, then this will be an important discovery. Obviously, the opposite claim is also true: it would be a challenge to find these characteristics of attention in groups of animals other than mammals.

It is of great importance to know the details of the attentional processes in humans. This is a dream of people belonging to many trades, from teachers and artists to advertising executives and presidential contenders: how to catch people's attention and to hold it. It is clear that the attentional processes are governed by the same rules in many mammalian species including humans. This include the generation of reactions to novel stimuli, selective attention paid to different stimuli, processes of inhibition and recovery which may lead to revival of attention and interest, and so on.

Let us consider a simple funny example. Nowadays pairs of lovers can present each others with velvet flowers provided with a button that controls an audiocassette. When the button is pressed, the happy lover will see the flower bowing and hear the magic words 'I love you' spoken by the partner's voice. After perhaps a hundred pressings of the button extinction will come, but an interval should see attention being restored. Isn't this a good chance to measure the degree of love in values of attention? I suggest that many quantitative characteristics could be measured here such as number of pressings till the first display of inhibition, the time duration which is needed for returning interest, the rate of increase of this time from one session to another, and so on.

Unfortunately none of present theories of attention is able to explain all the relevant experimental findings, although the development of these theories has led to the discovery of a wide range of experimental findings that now show the importance of attention in animal and human learning and cognition.

Chapter 7

Learning classes beyond 'simple' associative learning

The title of this chapter looks paradoxical because, as a matter of fact, the rest of this book is entirely devoted to forms of animal learning beyond 'simple' associative learning. The aim of this chapter is to give readers an estimate of the application of rules of associative learning and to review briefly the learning classes that will be considered further in this book, just to complete the list of learning classes that was given at the beginning of this Part.

Indeed, when analysing rules of associative learning, learning theorists consider some notable exceptions to the main principle of association formation. The fact is that on some occasions and in some species conditioning occurs at the 'wrong' time and with different probabilities for different stimuli.

7.1 Rules of 'simple' associative learning

Before considering some exceptions to the rules, let us first concentrate here on the main rules of 'simple' associative learning.

(1) Making associations is a matter of connecting events that occur together in time, and it is true that usually an increase in the interval between CS and US presentation slows the formation of associations.

(2) The speed or strength of learning increases with:

The intensity of the CS

The size of the reinforcer or US.

(3) When the reinforcer is withheld, the learned response declines in frequency or intensity (this is called extinction).

(4) All pairs of events can be associated with equal ease, e.g. it should be just as easy for a rat to associate the arrival of food with a light or a buzzer as with any other CS. This is called *equipotentiality*.

The first rule is now known to be a more general phenomenon, i.e. time is not necessarily the most important factor for learning. Even events far apart in time can be associated if there is a highly predictive link between them. The best example of this is in food aversion learning. The first experimental results in this field were obtained by Garcia and colleagues in their attempts to learn why are rats so hard to poison (Garcia and Koelling, 1966; Garcia *et al.*, 1972). In those experiments US = feeling ill, CS = novel flavour of food. The learnt response is avoidance of the novel food. The interval between tasting the food for the first time and feeling ill may be several hours for the rats. In the wild it is impossible to tell when the animal starts to feel ill. But food aversion learning can be studied experimentally. Illness can be induced at a set time after the animal has eaten a harmless but novel-flavoured food, e.g. by injecting lithium chloride. In rats, food aversion learning can take place with delays of up to 12 hours between consumption and the onset of illness. It was suggested that the US–CS interval was not as long as 12 hours due to lingering flavour of the novel food. But even if other (familiar) foods are eaten in the intervening period, aversion to the novel food is still demonstrated. Obviously such learning ability has a high biological importance. It indicates that it is the predictability between two events that is important in establishing learning. Predictability between US and CS is more important than the time interval between CS and US, although in many cases a short time interval will be strongly correlated with a high degree of predictability.

Other behavioural routines also need to be taken into account here. For example, rats show a cautious behaviour to novel foods. Usually, they will only nibble at any new food the first time it is encountered (Barnett, 1970). Then, if nothing happens, they may eat a bit more the next time, until, finally, it becomes a normal part of their diet. This coupled with the time lag on associations made between US and CS is why many rat poisons take much greater than an hour to have any effect.

Garcia and colleagues (Garcia *et al.*, 1977) found another effective application of the results obtained concerning slowed food aversion: they taught coyotes and wolves to coexist peacefully with sheep. Predators were made ill by feeding them chopped mutton that was wrapped in raw sheepskin and laced with lithium chloride. A dramatic effect was observed when the animals were allowed to approach live sheep. Rather than attacking them as they normally do, the wolves after their characteristic flank attack immediately released their prey.

Equipotentiality (the fourth rule listed above) is also problematic as a universal law of learning. It has been found in many studies that animals do not associate all events with equal ease (Mackintosh, 1974). Chickens will easily learn to peck a key to obtain food but take longer to learn to peck a key to obtain access to wood-shavings for dust bathing. They seem to find it easier to associate a food-related response with obtaining wood-shavings. For rats only five trials are sufficient to learn to run from one section of a box to another in order to avoid electric shock but hundreds of combinations are required on those occasions when animals must press a lever for the same purpose. For pigeons in the same situation it is much more difficult to learn to peck a key than to press a bar.

This is caused by the difference between displays of the animal's innate reactions of avoidance within a specific situation. Thus it is nearly impossible to shape avoidance in those experiments in which animals, say, rats must draw near a source of danger, for example to press a lever situated just under a bulb that lights up to signal an electric shock. It is easier to shape this association when the bulb is situated far away from the lever to be pressed (Biederman, 1972).

Returning to the set of examples concerning food aversion, rats associate a novel flavour with subsequent illness much more readily than an auditory–visual compound (Garcia and Koelling, 1966). Pigeons easily learned that a red light signals food, but they learned with difficulty when the same stimulus was used to signal an electric shock. Conversely, conditioning with a tone progresses more readily when it signals an aversive rather than appetitive US (Shapiro *et al.*, 1980).

The term *selective association* refers to this general finding that some CS–US relationships can be learned about more readily than others (Mackintosh, 1974). In the case of the Garcia and Koelling study cited above, taste is more likely to provide information than sound about whether or not a certain food is poisonous.

Selective associations may, at least in some species, be evident from birth. One of the best examples is the behaviour of newly hatched chicks of precocial birds. Chicks have an innate tendency to peck at small conspicuous objects such as coloured beads. When a new object is edible or positively rewarding, chicks will show enhanced pecking. However, when the object tastes bitter, chicks will subsequently learn to avoid similar beads even after a single experience. In this manner, chicks quickly make up a directory of edible, neutral and aversive objects encountering within days after hatching. This fact can be used as the basis for a variety of experimental tasks developed for studying memory formation in chicks (see Chapter 10 for details).

In experiments of Gemberling and Domjan (1982) rats were conditioned when they were only 24 hours old with either illness, induced by an injection of lithium chloride, or electric shock. The rats are unlikely to have learned anything during their first 24 hours of life. Instead, the most likely explanation for these findings is that rats are genetically disposed to learn about some relationships more easily than about others.

The concept of selective associations forms a physiological basis for the special form of learning, namely, *guided (selective)* learning, which means that species realise their innate predisposition to form definite associations in accordance with their ecological and evolutionary traits (Gould and Marler, 1984; Griffin *et al.*, 2002). We will consider the significant implications of this fact in Part VII.

Concerning the second and third rules listed above, we should note that relations between stimuli and reactions are not so simple and many questions concerning the nature of stimulus representation, attention, memory and learning are still not clear although much progress has been made. In contrast to theories of classical and operant conditioning, intervening variable theories proceed from the assumption that learning can involve knowledge without observable performance. This will become apparent if we consider such classes of learning as latent (exploratory) learning, insight and imprinting.

7.2 Latent (exploratory) learning

In his article 'A new formula for behaviorism', Tolman (1922) suggested the development of 'a new non-physiological behaviorism' aimed at the objective study of internal processes concerning learning such as motive, purpose, determining tendency, and the like. Tolman developed this concept in many experiments, and some of them led to the theory of latent learning. This theory describes learning that occurs in the absence of an obvious reward. Latent learning is 'hidden' internally rather than shown in behaviour.

For example, Tolman and Honzik (1930a,b) showed that rats could learn a route in a maze without obtaining reinforcement. Three groups of rats were trained to run a maze. The control group, Group 1, was fed upon reaching the goal. The first experimental group, Group 2, was not rewarded for the first six days of training, but found food in the goal on day seven and every day thereafter. The second experimental group, Group 3, was not rewarded for the first two days, but found food in the goal on day three and every day thereafter. Both of the experimental groups demonstrated fewer errors when running the maze the day after the transition from no reward to reward conditions. The marked performance continued throughout the rest of the experiment. This suggested that the rats had learned during the initial trials of no reward and were able to use a *cognitive map* of the maze when the rewards were introduced.

Indeed, what is learned in such a situation is not a series of S–R connections, rather, the organism learns 'what leads to what' and this may be considered a form of 'stimulus–stimulus' learning. These stimulus relationships get organised into a cognitive map that is obviously more than a series of individual routes to a goal. Once an organism has developed a cognitive map of its surroundings, it can get to a goal from any location. It follows the 'principle of least effort' by choosing the route requiring the least effort. This form of learning has been revealed in different species, from anthropoids to insects.

The term *latent learning* refers to only one phase of complex learning process. Thorpe (1963) preferred the term *exploratory learning* and defined it as the association of indifferent stimuli, or situations, those without patent reward. The motivation in latent learning situations seems simply be a desire to get to know the surroundings. Latent learning is undoubtedly very important for organisms in real life. The survival value of exploratory behaviour may not be obvious at first, but its adaptiveness becomes apparent when a need arises and is quickly met. After establishing the location of a commodity such as food or a hole, the animal can reasonably expect it to be there when it returns (Wallace, 1979). In a laboratory experiment by Metzgar (1967) two groups of mice were turned loose in a room with an owl. One group was given a few days to familiarise themselves with the room before the owl was introduced. The second group was put into the room at the same time as the owl. The mice that were familiar with the room fared much better – that is, fewer were caught.

A set of tests was elaborated by Reznikova (1981, 1982) for studying exploratory behaviour in ants in their natural surroundings. Ants were presented with an 'enriched piece of environment' which included several sorts of mazes, all lacking a food reward. These tests were aimed at comparing levels of routine exploratory activity in different species which reside in the same territories and belong to the same species community. Exploratory activity in ants was measured by recording the length of time spent by the ants within mazes as well as the numbers of visits. Ants of different species showed significant differences in intensity of exploratory activity. It turned out that explorative behaviour correlates with agility in hunting and searching for hidden victims in different ant species.

7.3 Insight

A 'what-leads-to what' *expectancy* is considered a basis for insight (Tolman, 1938). In the case of insight learning *confirmation of expectancies* leads to learning. Learning refers to cognitive knowledge generated by confirmed expectations, such as knowing how to navigate the maze or how to use things (such as to put sticks together for getting a banana) after exploring them.

As was mentioned in Part I, W. Köhler, one of founders of the Gestalt school of psychology, argued for the place of cognition in learning. In particular, he suggested that insight played a role in problem-solving by chimpanzees. Rather than simply stumbling on solutions through trial and error, the chimpanzees seemed to demonstrate a holistic understanding of problems, such as getting hold of fruit that was placed out of reach, by arriving at solutions in a sudden moment of revelation or insight. One of the best demonstrations of true insight in primates came from an experiment in which food was placed outside the arm's reach of a chimpanzee; the animal used short sticks that were placed within its cage to reach a longer stick placed outside, and then used the longer stick to reach the food. These results support the Gestalt assertion that it is the 'whole' rather than S–R elements that guide learning. A chimpanzee in such situations tries a number of possibilities mentally, and then tries it physically (see Fig. 7.1).

Fig. 7.1 A chimpanzee reaches a piece of food with the use of a stick. (Photograph by L. Firsov; courtesy of L. Firsov.)

In general, the learner ponders a problem thinking about possible solutions (this is called a pre-solution period) until one suddenly becomes apparent (insight).

There are several characteristics of insight learning:

- Transition from pre-solution to solution is sudden and complete
- Insight based on error-free performances
- Solutions gained by insight are retained longer
- Solutions gained by insight are easily applied to other problems (transposition).

For a long time, it was believed that insight learning is not possible below the primate level. Recently it has been shown that at least several bird species can learn this way and that 'folk physics for apes' (Povinelli, 2000) may be adapted for crows, finches and jays. This has encouraged investigators to take a look at insight learning in various other species (see Chapters 17 and 18 for detailed analysis).

7.4 Imprinting

Imprinting as one class of learning implies both features of conditioning and insight. One could name this 'ontogenetic insight', as during a particular short period of its ontogenesis an organism suddenly 'learns' what to do. At the same time imprinting implies features of both learning and instinctive behaviour. Imprinting occurs when innate behaviours are released in response to a learnt stimulus. This form of learning has been described as a rapid learning of certain general characteristics of a stimulus object. Most imprinting promotes survival of newborn animals and shapes their future breeding activities. There are still many discussions and criticisms of the imprinting paradigm. Many researchers believe that the imprinting phenomenon has too many special qualities for it to be considered just a special form of learning. The phenomenon of imprinting will be considered in details in Chapter 24.

CONCLUDING COMMENTS

In studying animal intelligence, we can discern basic forms of learning: some of them can be considered relatively simple (such as habituation and classical conditioning), some (such as imprinting and guided learning) have special qualities that make analysis of them difficult in isolation from innate forms of behaviour, and others (such as 'insight', 'latent learning' and social learning) require more cognitive explanations that will be given further in this book.

Although more than a century has elapsed since researchers turned from naturalistic studies of animal intelligence to experimental investigations and measuring characteristics of learning, there are probably no experimental paradigms in this field about which one can say that they have only historical importance. For example, habituation and conditioning still serve as the basis of experimental paradigms widely used by experimenters for studying abstraction and concept formation in a wide variety of species including humans. Relatively simple rules of operant conditioning can be used not only for shaping and governing complex behaviour but also for enabling organisms to exhibit creative abilities.

The whole picture of the interactions between different forms of learning in the subject's mentality is still far from completion. Even for 'simple' forms of learning its mechanisms and correlation with memory processes are still not completely clarified.

Part III

Past and future in animal life: remembering, updating and anticipation

. . . And suddenly the memory returns. The taste was that of the little crumb of madeleine which on Sunday mornings at Cambray my aunt Léonie used to give me, dipping it first in her own cup of real flower tea . . . At once I recognised the scent of all the flowers in our garden and in M. Swann's park, and all the water lilies on the Vivonne and the good folk of the village and their little dwellings and the parish church and the whole of Cambray sprang into being.

Marcel Proust, *Remembrance of Things Past*

The epigraph illustrates this mysterious storage and retrieval system that we call memory. A gap still exists between what we call memory referring to purely private events and what we mean by memory in its relation to learning and reasoning. In recent animal studies 'learning' is concisely defined as the acquisition of a skill, while 'memory' is the ability to retain that skill. In this Part we will consider experimental ways to solve a problem of how animals gain new skills and 'ideas' from experience, retain them over time and use for prediction of some events in the future.

For many years investigators of animal behaviour have been trying to clear the question of how organisms represent the external world. In particular, it is a very intriguing problem whether animals can operate with internal representations of stimuli received earlier, or whether they can only react to a real thing in sight. It is intuitively clear that when a dog returns to a place where it has buried a bone, it deals with an 'idea' of that bone. After leaving that bone, the dog could not see or

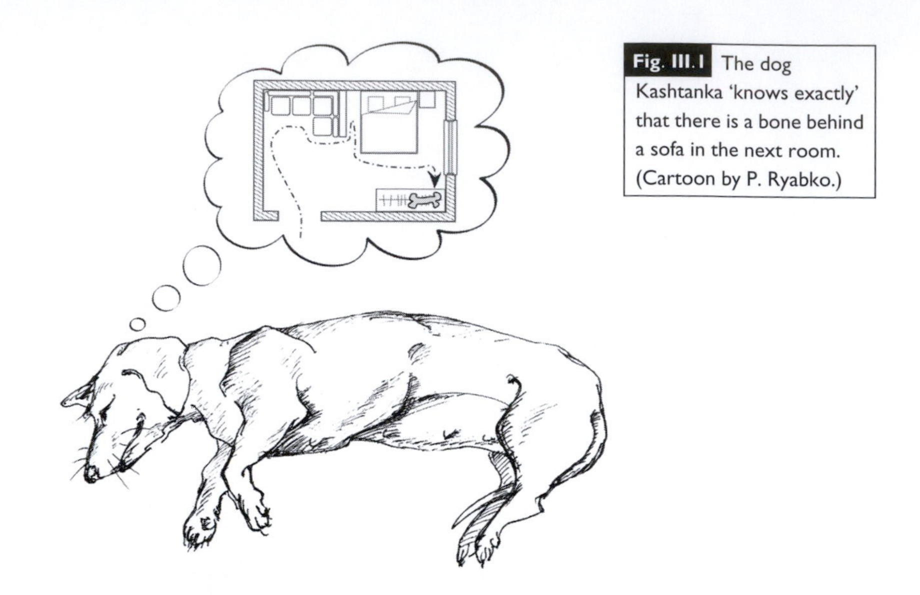

Fig. III.1 The dog Kashtanka 'knows exactly' that there is a bone behind a sofa in the next room. (Cartoon by P. Ryabko.)

smell this desirable thing; nevertheless the dog can dig up the bone many hours later. The Russian writer Anton Chekhov vividly described this in his novel *Kashtanka*. The dog Kashtanka 'knew exactly' that there was a bone behind a sofa in the next room whose door was closed, and she definitely intended to pick it up the first thing in the morning when her owner opened the door (Fig. III.1). Just like cognitive ethologists many years after him, the novelist used the mental states such as belief, desire and intention to describe the dog's mental capabilities.

Recent studies in animal social life have shown that in many species rapid learning of socially relevant information allows them not only to remember valuable items over long periods of time, but also to consider the future and plan ahead (see details in Part X). Many animals adjust their foraging strategies according to circumstances so as to optimise returns, and this implies that time and quantity are being calculated (Gallistel, 1990). Although it is still necessary to clarify the level of mental sophistication these strategic adjustments require, all these results in sum lead us away from monopolising cognitive capabilities for humans, in particular analytical skills based on recollecting memory and prediction.

Chapter 8

What is memory for an intelligent animal?

8.1 Kinds of memory

Much of animal behaviour seems to indicate the use of memory. Many animals develop an intimate knowledge of their local environment, such as where to find different food sources, where their home and shelters are and where danger lies.

Processes of *learning* and *memory storage* are tightly connected although they may be differently localised within the brain. The study of memory in animals looks at how information that has been acquired at one time (learned) influences behaviour at a later time and traditionally investigates how information is stored and retrieved.

Pearce (2000) points out that there are three main questions about memory in animals:

(1) How long can be the information retained?
(2) What type of information can be retained?
(3) How much information can be retained?

Most research on memory was historically connected with experiments on mammals or, at least, on vertebrates, although many animals with no cortex at all, and some without anything we would call a brain, can be classically conditioned. Moreover, we know now that excellent learning capacities and cognitive skills may be floated about eccentric nervous systems. It is worth noting that many facts concerning memory were first obtained from studies on humans and then applied to animals. To some extent the findings of the studies of memory in animals mirror those made in humans. Concepts of intelligence have fuzzy boundaries but one thing is clear to us: there is no straight correlation between powers of memory and intelligence in our species. There are many sorts of strange abilities associated with memory in humans. Idiots savants, a special class of retarded people, which includes autistics, are able to multiply two five-figure numbers in their heads. Another class of people called eidetics possess 'photographic memory'. The advantage of these capabilities seems to be enormous for our species but not for others. Later in this book we will meet non-human idiots savants within the classes of birds and mammals that take enormous advantage from their strange memory dimension.

Psychologists and memory researchers often divide memory into categories defined by the duration for which the memory is expected to last. The existence of distinct categories of memory was first noticed in humans with brain concussions who were not able to recall what happened just before an accident, but could remember what happened earlier.

Sensory memory refers to the fact that, after experiencing a stimulus, information about that stimulus is briefly held in memory in the exact form it was received, until it can be further processed. Typically, sensory memories may last only a few seconds before decaying – or being overwritten by new, incoming information. But, while they last, sensory memories contain detailed information: almost like an internal 'copy' of the stimulus, in perfect detail. For example, psychologists have assumed that there is a memory area where incoming visual information is stored as a picture or icon. This is sometimes called *iconic memory*. While visual information remains in iconic memory, an individual can answer detailed questions, such as what is the third row of numbers in a numerical display. Psychologists have assumed that there is also an *echoic* memory for auditory information (stored as an echo) and other buffers for information related to the other senses: taste, smell and touch.

Short-term memory refers to memories that last for a few minutes. Unlike sensory memory, which is stored in the exact form in which it was experienced, short-term memory has received some processing; thus, 'A' is stored not as a visual stimulus, but as an abstract concept of the letter 'A'. Short-term memory is of limited capacity, usually five to nine items. Beyond this capacity, new information can 'bump' out other items from short-term memory. This is one form of forgetting. Objects in short-term memory can be of indefinite complexity: thus short-term memory can hold several numbers, or several words, or several complex concepts simultaneously.

Working memory is sometimes considered a synonym for short-term memory. However, memory researchers often consider this a specialised term, which is conceptualised as an active system for temporarily storing and manipulating information needed in the execution of complex cognitive tasks (e.g. learning, reasoning and comprehension). Working memory is often described it as the 'blackboard of the mind' or 'online memory' (Goldman-Rakic, 1992). The term working memory is used to emphasise the fact that what is learned in one case may not be relevant on the next. There are two types of components: storage and central executive functions.

Speaking about human working memory, there are two storage systems within the model: the articulatory loop and the visuospatial sketchpad or scratchpad which are seen as relatively passive slave systems primarily responsible for the temporary storage of verbal and visual information (Baddeley, 1986).

Speaking about animal working memory in general, it should be emphasised that working memory is retained only long enough to complete a particular task, after which the information is discarded because it is no longer needed, or because it may interfere with the next task.

Long-term memory (reference memory) is memory that can last for years. It contains everything we know about the world, including semantic and factual information as well as autobiographical experience. In general, long-term memory is organised so that it is easy to reach a stored item by a number of routes.

Hebb (see Chapter 3) first distinguished between immediate memory (short-term memory, STM) and long-term memory (LTM). He suggested that STM consists of active networks of nerve cells that repeatedly excite one another. These reverberating loops of cells can maintain a stimulus trace in the brain from a few seconds to many minutes. What makes the patterns of cells form into loops is that the cell that originally brings the sensory information into the cortex is hypothesised to receive onto its receiving processes (dendrites) inputs from cells further along in the loop. As the output processes (axon) of these cells stimulate the first cell, the cell then sends back signals to the next cells. This ultimately causes the first cell to be restimulated. The excitation in the loop, if fairly weak, finally dies down; the loop then 'breaks' and the information is lost from the brain or 'forgotten.'

The development of LTM, called *consolidation*, is now known as a process of transfer from the short-term to the long-term system. It seems that short-term storage may be necessary before long-term storage can occur. During the last decades, a number of studies have revealed that LTM formation requires a *de novo* brain protein synthesis. Indeed, pharmacological experiments have shown that administration of protein synthesis inhibitors around the time of training impairs LTM. In contrast, STM is based on transient changes in synaptic morphology (for review, see Welzl and Stork, 2003).

The notion that it takes time to form a stable memory seems to capture a basic property of memory. That is, during the consolidation period after training, memories are susceptible to interference by shock, cooling or pharmacological manipulations and may be obliterated. This consolidation period might ensure that only the really important memories are stabilised (Menzel, 1999): it seems important to keep memory modifiable for some time, to erase it in case of contradictory experience, enhance it in case of affirmative experience, or to modify it according to other, already existing memories.

To analyse memory consolidation on a behavioural level, one can make use of the temporal characteristics of retention. Kamin (1957, 1963) found that retention of an avoidance response in rats follows a non-monotonic time course: retention is high immediately after training but then decreases to a minimum after 1 hour. Then retention increases again to reach a stable level after several days. This 'Kamin effect' (Denny, 1958) was subsequently found in many organisms, including humans.

A number of authors, beginning with Kamin, have suggested that the simplest explanation is to assume the existence of two independent and additive memory systems: one memory system that dominates retention immediately after training but then constantly loses impact and a second one that needs time to consolidate and is then increasingly responsible for retention. It is the 'secondary' rise in retention that is taken as a behavioural indication for memory consolidation. The functional background of such a system might be that animals cannot afford 'not to behave' until the stable, second memory is formed. Thus perhaps they transiently use an independent, potentially less specific, memory system. It has been demonstrated in different organisms such as honeybees, octopuses, goldfish, mice and humans that in conditioning procedures retention is often a non-monotonic minimum function (Gerber and Menzel, 2000).

Long-term (reference) memory is retained for longer periods and is used for completing successive tasks. There are several different ways to classify long-term memories according to their content. Cognitive psychologists and neuroscientists have divided memory into two broad classes or sets of strategies used by the brain to acquire information.

One set of strategies, termed *explicit* memory (or *declarative*, or *relational*), underlies memory of events and the circumstances of their occurrence. In human studies declarative memory is a term for information that is available to conscious recollection and verbal retrieval (i.e. it can be 'declared'). In forming and storing explicit memories, associations are done with previous related stimuli or experiences. Therefore, explicit memories can be remembered and recalled, and rely on previous experiences and knowledge.

The other set of strategies, *implicit* (*non-declarative*, or *procedural*, or *priming* memory) memory, encodes information about perceptual and motor skills, using non-cortical structures and requiring no conscious participation. Implicit memories cannot be looked up or remembered to be used for actions and reasoning. They consist of memories necessary for performing events and tasks, or for producing a specific type of response. Implicit memory is best demonstrated when performance is improved on a mechanical task. This type of memory is shown through activation of the sensory and motor systems needed to perform a certain task.

Humans with amnesia cannot form new conscious, explicit memories, while implicit memory is left intact. Many learning tasks require both memory systems. Explicit memory, like implicit memory, has a short-term phase that does not require protein synthesis and a long-term phase that does.

Declarative memory has been further subdivided into *episodic memory*, which is autobiographical information, and *semantic memory*, which is factual information about the world (vocabulary items, knowledge of what things are used for, memory of multiplication tables, etc.). The term episodic memory returns us to the epigraph from Marcel Proust. Episodic memory is the ability to 'reach back' into the past. More specifically, it is the ability to retrieve egocentric episodes as a temporal chain of life events.

There is some disagreement as to the nature of this division and the relationship between these memory systems, both in terms of function and the neural structures involved. It is generally agreed, however, that episodic memory is concerned with the conscious recall of specific past experience, whereas semantic memory involves the storage and retrieval of factual knowledge about the world (Griffiths *et al.*, 1999). This difference is often referred to in terms of remembering versus knowing: episodic memory is concerned with the remembering specific personal experiences, whereas semantic memory mediates what one knows about the world. As Griffiths *et al.* (1999) have it, remembering getting soaked in the London rain last Tuesday is an example of episodic memory, but knowing that it often rains in England is an example of semantic memory because it need not be acquired as a result of a personal experience of getting wet.

According to Endel Tulving's (1972, 1983) classical definition, episodic memory receives and stores information about temporally dated episodes or events, and temporal–spatial relationships among these events. Thus, episodic memory provides information about the 'what' and 'when' of events (temporally dated experiences) and about 'where' they happened (temporal–spatial relations). The different kinds of memory are linked to different levels of 'knowing': non-declarative memory is anoetic (non-knowing), semantic memory is noetic (knowing), while episodic memory is autonoetic (self-knowing). This suggests that episodic memory is critically dependent on the concept of self.

When presenting a new definition of episodic memory, Tulving and Markowitsch (1998) state that it possesses features that no other memory system has and is accompanied by a special kind of consciousness (autonoetic consciousness). This is different from the 'noetic' consciousness that is involved with the retrieval of declarative information. Such a distinction is based on the fact that human subjects can distinguish between recalling past personal experience and remembering an impersonal declarative fact. Remembering a specific event requires autonoetic consciousness, whereas knowing a fact is noetic in nature. Tulving and Markowitsch (1998) claim that episodic memory is what really sets human apart. Episodic memory allows human beings to travel back and forth mentally through time. It also gives us an awareness of ourselves and our state of existence within the dimensions of time.

Some recent studies, however, enable us to suggest that animals are capable, at least to some extent, of episodic memory. The fact is that most of the laboratory tests for animal memory that have been used up to date and which will be described in this chapter can be explained in terms other than episodic recall. A different strategy of testing whether or not animals are capable of episodic memory is to adopt an ethological approach, considering cases in nature in which an animal might benefit from the capacity to remember unique episodes that occurred in the past (Griffiths *et al.*, 1999). We will return to this problem and concerned experimental results in Chapter 22.

In general, the terminology concerning different kinds of memory is not completely settled yet. On the one hand, in some fundamental works concepts of explicit and implicit memory have been attributed to snails. On the other hand, sometimes in scientific literature there are no distinct boundaries between concepts of explicit and implicit forms of memory and explicit and implicit forms of knowledge. Besides, distinctions between explicit and implicit knowledge in non-humans have seldom been made yet. This is probably because explicit knowledge is typically defined in terms of consciousness, and most researchers are hesitant to attribute consciousness to non-human species. There are, however, no fundamental functional or biochemical differences between the nerve cells and synapses of humans and those of many more 'simple' organisms, both vertebrates and invertebrates such as snails and worms.

Many researchers believe that animals are capable of only non-declarative learning and memory, or at least that there is no way to test whether any of their learning is declarative because they cannot tell us what they remember, and that animals may know many facts that are important in their lives but do not know what they know (Cheney and Seyfarth, 1990). Recently some insights have come from different experimental techniques. These include, in particular, experiments with delayed reactions (see Chapter 10) and language-training experiments (see details in Part IX), in which apes, dolphins and African grey parrots have learned adaptations of human communication systems.

8.2 Different bodies, different memories

Post-training memory modulation is seen in many species, including bees, fish, birds and mammals, and this argues that the basic learning process was developed early in evolution. A comparative approach to understanding learning and memory can provide a broader perspective on how brains encode, maintain and retrieve information. We will consider several striking examples in order to imagine the complexity of the relations between brain and behaviour in different creatures who learn and remember.

We start with invertebrate representatives which differ from vertebrates in the evolution and functional organisation of their nervous systems (Sarnat and Netsky, 1981). These differences are so fundamental that many students of animal learning and memory argue that invertebrates lack the autoassociative mechanism, and hence lack episodic event memory. Indeed, for many invertebrates, a teaching input (unconditioned stimulus, US) must position its synapse on a single stimulus which is to be conditioned (CS). The teaching input conditions just that one stimulus. In the vertebrate, a teaching input (US) can spread long distances along a dendrite. As a result, the vertebrate teaching input can condition many stimuli simultaneously. The vertebrate mechanism is the one appropriate for pattern association, and for the storage and retrieval of episodic memories. Invertebrates are believed to lack episodic event memory.

As a behavioural example of this difference, one can contrast the ways in which vertebrates and invertebrates learn mazes. Both can learn simple T-mazes. But an ant taught a T-maze with one eye covered must relearn the maze when that cover is removed and placed on the other eye. No vertebrate would be hindered by such a trick. The vertebrate's general-purpose associative memory store is isolated from the individual sense organs. A vertebrate's memory of the maze is therefore a memory of a series of events which can be recalled at will, independent of sensory cues. Ants, like all insects, operate without benefit of such an abstract and centralised memory store (Wehner and Müller, 1985).

Despite some limitations of learning and memory in invertebrates, we should not keep our minds closed on the question of their intelligence. There are several marvellous compensatory mechanisms that allow at least some invertebrate species to memorise landmarks, remember and pass complex information, and do many other clever things. It is not without reason that Lehrer and Wehner published one of their projects devoted to social Hymenoptera (10-mg insects with a 0.1-mg brain) under the title: 'Mini brains – mega tasks – smart solutions' (see: Wehner *et al.*, 1996b; Lehrer, 1997).

Eight arms and one leg for juggling with memory

This section is devoted to molluscs, a seemingly 'primitive' animal group. But these creatures possess properties tempting for students of learning and memory.

The environment and life style of cephalopods means that they need to be capable of complex and flexible behaviour. As active predators, they need to explore, understand and remember their environment and the behaviour of other animals. They can solve problems as they remove a plug or unscrew a lid to get prey from a container. They use rocks and jets of water in a way that could be classified as tool use. They have been found to play with 'toys' and to have individual responses and individual temperaments (Sinn *et al.*, 2001; Kuba *et al.*, 2003).

Despite huge efforts devoted to the analysis of the nervous system of octopuses and other cephalopods such as squid, cuttlefish and nautilus, their brains and the limits as to what extent these animals are educable remains a mystery. The brains of cephalopods evolved entirely separately from the brains of the vertebrates, and they have an entirely different design. At the same time, at the fundamental level – cells communicating by chemical signals – the brain of the cephalopod is essentially the same as that of any vertebrate. Indeed, research on the squid's 'giant axon' has been an instrument in showing how nerves work throughout nature. Cephalopod neurons are closer to the vertebrate structure than that of a common invertebrate.

Cephalopods possess keen senses and are capable of extremely complex behaviour that requires a lot of neurons for processing. Their complex eyes, as large as car headlamps in some deep-water species, can distinguish detail as well as mammalian equivalents can. Cephalopods have highly developed senses of touch, taste and smell, and can detect gravity, a sense which is used in the coordination of muscles during movement. Division of labour between the central and peripheral nervous system greatly simplifies the movement control of flexible arms in the octopus. The peripheral nervous system is organised as an axial nerve cord composed of about 300 interconnected ganglia in two cerebrobrachial (axonal) tracts. The axons in the tracts carry sensory and motor information to and from a highly centralised brain. The brain sends global commands to the neuronal network in the arms to activate and scale their program variables. At the same time, the neuromuscular system of the arms does not require continuous central control. Movements resembling normal arm extensions can be initiated in amputated arms by electrical stimulation of the nerve cord or by tactile stimulation of the skin of suckers. A major part of voluntary movements of the arms is controlled by a pattern generator that is confined to the arms' neuromuscular system. The octopus also reduces the complexity of controlling its arms by using highly stereotypical movements. Due to the developed peripheral motor programme, the arm can be moved in any direction, with a virtually infinite number of degrees of freedom (Young, 1961; Wells, 1978).

Cephalopods use their brains for learning, not only for controlling eight arms. It was demonstrated as early as in 1950s and 1960s that octopuses can learn to distinguish between different shapes, orientations, sizes and degrees of brightness. In one experiment, Young (1960) trained octopuses to select between large and small squares, horizontal and vertical stripes, and black and white circles. He found that the animals could retain all three preferences at once. In other experiments, blinded octopuses learnt to distinguish between differently shaped objects using only their highly sensitive suckers. One octopus remembered the differences for 4 months (Wells, 1962). Long-term memory of associative learning has been revealed in cephalopods with the use of negative reinforcements such as a glass tube for cuttlefish and electric shock for octopuses (Boycott and Young, 1955; Messenger, 1973; Boal *et al.*, 2000; Agin *et al.*, 1998). Experiments with taste aversion in cuttlefish with the use of learning procedure in which the preferred prey was made distasteful by a bitter taste have clearly demonstrated that cuttlefish were able to learn that a particular prey is not acceptable food, even if they usually preyed on it, to recognise it and to avoid it for several days and as a result to eat a usually non-preferred prey (Darmaillacq *et al.*, 2004).

The remarkable memory of cephalopods can be understood in terms of the peculiarities of their brain structure. Cephalopods possess an organ analogous to the vertebrate hippocampus. This organ, the vertical lobe, appears to fashion autoassociation event memories, which cephalopods are known to store long term in the optic lobes. The octopus appears to use event memories in much the same way that vertebrates use memories recorded by the hippocampus. Field observations suggest that the octopus follows a detailed topographical map when navigating the coral reef near its den. *Octopus vulgaris* has also demonstrated the ability to retain arbitrary T-maze memories over a long period of time in the laboratory (Hanlon and Messenger, 1996). Recently the functional implication of the vertical lobe complex in learning and memory has been confirmed by the use of a metabolic marker, cytochrome oxidase (Dickel *et al.*, 2000).

The cellular and molecular mechanisms of LTM seem to be universal in vertebrates and invertebrates (at least in mollusc species). Insights in developing the theory of the formation of LTM have been obtained with the help of a close, but much simpler, relative of cephalopods, the giant marine snail *Aplysia*, which lends itself to experimental laboratory work as an ideal object for studying memory processes. This mollusc has only one 'leg' instead of the eight arms of the octopus. In his Nobel Lecture E. R. Kandel (2001) portrays *Aplysia* and describes the fundamental results that have been obtained in the study of memory processes with the help of this snail. *Aplysia* has 'only' 20 000 central nerve cells, and the simplest behaviours that can be modified by learning may directly involve fewer than 100 central nerve cells. In addition to being few in numbers, these cells are the largest nerve cells in the animal kingdom, reaching up to 1000 μm in diameter, large enough to be seen even with the naked eye. The cells can easily be dissected for biochemical studies and can readily be injected with labelled compounds, antibodies or genetic constructs, procedures which have opened up the molecular study of signal transduction within individual nerve cells. Kandel's research suggests that the cellular and molecular strategies used in *Aplysia* for storing short- and long-term memory are conserved in mammals and that the same molecular strategies are employed in both implicit and explicit memory storage. With both implicit and explicit memory there are stages in memory that are encoded as changes in synaptic strength and that correlate with the behavioural phases of short- and long-term memory. The short-term synaptic changes involve covalent modification of pre-existing proteins, leading to modification of pre-existing synaptic connections, whereas the long-term synaptic changes involve activation of gene expression, new protein synthesis and the formation of new connections. Recently clear evidence has been obtained that *de novo* protein synthesis is an essential and time-dependent event for LTM formation of associative learning in the cuttlefish.

Memory sits comfortably in bees' mini brains

For a long time many investigators of learning and memory argued that only 'higher' animals exhibit complex forms of learning and that these forms require neuronal organisations and neuronal mechanisms qualitatively different from those found in 'simple' animals. Recent studies on insects, especially on Hymenoptera, such as bees and ants, have demonstrated how excellent learning capabilities may be implemented on mini brains.

The honeybee *Apis mellifera* is a particularly useful animal for the study of learning and memory formation, because this insect exhibits easily manipulated feeding behaviour coupled with extremely high mnemonic fidelity. The size of the honeybee brain has allowed for electrophysiological analysis of the neural correlates of behaviour, sometimes with single-cell resolution, as well as identification of critical brain regions.

Experimental findings concerning how honeybees find a target go back to Romanes (1885). He put a hive in the basement window of a house with a large flower garden on one side and a lawn leading to a beach on the other. When he released foragers from this hive anywhere in the garden, they soon appeared back at the hive. When he released them on the beach, they did not return to the hive, even though many of the sites were closer to the hive than release sites in the garden. The contrast between homing performances from familiar versus unfamiliar territory implies that the homing was not mediated by a random search; rather, it was based on knowledge of the terrain. At the beginning of twentieth century, American scientist Charles Turner (1910, 1911; see Abramson, 2003) revealed that honeybees have 'ideas' about time and can readily distinguish colours and geometric patterns. Indeed, Turner's work on colour vision of bees and their recognition of patterns and shapes predated von Frisch's work and probably influenced the Nobel prizewinner's investigations.

The honeybee brain contains a pair of organs called 'mushroom bodies', due to their mushroom-like shape. They are the bee's primary organs for the acquisition of complex memories (Menzel, 1985). The mushroom bodies (MBs) or corpora pedunculata were first described in detail by Dujardin (1850). Forel (1874) developed a hypothesis of the close connection between brain structure and complexity of behaviour in different casts in Hymenoptera. Studying ants, he revealed that MBs are much larger in workers which perform different kinds of job in the colony, than in queens whose behaviour is more stereotypic and all the more than in males which can only fly and copulate. Workers are also not equal in their learning capability and development of MBs. It has been demonstrated in one of the 'cleverest' hymenopterans, red wood ants, that individual workers' abilities to solve Schneirla's maze correlate with MB sizes (Bernstein and Bernstein, 1969).

The MBs account for about a half of the whole volume of the brain in red wood ants and about a quarter in the honeybee whereas in such active hunters as water-tigers possessing complex behaviour, MBs account for only about one-twentieth of the whole volume of the brain. The honeybee's brain is as small as 1 μl containing 950 000 neurons (Witthöft, 1967). But the point is that the quantity of neurons does not make the cleverest organism. Neuronal activity and brain mechanisms in insects differ from that in vertebrates and this provides 'multiple memories' being implemented in mini-brains (Menzel, 2001). Possible regions of LTM storage in the MBs are labelled as median calyx (mC) and lateral calyx (lC). The microstructure of these regions bears a resemblance to that of the vertebrate cerebellum.

The brain of the honeybee has been studied for the last six decades using a standard technique of ablation established by Lashley (1950). Early experiments with surgical removal of MBs after classical conditioning to odours resulted in complete loss of conditioned responses for these stimuli whereas other vital functions remained (Voskresenskaya, 1957). Panov (1957) with the use of histological examinations demonstrated that MBs develop later than other brain structures in honeybees and that they possess the most complex structure. Recent studies based on the same technique have revealed more and more intriguing details concerning the role of different brain structures in olfactory and tactile learning in honeybees (Scheiner *et al.*, 2001).

Brain mechanisms of learning and memory are better studied in bees than in ants thanks to such a robust phenomenon in honeybees as reward learning based on olfactory and tactile stimuli. What is especially important for the elaboration of this technique is that some peculiarities of the bees' reactions make it possible to study this insect fixed in a tube like a Pavlov's dog fixed in a stall.

The preparation which is used to study reward learning in honeybees was introduced by Kuwabara (1957), who first studied colour learning in such a way, and then by his student Takeda (1961), who discovered that bees restrained in tubes form an association between an olfactory stimulus and a sucrose reward. Each bee is harnessed in such a way that it can move only its antennae and mouthparts (mandibles and proboscis) freely. The antennae are the main chemosensory organs. When the antennae of a hungry bee are touched with sucrose solution, the animal reflexively extends its proboscis to reach out toward the sucrose and lick it. Odours or other stimuli to the antennae do not release such a reflex in naive animals. If an odour is presented immediately before sucrose solution (forward pairing), an association is formed which enables the odour to trigger the proboscis extension response in a successive test. This effect is clearly associative and involves classical but not operant conditioning (Bitterman *et al.*, 1983). This redoubles the similarity of this preparation to Pavlov's experiments. The odour can be viewed as the conditioned stimulus (CS) and the sucrose solution as the reinforcing, unconditioned stimulus (US). Using this preparation, Hammer (1993) identified a single neuron that serves reinforcement during olfactory conditioning.

It turns out that in bees a single association of an odour and a sucrose reward will lead to a memory lasting for days. Three pairings of an odour and a reward lead to a lifelong memory. This is much faster and more reliable than in many vertebrates. Reward learning in honeybees initiates a sequence of memory phases that lead to long-lasting memory passing through multiple forms of transient memories. This has been called 'multiple memories' (Menzel, 1999). It is possible to distinguish between two forms of LTM: early LTM characterised by protein synthesis-independent retention, and late LTM characterised by protein-synthesis-dependent retention

(Menzel, 2001). In general, these experiments have revealed that in Hymenoptera, the MBs control complex behaviour, learning and memory and receive multisensory input.

An avian version of episodic memory

The title of this section is a paraphrase from Milius' (1998) paper about Clayton's investigations on scrub jays. Food-storing animals, and in particular scrub jays, are good candidates for studying animals' capability of declarative memory because they remember what they cached where and when based on a single caching episode. In the wild jays cache and recover many different food items during the autumn and winter months, and should therefore be able to remember which sites have been depleted by cache recovery and subsequently return only to those sites where their caches remain.

Clayton and her co-authors conducted some series of ingenious experiments on scrub jays *Aphelocoma coerulescens* basing on the food-caching paradigm as striking examples of how an understanding of the species and its natural history can be employed to develop novel approaches to the study of episodic memory in animals. Indeed, the memory capability of jays fulfils Tulving's classic behavioural criteria for episodic memory, and is thus referred to as 'episodic-like' (Clayton and Dickinson, 1998).

The experiments conducted on the episodic-like memories of food-caching birds were guided by two key features of their natural behaviour. Firstly, birds rely, at least in part, on memory to recover their caches. Secondly, in the wild some food-storing species including scrub jays cache insects and other perishable items in the wild as well as seeds. It may be useful, therefore, for them to encode and recall information about what has been cached when, as well as where. This enables experimentalists to capitalise on the jays' natural propensity to cache and recover perishable items in designing experimental test for memory of 'what, where and when'.

To test whether scrub jays are capable of episodic-like memory recall, birds were allowed to cache and recover perishable 'wax worms' (wax moth larvae) and non-perishable peanuts. The logic is as follows. Jays show a strong preference for caching, recovering and eating fresh wax worms when given both worms and peanuts. Worms decay rapidly over time, however, so that if worms are left for a period of 5 days or so they become rotten and unpalatable. If birds can remember when they cached as well as what they cached and where, then they should recover worms when they were cached just a few hours ago. They should avoid the worms, however, if the worms were cached several days ago and have had time to rot.

Sixteen hand-raised scrub jays were given worms and peanuts which they could cache in and recover from sand-filled plastic ice-cube trays, containing an array of 'cache sites'. Each tray was attached to a wooden board and surrounded by a visuospatially distinct structure of Lego bricks that was placed next to one of the long sides of the tray (Clayton and Dickinson, 1998). For each trial, the trays were unique and care was taken to ensure that the trays were not placed in the same location across trials so that the birds could not learn general rules about the contents of caching trays. Instead, they had to remember the 'what, where and when' of each individual caching episode.

The birds were divided into two groups, Degrade (D) and Replenish (R), which differed in whether or not birds had had the opportunity to learn that worms degrade and become unpalatable over time. In order to learn that worms decay, jays in the D group were given a series of pre-training trials in which they cached peanuts in one tray and worms in another tray, and then recovered their hidden caches from both trays either 4 hours or 124 hours later. Birds in the R group received the same treatments as those in the D group except that the old wax worms were removed and replaced by fresh ones just before the start of the cache recovery phase so that these birds never had the opportunity to learn that worms decay over time. The birds in the R group serve as an important control to test whether any switch in preference from worms to peanuts after the long retention interval can be explained in terms of either a genetic predisposition to prefer worms in some instances and peanuts in others, or simply that memories for perishable worms are forgotten more quickly than memories for non-perishable nuts. The results of the special test trials showed that birds in the D group preferred to recover worms after the 4-hour retention interval, but preferred to recover the peanuts and avoid the worms after the 124-hour retention interval.

As predicted, on test trials, birds in the D group reversed their preference from worms to peanuts at the long interval, but R group preferred the worms at both intervals. The switch in preference from worms to peanuts after the long interval required the birds to recognise a particular cache site in terms of both its contents and the relative time that had elapsed between caching and recovery. This result can only be explained by recall of information about 'what' items (peanuts and worms) were cached, 'where' each type of item was stored and 'when' (4 or 124 hours ago) the worms had been cached. Furthermore, the information was acquired as a result of single, trial-unique experience.

In further experiments researchers tested the jays' ability to recall specific past experiences during a cache recovery episode (Clayton *et al.*, 2003a,b). They concluded that scrub jays encode information about the type of food they store in cache sites. Besides, the birds can update their memory of whether or not a caching location currently contains a food item. They also can integrate information of the content of a cache at recovery with information about the specific location of the cache site. Taken together, these results fulfil the behavioural criteria for episodic memory in its classic definition: the jay remembers a series of facts about an object (the food item), a place (where they stored it), a time (how long it was since they stored the item) and an action (caching versus cache recovery) that allow the bird to subsequently recall that information and execute the appropriate behaviour. Each item could be considered as a semantic fact but when all the facts are integrated, the jay has sufficient information to isolate what was cached and what was recovered, where and how long ago: functionally, the animal has enough information to recall the episode of caching a specific item (Griffiths *et al.*, 1999).

One could call the scrub jay a member of the 'Caching Club'. This species shares with other animal cachers (such as several mice, squirrels and bird species) specific brain properties that could have evolved when there was pressure on memory and supporting brain structures. There is a speculation that spatial memory in birds and humans is hippocampus-dependent to a great extent (Pravosudov, 2003). There are some parallels which, of course, should not obscure many differences between the avian and mammalian hippocampus (see Macphail (2002) for detailed analysis).

Food-storing species have larger hippocampal volumes relative to telencephalon size than their non-food-storing counterparts in a wide variety of species. Volumetric differences are not accompanied by differences in cell density but rather by a greater number of cells as well quantitatively different cells such as larger immunopositive neurons. Comparative studies on songbirds have shown that seasonal peaks in food-storing correlate with seasonal changes in brain morphology. These seasonal changes in the hippocampus do not occur in non-food-storing species and are probably specific to food-storing birds (Krebs *et al.*, 1989). A striking example of difference in hippocampus sizes between food-storing and non-food-storing species is provided by two parid species, the storing marsh tit *Parus palustris*, and the non-storing great tit, *P. major*. The marsh tit, at 11 g, is a much smaller bird than the great tit (20 g), and possesses a telencephalon that is about 20 per cent smaller than the great tit telencephalon. But the hippocampus of the marsh tit is some 30 per cent larger than that of the great tit. Further support for a hippocampal role in food-storing is provided by studies that have found that cache recovery is disrupted by hippocampal lesions (Sherry *et al.*, 1989). It is important to note that hippocampal damage does not disrupt the tendency to cache food: it is the memory for the locations of the caches that appears to be disrupted.

A reasonable interpretation of the hippocampal enlargement seen in food-storing birds is that it is an adaptive specialisation that enhances the spatial memory that is served by the hippocampus. The experimental data have recently been reviewed by Macphail and Bulhuis (2001), who concluded, perhaps surprisingly, that the tendency to store food does not appear to correlate with spatial memory capacity in other contexts (see also Chapter 22). In other words, it seems clear that the hippocampus is in some way involved in food-storing and recovery, but it is not clear that this association strengthens the case for assuming a role for the avian hippocampus in spatial learning and memory.

Chapter 9

Chicks do not suffer from schizophrenia: a brief outline of brain mechanisms for processing and storing memory

The processes of learning and memory storage in some sense look like 'editing brains' and there seem to be a variety of ways of doing this: killing whole cells, disconnecting some interconnections between selected cells, creating additional synapses, and increasing and decreasing synaptic strengths. Memory thus may work not only by adding some new materials but also, in part, by eliminating some neural connections (Young, 1970; Calvin, 1989). Unlike other mental processes such as thought, language and consciousness, learning and memory storage has seemed from the outset to be readily accessible to cellular and molecular analysis (Kandel, 2001). There is a huge literature devoted to modern approaches and results obtained in this field. This chapter does not pretend to give a whole picture of how learning occurs as a function of the activities of the brain and its constituent neurones. I give only a brief review here leaving this theme with experts who professionally study these fascinating problems.

9.1 Searching for the spatial localisation of memory

Brain explorers feel their way like Tom Thumb and one of their ways of finding a pathway is to throw pebbles, or to find a correlation between a localised brain area and a definite problem solved by this area. One method for finding such correlations is based on ablation. For example, a series of experiments by Bingman and co-authors (Bingman *et al.*, 1984; Bingman and Able, 2002) exploited the remarkable ability of pigeons to fly home rapidly from both familiar and unfamiliar release points. It turned out, firstly, that when released from an unfamiliar site pigeons with hippocampal lesions show an initial orientation towards home that is no less accurate than that of control pigeons. The hippocampus is not, then, necessary for the efficient operation of the navigation system. Secondly, although hippocampal-lesioned birds successfully returned to the general region of their home lofts (indicating that their use of their compass was intact), they did not enter their home lofts even though they were in sight. It appeared from these results that hippocampal-lesioned birds were capable of homing successfully, but failed to recognise their home lofts. Subsequent studies found that if hippocampal-lesioned pigeons were given sufficient post-operative experience of their home lofts (7 days or more), then they did successfully re-enter their home lofts.

These findings indicate that the hippocampus plays a critical role in memory formation. Nevertheless, despite many studies, identifying the scope and nature of memory processing by this brain structure is still remains a challenge. Results from recent studies in human and non-human subjects have suggested that the hippocampal formation and related structures are involved in certain forms of memory (episodic and spatial memory) and contribute to the transformation and stabilisation of other forms of memory stored elsewhere in the brain (Nadel and Moscovitch, 1997). More recent experiments with pigeons of different ages and different deficits and noises in sensory inputs (clockwise, anosmic) demonstrated both the important role of the hippocampus in associability and memory storage, and a high level of plasticity in birds of essential functions such as homing (Ioalé *et al.*, 2000; Macphail, 2002).

In general, *plasticity* refers to the process of making long-term changes in the brain, particularly the changes that occur as a result of learning. It is known that during development neurons can change shape, location, function and patterns of interconnection. In the adult brain, neurons are less able to change location and function, but plasticity is still present.

Historically, there have been two opposed viewpoints about the extent to which learning and memory can be related to specific brain areas. The localiser (mosaic) theory position is that specific memories are stored in specific locations, much as mail is stored in pigeonholes in a post office (Garner, 1974; Swenson, 1991). This position was derived from the early successes of the great neuroanatomists, such as Paul Broca, in localising cortical functions. A less extreme version of this view was advanced by Pavlov. Pavlov suggested that the cortical projection areas associated with USs act as dominant foci; when excited, they radiate electrical excitation, which acts to attract the excitation produced by a CS, so that eventually the CS has the power to elicit the reflex. Thus, Pavlov saw classical conditioning as involving electrical connections between localised cortical projection areas (Schneider and Tarshis, 1975).

In contrast to this localiser theory, the Gestalt theorists advocated the holistic theory which states that the brain functions as a whole through electrical field forces generated by the brain during learning. Ross Adey, Wolfgang Köhler, and other Gestalt theorists followed Karl Lashley, an American physiological psychologist, who spent over 30 years trying to determine where, if anywhere, the engram might be localised. Lashley made deep cuts between cortical regions to disrupt the kinds of connections between the US-excited projection areas and the CS-excited projection areas postulated by Pavlov. This produced no losses of memory or learning ability. He then systematically removed various cortical regions from thousands of rats without destroying their ability to learn and remember, thereby disproving the localiser hypothesis. The holistic theory proposed that learning leads to changes in electric fields or chemical gradients, which were postulated to surround neuronal populations and to be produced by the aggregate activity of cells recruited by the learning process.

Based on the facts that memory deficits did appear in some of his subjects and that the extent of such deficits was related to the extent of the lesions not to their location, Lashley advanced his theory of mass action as a compromise between strict localisation theories and strict holistic theories. The *theory of mass action* states that within functional brain regions all parts of the region are equally effective in carrying out the function normally served by the entire cortical region. This theory received clinical support. Kurt Goldstein (1939), a Gestalt neuropsychologist, observed brain-damaged World War I veterans. He noted that patients with damage in the association areas adjusted by accepting that their overall intellectual abilities had been reduced. Luria, the Russian cognitive neuropsychologist, has also supported the mass function (mass action) concept by developing a series of testing strategies to evaluate the amount of intact function within a damaged region (Luria and Majovski, 1977). If some intact function remains, the patient can be trained to compensate for behavioural deficits. Training is possible because of plasticity, or flexibility, in brain functioning, which decreases with age.

While Lashley found no clear evidence of localisation of memories stored in the cerebral cortex, other researchers discovered a limited type of localisation. Diamond *et al.* (1958) found that cats were unable to remember an auditory discrimination after removal of their auditory cortex but were able to relearn the discrimination. This suggests that learning that is specific to a particular sensory modality - such as hearing - is usually stored only in the part of the cortex specialised for processing that type of sensory input. However, other cortical areas have some potential for storing that type of information and after damage to the primary memory area for a particular modality, these secondary brain regions do store this information.

One of the first series of experiments that demonstrated the role of a definite part of the brain in experiences with time was carried out by Jacobsen (1936). He showed that damage to the prefrontal cortex results in violation of delayed response behaviour. 'Lobed' monkeys (i.e. animals with surgically damage to the frontal lobe) coped with a simple discrimination task but could not solve a problem when requested to choose a cup under which a piece of food had disappeared before their eyes. It was, as it were, a case of 'out of sight - out of mind'. In early experiments with prefrontal cortex-ablated chimpanzees, animals that had been trained to use sticks for getting food were able to perform this action after surgical operation only if they kept their eyes on both the stick and the food simultaneously.

Later, when dealing with individual neurons became possible as well as mapping brain activity using magnetic resonance imaging, the behavioural method of delayed reaction has become a basis for studying the function of the frontal lobe, from single cells to how cells communicate with each other and to how animals behave. For example, the research of Goldman-Rakic (1995) on non-human and human primates included functional metabolic mapping, single-cell recording in behaving primates, and lesions and drug manipulations of behaviour in monkeys.

Experiments on monkeys have made it clear that a major contribution of the prefrontal cortex to cognition is the active maintenance of behaviourally relevant information 'online'. Experiments in many laboratories have verified that damage to the prefrontal cortex in humans and monkeys tends to produce impairments when available sensory information does not clearly dictate what response is required. For example, prefrontal lesions impair the ability to solve spatial delayed response tasks in which a cue is briefly flashed at one of two or more possible locations and the monkey must direct an eye movement to its remembered location. However, no impairment is observed if there is no delay and monkeys can immediately orient to the cue. Thus the prefrontal cortex seems critical when the correct action must be selected using recent memory and knowledge of the task (Funahashi *et al.*, 1993; Rainer *et al.*, 1998).

A correlation has been demonstrated between accuracy in performance of delayed response procedure and age-specific development of the frontal lobe, at least in primates. It turns out that in rhesus monkeys the ability to cope with such tests becomes apparent at the age of 2–4 months while in human infants it is after 8 months. Infants under 8 months, in which functional organisation differs very much from adults, are as bad at coping with the delayed response tests as monkeys with the surgically damaged lobe. In these cases behaviour is governed by current conditioning, not by mental representation. Both babies and prefrontal cortex-ablated monkeys repeat again and again the response that was reinforced earlier, instead of changing their reaction in accordance with the new information obtained. Even if the infants see a toy being moved into the left box from the right box, they insist on choosing the left one where the desired thing had been before (Goldman-Rakic, 1992).

It is worth noting that normal ageing, at least in mammals, is frequently accompanied by a decline in the cognitive capacities supported by the prefrontal cortex. For example, aged rats and monkeys display deficits in multiple testing procedures known to require the functional integrity of the prefrontal cortex. It was shown that ageing is not accompanied by significant neuronal loss; rather, intricate biochemical mechanisms affect the age-related decline in information processing capacities (O'Donnell *et al.*, 1999).

The non-human primate brain has been used as an animal model of human brain development and brain function. Comparative analysis of development has shown the close parallel between the developmental time-course of memory processes in infant monkeys and human infants. In human children procedural memory emerges very soon after birth, while associative memory emerges later, around the age of 4 or 5 years. Similar dissociation of memory processes also is found in monkeys. On tasks measuring procedural memory, monkeys as young as 3 months of age can perform as efficiently as adults, while associative memory develops later (see Bachevalier and Mishkin, 1989).

Results obtained in Goldman-Rakic's laboratory (Goldman-Rakic, 1992, 1995) showed that, among very dramatic aberrations observed in a schizophrenic brain, in which frontal lobe dysfunction is prominent, one of the main differences lies in the organisation of working memory. When performing delayed response tests, people suffering from schizophrenia usually repeated the answer they had given before although it was already clear that this answer had been wrong. The schizophrenic mind misapprehends events in the external world as a series of disconnected occurrences. Their behaviour in this context is governed not by a balance between current external information and mental representations but by online reactions to external stimuli.

Although the cause of schizophrenia remains a mystery, evidence suggests that it is at least 80 per cent heritable, stemming from complex interactions among several genes and non-genetic influences. Patients show abnormal activation of the prefrontal cortex, which is required for such 'executive' functions associated with the prefrontal cortex and working memory such as initiation and overall control of deliberate actions, goal-directed behaviour, attention, planning and decision-making.

Schizophrenia is a disease unique to humans. It is not that the prefrontal cortex is the part of the brain that really helps to separate man from beast, but that in animals the ability to solve problems requiring active working of the prefrontal cortex is limited by a species-specific level of cognition and guided by selective attention for stimuli and specific filters for perception. We will see in Part VII that cognitive skills may vary within a wide range among individuals belonging to the same species. Nevertheless, there is no doubt that humans are champions in cognitive abilities as well as in mental diseases. At the same time, cellular and molecular mechanisms of memory are common for many species and this gives researchers a great opportunity to discover universal laws of memory on animal models.

9.2 Becoming memories: consolidation

The hypothesis that new memories consolidate slowly over time was proposed some 100 years ago, and continues to guide memory research. In modern consolidation theory, it is assumed that new memories are initially 'labile' and sensitive to disruption before undergoing a series of processes (e.g. glutamate release, protein synthesis, neural growth and rearrangement) that render the memory representations progressively more stable. These processes are generally referred to as *consolidation*.

The famous researcher of biochemical mechanisms of learning and memory Steven P. R. Rose has found the young chick a powerful model system in which to study the biochemical and morphological processes underlying memory formation. In the reviews (Rose, 2000, 2005) he argues that chicks, being precocial birds, need to actively explore and learn about their environment from the moment they hatch; they learn very rapidly to distinguish edible from inedible or distasteful food, and to navigate complex routes. Training paradigms that exploit these species-specific tasks work with the grain of the animal's biology, and because this learning is a significant event in the young chick's life the experiences involved may be expected to result in readily measurable brain changes. At the same time, when working with chicks researchers have applied one of the few universal findings in studies of biochemical processes in memory formation, namely that long-term memory is protein-synthesis dependent (Davies and Squire, 1984).

The learning protocol that has been used in Rose's laboratory does not demand excellent cognitive skills from day-old birds. Nevertheless, their stable and well-working memory is amazing. Chicks are held in pairs in small pens, and pre-trained by being offered a small dry white bead; those that peck are then trained with a larger chrome or coloured bead coated with a distasteful substance. Chicks that peck such a bead show a disgust reaction (backing away, shaking their heads and wiping their beaks) and will avoid a similar but dry bead for at least 48 hours. However, they continue to discriminate, as shown by pecking at control beads of other colours.

This method enabled experimenters to identify a biochemical cascade associated with memory consolidation in the minutes to hours following training. Thus a change in some biochemical marker at a specific post-training time, occurring in trained compared with control chicks, might imply its direct engagement in memory expression at that time. Alternatively it could indicate the mobilisation of that marker as a part of a sequence leading to the synthesis of a molecule, or cellular reorganisation, required for the expression of memory. A similar argument applies to the timing of the onset of amnesia following intracerebral drug injection.

Detailed descriptions of the biochemical mechanisms of learning is available in books and papers generated by this research group and other specialists in this field (see, for example, Goelet *et al.*, 1986; McGaugh, 1989; Anokhin *et al.*, 1991; Damasio, 1994; Tiunova *et al.*, 1998; Freeman, 1999; Rose and Stewart, 1999). The training experience generates a sequence of rapid synaptic transients which provide a temporary 'hold' for the memory – the phases categorised as short- and intermediate-term memory. As well as forming the brain substrate of the remembered avoidance over this period, these transients serve two other functions. They initiate the sequence of pre- and post-synaptic intracellular processes which will in due course result in the lasting 'synaptic' changes presumed to underlie long-term memory, and they also serve to 'tag' relevant active synapses, so as to indicate those synapses later to be more lastingly modified.

The study of long-term memory has revealed the extensive dialogue between the synapse and the nucleus, and the nucleus and the synapse. In the long-term process the response of a synapse is not determined simply by its own history of activity (as in short-term plasticity), but also by the history of transcriptional activation in the nucleus.

Investigations and manipulations with the sequence of biochemical events occurring in the brain of a chick that faces a problem of avoidance of irrelevant food items have allowed the construction of an integrative picture of memory consolidation beyond cellular theories of memory formation, which is heavily based on Hebbian models. Memory appears to be not just a pre/post-synaptic event; rather, whether any particular experience is learned or not depends on a much wider array of neural and peripheral factors, humoral and perhaps also immunological. Furthermore the magnitude and diverse locations in space and time of the changes that have been found following training in such a simple learning task have demanded that researchers reconceptualise their model of memory storage, moving from a fixed and linear view of memory formation to a more dynamic concept, involving large ensembles of cells differentially distributed in space and time (Rose, 1993).

Chapter 10

Behavioural mechanisms of the experience of time

When my dachshund whelped, I decided to wire the five newborn puppies with common wonderful memories, just for fun. When they grew up enough to take solid food as well as their mother's milk, I snapped my tongue in a specific manner every time the puppies were fed from their common bowl. Now I have a secret password to enter the mind of each of them. Although they have grown up and become serious hunting dogs, they readily leave their owners to rush for my signal with infantile and pleased smiles on their faces. Thus I have shaped five 'Proustian' dogs who probably will keep these memories up to their old age. Everyone who deals with animals in any way has a collection of such anecdotes, but we need strong experimental evidence of how learned skills become memories in order to make judgements about animals' ability to navigate past and future in their life.

10.1 Travel into the past: delayed response behaviour

Delayed response behaviour is a behaviour that does not only depend on the current stimuli, but also depends on stimuli the subject had as input in the past. This pattern of behaviour cannot be described by direct functional associations between current input and output (Griffin, 2001).

In order to create situations in which animals could display their ability to operate with past and future fathomed, an experimental technique was elaborated by Hunter (1912) and later transformed by Tinklepaugh (1928, 1932) and Harlow with co-authors (Harlow *et al.*, 1932). This scheme was called a *delayed reaction test* (*delayed response procedure*, *delayed reaction time task*). The simplest task required the animals to discriminate between two cups placed in different locations, one of which had previously been baited with food.

The set-up described in Griffin (2001) after Tinklepaugh (1932) works as follows (Fig. 10.1). Separated by a transparent screen from the animal (at position p0), at each of the two positions p1 and p2 a cup (upside down) and a piece of food can be placed. At some moment (with variable delay) the screen is raised, and the animal is free to go to any position. Within this experiment the animal is able to observe the position of food, cups, screen and itself.

The following situations are considered:

Situation 1 At both positions p1 and p2 an empty cup is placed.

Situation 2 At position p1 an empty cup is placed, and at position p2 a piece of food, which is visible to the animal.

Situation 3 At position p1 an empty cup is placed and at position 2 a piece of food is placed, after which a cup is placed at the same position, covering the food. After the food disappears under the cup the animal cannot sense it any more.

Situation 4 At position p1 an empty cup is placed and at position p2 a cup and a piece of food are placed, in such a manner that the animal did not see the food.

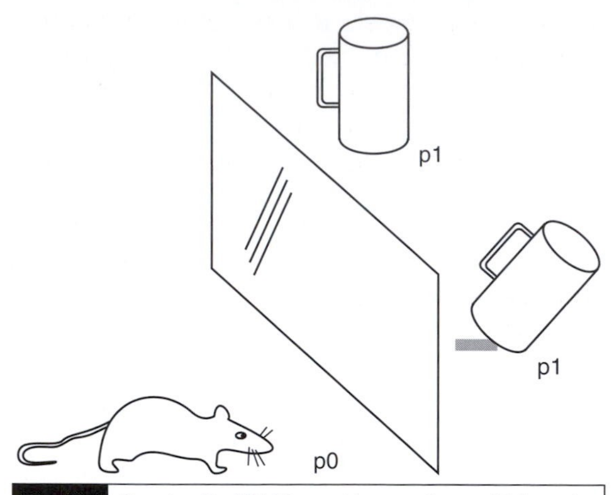

Fig. 10.1 Situation 3 of Tinklepaugh's experiment. (Adapted from Griffin, 2001.)

After the screen has been taken away, in situation 1 the animal, not having observed any food, will not show a preference for either position p1 or p2; it may even go elsewhere or stay where it is. In situation 2 the animal, observing food at position p2 and that no screen is present, will go to position 2, which can be explained as pure stimulus–response behaviour. In situation 3 the immediate stimuli are the same as in situation 1: no food is observed, and it is observed that no screen is present. Animals that react in a strictly stimulus–response manner will respond to this situation as in situation 1. Animals that show delayed response behaviour will go to p2, where food can be found, as has been observed in the past. In situation 4 animals will behave as in situation 1; however they can find food and eat it, and therefore they will be reinforced in successful behaviour.

This scheme of experiments has been applied to many species such as rats, dogs, cats, racoons, macaques, chimpanzees and human infants. It turns out that rats, dogs and cats orient their bodies towards the cup with food and keep themselves axial during the delay, that is, follow their own noses. Being disturbed by an experimenter, they fail to find food. Racoons are able to pass the test without resorting to 'following the nose' if the delay is short. Primates demonstrate delayed response within tests with relatively long delays; besides, they do not need to keep themselves axial, and moreover, they perform successfully even after being isolated in another room during the delay.

Tinklepaugh (1932) found that monkeys were very accurate in this task, but the accuracy declined when the task was made more complex by increasing the number of cup pairs. Two young chimpanzees coped with the complex task within the following scheme. The animals were led through six rooms and in each they were shown how the food was placed under one of two boxes. Then the animals were returned into initial point and allowed to go through all rooms but raise only one from two boxes in every room in order to obtain food. If the animal failed in one room, it was led into the next room without any reward. The accuracy demonstrated was from 88 per cent to 92 per cent successes.

The same subjects came through an even more complex test. An animal sat on a chair in a centre of a room where 16 pairs of boxes were placed around it. The animal could see how the food was placed by the experimenter under one box from each pair. Then the animal was led in turn to each pair of boxes and allowed to raise one of them. If successful, the animal returned to the chair, ate the food and then approached the next pair of boxes and so on, clockwise. The accuracy was between 78 per cent and 89 per cent, and this was higher than in human infants in similar tests. The chimpanzees remembered quite well what kind of food was placed under the box and demonstrated signs of disappointment and indignation when the preferred food was replaced by another kind of item or if a big item was replaced by a small one.

One of modifications of the delayed reaction test, known as *delayed matching to sample* (DMTS), has proved extremely useful for the study of animal memory. The procedure is as follows.

Subjects are presented with one of two stimuli at the beginning of a trial. In the case of pigeons this could be the illumination of a response key with either red or green light. After a while this *sample* stimulus is turned off and nothing is presented for a period known as *retention interval*. At the end of this interval, two different response keys are illuminated, one with red, the other with green light. These are referred to as the *comparison stimuli*. To gain reward, the pigeon must peck the comparison colour that is the same as the sample presented on the particular trial. Pecks to the other colour result in both keys being darkened and no reward being presented. After the completion of the trial there is a period in which nothing happens – the *inter-trial interval* – before the sample is presented for the next trial. In order to gain a reward, the subject must store information at the time the sample is presented and use it to select the correct response when the comparison stimuli are subsequently available (see Pearce, 2000).

Delayed matching to sample seems to be a rather difficult task for most animals. Subjects must first be trained to a simpler, *matching to sample* test with the single sample and two comparison stimuli presented simultaneously. Once they have learned to peck the comparison that matches the sample, the comparison stimuli are presented as soon as the sample is turned off. After considerable training, this 0-second retention interval is gradually extended, but rarely to long intervals. For example, most researchers use delays of 5 to 10 seconds when pigeons are used as subjects. In one study reasonably accurate performance in a delayed matching to sample test was obtained by pigeons with a 1-minute interval. This was achieved at the expense of considerable effort by both experimenter and subjects, as it required some 17 000 trials (Grant, 1976). In monkeys accurate retention was possible for 2 minutes, and with careful training this can be extended to 9 minutes (D'Amato and Worshman, 1972). In experiments with dolphins, accurate delayed matching to sample has been achieved with a retention interval of 4 minutes (Herman and Thomson, 1982).

Another well-known technique for studying delayed reaction is the Wisconsin General Test Apparatus (WGTA) elaborated by Harlow and co-authors (Harlow *et al.*, 1932; Harlow and Bromer, 1938). Since the 1930s, numerous versions of the WGTA have been utilised, including a recent computerised model operated by animals using touch screens or joysticks. This apparatus standardises situations and makes it possible to map brain activity during trials. WGTA is one of the basic techniques based on stimulus discrimination, which is still in use for a battery of tests, from simple discrimination to categorisation. As far as delayed response procedure is concerned, the basic task is the same as in Hunter's (1912) tests: an animal must remember a cue stimulus over a delay period and then make a behavioural response based on the cue. Despite the many modifications over the years, including increased automation, certain elements of the WGTA design remain standard. It is essentially a testing apparatus with: (1) a stimulus tray, (2) a food reward delivered under one to three objects, (3) the option to permit or deny a subject the opportunity to observe placement of the food reward under an object, (4) an observation interval in which subjects can see but not displace objects, (5) a subsequent interval in which subjects can obtain objects and food rewards and (6) a one-way screen for experimenter observation. This basic arrangement allows both the researcher and the subject to manipulate objects in a safe and controlled environment.

In a typical scheme, which is still used today (see, for example, O'Donnell *et al.*, 1999; Buckmaster *et al.*, 2004) a subject, for example, a monkey observes through a clear barrier while one of the two lateral wells of the WGTA stimulus tray is baited with a food reward. The wells are then covered with plaques identical in appearance and the barrier is raised immediately to permit a response. Many trials are provided per daily test session. The left/right position of the reward is balanced within a session. Usually animals are tested until they reach a 90 per cent correct level (nine or fewer errors in nine consecutive ten-trial blocks). Testing subsequently continues in the same fashion except that a 1-second retention interval is imposed by lowering the opaque door of the WGTA between the baiting and response phase of each trial. When the performance criterion has been re-achieved, the memory demands of the task are made progressively more challenging by imposing successively longer delays.

It is worth noting that a concept of secondary reinforcement (see Chapter 6) is used in many experimental schemes being implemented on WGTA. This is a so-called *go signal*. For example, if a required response is key-pressing, then, after a delay period a bridge stimulus is presented for an animal, say, white light is presented on the key as a go signal and the animal is required to press the key within 1 second. A typical 'go–no-go' procedure is illustrated in Fig. 6.1 where a pigeon is acting within a go–no-go conditioning chamber in Huber's laboratory.

Recent studies based on delayed matching to sample tests have helped to make the distinction between implicit and explicit processes in animals. For example, Hampton (2001) developed a new method in which two rhesus monkeys were trained in a match-to-sample procedure to report, by pressing the appropriate image on a touch-sensitive video monitor, whether they did or did not remember one of four visual patterns they had seen a short time before. In this test touching the correct image yielded the food, but pressing the wrong image produced no food. Before the apparatus presented the four images, of which one had been seen previously, the monkey had the opportunity to press either of two different images, one of which caused the test stimuli to appear, whereas the other avoided the test and yielded a food item that was less preferred than what would be delivered after a correct choice in the test. The monkeys learned to avoid the test when they did not remember the images well enough to believe they could make a correct choice. When the four images had been seen only 15 to 30 seconds previously they almost always chose to take the test, touched the correct image, and received a food reward. But when the interval was 2 to 4 minutes they made for more errors if they chose to take the test. These animals certainly seemed to know when they did and did not remember a visual pattern.

Another of Hampton's (2003) experiments was based on the following scheme. At the start of each trial the animal studies a picture, and a delay follows. In two out of three trials the monkey then gets to choose between taking a memory test and escaping from the test. In one out of three trials the monkey is only given the option of taking the test. Both monkeys tested were more accurate when they chose to take tests than when they had no choice about taking tests, indicating that they adaptively escaped from trials in which they had forgotten the picture. In order to rule out the use of cues other than the absence of a memory for the picture in controlling the escape response, Hampton presented monkeys with occasional probe trials on which no picture was presented for study. Since no picture entered the animals' memory on these trials, they should treat them like trials on which they had forgotten the picture. The monkeys were much more likely to choose to escape on trials in which no picture had been presented than they were to choose to escape from normal trials. These results demonstrate that monkeys know whether or not they remember a recently seen stimulus – a form of explicit memory.

Many reports of observed delayed response behaviour in experiments of different types, from simple to complex, have enabled researchers to assume that animals of the type studied maintain internal (mental) representations of the world state on the basis of their sensor' input, and that they make use of this world-state model (in addition to the actual sensory input that is used) to determine their behaviour (see Vauclair, 1996; Griffin, 2001).

Training subjects in more natural settings (e.g. using biological stimuli as discriminative cues) might yield more impressive results. It is well known that in more natural situations animals often amaze their trainers with their capability, speaking in anthropomorphic manner, to keep in mind many events and imaginations for a long time. Köhler and Yerkes gave many examples of how chimpanzees recognised their mates as friends after from 4 to 18 months of being separated. They also recognised their human friends after long intervals. For example, one ape after 4 years recognised a man who once restored it to health and fell on his neck affectionately.

10.2 Travel into the future: anticipatory coding and prediction

The famous innovative thinker and mathematician in Victorian Britain Charles Babbage said in 1840 that '... intelligence will be measured by the capacity for anticipation'.

The capacity to anticipate is necessary for survival, and contributes to the success of every organism. As a consequence, animals and humans behave as if the confirmation of an expectation makes the same anticipation more certain in the future. They behave in a predictive manner: what happened once will occur again.

Recently, a growing number of researchers have identified and emphasised the importance of anticipation as the basis of models of animal learning and behaviour. In this section we will concentrate on animals' capacity to predict future states in time in which their actions are embedded.

There is no doubt that species are not equal in their abilities to anticipate future events. If a mouse disappears into its hole it no longer exists for a snake, while a cat, for instance, remains in front of the hole and waits for the mouse to reappear.

Chimpanzees seem to be able to plan their behaviour in the light of future events which they can predict based on their personal experience. De Waal (1982, p. 192) reported an incident in the Arnhem Zoo where Van Hoff established the chimpanzee colony.

> It is November and the days are becoming colder. On this particular morning Franje collected all the straw from her cage (sub-goal) and takes it with her under her arm so that she can make a nice warm nest for herself outside (goal). Franje did not react to the cold, but before she can have actually felt how cold it is outside.

Another anecdote concerns Lucy, the chimpanzee reared in Temerlin's family. She often accompanied the family in their trips to a ranch. The ape was very much afraid of bridges, especially those of them that she considered unsteady. Long before coming near such 'bad bridges' she began to whimper, tremble with fear and grasp a driver by the hand in order to stop the car (Temerlin, 1975).

These and analogous naturalistic observations raise a question of how far animals can travel into the future.

Bischof (1978) and Bischof-Köhler (1985) suggest an explicit limit on the extent to which animals can represent the future. Their hypothesis is that animals other than humans cannot anticipate future needs or drive states, and are therefore bound to a present that is defined by their current motivational state. More recent, the so-called mental time-travel hypothesis (Suddendorf and Corballis, 1997) states that the cognitive time-window for non-human animals is restricted to the immediate past, the present and the immediate future. According to this hypothesis, animals are incapable of taking action in the present on the basis of either the recollection of specific past episodes (retrospective cognition) or the anticipation of future states of affairs (prospective cognition).

Some recent investigations question these assumptions about episodic memory and future planning. In fact, this is part of the general problem of whether our species is alone in a cognitive niche, a question that we will face more than once in this book. In this section experimental results on animals' capability of planning future events will be analysed in terms of cognition and adaptiveness.

In the scientific literature in general, *anticipation* refers to the attempt to predict the consequences of behaviour. Anticipatory behaviour refers to behaviour that is influenced by expectations about the future, such as future states of the environment, future actions or merely anticipations about the way things work in a given situation. Although it is natural to assume that at least some species of animals are capable of anticipating future events, it was believed for a long time that animals only possess planning for present needs, which is called *immediate planning*; planning for the future is called *anticipatory planning* (Gulz, 1991). Although it may appear obvious that Pavlovian and instrumental learning tasks would result in organisms learning to anticipate future events on the basis of current stimuli or present actions, respectively, the notion that animals can anticipate future events was strongly resisted by the majority of learning theorists from the early 1900s to the late 1960s (Grant and Kelly, 2001).

Some distinct behavioural evidences of expectation and participation came from Hunter's (1912) experiments, in which chimpanzees definitely demonstrated their disillusionment when more desired rewards were replaced in the boxes with less attractive items. More recent research in both Pavlovian and instrumental conditioning has shown that organisms have learned something in addition to, or, more likely, other than stimulus–response habits.

Particularly definitive evidence for anticipation has been provided by studies in which the value of the consequent has been modified after conditioning (Holland and Straub, 1979; Dickinson *et al.*, 1996). In these studies, initially a stimulus or an action had been established as a reliable predictor of an outcome (food pellets). Subsequently, the value of the consequent has been reduced (e.g. by satiating the animal on that consequent, or by making the consequent aversive by pairing it with induced illness). During testing in which the consequent is no longer presented, animals react to the predictor, whether stimulus or action, in a way clearly revealing that they had learned what is predicted by that stimulus or action. For example, rats that had initially reacted with agitated excitement to a tone paired with sucrose show little excitement to the tone following devaluation of the sucrose. Similarly, rats that had earlier eagerly pressed a lever to obtain sucrose pellets, will, after devaluation of sucrose, be reluctant to press the lever.

In the early 1980s, several investigators recognised that the information or code which mediates short-term retention in delayed matching to sample could have either a retrospective or prospective content (Roitblat, 1980; Grant, 1981; Riley *et al.*, 1981; Honig and Thompson, 1982). In particular, the animal could retain features of the sample stimulus during the delay, a retrospective or 'backward-looking' code. During testing, the retrospective code, in combination with the rules learned during simultaneous matching, would be sufficient to generate accurate choice performance. Alternatively, the rules learned during simultaneous matching could be activated during sample presentation to generate an anticipatory code. That is, the code could represent features of the correct choice stimulus, a prospective or 'forward-looking' code. This approach has generated experiments devoted to what is called *prospective memory* in animals.

Grant and Kelly (2001) reviewed a large number of findings including their own experiments designed to answer the question of whether an anticipation of a future action and/or event can function as an effective mediator of short-term retention in pigeons. They found that the literature provides only moderate evidence of the role of anticipation in mediation of short-term retention in these subjects. The strong evidence for anticipatory memory mediation comes from the study of Zentall *et al.* (1990) devoted to memory for spatial locations. They found that a delay interpolated after four of five locations had been chosen produced less disruption in performance than a delay interpolated after two or three locations had been chosen. This finding provides compelling evidence that, as the trial progressed, pigeons recoded from remembering previously chosen locations to remembering not yet chosen locations.

The scheme of the experiment was as follows. An array of five pecking keys was illuminated at the onset of a trial. Choosing (i.e. pecking) a particular key was correct if that key had not been chosen previously on that trial. An error was to choose a key which had been previously chosen on that trial, and resulted in a 2.5-second blackout after which the keys were again illuminated for a choice. A trial ended when all five keys had been chosen. The data relevant to the issue of prospective code content came from trials in which a delay was inserted at some point in the choice sequence. Across trials, the delay occurred after one, two, three or four correct choices had been made. The authors found that the delay caused more errors when it was interpolated after choice 2 than when it was interpolated after choice 1. This result suggests that the amount of information that the bird had to retain was greater after having made two choices than after having made only one choice. Hence, this result suggests that the animal begins the trial using a code with retrospective content; that is, early in the trial the animal remembers which locations it has previously chosen. Interestingly, however, a delay interpolated after choice 3 produced approximately the same amount of forgetting as a delay interpolated after choice 2, suggesting that the memory load was equivalent after either two and three choices had been made. Equivalent memory load after two or three choices could be obtained if pigeons remembered retrospectively after having made two choices (and hence remembered the two locations previously chosen) and remembered prospectively after having made three choices (and hence remembered the two locations not yet chosen). Further support for the notion that pigeons switched to prospective coding later in the trial was provided by the finding that a delay interpolated after choice 4 produced less forgetting than a delay interpolated after choice 2 or 3 and, moreover, produced about the same amount of forgetting as a delay interpolated after choice 1. This result is expected if the animal codes prospectively later in the trial because the memory load after choice 4 (i.e. one item) is equivalent to that after choice 1 and is less than that after choice 2 or 3 (i.e. two items). Hence, the results of Zentall *et al.*'s (1990) experiment provides evidence that codes having prospective or anticipatory content can mediate retention in pigeons.

A similar experiment employing rats was carried out by Cook *et al.* (1985). They investigated the content of the memory used by rats in mediating retention intervals interpolated during performance in a 12-arm radial maze. The delay occurred following either the second, fourth, sixth, eighth or tenth choice. A 15-minute delay had the greatest disruptive effect when interpolated in the middle of the choice sequence and less of an effect when it occurred either earlier or later. This pattern of results was obtained when both a free-choice or forced-choice procedure was used prior to the delay and regardless of whether post-delay testing consisted of completion of the maze or two-alternative forced-choice tests. Assuming that the disruptive effect of a delay is a function of memory load, this implies that the rats used information about previously visited arms (retrospective memory) following an earlier interpolated delay but information about anticipated choices (prospective memory) following a delay interpolated late in the choice sequence.

In Reznikova's (1983) experiments red wood ants showed a high standard of temporal interpolation in the 12-arm radial maze adapted for insects in such a way that the ants had to visit one of 12 cardboard strips sequentially in order to find a drop of syrup there. A 10-minute delay occurred after each choice. Each time the next arm (a strip) was baited so the ants were required to predict the appearance of the reward not on the strip they had visited before but on the next one. Nineteen series of trials were conducted with an interval of 2 days between series, and the bait ran through the whole cycle during each series. After the second series the ants' behaviour was not chaotic: they chose the strip where they had found syrup last time but having failed to find food, they proceeded to the next one. During the fourth series of trials the insects demonstrated that they can predict where the syrup ought to be. They visited the next strip ahead each time, not the one on which they had found the food previously. During the 20th to 24th series of trials the maze was shifted from a vertical to a horizontal position. The ants coped with this task successfully after the second series with the new position.

10.3 Foraging as soon as possible: impulsiveness and self-control in animals

Self-control has been defined historically as an intra-personal conflict: between reason and passion, between cognition and motivation, between higher and lower centres of the nervous system. Self-control is said to be the dominance of the former member of each of these pairs over the latter; impulsiveness, the dominance of the latter over the former.

In the last three decades these terms of human psychology have become frequently used in two fields of animal studies, namely, in comparative psychology and behavioural ecology. In animal studies *self-control* means choosing a larger delayed reward over a smaller immediate reward. The opposite of self-control is *impulsiveness*.

A behavioural theory of self-control versus impulsiveness is mainly based on binary-choice studies which are called self-control experiments (Rachlin and Green, 1972; Ainslie, 1974; Mazur, 1987; Bateson and Kacelnik, 1997). These experimental studies are mainly based on the delayed response procedure as well as on the Skinnerian concept of schedule of reinforcement (see Chapter 6). The behavioural self-control experiments are anticipated by typical Skinnerian shaping (sometimes called the method of successive approximation). During this training stage of the experiment a subject must learn which reaction would be counted as a 'choice': for example, pecking, hopping forward from one place to another, touching a figure on a screen and so on.

In recent laboratory experiments on self-control animal and human subjects are given choices between a small reinforcer available after a short delay and a larger reinforcer available after a longer delay. Preference for the smaller more immediate reinforcer is said to reflect sensitivity to reinforcement immediacy, sometimes labelled 'impulsiveness', whereas preference for the larger more delayed reinforcer is said to reflect sensitivity to reinforcer amount, labelled 'self-control'.

Figure 10.2 shows a single trial from such a study (Stephens, 2002). The subject waits for t seconds and then it is presented with two stimuli (say a red and a green pecking key for a pigeon). From previous experience it has learnt that the red key leads to a small–immediate consequence, and the green key leads to the large–delayed consequence. Typically, the experimenter designates a small proportion of the trials as forced-choice and no-choice trials, in which only one option is available, with the aim of ensuring that the subject has some experience with both options. After the animal makes a choice (say, it pecks the green key) it must wait for the programmed delay to expire before receiving the corresponding amount of food (in this example, it waits t_1 seconds to obtain amount A_1 after pecking the red key, but it waits t_2 seconds to obtain amount A_2 after pecking the green key). After delivery of the food the animal must wait another t seconds before the next presentation. The investigator can vary the amounts and delays, and record the proportion of choices made with each alternative. Motivation is typically high: in most experiments with pigeons, for example, the birds are very hungry, as they are deprived of food until they are at 80 per cent of their weight when allowed to eat during the experiments. The impressive strength of animal preference for immediacy has been demonstrated in many experiments: the first second of delay can cut value in half.

In principle, a technique of self-control experiments is a logically elegant way to ask an animal whether it is willing to wait, and under what conditions (Stephens and McLinn, 2003). Demonstration of decision-making in laboratory experiments should be derived from how animals solve their daily problems in real life and supplied by corresponding neural and sensory bases. No wonder that the self-control paradigm has become a useful tool for developing one of the fundamental concepts in behavioural ecology concerning how the timing and size of food

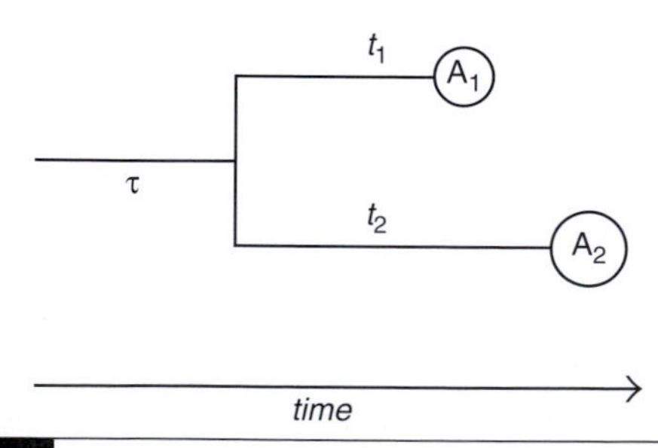

Fig. 10.2 Diagram of a trial in a self-control experiment. After waiting τ- time units the apparatus presents a choice between a small immediate option (A_1 delivered after t_1) and a larger delayed option (A_2 delivered after t_2). (Adapted from Stephens, 2002.)

gains affect an animal's fitness, and how the animal's actions determine magnitude and timing of fitness gains.

The main field within which parallels between self-control in comparative psychology and behavioural ecology work efficiently is foraging behaviour and corresponding decision-making. Results from experiments in laboratories indicate that animals' feeding decisions are guided by short-term considerations. In other words, animals prefer small, quickly delivered food rewards, even when they could do better (sometimes much better) in the long run by waiting for larger, more delayed rewards.

The difficulty in analysing data concerning impulsiveness and self-control is that animal psychologists and behavioural ecologists use different language when considering the problem of relevant choices in animal life. Some recent investigations have been aimed at integrating these approaches.

Within the framework of behavioural ecology, foraging theory (Stephens and Krebs, 1986; Krebs *et al.*, 1989) assumes a simple hierarchy of feeding decisions. It supposes that foragers first choose among habitats, then choose among patches within habitats, and finally choose among prey items within patches. These three decisions have usually been studied independently based on well-known ideas about diet choice and patch use (Pulliam, 1974; Charnov, 1976; Orians and Pearson, 1979). Stephens *et al.* (1986) have integrated these problems by considering patches as being nothing more than clumps of discrete prey, and they asked which prey should be attacked and in what order. As the authors noted, despite the simplicity of this approach, there is some confusion in the literature about how foragers should choose prey within patches. This problem directly concerns impulsiveness and self-control in animal life. Researchers have considered the *take-most-profitable rule* which claims that the relative profitabilities of the items encountered are sufficient to predict preferences, including long-term rate maximising, which is, in turn, based on the so called *marginal-value theorem* (Charnov, 1976), the best-known model of *optimal foraging* (for details see: Krebs and Davies, 1997).

Animal psychologists have proposed a concept of preference which is nearly identical to the take-most-profitable rule and which they call *momentary maximising* (Staddon, 1983). This concept claims that an immediate reward is fundamentally more valuable than a delayed reward and that animals are momentary maximisers, but that they sometimes take account of the long-term rate of gain.

Logue (1988) has suggested functional explanations for cross-species differences in self-control. One hypothesis could be called the metabolic hypothesis. It says that impulsiveness is adaptive for creatures with faster metabolisms – they need food now rather than later. The other hypothesis could be called the ecological hypothesis. This says that self-control or its lack has been shaped by the ecological conditions under which animals evolved.

It could be unadaptive for highly food-deprived animals to wait for a larger amount of food instead of choosing a smaller amount of food that is available sooner. Then one would predict that species with higher metabolic rates – animals for which there is a greater cost of waiting for food – should be less likely to show self-control in relation to food. This has indeed been shown in a comparison among three species: pigeons (with the highest metabolic rate and the most impulsiveness) rats (with the next highest metabolic rate and somewhat less impulsiveness) and humans (with the lowest metabolic rate and the least impulsiveness) (Tobin and Logue, 1994). In a further test involving macaque monkeys, it was therefore predicted that the macaques would show more self-control than rats, but less than humans, due to their metabolic rates falling between those of rats and humans. However, instead, the macaques showed more self-control than any other species tested (Tobin *et al.*, 1996). This may be due to the particular ecological niche in which these monkeys live. Wheatley (1980) has described the natural environment of this type of monkey as consisting of a constant year-round climate and fruiting of food trees. Similarly, Menzel and Draper (1965) found that chimpanzees would often pass up food that was easily accessible if there was a high probability that they would be able to find a greater amount of food at another location.

The world is a noisy place, as Stephens (2002) says, and short-sighted choice rules can lead to better long-term results in animal life because they provide a cleaner discrimination of delayed alternatives. This, in particular, enables us to suggest that in many situations the non-human subjects display impulsiveness not because they are incapable of self-control in principle but because testing procedures are not adequate.

Classic behaviouristic procedures sometimes reveal more or less adequately chosen strategies based on self-control. A scheme of experiments was suggested by Rachlin and co-authors (Rachlin and Green, 1972; Siegel and Rachlin, 1995). This scheme is as follows (see Fig. 10.3). In the first series of trials a pigeon is trained to peck a button to obtain food from a hopper. Then the pigeon is offered a choice between two buttons to peck. A single peck on the green button leads to an immediate reward of 2 seconds of access to food; a single peck on the red button leads to a 4 second blackout (a delay) followed by a reward of 4-seconds access to food. The first reward is called the smaller–sooner reward (SS), whereas the second reward is the larger–later reward (LL). The 'price' of SS is one peck, while the 'price' of LL is one peck plus a wait of 4 seconds. When offered such a choice, pigeons invariably choose SS and thus display impulsiveness.

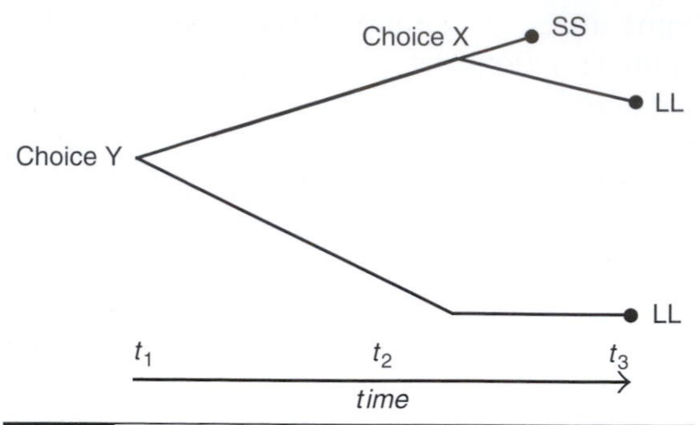

Fig. 10.3 Alternatives in an experiment by Rachlin and Green (1972) on commitment in pigeons. Pigeons choosing (choice X at time t_1) between a smaller–sooner (SS) reinforcement and a larger–later (LL) reinforcement strongly preferred the SS one. However, at a prior point in time (choice Y at t_2), pigeons preferred an alternative (the lower branch) that restricted their choice to LL only. (Adapted from Siegel and Rachlin, 1995.)

Stephens and Anderson (2001) consider the data from self-control experiments from the point of behavioural ecology and they analyse theoretically and experimentally the parallel between the patch exploitation by a foraging animal and its policy in making a choice in self-control experiments.

When a forager exploits a food patch it makes a decision that is analogous to that in a self-control experiment, because it must decide whether to spend a short time in the patch obtaining a small amount, or spend a longer time, obtaining more. Travel time from one patch to another is analogous to the inter-trial interval (ITI) in self-control experiments. Empirically it has been demonstrated in many behavioural experiments that ITI has very little effect. This dramatically contradicts some data of behavioural ecologists. Patch experiments consistently show animals spending longer to extract more when travel times are long (Stephens and Krebs, 1986). This empirical disagreement prompted the study by Stephens and Anderson (2001) devoted to testing foragers in equivalent patch and self-control contexts.

The subjects were six adult blue jays which were tested in two economically equivalent situations. The first situation followed the behaviouristic paradigm: jays made a binary choice between small–immediate and large–delayed options. The second situation was modelled on patch-use problems derived from behavioural ecology: the jays made a leave–stay decision in which 'leaving' led to small amount (two food pellets) in a short time, and 'staying' led to larger amount (four food pellets) in a longer time. Each context was tested at three different ITIs (30, 60 and 90 seconds).

The self-control situation was modelled as a typical self-control experiment. During the ITI, no stimuli were present and birds waited on the 'rear' perch. After the ITI expired, the computer switched on a pair of stimulus lights. The bird then chose a side by hopping forward. When the bird made its choice, the computer switched off the light signal that had not been chosen. The pellets were dispensed after the programmed delay if the bird was on the 'food' perch.

The 'patch situation' was modelled differently. Only the ITI phase was identical to that in self-control experiments. After the ITI expired, the computer presented the stimulus colour associated with the small amount on a randomly chosen side of the apparatus. Again, the bird must have been on the rear perch for this presentation to occur. When the bird hopped forward, the 'small colour' was switched off and the programmed delay began. When the programmed small delay expired, the computer dispensed two pellets and the colour associated with the large ('large colour') was displayed for 1 second and then switched off; the programmed delay to large ($t_2 - t_1$) began when the first two pellets were dispensed. If the bird was on the front perch when the programmed delay expired, then the remaining two pellets were dispensed. At any time after the initial hop forward, a hop on the rear perch cancelled the trial and started a new ITI, that is, the bird was free to 'leave the patch' at any time.

Self-control experiments demonstrated strong preferences for immediacy, and the observed outcome agreed with short-term rate maximising. In the patch situations, the results agreed more closely with long-term rate maximising, but the picture was complex. When the delay to small was 60 seconds, the ITI had no effect on preference, but the jays were more likely to choose the large outcome in the patch context. When the delay to small was 5 seconds, preference for large was greatly reduced. The most striking feature of these data, as the authors describe it, is the interaction between context and ITI. In a patch context, preference for large increased with ITI, while in a self-control context preference for large decreased with ITI.

Stephens and Anderson (2001) concluded that the birds use the same short-term rules in both situations, and this short-term rule approximates long-term maximising in the patch situations, but not in the self-control situations. This means that long-term rate rules and short-term rate rules accomplish the same thing in patch use, and possibly other situations. It is argued that natural selection may have favoured short-term rules because they have long-term consequences under many natural foraging circumstances.

A general question is still open, that is, how currency and context combine to shape decision rules.

Some answers could be obtained from experiments with honeybees. Stephens *et al.* (1986) asked a question: are forager's choices influenced only by immediate, relative pay-offs, as the take-most-profitable rule (and momentary maximising) predict, or do the values of the forager's options elsewhere affect preference, as the theory of optimal foraging predicts? This is the question of distinguishing between momentary rate maximising and long-term-average rate maximising. In experiments with honeybees the researchers arranged pairs of artificial flowers such that a close flower was less profitable and energetically less rewarding than a distant flower. Theoretic models allow the following predictions to be made. The behaviour predicted by the take-most-profitable rule is to pass by the close flower (since it always is less profitable) and to go for the distant flower. The theory of optimal foraging predicts that 'attacking close then distant' should occur at low habitat rates of energy intake and that 'attacking only distant' should occur at high habitat rates of energy intake. Investigators can therefore use the first flower visited to distinguish between the models.

In the experiment individually marked bees were trained to visit a long alleyway covered by nylon mesh. Bees extracted 'nectar' from artificial flowers of two kinds: 'open-faced' and 'doughnut'. Doughnuts flowers markedly increased the time taken to extract nectar. Manipulations with habitat richness, flower quality and several other characteristics led the experimenters to the conclusion that honeybees act like efficient shoppers in a market whose choices depend not only on what they see in the market, but also on what other markets have for sale. Honeybees do not simply take the most profitable flower available. The relative values (measured in time and energy) of the flowers offered affect honeybees' flower choice, but flower choice is also affected by how well the bees can do elsewhere in their habitat. The bees choose immediate gain by taking the close unprofitable flower when the habitat rate of energy gain was low, but they passed by the close unprofitable flower when the habitat rate of energy gain was high. The data obtained support long-term-average rate maximisation over momentary maximisation.

These findings are supplemented by recent study of Cheng *et al.* (2002). Three experiments based on different procedures showed that bees exhibit a relatively high level of self control. The experimental group was offered the classic choices of self-control on the first day. They preferred a large delayed reward over a small immediate reward. On the second day, they chose between a large non-delayed reward and small non-delayed reward. The seeming improvement from day 1 to day 2 is most likely due to practice at the task. That is because a similar pattern was found for bees offered both choices with no delay throughout (control). They too improved on day 2, to a similar extent. The results contradict the metabolic hypothesis, but a much clearer formulation of the ecological hypothesis is needed. As it stands at the moment, the formulation is too vague to allow precise quantitative predictions.

Honeybees have helped investigators to accept that momentary maximising cannot be a complete description of behaviour. From the human aspect, researchers have shown theoretically that taking the most profitable is sometimes the long-term rate maximising behaviour.

CONCLUDING COMMENTS

There is much work to be done to make our understanding of how the brain deals with past and future events more complete. We know now that very complex processes take place in the brain as well as in the adjacent organism when a newly gained skill becomes a memory. This occurs, for example, when a 1-day-old chick first distinguishes between edible and inedible items. At this stage of our knowledge, we may not consider a brain that learns and remembers as a 'black box' but rather as a 'Pandora's box'. An individual carries a brain so that the brain serves as a source of reward and punishment decoding memories whether appropriately or not. It seems that the correlation between power in memory and intelligence is more pronounced in non-human animals than in our own species. Animals do not use memory for eidetic tricks but rather for surviving and improving their lives. Natural selection should lead to differences in specialised learning and memory abilities between animals facing different vital problems such as decision-making, food searching and caching, landmark storage, information processing and social navigation. At the same time, there are no strong correlations between adaptive specialisation, types of nervous systems and memory capacities in living things. Exceptionally robust memory systems can be implemented in eccentric brains lacking both cortex and hippocampus. All these facts enable researchers to broaden comparative investigations on learning and memory to a wide variety of species in order to find the fine balance between memory and intelligence in nature and thus to better understand the advantages and disadvantages of brain design in our own species.

Part IV

Being in the right place at the right time: representation of space and objects in the animal mind

'. . . Would you tell me, please, which way I ought to walk from here?'
'That depends a good deal on where you want to get to,' said the Cat.
'I don't much care where –', said Alice.
'Then it doesn't matter which way you walk,' said the Cat.
'So long as I get somewhere,' Alice added as an explanation.
'Oh, you're sure to do that,' said the Cat, 'if you only walk long enough.'

Lewis Carroll, *Alice in Wonderland*

In Part III we considered animals' experience of time. Indeed, the separation of 'when', 'what' and 'where' more often takes place in laboratory than in the real world. In this Part we will concentrate mainly on how animals solve 'what' and 'where' problems within their timescale.

Many animals possess sophisticated navigation abilities and there have been a large number of studies devoted to investigating the phenomena and mechanisms of this amazing domain of behaviour (see, for example, Baker, 1984; Healy, 1998). For many species spatial navigation is one of the major problems to be solved every day for survival. Indeed, animals spend plenty of time in motion seeking food to eat, free space to settle in, a mate to have sex and offspring with, a tool to use, an enemy to fight and something to have fun with. Some species undertake extremely long-distance journeys such as the Arctic tern migrating from the North Pole to the South Pole, while others such as some invertebrate species do not move further away from their native place than a few centimetres. Distances are non-comparable, but there is one aspect that can easily be compared in many species, from ants to elephants. This concerns the feats of intelligence which can be equal in such different creatures. Even if you are not a long-distance explorer it could be difficult to find the right place and to come back again, unless you were equipped with something like Ariadne's thread or at least with good brains.

It is important to note that animals' mapping behaviour does not necessarily require cognitive abilities. Even 'simple' forms of learning are not always involved in the processes of navigation. At the same time, even 'simple' ways of navigation could be included in a process of cognitive mapping that requires the ability to form a 'bird's-eye view'of the environment based on sophisticated internal representation.

The hypothesis that animals can construct and use cognitive maps has a long and controversial history. Discussion still continues to sort out concrete questions such as 'Do insects have cognitive maps?' (Wehner and Menzel, 1990) and general ones such as 'Do animals have cognitive maps?' (Bennett, 1996), as well as between admission and denial of animals' capacities for mental representation of their spatial environment.

One of the basic components of intelligence is the understanding that there are invariant physical properties attached to objects in the world. The most commonly discussed aspect of the object concept is object permanence, the notion that an object still exists even when it is out of sight. Understanding how object permanence develops and works in animals and humans helps to bring together time-bridging capacity and space representation (mapping), the two important aspects of animal intelligence that will be discussed in this part.

Chapter 11

Navigation strategies in animals

11.1 Display of navigation in animals

Navigation can be defined as the process which enables a course or path from one place to another to be identified and maintained. This process is based on the capacity to plan and execute a goal-directed path (Gallistel, 1990). Behavioural and neurophysiological mechanisms of navigation in animals belong to one of the most fascinating and seminal fields, which is of great interest not only in the biological sciences but also in engineering and robotics. In this and in the next section we limit our consideration of the subject to a brief listing of displays and ways of navigation in animals, to create a sketch of the whole picture and to find room for intelligence in it.

Progress in technology has enabled students of navigation to observe animals' migratory repertoire with the use of short-distance radar recordings as well as long-distance satellite radio telemetry. These methods of observation in particular make it possible to look at migration trajectories from quite long distances and to observe a sequence of straight vector courses. Our knowledge about animals' navigating skills has essentially grown but many questions concerning the basic mechanisms of their travelling still remain obscure.

The most mysterious display of mapping behaviour in animals is *homing*. Homing refers to the ability of animals to return to their local place from any distance, and very often this journey involves travelling across unfamiliar territory. Typical but at the same time impressive examples of homing concern so-called *central place foragers* (Orians and Pearson, 1979). These animals undertake foraging trips which lead them away from their 'central place' such as the nesting site and then return back. For example, at the Crozer Islands in the southern Indian Ocean, wandering albatrosses *Diomedea exulans* leave their nests for foraging flights which take them over distances of hundreds or even thousands of kilometres. Finally, however, they return to their home island, a tiny speck within the vast expanse of water, with seemingly unerring precision (Jouventin and Weimerskirch, 1990).

Desert ants wander as far as 600 metres from their nest which perhaps demands the same level of reliability of navigation systems from these small creatures. The most amazing thing is that when an ant finds an edible thing such as a dead insect, it carries this thing straight home in order to feed the larvae there. In my experiments I used to give red wings of locusts to *Cataglyphis* ants, whereupon it was easy to observe them carrying these red 'flags' for very long distances (Reznikova, 1982). The ant's close relative, the honeybee, flies up to 10 000 metres from its hive and then returns home (Visscher and Seeley, 1982).

With many other examples of different species described in thousands of scientific papers, these pieces of animals' natural history enable us to consider the question as to what extent intelligence is involved into the process of homing. We will return to this problem later in the chapter.

Migration does not require great feats of navigation because it is not oriented towards a specific goal. Migration is defined as 'oriented, long-distance, seasonal movements of individuals' (Able, 1980). The role of learning could be rather modest in migration behaviour, at least in some species. In many cases of migration generations of the same species undertake almost the same journey. Gwinner (1972) suggested that the direction and duration of migration is largely under endogenous control and depends rather little on learning.

The migrating behaviour of the loggerhead sea turtles provides an instance of this type of migration (see Pearce, 2000). Loggerhead turtles hatch from eggs on the Atlantic beaches of Florida and then spend the next few years swimming in loops around the Sargasso Sea before returning to a beach in Florida. A series of experiments has revealed how newly hatched turtles are able to find their way to the Gulf Stream. Young turtles were placed into a large dish in which it was possible to measure their preferred direction of swimming (Lohman, 1992). When the experimental room was dark, except for the presence of a dim light source, the turtles were observed to swim towards the light. Apparently, newly hatched turtles emerge onto the beach at night and reflected light from the moon and stars makes the sea brighter than the land. By being attracted to light, the turtles are thus led towards the ocean. When the room was completely dark, the turtles had a tendency to swim head on into the waves, and also toward magnetic east. Once they have reached the ocean, the joint influence of these tendencies would then lead them to the Gulf Stream.

Far-travelling insects have even less chance to prove their free will. One of the most spectacular examples concerning large-scale insect migration is provided by the North American monarch butterfly *Danaus plexippus* (Brower, 1985; Wehner, 1997). From late August to early September, millions of monarchs leave their breeding sites in the eastern United States and Canada to migrate up to 3600 km to their overwintering sites in the high-altitude forests of central Mexico. Some individuals may travel 130 km per day. The return migration is completed by two or more short-lived breeding generations. The role of different sensory cues has been discussed in migrating movements of butterflies such as the use of magnetic fields and sun compass mechanism but it is most likely that these insects explore soaring using tailwinds. It seems reasonable that the overall migration pattern of monarch butterflies would not have evolved in the absence of the large-scale weather patterns prevailing above North America. Rapid microevolutionary change has occurred in monarch butterflies since they were introduced to Australia, together with their milkweed food plants. In contrast to their North American ancestors, the Australian butterflies have reversed the timing and direction of their migratory behaviour by 6 months and 180°, respectively (James, 1993; Wehner, 1997).

There is still a room for flexibility and decision-making for these migratory insects. They must actively embark on air currents by launching themselves into the air at the right time, and must exploit lift by soaring in thermals, but they should do so only when there is a wind in the appropriate direction. They must also select the right flight vectors and control the track vectors by visual contact with the ground. They also adjust their direction to compensate for wind drift.

In several bird species it has been revealed how learning interacts with endogenous processes in order to influence the direction of migration. At the same time the results of many experiments in which birds were removed a long distance from their home and then were able to return to the place where they were born still remain a riddle (see Able, 1980; Pearce, 2000 for details). As has been summarised (Ens *et al.*, 1990; Piersma, 1994), upon departure and en route the birds must make a number of deliberate decisions. They must decide when to depart, at what altitude to fly, how to adjust their air speed and what heading to keep relative to the wind. We do not yet have a relevant idea about how the birds switch from one decision to another and or how they integrate behavioural patterns into the whole process of migration. Finally, a 'simple' question remains unclear, namely, whether the birds are innately informed about what directions to steer and what distance to cover.

The second part of this question can be answered, at least partly, by the example of European warblers travelling from their Palaearctic breeding areas to various parts of Africa. The warblers exhibit a fairly sharp 'migratory division', with the western German and the eastern Austrian birds flying south-westwards and south-eastwards respectively. Studies of inexperienced, hand-raised birds tested in orientation cages at the time of day and year when they would normally migrate clearly show that the first-year migrants possess and use an innate vector programme. In cross-breeding experiments the first-generation offspring of mixed pairs of south-westwards and south-eastwards migrants choose migration directions that are in between of those of their parents (Helbig, 1994).

Although this is clear that the birds follow their endogenous vector programmes, these patterns seem to be rather flexible with respect to both migratory distance and direction. Within the last three to four decades, which have been characterised by progressively milder winter seasons, a fraction of the south-west migrating population has shifted its vector course and established new winter quarters in Britain (Berthold *et al.*, 1990). As in the example with monarch butterflies cited above, the microevolutionary processes interact here with some displays of behavioural flexibility.

11.2 Ways of navigation in animals

Both long-distance and short-distance travellers use multiple cues and a variety of ways for navigation. Here we will consider briefly the main ways of navigation in animals and illustrate them by several examples based on experimental examinations.

Olfactory navigation

The use of the sense of smell to find a goal is widespread in different taxa, of both invertebrates and vertebrates. One particular case of orienting by smell is the use of pheromones. The term *pheromone* was introduced by Karlson and Lüscher (1959) to refer to the chemicals that are used for communication, through the sense of smell, among individuals of a given species. Some animals possess scent glands that release pheromones. Pheromone trails can guide animals to food and to their nest sites, to detect boundaries of guarded territories of their neighbours, and to seek sex mates.

One of the most impressive examples of olfactory navigation concerns birds who gain information from atmospheric trace gases. Birds have been known for a long time to be weak users of olfactory cues. Human beings could be considered to be members of the same club. However, birds are the most prominent and best-known migrants and they possess one of the most intricate navigation systems which are probably based on using multiple sources. The use of olfactory cues may be built into an integrative 'cognitive map' in some bird species.

The question of whether birds can derive information on their current position relative to their home loft from airborne substances perceived with the sense of smell has only relatively recently been investigated. More than three decades ago, ten homing pigeons whose olfactory nerves had been sectioned were released 54 km west of Florence in Italy (Papi *et al.*, 1972), and these ten birds were pioneers who made a breakthrough in avian navigation research (for a review see Wallraff, 2004). Less traumatic methods of olfactory deprivation were then elaborated such as anaesthesia of the olfactory epithelia or occlusion of nostrils. Further studies based on olfactory deprivation have led to a suggestion that olfaction-dependent/independent homing is expressed differently in different species and that this perhaps reflects difference in the species' 'system profiles' of navigation. Fiaschi *et al.* (1974) applied olfactory deprivation to common swifts *Apus apus* and released them 47–66 km from home. Of 20 control birds 15 returned, but only three of 23 experimental birds; these three birds had lost their nose plugs. Anosmic European starlings, *Sturnus vulgaris*, were displaced over varying distances (Wallraff *et al.*, 1994). Up to 60 km, anosmia increased the time required for homing, but did not affect the percentage of birds that finally homed, whereas over 120 and 240 km return rates were drastically reduced. This increased effect of anosmia with increasing distance was similar to that obtained earlier in pigeons, but the starlings were successful in olfaction-independent homing within a larger radius around home. In principle, the range of olfactory navigation is not unlimited. It may extend over distances of some 300–500 km from home and seems dependent on particular geographical, mainly orographical, conditions.

Researchers have carried out many air filtration experiments with pigeons. They manipulated displaced pigeons by exposing them to portions of air to filter. In sum, these findings have shown that a pigeon decides in which direction to fly not primarily on the basis of olfactory inputs received at the moment of departure, but on the basis of inputs gathered over a longer period beforehand. Moreover, these experiments suggest that the pigeons develop an 'olfactory map' by associating current olfactory sensations with current wind directions. In a model aimed at describing the development of the pigeon's navigation system, the wind would have to be inserted as an essential link between compasses and grid map (Wallraff, 2004). We will return to these elements of pigeon's 'integral map' in the next sections.

Path integration (dead reckoning)

The term 'dead reckoning' has been borrowed from human navigation terminology: it was the method that was historically used by sailors to navigate across featureless open sea. More frequently it has come to be called *path integration* (PI) to reflect the assumption that the process takes place by the addition of successive small increments of movement onto a continually updated representation of direction and distance from the starting point. Darwin (1873) and Murphy (1873) hypothesised that path integration is based on the integration of interior signals. Path integration appears to operate in a great number of species with a fixed home base, during the exploration of a new environment or in commuting between home and familiar resource sites. It functions automatically and constantly, whenever the agent moves in space. Thus, a central place forager may, for instance, interrupt its excursion back home at any place and at any moment of its journey (Etienne and Jeffery, 2004).

Integrating definitions from recent literature, dead reckoning (path integration) can be defined as the process of estimating your position by advancing a known position using course, speed, time and distance to be travelled. In other words figuring out where you will be at a certain time if you maintain speed, time and course.

There is general agreement among those who study animal navigation that dead reckoning is fundamental to the navigation of animals ranging from molluscs and insects to humans. This mode of navigation is often considered rather simple. Cornell and Heth (2004) cite Herman Melville's poetic novel *Moby Dick* about reverting to elemental navigation such as dead reckoning after the ship's quadrant is destroyed by Captain Ahab's soul. At the same time, Cornell and Heth note that such a method of navigation without fixed references and landmarks has been especially intriguing to comparative and cognitive psychologists. As Pearce (2000) has pointed out, an obvious problem with navigating in this way is that once an error has entered into the calculations as a result of a faulty measurement, there is no means of detecting it and the navigator will have no indication of being lost.

Let us consider path integration in desert ants, as a good example. The organism here is the Sahara desert ant *Cataglyphis fortis*, which has been the subject of more than 20 years' detailed experimental analysis by R. Wehner's group (see, for example, Wehner and Srinivasan, 1981; Wehner *et al.*, 1996b). This ant is a solitary forager which scavenges for other arthropods that have succumbed to the physical stress of their desert habitat, and, what is important in the present context, does so by relying upon visual rather than chemical cues. Its navigation courses can be recorded in full detail because it walks rather than flies.

Dead reckoning in the ant and the honeybee is likely to be dependent on information acquired from the movements they make during their journey. How can insects successfully continuously compute the vector pointing towards their nests? In vertebrates this mode of navigation is said to be influenced by changes that take place in the vestibular system. In rats, for example, lesions in the vestibular system have been shown to disrupt their capacity for dead reckoning (Wallace *et al.*, 2002). It is known that arthropods successfully use path integration but it is still under discussion how they can estimate the distances travelled in a given direction.

What is clear now is that a common formula does not exist. For example, energy expenditure as a cue for distance estimation was proposed by Heran and Wanke (1952) on the basis of experiments with honeybees that were trained to forage on steep slopes. This hypothesis has been successfully revisited for bees (Goller and Esch, 1990), but not for ants as an additional load of up to four times the body weight did not affect the measurement of walking distance in the ant *C. fortis* (Schäfer and Wehner, 1993). The second way to measure the distance travelled is to do this by monitoring locomotor activity (the so-called idiothetic orientation). Experiments with spiders such as *Cupiennius salei* showed that they are able to evaluate travel distance on the basis of idiothetic cues (Seyfarth *et al.*, 1982). In experiments with the ant *C. fortis* Ronacher and Wehner (1995) tested the third hypothesis, namely that ants are able to use self-induced optic flow components to estimate travel distances.

The animals were trained and tested in transparent channels. One end of the training channel was situated near the ants' nest, and the ants were induced to enter the channel by a fence around the nest. The channel was equipped with stationary or moving black-and-white gratings of random dot patterns presented underneath a transparent walking platform. The patterns could be moved at different velocities in the same or opposite direction relative to the direction in which the animals walked. Experimental manipulations of the optic flow influenced the ants' homing distances. The use of different pattern wavelengths showed that distance estimation depended on the speed of self-induced image motion rather than on the contrast frequency. Experiments in which the ants walked on a featureless floor or in which they wore eye covers showed that they are able to use other cues for assessing travel distance. Hence, even though optic flow cues are not the only cues used by the ants, the experiments show that ants are obviously able to exploit such cues for estimating of travel distance.

Piloting with landmarks and use of geometric relations

Piloting refers to the act of setting a course to a goal on the basis of landmarks that are in a known relation to a goal. In many experiments many species have demonstrated their abilities to make use of landmarks to determine both the direction and the distance that they are going to travel.

The simplest form of piloting would be to navigate towards a feature that was located immediately by the goal – a beacon. But very often landmarks are not conveniently situated by a goal and their use then becomes more complicated.

Many elegant experiments concerning piloting have been carried out on insects. This tradition was started by Tinbergen. By means of simple and elegant tests he asked a digger wasp how it finds its way to an 'invisible' nest (simply a hole in sand). A mother-wasp digs a nest burrow in which she lays an egg. She covers the burrow entrance with pebbles, making it all but invisible, and then flies off in search of insect prey. Tinbergen and Kruyt (1938) arranged circles of sticks and pine cones around the nest while a wasp was digging it then displaced the circle while she was gone. The returning wasp landed in the centre of the displaced circle. She could not find her nest, even though it was only a metre away. This experiment demonstrated that the wasp located the entrance by reference to nearby landmarks, not by the sight or smell etc. of the nest itself. It is natural that when two or more landmarks are present, they create a geometric shape. Two landmarks create a line, three a triangle, and so on. Researchers from Tinbergen's group were the first to demonstrate that animals are able to detect this geometric information and to use it to define the location of a goal.

As a variant of the experiments with the digger wasp, van Beusekom (1948a, b) also constructed a circle of pine cones around the hole while the wasp was in it, but as soon as it flew away, he constructed either a square or an ellipse of cones around the circle. Upon returning, the wasp went to the circle rather than to the square, but it selected the circle and the ellipse equally often. This discovery indicates that during their initial flight around the hole the wasp remembers the shape created by the cones. Evidently the representation of this shape is sufficiently precise to permit a distinction between a circle and a square but not between a circle and an ellipse. Luckily nobody punished the wasp with an electric shock for its incapability of detecting such a subtle difference between geometric figures so it did not get traumatic neurosis like Pavlov's dog which also failed to make this distinction (see Chapter 6).

A series of experiments provides clear evidence that rats, pigeons, humming birds, hamsters and even newly hatched chicks are sensitive to geometric relations contained in the shape of their test environment and can make use of geometric information in piloting processes when they search for a hidden food (Cheng, 1986; Vallortigara *et al.*, 1990; Brown, 1994; Spetch, 1995; for a review see Cheng, 2005). For instance, Cook and Tauro (1999) investigated how the geometric relation between object/landmarks and goals influenced spatial choice behaviour in rats. Animals searched for hidden food in an object-filled circular arena containing 24 small poles. The results of tests with removed and rearranged landmarks that were configured in a square suggested that close proximity of objects to goals encourages their use as beacons, while greater distance of objects from goals encourages their use as landmarks.

Magnetic fields and compasses

A number of studies have experimentally demonstrated that the magnetic properties of the Earth provide a pervasive source of information for travelling both long and short journeys. In support of this proposal, pigeon homing is less accurate when a magnetic storm takes place (Gould, 1982b). It was assumed that birds were able to determine their global position by detecting features of the Earth's magnetic field, such as field intensity and inclination, for reading a bi-coordinate magnetic map based on these features. Besides, birds apparently use the sky and its celestial bodies as a source of information for navigation. To illustrate that the sun is an important element of homing the results of experiments with 'clock-shifted' pigeons can be cited. Pigeons with a phase-shifted circadian clock, released from home in any direction, deviate at departure from unmanipulated control birds by an angle roughly corresponding to the angle between the observed sun azimuth and that expected according to their shifted timescale (Schmidt-Koenig, 1960, 1979). Kramer (1953) suggested that birds must have some equivalent of a map as well as of a compass. Surely these are conveniently brief metaphors which do not precisely describe the system. The brief term 'map' refers to position-indicating factors while 'compass' refers to directional references. We must be aware that these terms are anthropomorphic. A bird hardly has a map at all and probably does not needs any idea about wide-ranging spatial structures. In other words, it does not necessarily imply that the bird has a representation of a coordinate system or any other wide-ranging spatial configuration in its brain. It merely means that we are dealing with environmental factors whose two-dimentional spatial order involves information on distant positions relative to a bird's home site. How a bird makes use of such information is not immediately implied (Wallraff, 2004). In particular, the sun-arc hypothesis suggested that birds could determine their position relative to home by comparing the actual movement of the sun along its arc with the remembered one at home. This suggestion reflects the paradigm of early research in animal navigation which represented birds as 'true astronavigators'. Since experimental investigations were carried out in the 1970s, it has become clear that birds are not astronomers in the sense that they can not use the sun for taking positional fixes anywhere on the surface of our globe; nor can birds use the stars for fixing their position at night (see Wehner (1998) for a review). Birds and other animals have not adopted a heliocentric or a general geocentric view of the sky surrounding them. Instead, they solve their navigation problems in more immediate ways in which signals from magnetic and celestial cues are included differently for different species and at different stages of their lifetime.

11.3 Redundant sources of spatial information in animals

Animals refer to different sources of information and use different sensory modalities in order to navigate successfully, and similar mechanisms may be involved in different taxa. Although comparative studies have clarified many problems of how animals find their paths in the world, many questions are still under discussion after more than a century of investigations. As Wehner (1998) notes, in spite of the impressive body of literature that studies of animal migration and homing have produced in the last decades, we seem only to have touched the surface of the navigation system at work.

There are several lovely objects for students of navigation such as pigeon, honeybee, the desert *C. fortis* ant, loggerhead sea turtle and some others that we have already met in the previous section and whose secrets of multiple ways of navigation have come to light little by little but still remain far from crystal clarity.

For example, so-called multi-coordinate grid maps were first proposed in 1882 by Viguier to explain long-distance navigation in birds. This question was reopened in the mid-twentieth century by the proposals that the grid map is derived from the Coriolis force and geomagnetic fields and that birds use the gradient of the sun's movement (Yeagley, 1947). In the 1950s and 1960s, the debate on birds homing was dominated by Matthews' sun navigation hypothesis (Matthews, 1955). More recent studies essentially corrected this hypothesis although it remained obvious that the sun is a key element of the homing processes in birds. Over 50 years, Kramer's 'map-and-compass concept' remained an accepted constant (Kramer, 1953). Further experiments added the use of olfactory gradients in birds as well as visual signals from the environment such as visual landmarks. Additional methods of navigation in birds are still being intensively investigated and discussed such as using ultrasound, polarised and ultraviolet light, and different sorts of landmarks (for reviews see Papi, 1986; Wiltschko and Wiltschko, 2003; Wallraff, 2004).

Another set of examples concerns the study of mapping behaviour in hymenopterans, which dates back to the classic experiments of Fabre (1879) and Romanes (1885) who displaced individually marked bees, wasps and ants to unfamiliar sites and tried to clarify details of their homing. More than a century of studies mainly based on different variants of displacement experiments have revealed that in bees and ants path integration employing a skylight compass is the predominant mechanism of navigation, but landmark-based information is used as well (Lewtschenko, 1959; Lindauer, 1963; Dyer, 1987, 1996; Wehner, 1998). Indeed, as Wehner and Menzel (1990) note, unexpected displacements in dark boxes carried by human experimenters have obviously not been an evolutionary force that has shaped the insect's navigation system. Nevertheless, plenty of displacement experiments have showed that the flexible use of vectors, snapshots and landmark-based routes enables insects' navigation systems to cope with many different ecological situations such as obscured sun, overcast sky, apparent movement of the sun and drift by wind. Hymenopterans demonstrate a great deal of flexibility switching between different sources of spatial information. Experiments by Chittka et al. (1999) showed that bumblebees *Bombus impatiens* are able to forage in complete darkness by walking instead of flying. Using infrared video, the researchers mapped walked trails. They found that bumblebees laid odour marks. When such odour cues were eliminated bees maintained their correct direction, suggesting a magnetic compass. They were also able to access travel distance correctly, using an internal, non-visual, measure of path length. The ability to switch between different orienting cues in response to changing circumstances (including complete darkness) were demonstrated in several ant species by Mazokhin-Porshnyakov and Murzin (1977), Dlussky *et al.* (1978) and Reznikova (1983). Indeed, ants amaze researchers by the diversity of ways they use orienting cues. For instance, *Camponotus japonicus aterrimus* marks stems in a fashion just like a dog by standing on tiptoe and moving the abdomen along the stem (Reznikova, 1981). Ants of different species were discovered to use different orienting cues when searching for food, homing and patrolling boundaries of defended territories (Reznikova, 1974).

It is interesting to note that hymenopterans switch between different ways of spatial orientation as their experience increases, in a manner analogous to the way that vertebrates switch navigational tactics with experience. For instance, naive giant tropical ants *Paraponera clavata* initially use chemical trails to find food sources and to return to the colony but switch to visual cues as they gain experience. Ants using visual cues run twice as quickly between sites as those following chemical trails (Harrison *et al.*, 1989). Laboratory experiments with the use of artificial flowers have demonstrated that naive bumblebees use relatively simple tactics of foraging; as their experience increases, however, they increasingly depend on their spatial memory for locating rewarded flower locations (Dukas and Real, 1993; Burns and Thomson, 2006).

Applying imagination about redundancy of information sources to water-dwellers, we can consider fish for which hydrostatic pressure acts as a global landmark or reference. Similar to compass direction or distal visual information, it provides information about the location of a goal, especially to an organism that must orient and navigate in three dimensions (Healy, 1998). Fish also use landmarks to orient and navigate. They can detect changes in landmark size and will modify their locomotor behaviour to integrate the change into an internal representation. If the water level changes, increasing hydrostatic pressure, the fish orient to a landmark, if present. If no landmark is present, the fish rely on an internal representation oriented to hydrostatic pressure (Cain and Malwal, 2002). Braithwaite and Girvan (2003) tested experimentally whether river three-spined sticklebacks are more adept at using water flow as a spatial cue than fish from ponds. Fish from two ponds and two rivers had to learn the location of a food patch in a channel where water flow direction was the only reliable indicator of the goal position. All fish were able to use water flow as a spatial cue but one of the two river groups was significantly faster at learning the patch location. When the task was reversed so that fish that had formerly been trained to swim downstream now had to learn to swim upstream and vice versa both river groups learned the reversal task faster than the two pond groups. In a second experiment, fish were given a choice between two different types of spatial cue: flow direction or visual landmarks (rocks, plants). A test trial in which these two cues were put into conflict revealed that the river population showed a strong preference for flow direction whilst the pond population preferred visual landmarks. Thus learning and memory are fine-tuned to the fish's local environment, and fish sampled from contrasting environments use different types of spatial information.

Surprisingly little is known about how mammals navigate under natural conditions and cope with given environmental constraints. Thus, field experiments in which positions of several feeding platforms as well as several artificial landmarks were manipulated revealed that free-ranging Columbian ground squirrels (*Spermophilus columbianus*) rely on multiple types of cues. Local landmarks were used only as a secondary mechanism of navigation, and were not attended to when a familiar route and known global landmarks (such as the forest edge or the outline of the mountains) were present (Vlasak, 2006).

Yet a fundamental principle of spatial orientation is that navigators utilise multiple and redundant sources of spatial information (Keeton, 1974; Cheng and Spetch, 1998; Jacobs, 2003). Compasses based on different cues interact in intricate ways; they are calibrated against each other, replace each other, and do so differently during successive stages of development (Wehner, 1998).

In this section we consider examples from two series that illustrate the principle of redundancy. The first series includes individuals being placed sequentially under different circumstances and thus forced to use different ways of navigation. The second series includes species as wholes which are characterised by using multiple navigation systems owing to peculiarities of their natural history.

But one example stands beyond the sequences mentioned above. In ant society perhaps the most original way of realising the principle of redundancy can be observed: the use of different cues is implemented in different family members. It is known that threading their way through stems of grass, these insects use multiple sources for navigation, referring to celestial cues, local landmarks of different sizes, and also pheromone trails (Rosengren, 1971). It has been shown that in steppe ants *Formica pratensis* individual members of each group that carry out their tasks on a local plot of a common feeding territory possess different preferences when choosing ways of navigation. Some ants use small local landmarks (such as pebbles) and experimental removal of small artificial beacons would be sufficient to confuse them. Other family members prefer bigger landmarks, from the size of a bottle to the size of a bush or, say, a man. And for other ants the shape of the wood against the skyline is sufficient to find their plots, so global changes of a scene would be needed to confuse their orientation. Finally, there are some ants that are guided by odour cues only and do not pay attention to any visual cues when looking for their destination (Reznikova, 1983, 2005).

More individualistic organisms than ants could consider different sources of information reliable for mapping depending on the context of their life. For example, experiments on laboratory rats indicate that they apply different systems of orientation depending on circumstances (Diez-Chamzio *et al.*, 1985). Rats were trained in a radial maze in a rather unusual way. They were placed at the end of one arm and then had to go to the end of another arm in order to receive food. The correct arm was distinguished by a sandpaper floor. Throughout the training, the landmarks provided by various objects in the room, and the shape of the room itself, were made irrelevant by rotating the maze from one trial to the next. In a second stage of the experiment, food was placed again at the end of sandpaper arm, but the maze was not rotated so that all the cues associated with the room were no longer irrelevant to the solution of the problem. Despite this change in training, subsequent testing revealed that rats had learned very little about the cues that lay outside the maze. In contrast, a control group, which received just the second stage, learned a great deal about the significance of the extra-maze cues. One way of summarising these results is to say that pre-training with the local landmark of sandpaper blocked the development of a cognitive map based on extra-maze cues in the second stage of the experiments (Pearce, 2000).

This example gives a flavour of the argument that the same navigator being placed in different situations is able either to be or not to be the constructor of a cognitive map. What system of navigation would be chosen is likely to be determined by such important factors as inherited predisposition for the use of definite methods of orientation as well as reliability of these methods in the current context of life. Tied to a wired programme of preferred ways of navigation, animals nevertheless demonstrate a high level of freedom in combining them. Recent studies have revealed a great deal of flexibility in the use of navigation systems in many species.

In order to judge whether a subject really prefers any particular system of orientation a so-called *disorientation procedure* (reorientation task) can be applied. When searching for hidden food, the subject is disoriented between acquisition phase and the test itself.

For instance, in Cheng's (1986) experiments rats searched for food previously hidden in one of the four corners of a rectangular apparatus. After animals were familiarised with the experimental environment, they were removed from the apparatus, disoriented within a closed box, and returned to the apparatus to search for food. In this reorientation task, the rat had to re-establish its position and heading before it engaged itself in goal-directed behaviour. To be oriented again, the rat could rely on the shape of the apparatus, on the patterns, on the odours, or on the brightness of the walls. In the cited experiments rats showed a high rate of searching both at the correct (reward) corner and at the rotationally equivalent opposite corner, and these two corners on the same diagonal are defined by the same geometric relation within the apparatus (length and width). This search pattern was constant, despite the availability of many other cues including strong distinctive odours and large differences in contrast and luminosity. This and other findings (Gallistel, 1990) suggest that geometric features of environment are spontaneously taken into account by rats and dominate over other landmarks.

Gouteux *et al.* (1999) have applied reorientation tests to rhesus monkeys and baboons and revealed more flexible behavioural tactics. Although primates prefer to rely on the large-scale geometric cues of the experimental rooms in order to reorient, they also used non-geometric information (a coloured wall). Surprisingly flexible ability to use both geometric and non-geometric spatial information to reorient has been found in pigeons and chickens (Vallortigara *et al.*, 1990; Kelly *et al.*, 1998).

Influence of ageing is significant in flexibility of navigation. Hermer and Spelke (1994, 1996), in a series of studies conducted with human children, have examined the use of a geometric module by toddlers. Children saw a desired toy that was hidden in one of the corners of a rectangular homogeneous experimental chamber. In one of the experiments, the chamber contained no distinctive landmarks. It turned out that when no information other than the shape of the environment was available, children searched equally often in the correct and in the rotationally equivalent corner. When non-geometric information (a blue wall or a pair of toys placed in the room) was added, children still divided their searches between two diagonally situated corners and seemed to ignore the added cues. Unlike children, human adults in the same experiments were able to use both geometric and non-geometric information to optimise their search.

11.4 Mapping in the context of natural histories

Some insights concerning the intricate mechanisms of spatial orientation and navigation came from the careful study of species for which some aspect of their natural history makes spatial navigation particularly important for biological success (Kamil and Cheng, 2001). We have already mentioned homing in birds and insects. Many bird species (among which pigeons have been best studied) are known to use multiple means of orientation such as magnetic fields, sun compass, mosaic of local landmarks and odour. The cleverest insects, hymenopterans, use in general the same set of ways of orientation. Here we consider several special examples concerning foraging processes in these two groups of organisms that require excellent spatial memory and orientation.

One amazing example of a navigating system is that of foraging hymenopterans. One explanation of how a honeybee finds a goal is that they follow a sequence of place-finding servomechanisms, following a vector to the vicinity of the target homing on a landmark located near the target and then flying in a stereotypical direction to match the image they see to a remembered image (Cheng, 2000).

As in many other animals, complex navigating behaviour in honeybees can be divided into several sequences that in turn are characterised by more or less specific features. Some of these features can be explained with the use of common properties of the insects' physiology, in particular, their visual systems. For insects, an image of a three-dimensional world on the retina is only two-dimensional. Whereas vertebrates have evolved several mechanisms for depth perception, such as stereotypic vision, convergence of the eyes or accommodation of the lens, most insects lack these mechanisms (Mazokhin-Porshnyakov, 1969; Mazokhin-Porshnyakov *et al.*, 1977; Frantsevich, 1993).

The first behavioural pattern the bee performs when it has found a novel and rich food source is the so-called circling behaviour. When bees depart from a novel feeding site for the first several times, they fly around it in wide and high circles. During these flights they memorise the landmark panorama around the feeding place (Frisch, 1967; Gould, 1986). The circling behaviour does not provide information about such details as shape and colour of the feeding site.

For these purposes other patterns serve, which were discovered by M. Lehrer and co-authors (Lehrer, 1987, 1993; Lehrer *et al.*, 1985, 1988; Srinivasan *et al.*, 1989) and subsequently called *scanning behaviour* and *turn-back-and-look behaviour* (TBL). Scanning behaviour is displayed by the bees when they leave a novel feeding site and try to remember its visual pattern. In experiments with the use of artificial black-and-white patterns the bees followed the contours of the patterns in front of which they were flying. This behaviour is effective for discriminating models that are placed on a vertical plane. In experiments in which depth cues were important as the main feature of the novel food source, the bees, upon leaving a reward box, turned around to view the entrance, approaching it again and again. For example, in one series of experiments bees were trained to collect food on a white paper 'meadow' surrounded by a white wall. Six black discs of different size, each carrying a drop of water, were placed flat on the ground, while one disc, placed on a stalk above the ground, offered a reward of sugar-water. The positions and the sizes of all seven discs varied, so the only cue that could be used by the bees was its height above the ground, which was the only parameter that was kept constant. In other experiments the bees were trained to collect sugar-water at an edge between two surfaces covered with different visual patterns. Colour learning tasks have been involved in other series of tests (Lehrer, 1996, 1998; Lehrer and Bianko, 2000). It has been revealed that the TBL phase of remembering spatial cues is based on the specific characteristics of the bee's eye. The bees use the speed of image motion in a variety of tasks involving small-scale navigation. These insects use different cues not only under different circumstances but also in different regions of the eye. For example, colour discrimination in the lower half of the frontal eye region is better than in the upper half, and estimation of the distance flown is not a function of the ventral visual field. At the same time, the bees effectively employ many options in a process of small-scale navigation depending on the context of their vital tasks.

Knowing different behavioural patterns and specificity of their visual system, we now can predict what the bee will do in different situations. For example, J. L. Gould and C. G. Gould (1995) observed that honeybees learn the colours of flowers only in the last 2 seconds before landing, and fix landmarks in memory only when they leave. A bee transported to a source of nectar, and later transported back to the hive, cannot find the place again. Even the location and appearance of their own hive is learned only when leaving for the first time each morning. If the hive is relocated after the first flight, the bee that leaves the relocated hive cannot find it again. As the authors sum up, the bees learn exactly what they were programmed to learn, exactly when they are programmed to learn it.

The distance information acquired during the TBL behaviour is used only in the initial phase of visiting the novel feeding site. Experienced bees arriving at a familiar food source use the size of the landmark as a cue to distance. Lehrer and Collett (1994) propose that cues to distance are particularly important in the initial phase of learning, because near landmarks are more useful for pinpointing the goal than are more distant ones. Thus, the insect must first of all determine which marks are near and should, therefore, be memorised.

It is easier for us to imagine a process of small-scale navigation during foraging trips in pedestrian hymenopterans. Recent studies have shown that ants take several 'snapshots' of a familiar beacon from different vantage points. This mechanism reminds of the TBL behaviour in bees and works as if the ants had acquired an eidetic image – a 'photographic snapshot' – of the landmark panorama around their nesting site and then move in the new area to match their individually acquired template as closely as possible with their current retinal image (Wehner, 1997).

As has been found by Judd and Collett (1998) from their laboratory experiments on red wood ants presented with a sucrose reward placed near artificial landmarks (black cones), foraging individuals take several snapshots of a familiar beacon from different vantage points. An ant leaving a newly discovered food source at the base of a landmark performs a tortuous walk back to its nest during which it periodically turns back and faces the landmark. The ant, on revisiting the familiar landmark, holds the edges of the landmark's image steady at several discrete positions on its retina. These preferred retinal positions tend to match the positions of landmark edges that the ant captured during its preceding 'learning walks'. So, as Srinivasan (1998) puts it, red wood ants match as they march during their foraging journeys.

Another biologically significant situation, in which spatial navigation plays a central role, is the recovery of stored food by several species of birds and small mammals. These animals are highly dependent on stored resources for survival and reproduction. We have already met with these masterly creators/detectors of treasures in Part III and will return to them in Part VII to investigate constituents of their special education. Here a central problem is what special means these animals use for spatial navigation in order to recover the thousands of caches hidden by themselves.

Shettleworth and Krebs (1982) carried out experiments with marsh tits. These birds hide seeds in tree branches, and are able to remember months later the location of thousands of hidden seeds. The seeds are found using a memory for spatial coordinates. When landmarks (for example rocks) are moved, the birds look for seeds in the wrong places; they search where the seeds would be if they had been moved together with the rocks.

Researchers carried out long-term experiments with another cacher – Clark's nutcracker (*Nucifraga columbiana*). In an autumn, with a good pine seed crop, an individual nutcracker will cache tens of thousands of pine seeds in thousands of locations, subsequently returning to them throughout winter and spring (Tomback, 1980). Birds recover their caches with great precision. Many field and laboratory studies have strongly supported the cache site memory hypothesis for the accurate cache recovery of nutcrackers (reviews in: Vander Wall, 1990; Kamil and Balda, 1990). What makes this story puzzling is that many caches are located relatively far from large landmarks. As far as small objects or marks on the ground were concerned, experimenters found no evidence of their use by birds. It sounds natural that nutcrackers should ignore small landmarks because these markings can change dramatically with the changes of seasons. A major puzzle for a student of animal navigation is how a nutcracker can possibly achieve the precision required to relocate a small cache while digging with its small-diameter beak (Kamil and Cheng, 2001)?

Kamil and Jones (2000) have demonstrated that nutcrackers use spatial memory to retrieve their caches with accuracy and precision and that the birds can learn directional relationships between a goal and two landmarks. They tested birds in a rectangular room. In one of the settings of their experiments, birds were presented with two landmarks with varying distances between them. The birds always buried a seed at a third point relative to two landmarks. The distance between the goal and the landmarks varied with the inter-landmark distance so as to maintain constant directional relationships: the seed was always buried north-west of one landmark and south-west of the other. The birds thus demonstrated an ability to use directional relationships. Another series of experiments revealed the evidence that the direction from the goal to a landmark is a more potent cue than the distance between them. Two groups of birds were trained to find the third point of the triangle. The goal location was defined by two landmarks whose inter-landmark distance varied from trial to trial. For one group, the third point of the triangle was defined by bearings. The goal was always buried at the intersection of two fixed bearings whose value was constant (requiring the goal–landmark distance to vary). For the second group, the goal was always buried at the same distance from each landmark (requiring the goal–landmark bearings to vary). The constant-bearing group learned to solve this problem much more rapidly than the constant-distance group. Under these conditions, bearings provided a more useful cue to location than distance did.

Basied on these and many other experimental results and on computer simulations Kamil and Cheng (2001) have come to the multi-bearings hypothesis of navigation in nutcrackers. These birds use the metric relationships between a goal and multiple landmarks. They prefer to bind the direction up with the distant landmarks rather than with the nearest one. The researchers propose that nutcrackers use a set of bearings, each a measure of the direction from the goal to a different landmark, when searching for that goal. Increasing the number of landmarks used - within certain limits - results in increasingly precise searching.

Once more impressive example of essential function of spatial representation concerns meerkats, cooperatively breeding mongooses (see Part X for details) that live under high predation pressure, and thus need to know which of more than a thousand bolt-holes on their territory is closest for finding shelter from predators quickly. Manser and Bell (2004) used observations and manipulation experiments to investigate how meerkats succeed in finding shelter immediately when an alarm call is given. Experimenters played back alarm calls to foraging meerkats and dug new bolt-holes and covered existing ones to see whether location or other cues were used. Meerkats almost always ran to the bolt-hole closest to them. This was not done by a simple rule of running back to a bolt-hole they had just passed, nor by escaping in a random direction and finding a bolt-hole by chance. They nearly always ignored the bolt-holes that researchers dug but ran to those they had covered up. Adult animals seemed to have an accurate knowledge of the distance and direction to the closest shelter in relation to their own position in their territory at any time (Fig. 11.1). Although meerkats sometimes failed to find shelter, they still appeared to know a large proportion of the bolt-holes available in their territory, which still leaves several hundred or more locations to remember. This ability to memorise many locations might, as the authors suggest, be comparable to the skills of food-storing birds.

Fig. 11.1 Meerkats seem to have an accurate knowledge of the distance and direction to the closest shelter in relation to their own position in their territory at any time (Manser and Bell, 2004). (Photograph by L. Hollén; courtesy of L. Hollén.)

Chapter 12

To what degree is mapping cognitive in animals?

12.1 Cognitive mapping as a methodological problem

Long-term experiments with rats navigating mazes led Tolman to suggest that

> something like a field map of the environment gets established in the rat's brain . . . And it is this tentative map, indicating routes and paths and environmental relationships, which finally determines what responses, if any, the animal will finally release
>
> (Tolman, 1948, p. 192)

This concept was founded on Tolman's general idea that animals do not merely base their actions on specific stimulus–response associations, but that they also internally reorganise acquired spatial information to form cognitive representations of the environment (Tolman *et al.*, 1946a, b). One important property of such representations is that they allow animals to react to stimuli that are not immediately present because the relationship of such stimuli to those actually perceived is maintained in a cognitive representation, in other words, a map. Once an animal has built a map, it can use the information to solve a variety of problems. If animals are introduced into an environment from a new starting point or if the previous path is blocked, they will be able to deduce the most direct trajectory from that place to a known goal.

Having been 'officially introduced' by Tolman in the 1940s, cognitive maps came into fashion for behavioural neurobiologists and experimental psychologists in the 1980s and, as Wehner and Menzel (1990) comment, it is not astonishing that recently even insects have been claimed to possess cognitive maps. As these authors note, any such claim depends crucially on how a 'cognitive map' is defined in operational terms. Ants ran parallel ways in labyrinths with rats as early as the 1930s and these two groups of inhabitants of underground secret passages yielded to no one in this respect (Maier and Schneirla, 1935). Anyway, we are going to face some controversy when consider cognitive mapping in insects.

The main problem to be discussed is to what degree mapping is cognitive, or, in other words, whether intelligence is involved in the process of mapping. In order not to be lost in the labyrinth of wild mapping, let us start with several examples before entering into details of modern theories of cognitive maps.

In his pioneering experiments with map-like behaviour in chimpanzees E. Menzel (1973a, b) began straight away with a problem of rational choice of multi-destination routes. He carried young chimpanzees around while he hid pieces of fruit at 18 different sites in their enclosure. The animals could see where the experimenter hid the pieces of fruit. Then they were placed in a dark enclosure for 2 minutes. When the animals were released, they retrieved food from all 18 sites in a single rapid foraging expedition. Menzel supposed that the apes tended to use a 'least-distance' strategy to retrieve the hidden food and that this was based on cognitive mapping.

Experiments with radial mazes that have been carried out in order to establish how much information animals can remember also lead to discussion about cognitive mapping. Researchers have tried mazes with from eight to 24 arms to test rats, which easily remember the position of the rewarded arm (Olton, 1978; Roberts, 1979). After a while the rat would be able to empty the maze of treats without going down the same arm twice. It did not follow a set pattern and the route it took could not be predicted. At first Olton (1978) suggested that the rat left an odour marker at the entrance to each arm, but the researcher conducting the experiment eliminated that possibility. The clue came when the maze was rotated through 90°. The rat became disorientated, entering tunnels that it had already been travelled down because their position in relation to the objects in the laboratory had changed. The test demonstrated that the rat could 'picture' locations of the tunnels it had already entered and integrate them into cognitive map. It thus was able to reflect on its recent experience in order to find food with the minimum of effort.

Another set of maze experiments that presented evidence of a cognitive map in rats was the work of Morris (1981) using a swimming pool. He trained rats in a circular pool of opaque water, from which they could escape by climbing up onto a platform that was submerged 1 cm below the water level and thus could not be seen by the animals. The platform maintained a fixed position with respect to landmarks or objects in the experimental room. It turned out that rats can swim directly even towards a platform that is completely invisible. The results showed that the rats learned how to find objects that they could not see, smell or hear, locating their position based on their spatial relationship to landmarks in the room.

There is extensive evidence that rats and other laboratory animals such as pigeons and goldfish use complex relationship of landmarks to find a goal (Gould and Gould, 1994; Rodríguez *et al.*, 1994). Nutcrackers use an original form of space navigation when relocating their hidden food: the birds act as if they cover an open space with 'mental triangles'. They seem to form representations of the locations of cache sites, each based upon the relationship between the goal and surrounding landmarks; thus the nutcrackers probably do use cognitive maps (Kamil and Cheng, 2001).

Subsequently it has become an issue of great controversy whether or not animals construct 'true cognitive' maps. Perhaps Noah faced the same problem when filling his Ark: the more claimants, the more problems. Even for a strong lobbyist of intelligence in animals it is somehow difficult to imagine that a goldfish is really able to mentally build up a global representation of space.

The cognitive map can be operationally defined as novel route construction that cannot be explained by orienting either by pure path integration or by the use of beacons coincident with the goal (Jacobs and Schenk, 2003). O'Keefe and Nadel (1978) were the first to present a formal theory of the cognitive map. They define the cognitive map as 'the representation of a group of places, some related to others by means of a set of rules of spatial transformation'. When an animal enters a new environment, it forms a representation of the spatial relationship maintained by landmarks or other cues that it perceives from its position. Starting from this initial place representation, as the animal moves around the environment, it begins to incorporate new place representations into a cognitive map. This occurs because of the information coming from the sensory and motor systems with respect to the different distances that the animal perceives; these systems enable space to be represented in a relative way, with reference to the animal. O'Keefe and Nadel called this space relative or egocentric space. This theory explores the idea that navigation based on a cognitive map ('cartographic learning') occurs in an all-or-nothing way and that the hippocampus is the actual structure that is responsible for cartographic learning.

In the 1990s, Gallistel suggested a more tolerant theory of cognitive maps based on the claim that any orientation that includes implicitly distances and directions is evidence of a cognitive map ('metric map'). He defined a cognitive map as 'a record in the central nervous system of macroscopic geometric relations among surfaces in the environment used to plan movements through the environment' (Gallistel, 1990, p. 103). Since an animal can only perceive a part of its environment from any one vantage point, the construction of a cognitive map requires the integration of positional information derived from different views of the environment made at different times (Gallistel and Cramer, 1996). The construction of a map for navigation involves combining two sorts of position vectors: egocentric vectors which specify the locations of landmarks in a body-centred coordinate system and geocentric vectors which specify the animal's position in a coordinate system, defined with respect to animal's body. The geocentric vectors are calculated by path integration and specify the animal's position in a coordinate system based on the path's starting point.

Recently Jacobs and Schenk (2003) 'unpacked' the cognitive map and proposed the *parallel map theory* (PMT) which is founded on the premise that there is not one map but several: the integrated (or cognitive) map and its two components, the bearing map and the sketch map, which when integrated can create the novel short cut. The *bearing map* is the multi-coordinate grid map, derived from sources of distributed stimuli such as gradients. The *sketch map* encodes and stores fine-grained topographical data and is constructed from the memory of the positions of unique cues. In mammals the bearing and sketch maps are mediated by independent structures of the hippocampal formation, but the authors consider other structures in other taxa including invertebrates mediating these functions.

The bearing and the sketch maps thus represent two classes of maps. They are also constructed from two classes of cues, defined as directional and positional. Directional cues can include distributed cues such as gradients of odour, light or sound, as well as compass marks, distant visual landmarks which provide directional but not distance cues. Positional cues are discrete and unique objects, often near the goal, that can be used to estimate distance to the goal accurately. These maps differ in their stability and plasticity. The bearing map, once created, serves as a scaffold for the localisation of positional cues in absolute space or for global positioning. The sketch map, in contrast, is not a singular representation but a type of map. Sketch maps may coexist as a population of a topographic representation of local spaces, where each sketch map encodes the navigator's position within a specific panoramic array. Such maps arise from disjoined exploration, where a navigator may experience several local areas without recognising their directional relationship to each other, i.e. without forming an integrated map, and encodes them into separate sketch maps. Unlike the bearing map, where position can be deduced by estimating the rate of change in a gradient, a sketch map requires significant spatial memory to encode the features of individual landmarks in an array.

Jacobs and Schenk (2003) consider a set of examples in which the bearing map overrides the sketch maps. This concerns navigation in clock-shifted birds. Because the sun's movement is a distributed cue, the bearing map also incorporates data from the sun compass. With the clock shift, the pigeon shows a directional shift, even when the topographic information is not changed. The same phenomenon has been observed in small-scale orientation in other contexts: despite the presence of well-learnt positional cues, clock-shifted scrub jays, nutcrackers, pigeons and black-capped chickadees persist in searching for hidden food in the predicted compass deviation.

With a concept of an 'unpacked' cognitive map in mind, we can give more parsimonious explanations of some cases of map-like behaviour in insects and birds. Wehner considers that insects such as ants, bees and wasps possess intricate navigation systems and remarkable memory stores but lack 'true' cognitive maps (Wehner and Menzel, 1990; Wehner, 1999). Instead, mapping behaviour in insects can be attributed to snapshot matching. This strategy might be quite sophisticated in so far as landmarks positioned at different distances can be disentangled by motion parallax, and snapshots of the same scene might be taken from different points. According to Wehner (1997), insects do not incorporate their routes into a map-like system of reference. They cannot take a bird's-eye or even bee's-eye view of the terrain over which they travel; instead, they assemble the necessary information piecemeal over time. For insects, it might be a safer and more robust strategy to rely on sequentially organised, gazetteer-like memories rather than to encode the spatial relations among a multitude of similar sites and routes in a large-scale mental map.

It is still possible, however, that hymenopterans like ants and bees do possess a cumulative global path integration memory reflecting long-term experience of where abundant food is to be found (Collett *et al.*, 2003).We should not to discard insects as lacking at least a map-like organisation of spatial memory. Recently by using harmonic radar that monitors the flight path of an insect carrying the transponder antenna over a distance of up to 900 metres, R. Menzel *et al.* (2005) reported the complete flight paths of displaced bees. A sequence of behavioral routines become apparent: (1) initial straight flights in which they fly the course that they were on when captured (foraging bees) or that they learned during dance communication (recruited bees); (2) slow search flights with frequent changes of direction in which they attempt to 'get their bearings'; and (3) straight and rapid flights directed either to the hive or first to the feeding station and then to the hive. Two essential criteria of a map-like spatial memory are met by these results: bees can set course at any arbitrary location in their familiar area, and they can choose between at least two goals. This finding suggests a rich, map-like organisation of spatial memory in navigating honeybees.

Recent experimental data on pigeon homing along highways in Europe (Lipp *et al.*, 2004) demonstrated that in pigeons raised in an area characterised by navigationally relevant highway systems, mapping includes different stages. During early and middle stages of the flight, following large and distinct roads is likely to reflect stabilisation of a compass course rather than the presence of a mental road map. A cognitive (road map) component manifested by repeated crossing of preferred topographical points, including highways, is more likely when pigeons approach the loft area.

Indeed, there is much work to be done on map-like behaviour in different taxa. Theories and models of cognitive maps still cannot reasonably explain the observed behavioural flexibility, for example, planning new trajectories, selecting a novel route to a goal and making a detour around an obstacle. It seems that at this stage an adequate experimental method rather than additional models would be of great help. In the next section we will consider several sets of examples linked to experimental methods which provide a perspective for investigation of map-like behaviour.

12.2 Adjusting the track to the goal: short cuts and detours as elements of cognitive mapping in animals

In Tolman's concept of the cognitive map one of the key elements is the ability to construct new short-cutting routes. This behaviour is not so simple and requires the ability to link local scenes without any experience of the direct route from one point to another.

Besides demonstrating new short cuts themselves, another possible experimental strategy that can demonstrate the operation of a cognitive map is a so-called *detour test*. Detour behaviour demonstrates the ability of an animal to reach a stimulus (goal) when there is an obstacle between the subject and the stimulus. Detour tests allow researchers to integrate experimental approaches aimed at studying how animal intelligence is displayed in time and space. Indeed, the detour test is of particular interest for comparative cognitive research as a peculiar and quite naturalistic example of 'delayed response' (see Chapter 10). In the absence of any local orienting cue emanating from the goal, detour performance should require the maintenance of a 'memory' of the location of an object that has disappeared, As we will see in Chapter 13, developmental psychologists have thoroughly investigated these abilities in human infants (Piaget, 1954). In order to test how animals master the detour task, experimenters first show a goal to an animal and then place the subject behind a barrier some distance away. If the subject selects the correct, shortest route to the goal, then this may be because it possesses a cognitive map of the problem area and uses it to determine how to respond.

Short-cutting

Tolman's interest in novel short-cutting as an essential part of cognitive mapping has been developed in many experiments. It is worth noting that in this local field of studying cognitive mapping, the general problem of parsimonious explanation reproduces itself at a small scale: some authors claim that the results provide clear evidence of cognitive maps while others suggest simpler alternatives to explain the behaviour. Many examples of novel short-cutting have been described but it is still open to discussion whether they unambiguously indicate cognitive maps or could be explained as movements towards familiar landmarks seen from a new angle.

For example, a useful demonstration of short-cutting behaviour is reported by Chapuis and Scardigli (1993). In one of their experiments, a hamster was placed into Chamber A from the apparatus shown in Fig. 12.1. An animal had to pass along the route identified by the dotted line to obtain food in Chamber E. Locked doors prevented the hamster from deviating from this route. After a number of sessions of this training, all the doors in the apparatus were unlocked except the one that allowed the animal to leave Chamber A on its normal route to food. If hamsters are able to calculate the shortest detour to food, then they should pass along the route marked by the solid line in Fig. 12.1. For the three test trials that were conducted, hamsters selected this route significantly more frequently than would be expected on the basis of chance. It is important to note that during the training phase, the maze was rotated relative to the room from one trial to the next. In addition, the chamber that served as the start of the route (Chamber A) was varied from trial to trial. The success of the animals in solving this problem was thus not due to them orienting towards landmarks that lay beyond or within the maze. A more plausible explanation is that the hamsters formed a map of the shape of their route during the training stage, and then used this map to plot their course to the goal on the test trial (Pearce, 2000).

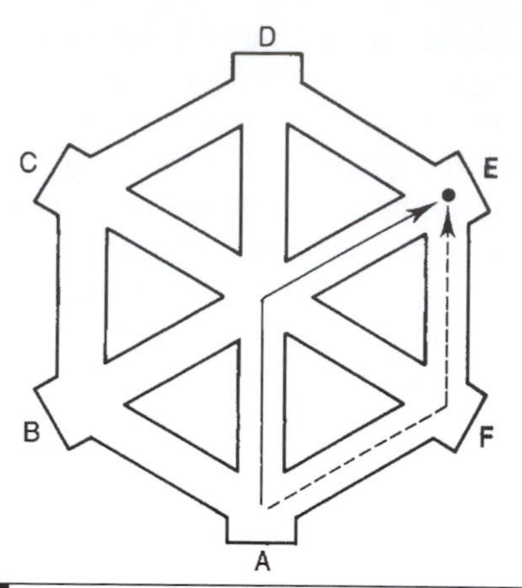

Fig. 12.1 A plan of the circular maze used by Chapuis and Scardigli to investigate the way in which hamsters solve detour problems. (adapted from Chapuis and Scardigli, 1993.)

Bennett (1996) argues that in this and other experiments path integration could be use to perform the short cut. As was described above, path integration, or dead reckoning, is the mechanism of integrating distance and direction while moving that allows an animal to make a straight line return to the starting point. Several experiments have shown that it can be performed without access to any previously seen landmarks, acting through either optic flow or internal acceleration detectors. Path integration then does not require memory, special or otherwise, of previously seen landmarks. Thus, the possibility that animals make short cuts using path integration must be eliminated before one can conclude from novel short-cutting that an animal has a special form of landmark memory in the form of a cognitive map.

One more example, concerning novel short-cutting in honeybees, demonstrates the importance of methodological details for travelling in this field of cognitive mapping. The question of whether bees can take novel short cuts between familiar sites has been central in the discussion about the existence of cognitive maps in these insects. The failure of bees to show this capacity could be a result of the training procedure, because extensive training to one feeding site may eliminate or weaken memories of other sites that were previously learned. Menzel *et al.* (1998) investigated this problem by rewarding honey bees at two feeding sites, one (S_m) at which they could eat in the morning, and the other (S_a) at which they could feed in the afternoon (Fig. 12.2). The bees were then displaced to S_a in the morning and to S_m in the afternoon either from the other feeding site or from the hive. Bees were also displaced to two novel sites, one at a completely unfamiliar location (S_4) and another that was located halfway between the two feeding sites (S_3) Bees displaced from either of the feeding sites never took novel short cuts; instead, they used the homeward directions that would have been correct had they not been displaced. Bees caught at the hive entrance, however, chose the correct homeward direction not only when displaced to both feeding sites, but also when displaced to S_3, although not from S_4. Control bees that had been trained to only one of the feeding sites were not able to travel directly home from S_3, excluding the possibility that bees use landmarks close to the hive.

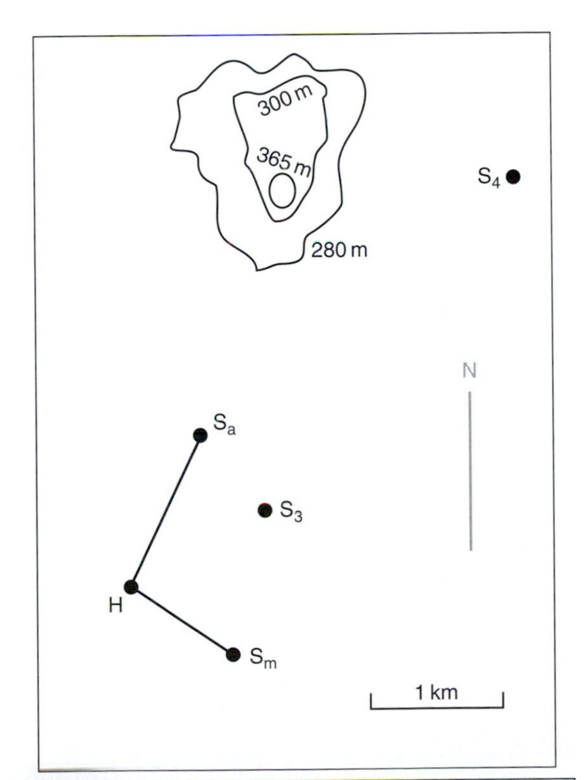

Fig. 12.2 Route integration in honeybees. (Adapted from Menzel *et al.*, 1998.)

These results support the hypothesis of taking a novel short cut by activating two vector memories simultaneously and meet the basic requirement to prove the existence of a cognitive map. However, the authors assessed their results rigorously. They support the notion that what their bees do is an instance of basic cognition, that is, an activation of separately acquired memories and their application in a novel adaptive sense. Whether such a memory organisation is classified as a cognitive map depends on how rich the map is under the concrete circumstances. The experimenters suggest that the inference of a map would be premature if the bees' map only allows them to navigate in one direction (home), and only in some motivation state (hive arriving but not feeder-related motivation). Indeed, there are supportive data concerning bees and wasps which memorise landscape hallmarks when searching for a way home and do not use this information when they search for food source; these insects have to learn afresh under the new circumstances and with a another motivation (Mazokhin-Porshnyakov and Kartsev, 1979, 2000; Kartsev, 1990; Kartsev *et al.*, 2005). More data are needed to demonstrate whether multiple vectors between landmarks, feeding sites and the hive could be stored together with the particular sensory properties of these sites. Anyway, a window of opportunity is still open for honeybees together with other insects to demonstrate that their mentality is rich enough to construct a cognitive map.

Detour tests

Detour tests can be considered parental for a theory of cognitive map. Detour behaviour has been widely studied in many species (review in Chapuis, 1987; Vallortigara, 2004). In this issue we will consider several concrete examples of experimental investigations.

Köhler (1925) was one of the first to study the detour problem. In his experiments a subject was placed behind the bars and could see food being placed on the other side. When dogs were tested, their behaviour depended very much on the distance between the food and the barrier. If the dog was some distance away, then it would immediately run in a smooth loop around the barrier to collect the food. When the food was placed behind a fence, the dog would run directly towards it and stay there even though the food was inaccessible. Köhler tested chickens in a similar way, but they rarely solved the problem. Instead, they persistently tried to reach the food by pushing through the barrier. The chickens were successful only when their attempts to pass through the fence led them by chance to pass around it. Köhler suggested that chickens are not capable of solving detour problems. However, subsequent studies have challenged this view. For example, Etienne (1973) presented 6-day-old chicks with a mealworm that disappeared behind one of two screens. She found that, after repeated testing, all chicks developed searching behaviour behind either screen and a minority of the birds (24 per cent) spontaneously learnt to orient their delayed response directly to the correct screen.

During the last few decades it has been demonstrated that it was not so much cognitive deficit in animals as deficit of motivation and biological significance of stimuli in early studies of animals' detour behaviour that was responsible for their failures. For instance, young chicks provide experimenters with the unique opportunity to test a highly motivated animal that is eager to follow a moving object that was imprinted a few hours after its birth as a 'mother'. This may be a ball, a cardboard triangle or something similarly far from a real bird (see details in Chapter 24). Regolin *et al.* (1995) tested 2-day-old chicks in a detour situation requiring them to abandon a clear view of a desired goal (a small red object on which they had been imprinted) in order to achieve that goal. The chicks were placed in a closed corridor, at one end of which there was a barrier with a small window through which the goal was visible. Two symmetrical apertures placed midline in the corridor allowed the chicks to develop strategies to pass round the barrier. After entering the apertures, chicks showed searching behaviour for the goal and appeared able to locate it, turning either right or left depending on their previous direction of turn. Thus, in the absence of any local orienting cue emanating from the goal, chicks showed an ability to understand the continuing existence of an object that had disappeared and to represent its spatial localisation in egocentric coordinates.

Recent comparative study in three species of birds, belonging to different eco-ethological niches, allows for a better understanding of the cognitive mechanism of such detour behaviour (Zucca *et al.*, 2005). Young quails (*Coturnix* sp.), herring gulls (*Larus cachinnans*) and canaries (*Serinus canaria*) were tested in a detour situation requiring them to abandon a clear view of a biologically interesting object (their own reflection in a mirror) in order to approach that object. Birds were placed in a closed corridor, at one end of which was a barrier through which the object was visible. After entering the apertures, birds could turn either right or left to re-establish social contact with the object in the absence of any local sensory cues emanating from it. Quails appeared able to solve the task, though their performance depended on the type of barrier used, which appeared to modulate their relative interest in approaching the object or in exploring the surroundings. Young herring gulls also showed excellent abilities to locate spatially the out-of-view object. Canaries, on the other hand, appeared completely unable to solve the detour task, whatever barrier was in use. It is suggested that these species differences can be accounted for in terms of adaptation to a terrestrial or aerial environment.

Pongrácz *et al.* (2001) have exploited dogs' attachment to their owners in order to estimate limits of social learning (see Part VIII) with the use of detour tests. The experimenters used a V-shaped fence, 1 m high, with sides 3 m long. The task for the dogs was to get to a piece of food or their favourite toy by detouring along the fence. Thirty dogs were divided between two equal groups that differed in the direction of detour necessary for reaching the target. For the inward detour group the experimenter placed the target behind the V-shaped fence near to the inner side of the angle. After six trials the position of the target and the dog were reversed (outward detour). In this seventh trial the dog was positioned inside the fence and the target was placed at the outer side of the intersecting angle. Dogs in the outward group were exposed to the same test procedure but in the reverse order. As one of the main characteristics, the dog's latency to obtain the target was measured; this was defined as the time elapsed between the owner releasing the dog from the leash and the dog taking the target in its mouth.

The results showed that the dogs performed differently depending on their position with regard to the fence. Learning the detour seemed to be a difficult problem for the dogs from the inward group. In contrast, dogs in the outward group mastered this task much more easily; latencies were significantly shorter after the first trial. The direction of the detours along the fence in the outward group showed strong concordance with the direction of the first successive detour. The inward group lacked such a concordance. This could indicate the difficulty of this task for the dogs; these dogs might have got behind the fence by chance during the first trial, which made it more difficult for them to remember the direction in which they started.

The dogs displayed relatively conservative strategies in mastering detours. Even after several consecutive trials animals could not improve significantly if they started from outside the fence. Dogs were apparently unable to transfer their experience of mastering the task from inside the fence to a reverse detour. Pongrácz *et al.* (2001) have demonstrated that the dogs were not flexible in performing detour tests and that they coped with the problem more easily starting from inside the V-shaped fence than outside it. A possible explanation for this asymmetry lies in stored individual experience that is, in many respects, similar in domestic dogs. It could be, as the authors suggest, that dogs might more often encounter situation in which they had to get out from somewhere (e.g. from a garden), rather than get inside it. Furthermore, in the outward detour there is less ambiguity, that is, any exit is successful. It is also could be that dogs in the outward group had to walk mainly tangentially to the target, but the inward detour needed a long walk away from the target at first. One could also argue that the inward detour situation might generate higher levels of neophobia in the dogs than the outward detour task. However, the fence was an equally strange obstacle for the dogs in both situations. Furthermore, dogs in the inward detour group tried strongly to obtain the target, barking at it and sometimes trying to dig under the fence.

Indeed, it was not without reason that Thorpe (1950) estimated the capability of digger wasps to solve detour tasks as higher than that of dogs. These animals brave fences without wasting time in barking. The most astonishing thing is that the insects demonstrated ability to transpose a picture of their place from a bird's(wasp's)-eye view to a 'pedestrian's'-eye view. In 1950 Thorpe did some experiments which suggested that digger wasps rely on a cognitive map of their large-scale environment to find their way to the vicinity of their nest from wherever they happen to capture their prey. He worked with the species of digger wasp that preys on caterpillars too large to be carried home in flight. A female of this species drags her prey home across the gravelly ground. In his experiments Thorpe manipulated a wasp dragging its prey homeward. He placed obstacles in its path that forced the wasp to deviate from its course. As soon as the wasp had cleared each obstacle, it resumed its course. This means, as Gallistel (1998) stated, that the wasp was not marching blindly along like a wind-up toy tank, without regard to where the marching was actually getting her; instead, when forced to deviate from the course, she corrected for the deviation she had made.

Extrapolation

Animals' capacity to make detours could be considered a particular case of a more complex ability to extrapolate the trajectories of a moving objects. Thompson Seton (1898) described a raven that lost a piece of bread over a river. Being entrained by a stream, the bread disappeared within a tunnel. After a short glance into the tunnel, the raven flew round it, waited till the bread appeared and picked it up. Every happy owner of a puppy or a kitten can add personal anecdotes to this story. Usually young dogs and cats do not attack a cupboard when a ball rolls under it; instead they prefer to wait for the ball at the other side extrapolating the trajectory of their favourite toy.

In his experiments with chimpanzees Köhler (1925) simulated such a situation. For example, in one of experiments a basket containing fruit was fixed by a long rope. An experimenter set the basket in oscillatory motion in such a way that at a certain moment it passed near one of a couple of rafters. As soon as the animal caught sight of this, it jumped on the rafter and then waited for the basket with its hands extended. In the second experiment a chimpanzee entered a room in which a window was closed by shutters. Before the animal's eyes the experimenter opened the shutters, threw a fruit out and closed the shutters again. Instead of attacking the window, the chimpanzee immediately started to move in the opposite direction. It ran out into the garden and began searching for the fruit under the window. In both situations the chimpanzee's behaviour could be explained by animal's ability to grasp a problem and extrapolate trajectories of desirable things when they are moving.

The term *extrapolation* for explaining of animal behaviour was first suggested by Matthews (1955) who hypothesised that pigeons extrapolate the movement of the sun for navigation. L.V. Krushinsky (1958, 1977) was the first to consider extrapolation to be a key element of reasoning in animals. Krushinsky and his colleagues and students carried out many experiments on different species estimating their ability to extrapolate trajectories of objects disappearing from the animal view (corridor test, screen test, etc.: Krushinsky, 1990).

The tests in these series differ from the detour tests in which a subject can observe only a segment of the trajectory of the desired object (Fig. 12.3). For example, in a series of 'corridor' tests an animal needed to extrapolate the trajectory of a bowl with food moving in the opposite direction relative to the animal's view. First, on the path to a tunnel a bowl was open so that it could be followed and food taken from it. Then it went into the opaque tunnel which was immediately shut in order to prevent the subject from seeing the following movement of the bowl. These experiments were conducted with rabbits and several avian species. Rabbits, hens, ducks and pigeons persisted in searching for the bowl at the place where it had disappeared just before, while corvids manifested the ability to extrapolate. It was clear at the second stage of experiment when the tunnel contained two bends with a gap between them. Magpies, after they caught site of the bowl in the gap, hurried to the end of the second bend where the bowl was expected to appear. Standing there in a tense posture, they were waiting for food.

The 'screen' tests were so devised as to provide the subject with information about the

Fig. 12.3 A fox engaged in an extrapolation test: a screen with complicated device. (a) The fox is going out from the start, from where it had the opportunity to observe the trajectory of a bowl; (b, c) the fox passes round the corner without entering the wrong chambers; (d) the fox finds the bowl and eats the food. (Photographs by O. Bolovinova from the archive of Krushinsky's laboratory; courtesy of Z. Zorina.)

Fig. 12.3 (cont.)

trajectory of a desirable stimulus (such as a trough with food or a toy) by the possibility of looking at the stimulus through a narrow vertical slit in the centre of the opaque screen. When food served as a stimulus the animal was allowed to eat from the trough through the gap in the screen and then the filled trough was moving in one direction while an empty trough was moving in the opposite direction in order to prevent an animal to orient by sounds. These directions changed in different trials. The subject had to solve a problem based on its observations on how the trough moved when it could be seen through a narrow vertical slit. To solve this problem an animal must realise that the food bait, which has disappeared from its view, continues to move in the same direction as before, i.e. to perform extrapolation of the movement direction of the invisible food bait and, based on this knowledge, to go round the correct side of the screen. Therefore, the animal had to decide from what side it should run round the screen in order to obtain food. Experimenters complicated the screen tests by adding screens. In a simple case the added screen forced the animal to deviate from the course toward the target by moving

Fig. 12.3 (cont.)

athwart the traffic route and then to resume its course. In more complex cases the animal had to run round several added screens. This task requires the ability not only to extrapolate but also to sum vectors.

After the first tests on dogs, a wide range of species was studied. Comparative studies revealed distinct tendencies concerning species-specific abilities to extrapolate in animals. They ranked as follows: voles and mice displayed almost complete inability to solve even simple extrapolation tasks; then followed rats, rabbits, cats, dogs, foxes, corsacs, polar foxes, racoons, wolves, dolphins and apes (Krushinsky, 1965; Firsov, 1977). Among avian species only corvids manifested a high level of capability while chickens, ducks, pigeons, kites and the honey buzzard did not cope with extrapolation problems. In some birds the number of correct choices gradually increased; that is, they become trained for performing these tasks. Surprisingly, reptiles (including three species of tortoises and one species of lizards) were champions in many tasks. Similar experiments were conducted on human infants whose attention was called to a toy in motion. It turned out that 2-year-old children behave a good deal like rabbits. They were eager to get the toy and refused to leave the place where it disappeared. Six-year-old children were able to solve the most complex tasks in a majority of trials, and they reached the 100 per cent level of capability by 7.5 years.

Recently detour tests for measuring animals' ability to extrapolate the trajectory of a shifting object have been implemented on a computer for those animals that could watch the movements of a dot on a monitor and react in an adequate manner. Both apes and monkeys displayed the capability of extrapolating the trajectory of a dot on a monitor with the same level of accuracy as adult humans. It is interesting to note that when primates learn to mentally track a dot moving on the monitor, say, from left to right, they easily transfer this accomplishment to doing the same with a dot that moves up and down (Washburn and Rumbaugh, 1992).

12.3 Is a treasure map cognitive? Just ask a wild explorer to inform conspecifics

One possibly effective method for revealing cognitive mapping in animals is to ask them to transfer information to each other about how to reach a 'treasure' by a short route. This ability needs special properties of intelligence in general and 'social intelligence' in particular. That is, the first problem is whether an explorer and a recipient are able to encode the transferred information. When Little Red Riding-Hood explained to the Wolf how to reach Grandma's house she naively described her own path. The Wolf definitely possessed a cognitive map and this enabled him to convert Little Red

Riding-Hood's naive description into an optimal route and thus to run up to the goal much faster than had been expected.

It seems that only a few species can be candidates for participation in such a project. To join this crew, animals must be not only clever but also 'altruistic' enough. The relevant species are those that display a tendency to share food with members of their society. Moreover, they should possess a reliable means of communication to do this. Menzel's experiments with chimpanzees cited above could serve as a good example. In a series of experiments apes not only successfully found hidden food and toys but also informed their mates about the hiding-places. What is of most interest is that animals who were 'informed' came by most direct trajectories, and not by being escorted by knowing leaders. It still remains a challenge to reveal what means of communication chimpanzees use in such situations. We will discuss problems of communication and characteristics of animal communities in Parts IX and X. Here two examples will be considered concerning social hymenopterans whose 'altruistic' nature is above suspicion.

Bees in the middle of the lake

Bees navigate using 'maps' with landmarks, as well as orientation from the movements of sun and from magnetic data. When captured and transported to a new area within their foraging territory, they are able to return directly to their hive, or to a coveted source of sugar-water. The sophistication of the bee's cognitive map is shown in an experiment in which a researcher placed a source of sugar near the shore in a lake. The bees soon learned its location and communicated it to the other bees in their hive. Little by little the researchers placed the sugar further into the lake. Bees transported to this sugar source later communicated their find to their hive mates, but the other bees apparently did not believe the story, since none reacted. The researcher explained that normally a bee would not expect to find a sugar source in the middle of a lake, and so the bees must have thought the bee's story too fantastic. This experience suggests that bees possess fairly sophisticated cognitive maps that permit them to join different kinds of information in order to draw conclusions (Gould, 1986).

In the experiments of Dyer (1991), foragers were trained to collect food from a boat in the middle of the lake. On returning to the hive they were unable to attract any recruits with their dance to indicate were the food was located. When the boat was moved to the shore station, however, returning foragers were able to mobilise recruits to collect food. According to Gould (1984), the explanation of these findings is that bees possess a map of the area around the hive, and when the direction and distance of food is communicated to them by a forager, they identify its position on this map. Because food, for bees, is never found in the middle of lakes, it would be foolish to leave the hive when the map indicated this to be its predicted location, On the other hand, when food is indicated as being on the far side of the lake, then it makes good sense to search for it. Although the existence of cognitive map in bees is still under discussion, and Dyer himself named his 1991 article 'Bees acquire route-based memories but not cognitive maps in a familiar landscape', these data provide an argument that bees use cognitive mapping in their social life.

Ants on the Cartesian grid

In the experiments of Reznikova and Ryabko (1994, 2001) red wood ants were requested to pass to their nestmates information about coordinates of objects in a lattice shaped as different variants of Cartesian plane grid (Fig. 12.4). Group-retrieving ant species have a similar system to honeybees' dance language, in that scouts transfer symbolic information to foragers based on distant homing. This sophisticated communication will be considered in detail in Chapter 34. What is important here is that in these experiments scouting ants passed the information concerning coordinates of a trough with syrup on the lattice and foragers reached the 'node' with the trough each by its own pathway with the rate which essentially differed from those in control, naive ants. It is important to note that each

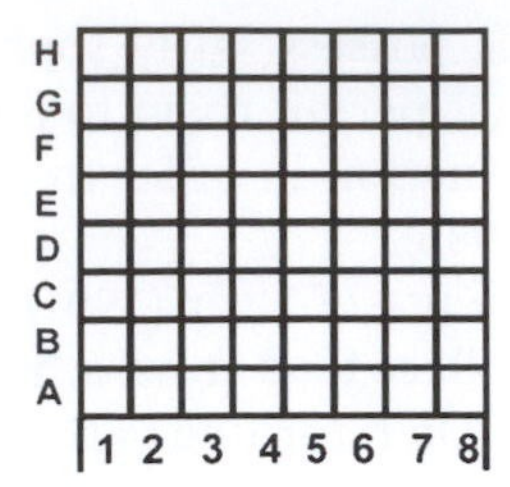

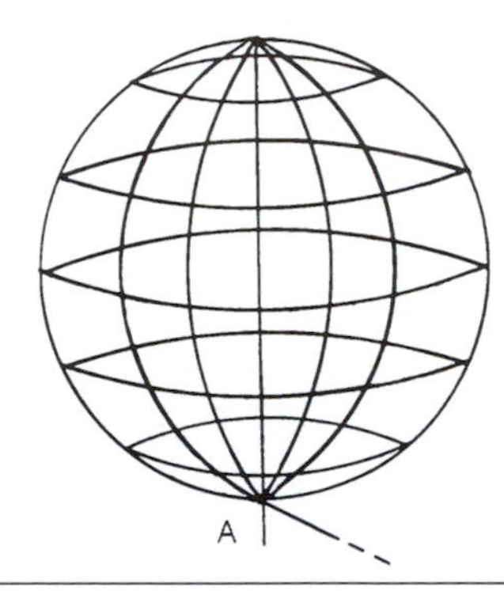

Fig. 12.4 The Cartesian coordinate set-ups ('lattice' and 'globe') used by Reznikova and Ryabko (2001) to study ants' ability to memorise and transmit information regarding the location of a trough containing food on the plane.

series of experiments was preceded by a training stage as long as 4 weeks, during which the ants explored set-ups and got accustomed to the work regime. This enables the authors to suggest both scouts and foragers use representation for successful coding of information concerning coordinates of objects.

In the cited experiments ants lived in a laboratory arena, in a transparent nest. The arena was divided into two sections: a smaller one, containing a laboratory nest, and a bigger one with an experimental system (see details in Chapter 34). Both sections were connected by a plastic bridge that was from time to time removed to modify the set-up or to isolate the ants. Each of the colonies consisted of about 2000 ants. All experimental ants were marked with coloured labels and fed with sugar syrup only in the experimental set-up once every 2–3 days. As soon as a scout found food, it informed the members of its foraging team. After that the scout was removed and the researchers watched its team. The experiments were so devised as to eliminate all possible ways that might be helpful in finding the food, except distant homing, i.e. information contact with a scout. During contact between a scout and foragers in their nest, the experimental apparatus was replaced by a similar one with all troughs empty to avoid the use both of an odour track and the food odour.

The Cartesian coordinate set-up is a flat grid of thin metal rods (8 by 8, 6 by 6, 5 by 5 lines: Fig. 12.4). Every set-up was fixed so that the ants could only reach the starting point of the maze across a small bridge. Further movements were possible only along the lattice rods. During the experiments the trough containing food was placed at various nodes of the lattice, while on the other nodes the empty troughs were placed. In total, five scout–foragers groups worked in two set-ups. The teams abandoned the nests after they were contacted with scouts and moved towards the trough 40 times (note that the scouts were specially removed). In 31 cases the team found the correct path to the trough in less than 5 minutes.

In order to test whether the foragers actually are guided by the information obtained from the scouts, a series of control experiments was carried out in which all of the ants were given the opportunity of access to an 8 by 8 lattice, on one of the nodes of which the trough with syrup was found, and were able to walk about it for 15 minutes (let us recall that in the experiments the scout was purposely placed on the trough) (see Reznikova and Ryabko, 2001). A result in which during the allotted time (15 minutes) at least one ant found the food, was considered positive. No one forager found food in 24 cases out of 30.

Two statistical hypotheses were considered:

H_0: the probability of finding the food in which the foragers obtained the information from the scouts is the same as in the experiments with free search for the food

The alternative hypothesis H_1: the probability of finding food by 'directed' foragers is greater than in cases of free search for the needed point.

In order to test H_0 against H_1 the χ-square test was used. Computations show that hypothesis H_0 is rejected in favour of H_1 at a significance level of 0.001. Thus, it follows from the data presented that ants are capable of transmitting the information about the coordinates of objects.

These results provide evidence that red wood ants are able to memorise and transmit information regarding the location of a trough containing food in a plane. At this stage the authors do not propose any hypothesis about the mechanisms of coding or representation of such information in ants' communication.

Chapter 13

'Object permanence' in animals

13.1 Experimental paradigm to study object permanence

Indeed, when happy parents play hide and seek with their newborn infant and discover that their child, when it matures, becomes aware of where a hidden - and even rehidden several times - toy can be found, they need not to refer in an academic manner to Piaget's developmental theory. Meanwhile, in accordance with the Piagetian paradigm, an individual, be it animal or human, who understands object permanence, knows (1) that objects are separate entities that continue to exist when no longer in view and (2) that these items have physically distinct identities and properties, fixed in time and space, that are unaffected by simple movement (Piaget, 1932). According to Piaget, such abilities reflect a capacity to form a mental picture of an object, and to understand that its existence continues independently of the observer - fairly sophisticated abstract reasoning. This understanding is not a basic inherited property of a living organism. In his experiments with children, Piaget showed that newborn infants lack object permanence, that the concept develops in six major stages, and that each stage represents a different level of understanding.

At Stage 1 subjects do not search and even look for an object they have watched disappear from sight. As subjects mature, eventually, they pursue such an object: at Stage 2 by tracking its movement; at Stages 3 and 4, respectively, by recovering a partially and a fully hidden item exactly where they see it disappear. At Stage 5, subjects retrieve an item that has been hidden successively in several locations - hidden, exposed and then hidden again several times. At Stage 6 a subject musters invisible displacement. For example, when an object is hidden in a container, the container is moved behind another container which hides the first one from view, and the item is transferred to the second container. A subject is shown that the first container is empty and is able to infer where the item is now located. In experiments various controls are used to prevent subjects from responding to odour cues or hand or eye movements by the experiment. According to Piaget (1937), Stage 6 object permanence indicates the individual's ability both to assimilate and to use available environmental information.

In humans, development of object permanence seems to occur over a child's first two years. This process correlates with brain maturation and language development. Acquisition of words representing the concept of disappearance appears to be correlated with reaching Stage 6 object permanence. Many data obtained on children suggest that complex object permanence tasks require particular types of cognitive ability such as handling abstraction and self-regulating behaviour (Gopnik and Meltzoff, 1986; Adrien *et al.*, 1995; Tomasello and Farrar, 1986).

For animals, the extent of development of object permanence and how it should be tested are less clear and intensively discussed recently (Chevalier-Skolnikoff, 1976; Pepperberg, 1999).

Building on Piaget's concept, Uzgiris and Hunt (1975) elaborated relevant tasks for the comparative study of object permanence which have been applied for a wide variety of species. Uzgiris and Hunt's procedure includes the scale that directly measures competence level of a subject via 15 tasks of increasing difficulty; success in specific subsets of tasks correlates directly with Piaget's six stages.

Tasks 1 and 2 examine a subject's visual capacity to follow a slowly moving and then a disappearing object (Stage 2). A subject succeeds in Task 1 (arc track) if it tracks an object (determined by eye gaze, head position) smoothly through a complete arc. To succeed in Task 2 (object disappearance), a subject must remain gazing at the point of disappearance or return its glance to the starting point.

The next set of tasks involves simple visual displacement. Tasks 3 (partial hiding) and 4 (complete hiding) test Stages 3 and 4, respectively: the ability to retrieve an object that is partly and then completely hidden under a single cover. In Task 3, a subject can succeed by pulling either the item or the cover. In Task 4, a subject must remove the cover; pulling an item from underneath the cover does not count.

Stage 5 requires success in many different tasks. Only the beginning of Stage 5 is tested with Task 5 ('two sites, visible'). In Task 5, two hiding sites are available; an object is visibly hidden in the same place for two or more trials before being hidden in the second site; the task tests whether a subject recognises when the site changes. Young children commonly fail: this failure is called 'A-not-B error'. In their bird studies, researchers used differently marked covers, placed the covers in front of birds, and then hid the object.

Tasks 6 and 7 are also part of Stage 5. In Tasks 6 (two sites, alternately visible) and 7 (three sites, alternately visible), the hiding place varies respectively between two and among three covers. These tasks test how well subjects track additional slight complexities in hiding. In both tasks, subjects must search only under the correct cover.

Tasks 8 and 9 involve more complex manipulations of hidden object, but still test Stage 5. In Task 8 (three sites, successively visible), an object, in full view in the investigator's hand, is passed under successive covers until it is hidden in the designated site. The task tests a subject's ability to track object movement and search where the item has most recently been seen. Success requires searching only the final location; searching the other locations, even if the search replicates the order of hiding, is incorrect. The final Stage 5 tests, Task 9, measures persistence in obtaining the hidden item; a subject must remove three superimposed covers to obtain the object.

Stage 6 also includes several tasks, and all must be successfully completed for full competence. Many subjects succeed on only some tasks, and thus do not achieve human levels. All Stage 6 tasks prevent subjects from directly viewing the transfer of the hidden item between covers: subjects must infer the occurrence of such a transfer. Inferential, rather that perceptual, abilities are thus tested (Pepperberg and Funk, 1990). The first Stage 6 tasks, Tasks 10–13, are similar to Tasks 4–7, except that the object is not seen during transfer. The item is, instead, hidden in a small container or in the experimenter's hand – the 'implementing cover': the subject sees the implementing cover, but not the item, move to one other screen (Tasks 10, 11) or a choice of two (Task 12) or three (Task 13) other screens, and the experimenter surreptitiously transfers the object. Only then is a subject shown that the implementing cover is empty. To succeed, a subject must search directly under the correct screen. Task 14 (successively invisible) is similar to Task 8, but the object is not seen after it is placed in the implementing cover, which the investigator then moves between screens. To succeed, a subject may search either the last cover in the past or all covers in order of hiding.

Task 15 is a trick. An item is hidden, as in Task 14, but under an intermediate screen as the investigator's hand continues to move under the other screens; the subject is led to believe the object is under the final screen. The task thus tests whether a subject goes to where the object logically should be found.

Task 16 is called 'substitution' or 'violation of a rule' (LeCompte and Grach, 1972; Pepperberg *et al.*, 1997). This tests whether the subject has a representation of the specific hidden object and does not simply search for 'something of interest'. For testing this ability, the experimenter repeats Task 14 but surreptitiously hides an item less favoured than one that the subject initially observed. In fact, Tinklepaugh (1932) used this technique in his experiments with chimpanzees in the delayed reaction test.

Task 17 is also called Task S, as a form of shell game. This technique has been applied to infants (Sophian, 1985), as well as to cats (Doré *et al.*, 1996) and birds (Pepperberg *et al.*, 1997). An item is visibly hidden under one of three covers and the experimenter then visibly exchanges the position of this cover with one or two others. Success in Task 17 may or may not be an acceptable criterion for Stage 6 because it also tests attention, different types of memory and spatial cognition, but the task provides data for cross-species comparison.

13.2 Walnut-sized brains master object permanence tasks: insights from grey parrots

Let us consider the detailed experimental study on grey parrots carried out by Pepperberg and her colleagues (Pepperberg and Kozak, 1986; Pepperberg, 1990, 2002; Pepperberg *et al.*, 1997). Studying parrots, researchers primarily used all 17 tasks described above. To hide the objects, experimenters used plastic boxes, cups, toy barrels and crumpled pieces of paper towel and construction paper to hide the objects. The items hidden included small toys from the parrot's play sessions and a few foods.

The first series of experiments was conducted with Alex, the most famous grey parrot in the world, whom we will meet in Chapter 31 devoted to language-training experiments. Alex possesses outstanding language-related skills but here we are interested in his ability to solve object permanence problems. Alex was 8.5 years old when researchers began their object permanence studies, so they could see what study of object permanence an adult grey parrot might achieve (see Pepperberg, 1999 for details).

Alex succeeded on all 15 stages and these were the first data to show avian competence at such a high level. Also, he reacts in some tasks in an interesting, 'Alex-specific' manner that was caused by his specific experience gained from language-training experiments based on direct intelligent dialogue between him and experimenters about the nature of things (see Chapter 31). For example, being presented with Task 5 with two cups, Alex immediately said 'Two'. People told him that yes, there were two cups, and asked him to get nuts. In two trials with nuts hidden under yellow cup, Alex chose the red, experimenters switched to red, he chose the yellow. The most conservative interpretation was that Alex could not move beyond Stage 4, but the researchers suspected that his failures might be due to inexperience in two-choice tests with physical objects. Until then, Alex's studies had never involved physical choice between objects. During numerous language-training tests Alex was commonly shown one object and asked for information about that item; choice was among numerous possible vocalisations, not between physical items. To see whether Alex's Task 5 failure could be due to lack of relevant experience rather than a conceptual deficit, the experimenters proceeded to Task 9, which did not involve multiple choices. Success in more advanced tasks implies understanding of previous ones (Uzgiris and Hunt, 1975). The parrot immediately removed three superimposed covers, thus suggesting that his Task 5 failure was not conceptual.

At final stages, occasionally, at the end of the session, rather than choose a screen himself, Alex requested that people have to 'go pick up cup'. On Task 15 he not only immediately overturned the cover where the object logically should have been hidden, but also emitted 'Yip' (which he uses when startled) after his obvious failure to obtain the expected reward. On subsequent Task 14 trials, he occasionally knocked over all covers simultaneously with a single motion of his head, often before the experimenters completed the displacement. Such occurrences were listed as mis-trials but, as Pepperberg writes in her book (1999), the experimenters could not help thinking that Alex had devised a 'Gordian knot' approach to solving this task.

Researchers knew now from experiments with Alex that adult grey parrots are capable of completing complicated object permanence tasks. Their question now was: 'How long would it take a baby grey parrot to progress through all six stages?'

The baby grey parrot Griffin, who entered the research team when he was barely 2 months old, helped Pepperberg to answer that question. Though Alex had participated in a previous study, he was included in one task and several control trials for this project. At 8 weeks old, possibly younger, Griffin was capable of Stage 2 object permanence. He also searched for hidden objects at 8 weeks (Stage 3). At 9 weeks he reacted to the 'object loss' (i.e. hiding of the object) which indicated Stage 4, but could not retrieve it. He did not complete Stage 4 until he was 15 weeks old. It is possible that he initially was not physically capable of removing the cover or that he was distracted. He was learning to fly and spent a lot of time practising during the testing sessions. Griffin reached Stage 5 at 18 weeks, and at the ripe old age of 22 weeks, he graduated to Stage 6.

In the shell-game test (Task 17) Griffin never hesitated and seemed to track the experimenter's hand very closely, and he almost always knew where the object was. Alex, who was then 19.5 years old, also passed this test easily. That suggests that grey parrots, unlike dogs and cats but like humans and great apes, develop a robust sense of object permanence.

Were the grey parrots simply 'following the action' as the experimenter moved objects and containers? Were they simply focusing on the 'arena of activity'? Or did they actually search for the object? This question was examined with both Griffin and Alex passing through Task 16.

This task used successive invisible displacement in which Griffin sees the object that is initially hidden but cannot see that object transferred to several successive containers. He must infer the location of the object. It is worth noting that the grey parrots behaved very much like the chimpanzees in Tinklepaugh's (1932) experiments in the delayed reaction test (see Chapter 10). They reacted with surprise and anger when the hidden item is other than the expected one. The standard argument is that animals have a representation of what is hidden, and react to the difference (the so-called *cognitive dissonance*) between the observed item and its representation (Pepperberg *et al.*, 1997; Pepperberg and Lynn, 2000). For example, experimenters showed Griffin a cashew but then hid a bird diet pellet. Griffin immediately uncovered the last box, turned it over and discovered the pellet. Then he turned over all the other (empty) boxes, found nothing and ran to the experimenters. This was repeated with the same results. But when they really hid the cashew, Griffin found it immediately and didn't bother searching any further. When Alex found the pellet, he stared down the experimenters and narrowed his eyes to slits. The experimenters knew well that this slit-eye look is a final warning before Alex bites if the cause of anger is not removed. When they repeated the test Alex banged his beak on the table.

What do results in the tests on object permanence mean in the big picture of the parrots' life? Until someone is able to study wild African grey parrots comprehensively from infancy to adulthood, we may never have a definitive answer. But Pepperberg did get some clues from parrot breeders. Apparently Stage 2 is demonstrated by very young parrots: nestlings follow feeding syringes at 2–3 weeks, essentially as soon as their eyes open. By 18 days, they follow syringes through a 180-degree arc; by 20 days, they track until a syringe goes out of sight and cheep as soon as it disappears. Because a baby grey parrot in the nest must compete with its siblings for parental attention (food), the ability to track the food source enhances survival.

13.3 Comparative studies of object permanence

Tests based on Piagetian staging have been tried on a wide variety of animals. Speaking as to who are the champions, it may be postulated that most never make it to Stage 6: monkeys, cats, doves, chickens and hamsters never figured it out. Only great apes, parrots and possibly dogs passed the corresponding tests. It is interesting to note that great apes, human infants and psittacids (parakeets and grey parrots) make A-not-B errors, while dogs, cats and magpies manage this test (Wood *et al.*, 1980; Funk, 1996). Different interpretations compete based on ideas about differences in the functional organisation of memory in these organisms (Diamond, 1990); somehow or other, this is an intriguing grouping of species. In this context, it is worth noting that black-billed magpies *Pica pica*, food-storing passerine birds, do not make A-not-B errors, like dogs and cats, but contrary to parrots and humans (Pollok *et al.*, 2000).

The age at which different stages of object permanence develop and ultimate competence varies with species (Gagnon and Doré, 1994; Pepperberg, 1999). This variation may reflect different rates of physical or cognitive maturation, which may involve genetic predisposition, that is, evolutionary pressures exerted over time (see Part VII for details), environmental effects, or some combination of factors.

Great apes generally reach the earliest stage more rapidly than humans (i.e. Stage 3 by ~4 months), plateau temporally at Stage 5, and reach Stage 6 about the same time as humans (Redshaw, 1978; Mathieu and Bergeron, 1981). Monkeys develop faster than apes, but slow down as tasks increase in difficulty and fail Stage 6 tasks (Spinozzi, 1989; De Blois *et al.*, 1998). At the same time, at least one New World monkey species, cotton-top tamarins (*Saguinus oedipus*), possess Stage 6 object permanence capabilities. In a situation involving brief exposure to tasks and foraging opportunities, tracking objects' movements and responding more flexibly are abilities expressed readily by the tamarins. Cotton-top tamarins were tested in visible and invisible displacement tasks. All subjects performed at levels significantly above chance on visible ($n = 8$) and invisible ($n = 7$) displacements, in which the tasks included tests of the perseverance error, tests of memory in double and triple displacements, and 'catch' trials that tested for the use of the experimenter's hand as a cue for the correct cup. Performance on all nine tasks was significantly higher than chance level selection of cups, and tasks using visible displacements generated more accurate performance than tasks using invisible displacements. Performance was not accounted for by a practice effect based on exposure to successive tasks (Neiworth *et al.*, 2002).

Cats and dogs develop more rapidly than monkeys, reaching Stage 5 at 5–7 weeks. Cats, however, develop no further, but dogs muster standard invisible displacement after several additional months. Dogs appear to fail Task S (Doré *et al.*, 1996).

Temporary plateaux in mastering object permanence tasks correlates with behavioural and cognitive milestones in different species. Dogs' Stage 5 plateau occurs at weaning (Gagnon and Doré, 1994). Children's final stages of object permanence emerge only as words involving movement and disappearance are acquired (Tomasello and Farrar, 1986).

It is important to note that, as in many other learning studies, animals display essential individual variability in succeeding at Piagetian tasks. For example, in experiments of Mendes and Huber (2004), a series of nine search tasks corresponding to the Piagetian Stages 3–6 of object permanence were administered to 11 common marmosets (*Callithrix jacchus*). Success rates varied strongly among tasks and marmosets, but the performances of most subjects were above chance level on the majority of tasks of visible and invisible displacements. Although up to 24 trials were administered in the tests, subjects did not improve their performance across trials. Errors were due to preferences for specific locations or boxes, simple search strategies, and attention deficits. The performances of at least two subjects who achieved very high scores up to the successive invisible displacement task suggest that this species is able to represent the existence and the movements of unperceived objects.

Dimensionality problem-solving

For studying the ability of organisms to form a mental picture of an object Krushinsky (1965) developed the method of examining geometric reasoning which involved 'dimensionality tests': testing the ability of animals to solve problems based on the fact that voluminous food bait can only be placed into a three-dimensional and not into a flat object. In principle, this task satisfies the requirements of Stage 6 according to Piaget and it is very much like Task 10 described above (a hidden object is 'packed').

Krushinsky arrived at this idea when observing the behavioural tactics of his hunting dogs and playing with them. For example, the fox terrier Afka was very much motivated to fetch a thrown stick but he did not react to a thrown rolled coat. The owner packs a stick into a coat behind his back and throws it with his left arm keeping his right arm behind his back. The dog does not rush for a thrown object, stands in front of the owner and looks up at him. The owner then extends his right arm in order to demonstrate to the dog clearly that both hands are empty. Immediately Afka rushes to the rolled coat, rakes over it, takes the stick out and brings just the stick back to the owner. In this situation, Krushinsky said, the dog must form a mental picture of the object that includes such object's properties as to keep its permanence and to be put into another three-dimensional object and thus move with it being hidden inside (*holding capacity* property). He elaborated the experimental scheme based on *geometric reasoning* that included dimensionality tests (Krushinsky, 1977, 1990). To solve the dimensionality problem successfully, that is, to represent that, say, a bait, which can no longer be seen, does not disappear at all (object permanence) but can be put into another three-dimentional object and move with it (holding capacity property), subjects must be able to perform at least the following operations:

(1) To evaluate and then remember the geometric parameters of a bait and other objects.
(2) To compare metric characteristics of the bait and other objects and decide where the bait is hidden.
(3) To remove the voluminous object and take possession of the bait.

To test these abilities, the dimensionality test was carried out as follows. A food-cup was shown to a subject, and then it was separated from the subject by an opaque screen. Behind this screen (i.e. out of sight of the subject) a food-cup was put inside a three-dimensional object (for example a cube) and then placed onto one of the two demonstration platforms. A two-dimensional flat object was placed onto the second platform, and this two-dimensional figure was the frontal projection of the three-dimensional one (a square in this case). Then the opaque screen was removed enabling the subject to see both objects (two- and three-dimensional) moving in opposite directions. Both objects revolved around their own axis so that a subject could evaluate their dimensional properties. In all tests about 30 pairs of different objects were used. In order to ensure the maximal possible novelty of each trial, all three-dimensional objects as well as the respective two-dimensional ones differed from each other in colour, shape, size and structure (Fig. 13.1).

Although it was Afka the dog who put ideas about animals' holding capacity into the experimenter's mind, dogs commonly did not grasp a logical sense of the dimensionality test. On average, dogs need about 40 trials to learn that the right choice is the choice of a three-dimensional object. If the conditions to which the dog is accustomed are changed, for example, if a little object is raised above a platform, the dog has to learn how to solve the problem afresh. Unlike dogs, wolves, monkeys and bears do not need preliminary training for choosing volumetric objects and they are not put off by any changes of conditions.

Fig. 13.1 The wolf chooses a bait that has volume, rather than a flat one. (Photograph from the archive of Krushinsky's laboratory; courtesy of Z. Zorina.)

In experiments with dolphins a ball was used as the attractive bait. The ball was hidden inside a volumetric object, while a contrasting object was a flat one. A dolphin had to press a bar in order to move one of two objects. A three-dimensional object thus overturned, and the ball slipped out so that an animal could play with it. The dolphins successfully solved this problem (Krushinsky, 1977, 1990). Zorina (1997) has applied the dimensionality test to crows, rooks, jackdaws, ravens, magpies and a jay. Successful testing of Corvidae for their ability to solve the dimensionality problem suggested that they are also capable of solving other problems based on the operation of the representation of object permanence and the notion of the holding capacity properties. For this purpose, a new test was designed in which two three-dimensional objects were used differing substantially in their volumes. Both objects could hold other objects, but only one of them was large enough to contain the food-cup. To solve this problem, the birds must not only evaluate both objects with regard to their dimensionality, but also make a quantitative comparison of their sizes. In this context, the test with two voluminous objects may be regarded as a combined one requiring the operation on two attributes of stimuli, that is, geometrical and quantitative. While solving this problem, individual birds showed considerable variability in their behavioural patterns. A least, in seven out of 20 birds the choice of the larger voluminous object exceeded 70 per cent, with an average of 87 per cent. This suggests a relatively high potential of Corvidae in solving geometric and quantitative problems. It is interesting to note that the more successful were individuals in the dimensionality tests, the better were the results of the test with two voluminous objects. At the same time, since this problem was correctly solved by a relatively small proportion of birds, one may conclude that for birds this represents a greater degree of difficulty than the common dimensionality test.

CONCLUDING COMMENTS

Spatial intelligence in animals is based on the use of multiple cues and they use a wide variety of methods for navigation. Many species learn the same or slightly changing home ranges during their whole life, whereas others undertake risky voyages into unexplored lands, at distances from several centimetres to hundreds of kilometres.

There is a growing body of evidences that animals can solve spatial problems without formation of 'truly cognitive' maps. Moreover, it is not necessary that human navigation depends on internalised versions of geographical maps. Instead, human spatial intelligence can be based on similar mechanisms to those described in many other species, such as path integration, place recognition and reorientation. What is important to note is that all these mechanisms have cognitive implications. Animals relying on path integration continuously update vectors describing their distance and direction to one or more significant locations as they move. Animals using place recognition form snapshot views so that the visual image of the environment is about the same each time an animal approaches the specified location. Reorientation is usually based on memory about geometric features of the environment, and can also be influenced by non-geometric characteristics.

In some species the use of geometric information in piloting processes, as well as their abilities to memorise many locations and landmarks, become apparent early in ontogeny, and can be considered part of essential adaptations. At the same time, a great deal of flexibility is demanded from space explorers in order to navigate changeable environment, in particular, to select novel routes and to apply short-cutting when making detours around obstacles. A portrait of an intelligent space explorer can be complemented by the ability to extrapolate trajectories of moving objects and development of object permanence, that is, forming mental pictures of movable and changeable things. Many species as well as human infants have serious limitations in their capability for object permanence and geometric reasoning, so wide comparative studies are required to complete our knowledge about the specificity of space representation in wild and human minds.

Part V

Experimental approaches to studying essential activities of animal intelligence

Good God! I should have known it! Destiny will play her little tricks, and all jokes have their cosmic angles. He is a psychologist! Had I given it due consideration, I would have realized that whenever you come across a new species, you worry about behavior first, physiology second. So I have received the ultimate insult – or the ultimate compliment. I don't know which. I have become a specimen for an alien psychologist!

James McConnell, *Learning Theory*

The main purpose of this Part is to explore the question to what extent living things understand their relationship with the world and to what extent we can measure their understanding. Binet and Simon (1905), the pioneers of intelligence testing, considered that 'to judge well, to comprehend well, to reason well, these are the essential activities of intelligence' (I refer to Gottfredson (1998) to complete a picture of how this method has been developed for studying humans). In this Part we will be trying to clarify what 'well' means in this context. To do this, we need to summarise experimental methods elaborated for studying displays of intelligence in animals. As we will see in Chapters 14 and 15, experimental paradigms for testing reasoning in animals are mainly based on a procedure of differentiation. Despite the methodological simplicity of this procedure, it allows the examination of rather complex forms of reasoning and comprehension, right up to abstraction and categorisation.

A central problem of this Part is whether animals are logical and whether they possess conceptually mediated behaviour that permits them to judge about relations between things and to adjust their behaviour to novel objects and events by virtue of membership in an already familiar class. To approach this problem, we will consider animals' abilities for categorisation, concept formation, abstraction and classification, as well as for matching the relation between relations, in other words, for analogical reasoning. Among others, such concrete features of intelligence will be analysed as imagery, i.e. the ability for mental representation which implies generalisation of a familiar image to its visual transformations, as well as the ability to put objects in order and to memorise them as a whole sequence. All these advanced types of learning can be considered as highly developed cognitive skills in animals.

Early attempts to measure animal intelligence go back to the beginning of the twentieth century. Descriptions of the experimental results on learning in rats obtained with the help of the first laboratory labyrinth (Small, 1900, 1901) enable students of animal learning to use mazes for the comparative study of intelligence in animals. Thorndike's puzzle-box and Skinner's chamber met with enthusiasm from comparative psychologists in their attemps to estimate the limits of intelligence in different species (see Parts I and II).

The theory of association has served until now as one of the first tools for evaluating intellectual skills in animals. As we saw in Part II, although this theory is still topical it is not sufficient for solving many problems when studying animal intelligence. Alternative sets of experimental schemes have been elaborated for advanced intellectual skills such as concept formation, categorisation, cognitive mapping and so on. At the same time, we will meet devices of early behaviourists that appear in modern experimental schemes aimed at measuring the highest forms of intelligence such as categorisation and concept formation. For example, so-called concurrent discrimination is based on shaping an operant response to only one of a set of stimuli from more than one set of stimuli concurrently. Another example concerns concept learning. The main experimental scheme is based on discrimination procedure, the basic element of Pavlov's and then Skinner's experiments (see Part II).

The general problem is that when studying complex levels of animal intelligence, we are dealing with internal representations hypothetically built by the animal and thus face the necessity of recording any relevant behavioural reactions that could reflect this process. The concrete problems concern the difficulties of interpretation of obtained data. For example, when a subject has to choose a definite picture it is sometimes difficult to decide whether concept formation or stimulus generalisation is responsible for the result. Or when a subject finds the goal in a maze, it is difficult to decide whether it forms a cognitive map of the maze or whether it reacts to a sequence of stimuli. In other words, it may be difficult to distinguish between 'simple' associative learning and 'complex' forms of learning and cognition. Nevertheless, gradual iteration of experimental methods has led to impressive results concerning some aspects of animal intelligence.

In order to estimate the limits of animal intelligence, Thomas (1986) has proposed a hierarchy of learning skills that could be used as an

Table V.1 | *A hierarchy of learning abilities (Thomas, 1986)*

Level	Description
1. Habituation	Learning not to respond to a repeated stimulus that has no consequence
2. Classical conditioning	Making reflex responses to a new stimulus that has been repeatedly paired with the original innate stimulus
3. Simple operant conditioning	Learning to repeat a voluntary response to obtain reinforcement
4. Chaining operant responses	Learning a connected sequence of operant response to only one of a set of stimuli for more than one set of stimuli concurrently
5. Concurrent discriminations	Learning to make an operant response to only one of a set of stimuli for more than one set of stimuli concurrently
6. Concept learning	Discrimination learning based on some common characteristic shared by a number of stimuli
7. Conjunctive, disjunctive and conditional concepts	Learning of concept involving a relationship between stimuli of the forms 'A and B', 'A or B' and 'if A, then B'

'index of intelligence' by determining how far up the hierarchy an animal is capable of performing. As one can see from Table V.1 below, the hierarchy has eight levels, starting with habituation (level 1) and classical conditioning (2) and progressing through complex conceptual learning involving logical operation (7).

Half these criteria have been analysed in Part II. Here we will consider experimental approaches for studying animal abilities to catch the meaning of problems to be solved, from relatively simple tasks to very complex ones that request navigating cause–effect relations. These abilities concern levels 5–7. It is worth noting that recently a new experimental paradigm of language training has been elaborated that allow asking animals such as great apes, grey parrots and dolphins directly about their understanding of many things (Gardner and Gardner, 1969; Premack, 1971, 2004; Herman, 1986; Pepperberg, 1987, 1999). In this Part we consider this approach very briefly and return to its detailed description in Part IX.

Chapter 14

Conditional discrimination as a basic technique for studying rule learning

14.1 Experimental paradigm of discrimination learning

Experimental studies of conceptual abilities in animals demonstrate the substantial range of these abilities as well as their limitations. Such abilities range from categorisation on the basis of shared physical attributes, associative relations and functions, to abstract concepts as reflected in analogical reasoning about relations between relations (Thompson and Oden, 2000). Many experimental procedures for studying species-specific and age-specific abilities to judge about things and relations are based on *discrimination learning*. It is a general and already trivial observation that discrimination is a basic element of decision-making in the animal kingdom. Organisms must respond appropriately in an enormous number of stimulus situations and learn which stimuli are significant for their goals and which are not. Relatively simple methods based mainly on discrimination learning serve for solving complex problems concerning the way in which animals represent knowledge. Moreover, the discrimination of presumably more abstract relations commonly involves relatively simple procedural strategies mediated by associative processes likely shared by many species.

When discrimination learning is examined in laboratory, the subject is reinforced to respond only to selected sensory characteristics of stimuli. Discriminations that can be established in this way may be quite subtle. Ways in which animals solve discrimination problems have been the focus of interest since early experiments conducted by Kinnaman (1902), Pavlov (1917, described in Pavlov, 1927) and Köhler (1918). The paradigm of discrimination learning appears to be fruitful for studying animal intelligence because simple discriminations lie at the base of concept formations of the highest order, such as 'sameness' or the 'relation between relations' and thus can be considered 'atoms of cognition'.

The Skinnerian technique of discrimination learning (*discriminative operant*) involves the use of environmental cues to signal the availability or unavailability of positive and negative reinforcement as well as punishment (see Chapter 6). For example, for a horse a trial begins when an experimenter brings the horse to the entrance of the testing stall and lets go of its halter. The horse enters the testing stall unescorted, faces the stimulus wall, and selects one of the stimuli by pushing a panel by its nose. When the correct stimulus panel is pushed it opens inwards, and the subject is allowed access to a food bowl. The incorrect stimulus panel is locked from behind and would not move if pushed, although the bowl behind it also contains the reward.

Another technique that was used for many years to study discrimination learning is a jumping stand. This apparatus is known as the *Lashley jumping stand*: (Lashley, 1938). The discriminative stimuli are situated behind ledges that the subject must jump onto from the central platform. If the subject jumps onto the ledge in front of the correct stimulus, then it can push the stimulus card over and gain access to a reward. The position of the discriminative stimuli is normally varied randomly from trial to trial, which ensures that subjects learn that food is behind a certain stimulus, rather than in a particular location.

In experiments with non-human and human primates *Wisconsin General Testing Apparatus* (WGTA) described by Harlow (1949, 1959) is used for a battery of tests including simple and reversal discrimination, matching to sample and the so-called learning set, that is, animals' ability to 'learn to learn'. We may, for example, try to teach an animal to systematically switch to the other stimulus after every successful response, a discrimination procedure that requires multiple problems and a transfer set-up. Such work on *learning sets* (also called *learning-to-learn*) has been done by Harlow (1949), and we will examine it more closely in Part VI. Another example of a complex task which is implemented on WGTA is the Wisconsin Card Sorting Task, a test of the ability of human subjects to flexibly alter their responses to the same stimuli. The sorting rule varies surreptitiously every few minutes and thus any given card can be associated with several possible actions; the correct response is dictated by whichever rule is currently in effect. Impairment on this task is a classic sign of prefrontal cortex damage in humans (Milner, 1963), and marmoset monkeys with prefrontal lesions are impaired on analogous tasks (Dias *et al.*, 1997)

The ways in which subjects are presented with stimuli and reinforcement differ in accordance with different schemes of experiments. If there are two or more stimuli present only one of which needs to be responded to, then we have what is termed a *simultaneous discrimination*. The task in a simultaneous discrimination is to determine which stimulus to respond to amongst the various stimuli present (Fig. 14.1). Normally, the animal will initially respond to both stimuli, and thus discover that one of these consistently fails to yield the desired outcome. Over the course of training, then, we will see responding first rise to both stimuli, and then drop off to the S− (a negative stimulus). This is a use of normal *discrimination training*. If many stimuli are present from which the subject has to make a choice, the task is termed *concurrent discrimination*. In a variant of simultaneous discrimination procedure, a technique termed *errorless discrimination* training, however, we typically start with the S− being so reduced in intensity compared to the S+ (a positive stimulus) that it is virtually non-salient (here + means that selection of this item is rewarded and − means this item is not rewarded). Over the course of a large number of trials, we gradually increase the intensity of S− until it equals that of S+. If this is done slowly enough, then we may discover that our subject has never actually made a false response to the S− (hence the name errorless).

An alternative approach is to present just one stimulus at a time. In this case, we have a *successive discrimination*. Successive discrimination techniques open up several different possibilities. If the successive discrimination is like the simultaneous discrimination in that only one stimulus is associated with a reinforcement, then we have what is termed a *go–no-go* training procedure, with rate of response serving as the measure of learning. Subjects can be trained to respond differently to different stimuli in a choice situation. Typically, the trial is initiated by the appearance of the discriminative stimulus. For example, in experiments with pigeons, a peck at one of two stimuli illuminates the choice keys signalling the opportunity to make a choice. As was described in Chapter 10, in this procedure the intermediate stimulus serves as a 'go' signal. In this case, you go (respond) when the correct stimulus (the S+) is present, but do not go (withhold a response) when any other stimulus is present. In successive discriminations, however, we also open up the possibility of requiring several different responses, depending on which stimulus is present. A triangle, for example, could be used to signal pressing a left-hand button, whereas a circle could be used to signal pressing a right-hand button. In the simplest version of this situation, a reinforcement can be obtained on each trial (assuming you are using a continuous reinforcement schedule) so long as the animal knows which response to make to which stimulus. This type of set-up is referred to as a *choice situation*. Unlike the simultaneous condition or the go–no-go situation, each stimulus in the choice situation could, in principle, be associated with the same degree of excitation or inhibition. Or to put it another way,

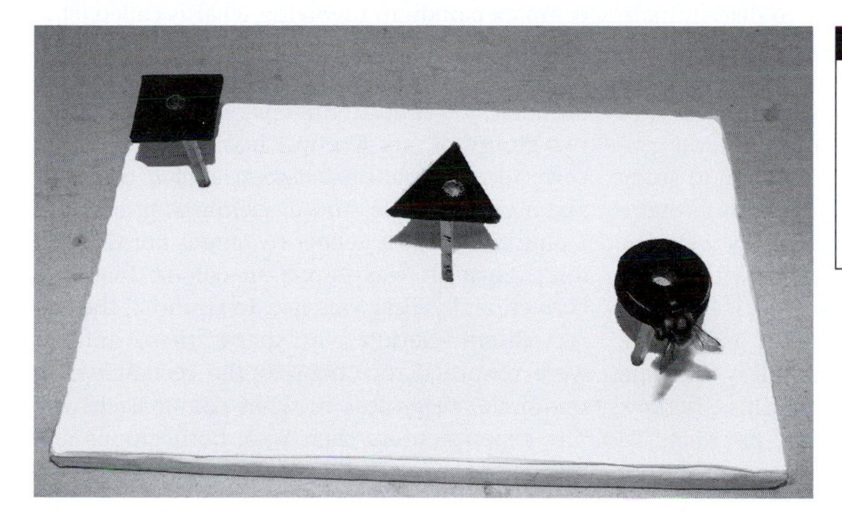

Fig. 14.1 A bumble-bee engaged in a task of simultaneous discrimination of geometric stimuli in Reznikova's laboratory. (a) A flight arena is connected with nests of bumble-bees; (b) A bumble-bee in the flight arena chooses the relevant geometric shape. (Photographs by A. Chernenko.)

in a choice situation, the stimulus acts as an occasion-setter indicating which response is appropriate.

Conditional discrimination involves successive discrimination. In conditional discrimination, at least two stimuli (or stimulus dimensions) are typically present. The reaction to one will depend on the presence of the other. For example, in an experiment by Nissen (1951), chimpanzees were trained to discriminate between a large square and a small square. Only one of these yielded reinforcement, but which one that was depended on a second stimulus characteristic. Because this was a successive discrimination procedure, each trial involved just one of the four complex stimuli below:

Stimulus	*Outcome*
large black square	reinforcement
small black square	no reinforcement
large white square	no reinforcement
small white square	reinforcement

In this example, if the squares were black then large was the S+ and small the S−; but if the squares were white, this assignment was reversed. So, colour in this case was the occasion-setter informing the animal about the meaning of size.

Assigning a *transfer tasks* to the animals, we ask whether or how prior discrimination training on one problem affects later discrimination training with another problem. Sometimes problems are very similar. A common task involves a *reversal shift*, in which a subject learns to do the exact opposite of what it did earlier. For instance, if it was trained to respond to a square but not a triangle, in a reversal shift, it would have to respond to the triangle, but not the square. Sometimes the problems are very different, and sometimes the training involves combining problems. In a technique called *acquired distinctiveness of cues*, for example, we try to speed up the process of discrimination by compounding the two stimuli we want the animal to discriminate with very different stimuli that we know from earlier training are easily distinguished.

A way of looking at the ease and success of discrimination training involves *matching-to-sample procedure* (MTS). This is a method to study how animals learn not only about stimuli themselves but about their relationships as well. For this technique a subject is presented with a sample stimulus, and, a short while after it has been turned off, two comparison stimuli are shown. When one of the comparison stimuli is the same as or identical to the sample stimulus, the task often is called an *identity matching procedure*. To gain a reward, the comparison stimulus that matches the sample must be selected. If a subject can also perform adequately when novel stimuli replace the training stimuli, the performance is often called *generalised identity matching* or *true matching* and taken as evidence that training generated an identity concept.

14.2 Discrimination and reversal shift

To train animals in discrimination shift, experimenters first suggest to them a simple discrimination between pairs of stimuli. For example, in the experiments of Dias *et al.* (1997) marmosets were trained on a simple visual discrimination consisting of either a pair of blue-filled shapes ('shape' group) or a pair of black lines ('line' group) presented randomly and simultaneously on two test boxes positioned on the far right and left of the test apparatus. A response to either 'correct' shape or 'correct' line resulted in removal of the transparent screen, allowing the monkey access to a food item hidden within the test box, whereas an incorrect response resulted in the replacement of the opaque screen and no reward. Once the correct reaction was attained on the simple visual discrimination, on the following session the reward contingencies were reversed such that the stimulus that had been negatively correlated with reward was now positively correlated with reward and vice versa.

Mackintosh and Little (1969) examined a more complex paradigm involving what is called *intradimensional* (IDS) and *extradimensional* (EDS) shifts. They started by training animals to make a discrimination on multi-dimensional stimuli. From two groups of rats, Group 1 had two problems to solve: discriminating between a blue rectangle and a yellow circle, and discriminating between a blue circle and a yellow rectangle. For Group 1 discrimination was based on colour: blue was correct, and yellow was not. In Group 2, the correct discrimination was shape: these animals were rewarded for choosing the rectangle over the circle, regardless of what colour each was. The experimenters then took both groups and moved them to the new set of problems. These involved different shapes than before and different colours. There was a red plus-sign and green triangle in one test (plus is rewarded) and green plus-sign and red triangle in another test (again, plus serves as S+). The dimension or relation of shape is still relevant: the animal needs to respond to the plus-signs and avoid the triangles. For the animals faced with this set of problems, whether something is red or green is irrelevant. For Group 2, the transfer problem is an IDS: the dimension stays the same, though new stimuli are presented. But for Group 1, the transfer problem is an EDS: they need to move from responding to the dimension of colour to responding to the dimension of shape. Reversal shifts are a type of IDS, and non-reversal shifts may be regarded as a type of EDS. The ability to solve IDS problems supports a claim of learning about abstract dimensions such as colour or shape, rather than physical values such as yellow or triangular.

When only one stimulus from many stimuli is associated with reinforcement, then *concurrent discrimination* is examined. For example, collaborating with the young chimpanzee Ioni, Ladygina Koths (1935) decided to skip the procedure of training to match pairing stimuli and proceeded with testing the concurrent discrimination. Ioni was presented with groups of eight figures among which he had to select one that matched the sample. Ioni was able to discriminate between 13 plane geometric figures and 10 volumetric figures (see Fig. 6.3).

Giebel (1958) reported that a horse learned to discriminate the correct stimulus in 20 pairs of visual patterns concurrently, which ranks it high among other animals that have been tested. Thomas (1986) compiled data on concurrent discrimination learning for various fish, reptiles, birds and mammals, including mice, rats, zebras, donkeys, horses and elephants. Of these species tested, only the elephant was able to successfully complete as much concurrent discrimination as the horse.

Chimpanzees were champions for concurrent discrimination in a study by Farrah (1976) on what he called 'Picture memory' in apes. In these matching-to-sample experiments each discrete trial began with one of 24 possible visual samples, varying in form and colour and being presented on a black-lit key. Immediately after the sample was terminated one stimulus was presented on each of four black-lit keys arranged side by side in a line horizontally below the sample key. The chimpanzees were rewarded if they pushed the key on which the displayed stimulus was the same as what had appeared on the sample key. Importantly, although the four stimuli varied across the 24 possible matching problems, the same set of three incorrect stimuli appeared with each correct 'match' on any given problem. The terminal performance levels of all three chimpanzees exceeded 90 per cent of correct matching responses, suggesting perhaps that they had learned a generalised matching concept. It is interesting to note that on the basis of this experimental paradigm, experimenters have since revealed extraordinary abilities of chimpanzees to remember faces of conspecifics and to make judgements as to their kinship (see details in Part VIII).

14.3 Conditional discrimination and rule learning

In a wide sense, conditional discrimination is discrimination in which the reinforcement of responding to a stimulus depends on or is conditional upon other stimuli. Conditional discrimination is one of relevant methods for empirically determining whether ordered pairs of events or interrelated stimuli make up an equivalence class. Using matching-to-sample procedures, relations are established with explicit reinforcement contingencies in stimulus pairs. The function of one stimulus depends on the nature of another stimulus. Thus, the S+ would be $S-1$ and not $S-2$ when both are presented with one contextual stimulus $S-x$, but the S+ would be $S-2$ and not $S-1$ when both are presented with another contextual stimulus $S-y$. This technique involves different sets of tests from simple to very complex ones that have been applied successfully for studying intelligence in animals as well as for the diagnosis of normal ageing and mental disorders in humans.

The use of complex tests on conditional discrimination in animals demands infinite patience from experimenters. Nissen (1934) managed to teach a chimpanzee how to solve the problem as follows: choose the smallest from two squares if they differ by only one characteristic (colour, design or border) but choose the biggest one if they differ by two characteristics. The experimenter did not assert, however, that the ape defines a problem using the same concepts as humans. However, it is remarkable that Frank, the chimp, reached a level that exceeded 70 per cent of correct answers after 1700 trials. Later Warren (1965) applied the matching-to-sample procedure in order to train a chimpanzee to choose a stimulus that differs from a sample by shape if stimuli were presented on white background and by colour if stimuli were presented on black background.

Like many compound tests, tasks of conditional discrimination should involve multiple controls, and results do not always reach a satisfactory point. Let us consider classical experiments of Lashley (1938) with three rats as subjects. The apparatus was a two-alternative jumping stand. The first task for animals was to choose an upright triangle rather than an inverted triangle when both were presented on a black background. In the second problem, rats learned to choose an inverted triangle rather than an upright triangle when both were presented on a striped background. In subsequent mixed trials, rats reliably (>90 per cent) chose the appropriate triangle based on the nature of the background. What are the rats learning in this situation? Should we say they are learning and applying a generalised 'if–then' rule? By asking if the rat is learning a generalised if–then rule, we are asking whether the rat is learning 'if the background is black, then A and not B are correct, and if the background is striped, then B and not A are correct'. On the other hand, are they learning to respond to a specific combination of stimulus plus background? Do they learn simply to pick the compound of upright triangle/black background and the compound of inverted triangle/striped background, as independent problems? This question can be answered by training rats in a series of conditional discrimination problems. A pair of stimuli should be presented, first on a black background. One stimulus would be correct and the other incorrect and the speed with which the rat acquired the discrimination would be measured. Then the same pair of stimuli would be presented on a striped background, now with the formerly incorrect stimulus being correct, and vice versa. If the rat was learning a generalised if–then rule, one would predict that after the rat had learned some number of these sorts of problems, whenever it was presented with a new problem it would learn at a faster rate when the background was striped than when it was black. For example, if the rat was learning an if–then rule, one would predict the rat would choose the correct alternative on the very first trial of the second exposure (i.e. with the striped background) with a very high probability. In contrast, if the rat was learning to respond to a specific combination/compound of stimulus plus background, one would predict that problems on the second exposure would be learned at the same rate as those on the first exposure.

In further tests, Lashley trained rats with four new problems (large circle versus small circle; + versus ×; circle versus vertical bar; star versus square). One stimulus was correct when both were presented on the same black background. The other stimulus was correct when both were presented on the same background with stripes. The question was how fast the rats would learn the second exposure to the problem, when the stimuli are presented on the striped background. The rats learned the discrimination with the striped background at more or less the same rate as they had learned the original discrimination with the black background. There was no evidence of a generalised tendency to reverse and choose the alternate stimulus on the first trial with the striped background. So the inference could be drawn that rats learn to respond to a specific combination of stimuli (e.g. a compound of upright triangle/black background; or a compound of inverted triangle/striped background), rather than a generalised if–then rule.

Development of laboratory studies on higher-order conditional discrimination provides evidence of behavioural flexibility in non-human animals, together with limitations in some species. Members of several species such as rats (Preston *et al.*, 1986), pigeons (Santi, 1978; Edwards *et al.*, 1982) and monkeys (Fujita, 1983) appeared to take multiple-rule strategies. For example, in the experiments of Nevin and Liebold (1966), pigeons could solve the multiple discrimination task following four rules: 'if red is presented in the lit chamber, then peck red', ' if green is presented in the lit chamber, then peck green', 'if red is presented in the dark chamber, then peck green', 'if green is presented in the dark chamber, then peck red'.

To further explore the possibility of conditional rule learning in animals, Nakajima

(1997, 2001) devised another method of testing. Pigeons were tested in a three-key operant chamber, where a correct response (a left- or right-side key peck) depended on three preceding events. Pecking the left or the right key was followed by a food reward according to the colour (amber or blue) of the centre key, presence or absence of flashing of the three keys with green colour prior to the centre colour presentation, and the house-light illumination condition (dark or light) of the chamber.

The problem posed was: can pigeons learn a hierarchical conditional rule or do they solve the task using other strategies?

With eight pigeons, seven out of eight possible types of trials were trained in a step-by-step fashion, and then the remaining trial type was tested. All birds responded correctly to the training test types, but their poor performance in the untrained test type indicates that they solved the training task by rote-learning of the individual trial types. Their test performance enabled Nakajima (2001) to suggest that the birds probably had learned to configure each sequence of events into a unique stimulus to respond properly during the training, and their test performance reflected generalisation from that learning. Another possibility is that the birds had learned seven rules of event sequences (multiple-rule strategy), and in testing they resorted to the rule that was most similar to the test type. In sum, pigeons did not derive an adequate response to a novel trial type from the familiar trial types by completing the hierarchical structure. It remains unclear whether this failure reflects the limitation of the birds' cognitive abilities or an unfavourable procedure for testing such abilities.

Chapter 15

Categorisation, abstraction and concept formation: are animals logical?

In defence of animal reasoning, McGonigle and Chalmers (1992) exclaim in the title of their paper 'Monkeys are rational!' This echoes an earlier paper which, in its title, asks 'Are monkeys logical?' (McGonigle and Chalmers, 1977). Then '... The Paleological Monkey and the Analogical Ape' appeared in the title of Thompson and Oden (2000), who in turn are referring to 'palaeologicans' in the sense enunciated by von Domarus (1944) in his interpretation of reasoning by schizophrenics: whereas a normal person accepts identity upon the basis of identical subjects (i.e. conceptual equivalence), a palaeologican accepts identity basing upon identical predicates (i.e. shared features). Of course, Thompson and Oden (2000) do not claim that monkeys are schizophrenics; they do claim, however, that monkeys discriminate categorical equivalence classes on the basis of perceptual identity of features, or some combinations, possibly configurable.

In fact, there is a wide range of ideas concerning *concept formation* in non-human animals. In terms of operant learning, the formation of discrimination is based on a class of stimuli such that an organism generalises among all stimuli within the class but discriminates them from those in other classes. Such classes play much the same role in the analysis of discriminative stimuli as operants do in the analysis of response classes. Following a pioneering experiment by Herrnstein and Loveland (1964), much work was concentrated on experiments on discrimination between sets of stimuli. The stimulus sets are usually defined in terms of human concepts, e.g. person vs. non-person, fish vs. non-fish, or artificial concepts defined by specified multiple features. Most discussion has centred on the question of whether animals need to possess concepts in order to perform concept discriminations, and what does it mean for an animal to 'possess a concept'. It is important to note that in order to dismiss simple *stimulus generalisation* (see Part II), it must be demonstrated that stimuli to be classified in the same category certainly differ from one another (Vauclair, 2002).

Cognitive ethologists concentrate on what Thompson and Oden (2000) call *conceptually mediated behaviour* which permits animals to adjust their behaviour to a novel object or event by virtue of its membership of an already familiar class. To infer that an animal has a concept one must provide evidence that it applies the same judgement in the form of an explicit response rule or cognitive operation to objects or events that are perceived to be common members of the same physical or relational class. The measure of conceptual categorisation is based on experimental procedures that match different levels of complexity in animal conceptually mediated behaviour.

15.1 Acquisition of the same/different concept in animals

It is an advanced intellectual feat for non-human animals to detect the sameness or difference of a collection of stimuli and to make distinctively different responses to the relation 'the same' and 'different'. An even more advanced feat would be for animals to match the relation between relations, in other words, to exhibit the essence of analogical reasoning (Delius, 1994; Premack, 1983). According to the experimental protocol of same/different discrimination, the subject has to respond 'same' when two or more stimuli are identical and 'different' if one or more of the stimuli is different from others. This is the basic element of many cognitive tasks. After learning same/difference discrimination, the degree to which this behaviour transfers to novel situations having same and different relations is taken as evidence of concept formation (same/different (S/D) concept). Using the same/different choice task, it has been found that pigeons, parrots, rhesus monkeys, baboons and chimpanzees are capable of learning and applying the same/different concept across a wide variety of simultaneously presented visual elements. An advanced variant of the same/different choice task is *multi-dimensional scaling* (MDS). As many other techniques which are applied to studying animal intelligence, multi-dimensional scaling has its origins in psychometrics (Richardson, 1938; Torgerson, 1958). Animals from widely different species appeared to acquire the same/different concept comparing complex stimuli of multiple components.

For example, Cook *et al.* (2003) tested pigeons using go–no-go discrimination, in which alternating sequences of either same (AAAA ... or BBBB ...) or different (ABAB ... or BABA ...) photographic stimuli were presented within a trial. At any one time only a single item was visible, thereby eliminating any perceptual features related to element simultaneity as a basis for learning or transfer. Pigeons were first trained to peck at a white warning signal. Once consistent responding was established, sequences of stimuli were introduced. Each of these trials started with a single peck to the warning signal, followed by either identical or different stimuli for 20 seconds. The stimuli tested during discrimination training consisted of combination of two elements – a small object figure centrally located on a naturalistic background. In one test the figure elements consisted of pictures of six common objects (soccer ball, bell, key, cup, teddy bear and phone). The background elements consisted of pictures of six landscapes. In other tests the number of figure elements was increased up to several tens and the stimuli varied in size. Separate tests were performed with video stimuli. Each test involved the successive presentation of different events. Based on many series of alterations and variants of combinations of stimuli, these experiments provided evidence that pigeons can differentiate same and different sequences based on the alternation of only two picture stimuli. Birds can discriminate change in colour, grey-scale and video stimuli, although stimuli with smaller differences proved difficult for the pigeons to learn. The discrimination was maintained when tested with large numbers (55 figures) of randomly combined photographic stimuli of two different sizes.

In another series of experiments concerning pigeons' acquisition of 'sameness' based on multi-dimensional scaling (Wasserman *et al.*, 1995; Young and Wasserman, 1997), subjects were first taught to peck one button when they viewed an array of computer icons that comprised 16 copies of the same icon and to peck a second button when they viewed an array that comprised 16 different icons (a same/different discrimination task). The pigeons were later tested with new same and new different displays that were created from a second set of 16 computer icons that had never before been shown during discrimination training. Accuracy to the training stimuli averaged from 83 per cent to 93 per cent, and accuracy to the testing stimuli averaged from 71 per cent to 79 per cent; in each case, choice accuracy reliably exceeded the chance score of 50 per cent. Such robust discrimination learning and stimulus generalisation attest to the pigeon's acquisition of an abstract same/different concept. Baboons were similarly trained and tested with the same visual stimuli. Accuracy to the training stimuli averaged 91 per cent correct, and accuracy to the testing stimuli averaged 81 per cent (Wasserman *et al.*, 2000).

A proper assessment of multi-dimensional scaling requires not only that subjects conceive that different objects have common class attributes but also that the subjects can discriminate among individual members within a category. For example, if pigeons get food in the presence of one random set of photos and do not get food in the presence of another set of photos, the birds may respond equally to all the 'food' items and withhold responses to all the 'no-food' items. This presumably does not mean that pigeons find the photos within each group perceptually similar to each other, and dissimilar to those in the other group (Blough, 2002). So the domain of multi-dimensional scaling in animal mentality is closely related to categorisation.

15.2 Categorisation in animals

Experimental work in the field of animal categorisation is aimed more at studying to what level animals are able to follow 'human' rules in grouping objects than investigating how far in the animal kingdom does this mode of conceptualisation extend. A useful general framework for the investigation of categorisation in animals was provided by Herrnstein (1990) who described categorisation abilities in five levels of increasing abstractness, including (1) discrimination; (2) 'categorisation' by rote; (3) open-ended categorisation (namely, category formation resting on a perceptual similarity between objects that belong to a given class); (4) conceptual categorisation; and (5) abstract relations. Herrnstein (1990) used two criteria to define conceptual categorisation (level 4). The first criterion is met when a rapid generalisation about members of a class of items is observed. The second criterion, which is related to conceptual processing, implies categorisation abilities that go beyond perceiving a similarity between exemplars or a class. Thus, level 4 is more complex than open-ended classification, the latter being related to the use of perceptual dimensions of stimuli. To perform this, a subject has to discriminate polymorphous stimuli, i.e. stimuli for which no single feature is either necessary or sufficient to determine category membership. Level 5 is attained when a subject is able to use abstract relations not only between objects but also between concepts, such as in conceptual matching or in conceptual identity (for example, the mastery of 'sameness' relations).

In their attempts to investigate animal mentality, experimenters have tested ideas about what is called *prototype effect*, which was initially reported in studies on humans (Rosch, 1973). This effect is expressed by a better categorising performance with prototypical stimuli representing the central tendency of the category than with other, less typical exemplars. For example, humans think that a sparrow is a better exemplar for the 'bird' category than an ostrich. In fact, the prototype effect is only one of three theories of categorisation that have been elaborated in human psychology. These are the *exemplar*, the *feature* and the *prototype* view as three types of representation.

Evidence for the capacity to perform the first three levels of categorisation is abundant for several animal species. It is, however, much less clear concerning levels 4 and 5 (for review, see: Huber, 2000; Thompson and Oden, 2000; Vauclair, 2002).

Several experiments were conducted with baboons in order to assess the abilities of these monkeys to discriminate objects on the basis of their membership in a category and to study the nature of the representations of categories the baboon formed (Vauclair and Fagot, 1996). A video task required the baboons to manipulate a joystick that controlled the movements of a cursor on a screen. The subject was required to manipulate the joystick so as to touch with the cursor a response stimulus that matched the sample stimulus on an arbitrary (experimenter-defined) basis. In one set of experiments the baboons were tested to categorise characters displayed in various typefaces. For this purpose, the baboons were first trained in a symbolic matching-to-sample task with 21 different fonts of the characters B and 3 as sample forms, and colour squares as comparison forms. After training, novel fonts were displayed. This task can be traced back to the basic concept of Gestalt psychologists who investigated humans' ability to grasp whole figures (such as letters) from their fragments. Baboons showed positive transfer of categorising performance to the novel stimuli of the characters used in the original training. These results demonstrate that the original learning was not achieved by rote, because in that case the animals would have demonstrated no transfer to the novel typefaces. Thus the baboons' performance indicated that these monkeys were able to exhibit level 3, i.e. open-ended categories in Herrnstein's (1990) sense.

Identical polymorphous stimuli were presented to humans and baboons in a symbolic matching-to-sample task (Dépy *et al.*, 1997). Subjects were trained to classify two out of three feature stimuli (colour, shape, position), and then to assess transfer of performance with the prototypes of each category. Whereas human participants solved the task in a propositional way, the baboons did not extract the prototypes; instead they used a mixed procedure that consisted of memorising salient cues between stimuli or specific associations between exemplars and response associations.

In series of experiments by Bovet and Vauclair (1998, 2000, 2001), olive baboons living in small social groups in an outdoor enclosure were individually trained and tested on the natural category on food versus non-food with real objects with the use of an adapted version of Wisconsin General Test Apparatus. The apparatus was made of a vertical wooden board comprising a one-way screen, a horizontal board to present the stimuli behind a plexiglas window and two openings, one for each of two ropes. When the experimenter placed one or two objects on the board, the subject had to respond by pulling one of the two ropes, according to the categories or to the relations presented. A food reward was provided when the baboons' response was correct. In each task, the baboons were trained with two stimuli (objects or pairs of objects) and when they succeeded, new objects were presented in order to assess transfer abilities. Four baboons were first trained to categorise two objects, one food and one non-food; then 80 other objects (40 foods and 40 non-foods) were presented and the response to each object was recorded. The baboons showed a high and rapid transfer of categorising abilities to the novel items. A similar performance for vervet monkeys was described by Zuberbühler *et al.* (1999). The set of data in which various modes of picture presentations were used further demonstrated the abilities of the baboons to relate real objects to their pictorial representations. Using the procedure of successive simple discriminations, the experimenters demonstrated the monkeys' abilities of categorisation corresponding to the level 5 of Herrnstein's classification scheme. In a first experiment, the monkeys had to judge two physical objects as 'same' or 'different' (perceptual identity). For example, they were required to judge two apples as being the same, or an apple and a padlock as being different. In a crucial test of conceptual identity (corresponding to Herrnstein's level 5), the baboon, had to combine their previously acquired skills in order to classify as 'same' two (different) objects that belonged to the same functional category (food or non-food) and apply that learning to new exemplars. For example, they had to classify as 'same' an apple and a banana, or a padlock and a cup, and as 'different' an apple and a padlock. The monkeys attained a high level of performance at the end of the experiment with totally novel objects (i.e. objects novel in the task but left in the monkey's enclosure before the experiment). Such results demonstrate the mastery of the same/different concepts and the ability to conceptually judge as same or different objects in the previously learned categories (Vauclair, 2002).

The outstanding capacity of pigeons for categorisation still remains a mystery. In fact, it was Herrnstein's work on pigeons that initiated investigations on categorisation and concept formation in animals. Herrnstein and Loveland (1964) trained pigeons to peck for food at any of a number of photographic pictures that contained a person or people somewhere in the picture. There was a wide variation in the appearance of the people shown in these slides – e.g. their number, orientation, size, colour of their clothing, etc. Other slides that did not contain a person in them were also shown to the pigeons; when these appeared, pecking at them did not produce food. With training, pigeons soon came to peck quickly and rapidly at the 'people' slides and not to peck at the 'non-people' slides. Similarly, when shown a new slide without a person in it, the pigeons either did not peck at it or did so much less frequently than at a 'people' slide, again despite never having experienced non-reinforcement with that particular slide. These latter results represent crucial data for inferring categorisation: the pigeon's explicitly taught behaviour generalised to new instances of the 'people' and 'non-people' groups (or 'tree' and 'non-tree' in other experiments).

This type of experiment has received considerable attention not only because the results suggest that pigeons possess an ability that transcends the discrimination of simple stimulus dimensions such as wavelength, intensity or frequency, but also because they imply that the pigeons' classification behaviour is mediated by abstract, or conceptual, rules, and therefore resembles the cognitive solution accomplished by humans.

An even closer analogue to human categorisation by appearance was developed by Wasserman and his colleagues (Bhatt *et al.*, 1988). They applied the pigeon version of the 'name game', which parents often use to teach their young children the names of different objects from a number of perceptually distinct categories. On each training trial in the pigeon version of the name game, hungry pigeons were shown a photographic slide on a centre viewing screen that depicted one of a variety of examples drawn from four different categories: cats, chairs, cars and flowers. After pecking at the slide appearing on the centre screen, pigeons could then obtain food by 'naming/labelling' it appropriately. This meant pecking one of four different keys located at the vertices of the viewing screen. Thus, if a picture of a flower appeared on the centre screen, pecking the top left key produced food; if a picture of a car appeared, pecking the top right key produced food, and so on. Like the pigeons in the Herrnstein and Loveland (1964) study, these pigeons readily learned to correctly categorise the various pictures shown to them in each training session. In this case, they quickly learned to confine their pecks to one key when shown any instance of a cat, to another key when shown any instance of a flower, and so on. The results suggest that the pigeons had grouped the various slides together by appearance. Things that looked alike were treated alike. The four different, explicitly taught pigeon 'names' generalised immediately and appropriately to novel instances of each group that were later shown to them during a subsequent generalisation test. To distinguish between 'accident' and true categorisation, researchers have compared pigeons' performances on the task with the performances of other pigeons that have been shown exactly the same set of photographic slides but have been taught a 'pseudo-category' task (Wasserman *et al.*, 1988). In this task each picture is again associated with one of the four possible responses but different pictures within a particular group (e.g. different cats or different flowers) are associated with different responses. For example, the pigeon might receive food for pecking the top left key after seeing one of the flower pictures, but the bottom right key after seeing a picture of a different flower, and so on. Pigeons learn this task, too, but they do so much more slowly and they are not as accurate in their choices as the 'true category' pigeons.

Some experiments have demonstrated that pigeons are able to use multiple facial features to discriminate human faces (Jitsumori and Yoshihara, 1997). Researches have also examined prototype effects in pigeons by using human faces, i.e. naturalistic visual stimuli, created by the morphing technique. A physical prototype created by averaging training exemplars is assumed to correspond to a prototype abstracted by categorisation training. With the categories constructed to mimic family resemblance of natural categories, pigeons showed clear evidence of prototype effects (Makino and Jitsumori, 2001). Pigeons trained to discriminate typical and atypical exemplars of a category showed multiple prototypes of the sets of atypical exemplars, each of which were perceptually disparate to one another (Jitsumori, 2004).

Huber (2002) suggests that pigeons use characteristics of objects that differ from those of humans', such as tiny details of texture and shape. Texture is mostly related to the surface properties of aerial photographs, landscapes or industrial materials. Huber and colleagues (Huber, 1995; Huber *et al.*, 2000) used human faces as stimuli and the concept 'sex' to define class membership. Experimenters compared the classification performance of pigeons presented with different versions of the same set of stimuli

Fig. 15.1 Experiments in Huber's laboratory for studying the classification performance of pigeons: a pigeon is presented with different versions of the same set of stimuli. (Photographs by U. Aust; courtesy of L. Huber.)

(Fig. 15.1). The stimuli could be distinguished according to their texture and shape information, and were derived from laser-scanned models of the faces of 100 men and 100 women. The faces were free from any kind of accessories such as glasses or earrings. The men were carefully shaven and the hair on the head was digitally removed from the three-dimensional models. The 200 faces were randomly divided into two sets (A and B), each containing 50 male and 50 female faces. Group O were shown the original images, while Groups T and S were shown images only after they had been subjected to a technique described in Vetter and Troje (1997) which involved separating the texture and shape components of each image. Group T was shown images generated by combining the original texture of each face with an average shape. This yielded an image set that varied with respect to texture but not shape. Group S was shown images generated by combining the original shape of each face with an average texture, which yielded an image set that varied with respect to shape but not texture. The results of this experiment indicated that Groups O and T learned very quickly and accurately to discriminate faces, whereas Group S failed to do so. These results suggest that pigeons are extraordinarily sensitive to texture differences, but that they find it very difficult to discriminate shapes.

The stimulus parameters that controlled the performance of successful subjects remain to be determined. Male and female faces differ both in average size and in average intensity. Female faces are generally smaller and brighter than male faces. Therefore, the experimenters computed the rank correlation between pecking rate to individual faces and either the average size or the average intensity of these images. The four parameters that describe the texture of images (energy, contrast, entropy, homogeneity) as well as the three components that describe their colour (red, green and blue) were also quantified. The correlation was computed separately for male and female faces. The pecking rates of almost all Group O and Group T subjects correlated significantly with intensity, but not with any of the other texture parameters or size. Pigeons assigned to these groups appeared to use the intensity of faces as a cue to discriminate between male and female faces. Computer analysis of the images revealed a subtle difference in colour between male and female faces; male faces are redder than female faces, while female faces are more blue and green than male faces. It is likely that pigeons use their extraordinary physiological capacity for the perception of colour. These findings raise question of whether animals might sort the complex objects of the natural environment, even the so-called higher-order concepts like 'persons' or 'fish' by fixing on some specific, single feature. Huber (2002) concludes that although pigeons have strong resources for learning specific exemplars, and display surprising cognitive capacities, neither categorisation in terms of exemplar memorisation nor in terms of abstract concept formation is plausible; it seems reasonable that representations of stimuli at the level of relationships between two or more arrays are at the limit of their cognitive capacity.

We are obliged to leave open the question of whether categorisation research with pigeons has demonstrated their extraordinary cognitive skills or 'only' their capacity to exploit very subtle variation in the texture of natural objects. Nevertheless, advanced forms of categorisation are currently being intensively investigated in many species including small-brained but specifically gifted creatures such as pigeons.

15.3 Abstraction in animals

Numerous experiments have demonstrated that animals have the ability to make abstractions. They are capable of recognising the essential features common to different phenomena and thus by abstracting certain relevant characteristics to arrive at concepts.

In Révész's (1924) pioneering study chicks learnt to peck at the smaller one from two geometric stimuli such as a circle, a square or a triangle. Chicks responded to sizes of stimuli irrespective of their geometric shape. Similar experiments were conducted by Protopopov (1950), Hilchenko (1950) and Markova (1962) with dogs, monkeys, chimpanzees and human infants. For example, in one of these experiments baboons had to choose one of two boxes with figures attached to lids. A reward was always placed inside the box to which a smaller square was attached. The boxes were situated in different places in order to avoid the subject using landmarks to orient. When the animal managed to choose the smaller square, it was tested with figures that had never been seen in experiments. Monkeys insisted on choosing the smallest figures irrespective of to their shape and colour. In a special series of tests the subjects were shown volume figures which at the second stage were replaced by their plane projections. In some experiments plane figures were subsequently replaced with their outline and finally with dotted lined images. In general, monkeys managed all these tasks but the less the figures' style resembled the initial one, the worse results were achieved.

Experiments conducted by Rensch and Dücker (1959) with a civet cat revealed a considerable ability to sift various sensory impressions for certain characteristics essential to the significance of the whole (generalising abstraction). The animal was trained to distinguish between two parallel semicircles (meaning 'food') and two straight lines (meaning 'punishment'). It was then presented with increasingly complicated patterns in which these two recurred in modified form. The cat showed that it could eventually distinguish between the concepts 'bent' and 'straight'. It also, in similar fashion, formed the twin concepts 'equal' and 'unequal'.

Many modern experiments on abstraction and concept formation are derived from the method used by Neet (1933), Gellerman (1933) and Harlow (1949) to show that human infants, chimpanzees and rhesus monkeys (but probably not rats) are able to form abstracts concepts, such as a *concept of triangularity* (Fig. 15.2). The same experimental paradigm has been repeatedly applied with wide variety of species.

For example, in the experiments of Sappington and Goldman (1994) with Arabian horses, the stimuli were grouped into six problems in such a way as to bring animals to shaping a concept of triangularity. Problems 1 to 4 involved simple discrimination learning with only two stimulus patterns: correct and incorrect. Problems 5 and

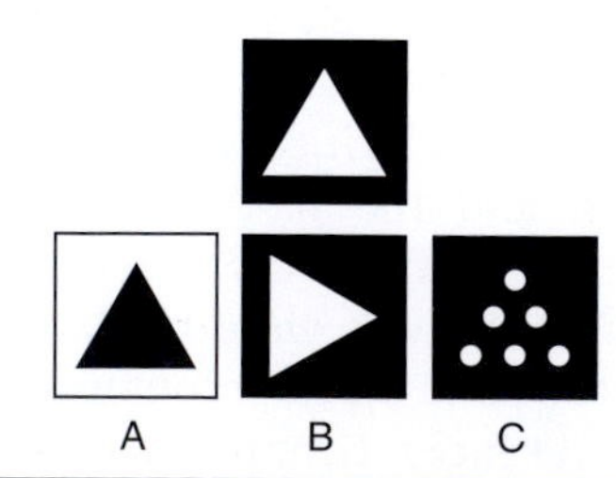

Fig. 15.2 A scheme depicting the 'concept of triangularity'. Rats trained to react to the uppermost triangle then react at random to any figure from the lower line; chimpanzees react to A and B, but not to C; human children 2 years old recognise all triangles.

6 each had three different stimulus patterns that were variations of triangles which were always correct, and three stimulus patterns that were non-triangular shapes which were always incorrect. On each trial, one card from the set of correct stimuli was randomly paired with one card from the set of incorrect stimuli. These two problems thus involved the concept of triangularity. Of these two, Problem 6 was a stricter test of the concept of triangularity, as it involved three triangles never before seen by the subject paired with three new, non-triangular shapes. The results clearly demonstrated that at least some horses were able to form the concept of triangularity. One of four subjects completed both Problems 5 and 6 at a criterion level of 85 per cent for two consecutive sessions, while another was performing at more than 80 per cent correct on Problem 5, the first test on the triangle concept. Although both subjects did not reach the point at which they responded correctly to a novel triangle the first time it was presented, which would be the strictest test for concept formation, the fact that they could learn relatively quickly that one of several triangles was correct regardless of what it was paired with suggests an ability to respond to a common characteristic of diverse stimuli (Sappington and Goldman, 1994).

In the experiments of Delius and Habers (1978) pigeons had to conceptualise symmetry. Birds in two groups were given pairs of stimuli that differed by several minor attributes. For one group only symmetric stimuli were rewarded, irrespective of other characteristics such as shape, ornament and colour, while for the second group the asymmetric stimuli were reinforced. When a level of 80 per cent of positive responses was reached with one pair of stimuli, the experimenters presented to the birds another pair of stimuli which consisted of unfamiliar items, again differing by the characteristic of symmetry for one group and asymmetry for another group. When the pigeons grasped regularity within a context of a given problem, they accurately chose only symmetric (or asymmetric in the other group) stimuli irrespective of other attributes. In testing experiments baboons were demonstrated to be able to categorise spatial relations such as 'above' and 'below' as well as 'short' and 'long' distances (Dépy *et al.*, 1999).

Abstraction was achieved by honeybees in the experiments of Mazokhin-Porshnyakov (1969, 1989) in which insects conceptualised paired/non-paired objects and solved some other complex tasks. The bees started from simple tasks in which they had to ignore irrelevant experimenter-dependent ('conventional') features of stimuli such as geometric shape in favour of relevant ones such as sizes. The method of these experiments was as follows. A feeder (rotating table) was situated at a distance of several metres from the hive. An experimenter placed a trough (watch crystal) with a drop of syrup to starting foragers at a bee entrance. When an interested bee began to eat, it was labelled with paint and then transported to the feeder just on the trough. There was a set of pictures covered by a glass on the feeding table. Rotation prevented the bees from using landscape landmarks for determining the location of the food. At the training stage of the experiments the bee was presented with the trough filled with a drop of syrup placed on the rewarded figure while troughs filled with water were placed on other, non-rewarded figures (Fig. 15.3). In a relatively simple task the same stimuli were used as in previously described experiments with vertebrate animals. Bees had to choose the smallest figure irrespective of its shape and colour. After several visits the bees memorised the rewarded figures. In order to test this, a series of examinations was performed during which the bee was presented with all the troughs filled with water in order to avoid possible use of the smell of sugar. The main part of the experiment – as in many procedures already described – was the use of novel stimuli which previously had not been seen by the subject. If the bee continued to choose the smallest figures from the new sets of stimuli, the experimenter concluded that the insect had grasped the concept of the task and marked out the main (experimenter-dependent) property (i.e. size) of the stimuli independently of accessory properties (such as shape and colour).

Fig. 15.3 Experimental scheme elaborated by Mazokhin-Porshnyakov for studying insects' capacity for abstraction. (a) A feeder for the experimental study of complex discrimination in bees and wasps; (b) a wasp engaged in a complex discrimination task. (Experiments and photographs by V. Kartsev.)

The bees also managed more complex tasks such as conceptualising 'novelty of coloration' or 'dichromatic/monochromatic'. Here are two examples of the most complex problems solved by the most gifted members of the bee family. Both problems are concerned with choosing one out of a pair of compound figures, namely, chains of small pictorial elements. The conventional feature of the stimulus in one task was location of a single black element at the end of the chain. The pictorial elements varied in different tasks while their numbers varied in different chains. The second problem was to choose a chain consisting of paired elements versus non-paired. After 100 visits to the feeder the foraging bee learned that the food was placed on the picture with the chain consisting of paired elements while the trough which was placed on the picture with non-paired elements was always empty. After several error-free solutions the bee was presented with chains consisting of new elements. Choosing of paired chains consisting of different novel stimuli in tens of trials enabled experimenters to conclude that honeybees can master the 'paired/non-paired' conceptualisation.

15.4 Animals' natural concepts: classification at different levels of abstraction

As we have seen in previous chapters, animals of different species have shown some evidence of the acquisition of *natural concept formation*, that is, they can generate concepts which represent real objects and events. The ability to create concepts is closely related to the capacity for abstraction and classification, and demands generalisation of knowledge about the whole category as opposed to memorisation of specific exemplars.

When the Tahitians first saw a horse, introduced to the island by de Bougainville's French sailors, they immediately classified it in terms of the mammal they knew that most resembled it . . . a pig. It was obviously closer to the horse than the other two mammals of which they had any experience: the dog and the Polynesian rat. Very sensibly, they referred to the horse as a 'man-carrying pig'. This was a sophisticated attempt to understand the strange beast in light of their experience of similar beasts. There is no reason to suppose that animals do not do likewise (Goldsmith, 1998).

It is intuitively clear that the acquisition of concept formation in animals should be limited. Abstract concepts are defined by logical rules rather than perceptual features. Human infants are able to determine category membership at an early age. When they address a cat, or a cow, or a dog with the word 'bow-wow' they apparently have in mind common features of these animals. Parents will later suggest the idea to the child about that neither a cat nor a cow says 'bow-wow' and that the cat differs from the two others. Nevertheless these pedagogical efforts are only aimed at teaching the child to select classes of objects more precisely. The tendency to classify objects is innate to human beings. The ability to select character features and to group animals, plants and many other things as classes of objects makes the process of cognition more parsimonious and effectively improves our adaptiveness.

It is of no surprise that tests of the ability to make use of categories are widely applied for diagnosing mental diseases as well as local damages of the brain. Age psychologists also routinely use these tests. For example, psychologists assume that there is a possible developmental shift from similarity-based to theory-based categories in 2-year-old children. At this age, children are likely to realise that category membership can be a better basis than perceptual appearances for drawing certain inferences. Children can detect the category membership of some novel pictures on the basis of subtle perceptual cues. Being presented with pictures, 2-year-olds recognise a lamb-like dog as a dog on the basis of its nose, feet and tail, despite its overall colour and fur (Anglin, 1977; Gelman and Coley, 1990). An interesting situation has been described by Makarova (1990), a teacher in an art school for children. She presented a 2.5-year-old boy with a box filled with different buttons. The boy began to play with the buttons; spontaneously, he began to sort them by size and then by colour but very soon he got confused and raked all the buttons together. To the teacher's surprise, the boy then quickly found the way to sort the buttons: he put buttons with two holes to one side, and buttons with four holes to the other side.

In the early 1920s Köhler and Ladygina Koths demonstrated that chimpanzees can make out things in pictures quite well, which was the necessary prerequisite for studying non-human capacity for classification with the use of pictorial stimuli. The Kellogs, who reared the chimpanzee Gua together with their little son Donald, showed that the chimpanzee could recognise pictures and point at them on a request by her 'parents' at an earlier age than the child could (Kellog and Kellog, 1933). For example, 17-months-old Donald and 15-months-old Gua were shown paintings of a cup, a dog, a shoe and a house. Donald pointed at the dog in response to a request 'Show bow-wow' but he did not know the names of the other three things. Gua pointed at the dog, at the cup and at the shoe. Like Ladygina Koths' (1935) pupil Ioni, Gua found difficulty in recognising things whose sizes were dramatically changed in pictures in comparison with the originals (such as the pictured house in comparison with the real one). It is interesting to note that during his leisure hours Ioni sorted brightly coloured pieces of silk by colour; for example, he grouped yellow pieces and red ones separately.

Recent studies have examined concept discrimination learning in animals at various levels of abstraction. As concepts become more abstract, exemplars within the category share fewer features in common. At the concrete level exemplars within a concrete category may share more readily perceivable attributes such as size, shape and colour. More abstract categories contain members that may be difficult to discriminate on the basis of sensory properties (for example, all members of the category 'animals' share the ability to breath and reproduce).

Roberts and Mazmanian (1988) examined concept discrimination learning in pigeons, squirrel monkeys and humans. At the most concrete level, subjects were asked to select photographs of kingfisher birds from photographs of other bird species. All the kingfisher photographs shared readily distinguishable features such as size, colour and shape that were not shared by members of the non-reinforced category. All three species tested easily learned this discrimination. At an intermediate level of abstraction, subjects were asked to select photographs of birds from various other animal species. The bird members of the reinforced set shared several features but there was more variance between them. At the most abstract level, subjects were asked to select animal photographs from those of non-animals. Pigeons and monkeys learned the most abstract category more easily than the intermediate discrimination. There was some evidence that intermediate discrimination was easier to learn if non-animal photographs comprised the negative category. It is possible, however, that pigeons and monkeys were not necessarily operating on the basis of conceptual processing but may have been relying on the perceptual features of the stimuli – like, as Huber (2002) suggests, pigeons in the Herrnstein and Loveland (1964) study.

Vonk and MacDonald (2002) conducted experiments with a young gorilla Zuri based on tasks analogous with those of Roberts and Mazmanian (1988) and intended to represent the same three levels of abstraction. Zuri was presented with photographs on a touch-screen computer. The experimenters were trying to eliminate the possibility that their subject was using irrelevant aspects of the photographs to make the discriminations, both by varying as many aspects of the pictures belonging to the same categories as possible, and by detailed analysis of Zuri's errors.

Four discrimination tasks were considered to be concrete discriminations with the use of many pictures: (1) gorillas or orang-utans versus humans; (2) orang-utans versus other primates; (3) gorillas versus other primates; (4) orang-utan colour test. Zuri was asked to select orang-utan photographs by attending to a single feature: their reddish colour. In order to illustrate the fullness of classification, the second task (orang-utans versus other primates) consisted of three sets of orang-utans photographs versus other primates, over a total of 60 photographs. The orang-utan photographs included subjects at various stages of development and of both genders (Zuri had never seen an orang-utan before). The photographs of other primates included a wide range of primate species including gorillas. Zuri showed a high degree of transfer to novel photographs of both gorillas and orang-utans when these photographs were contrasted with photographs of humans. She was generally not distracted by irrelevant features of the stimuli, such as orientation, size or gender. In order to avoid the possibility that Zuri discriminated the two sets of exemplars by attending to specific features, rather than by attending to a general concept of 'gorilla' or 'orang-utan' (for instance, she might have learned to 'choose a black face' or 'avoid a white face'), the experimenters tested her performance on photographs that might be expected to be difficult had she been using simple cues. For instance, the transfer set included a photograph of an albino gorilla with a pink face. At this stage of concrete discrimination as well as at two other stages (intermediate discrimination such as primates versus non-primates including reptiles, insects and so on; and abstract discrimination such as animals versus non-animals), the experimenters were trying to distinguish between the use of a perceptual and a conceptual strategy. The young gorilla seemed to have the most difficulty with the intermediate discrimination. The authors acknowledge that the ease with which Zuri learned the most abstract discrimination was partially due to a generic 'learning to learn' phenomenon, because this was the last discrimination tested. In sum, acquisition and transfer involving abstract stimulus sets suggest a conceptual basis for categorisation in gorillas.

Savage-Rumbaugh *et al.* (1980) demonstrated that chimpanzees are able to sort real objects or pictures into two trays according to their belonging to 'food' or 'tool' categories. The chimpanzees were able to perform this task with great accuracy, even with objects that had not been used during training. In Part VIII we will see how language-trained primates perform categorisation on the basis of these language-based skills.

15.5 Mental representation (imagery)

Mental representation is one of the central paradigms in cognitive psychology. The ability for mental representation implies generalisation of a familiar image to its visual transformations. In terms of an experimental procedure, subjects prefer transformed (say, the mirror image or the left–right transformation) of the rewarded stimulus or group of stimuli to a neutral stimulus when the learned stimuli are absent. Being confronted with these transformations, some animals respond to them as if they were the original learned stimuli, thus showing a capacity for flexible stimulus use. This kind of ambiguity is associated with cognitive processes such as *mental rotation*, the capacity of mentally changing the orientation of an image in order to reassess it from a new imaginary perspective. It is assumed that the underlying mechanisms allowing such generalisations have an important cognitive basis. For instance, in generalising to a rotated image, mental rotations of the learned image are assumed to occur in the brain. A combination of several techniques devised by psychologists for studying mental manipulations with images has been extended to develop an imagery paradigm for animals. Processes of imagery in animals could be based on what is called *detached representation*. A chimpanzee that walks away from a termite hill to break a twig in order to make a stick for fishing termites has a detached representation of a stick and its use.

Operationally, imagery refers to processing visual information not currently before the subject, that is, the processing of representation of visual stimuli. This capacity of organisms is closely related to extrapolation and stimulus generalisation; hence the resemblance of the experimental procedures used to investigate them. An integral part of the method for testing imagery in animals is to induce transformation of the remembered stimulus.

Neiworth and Rilling (1987) have suggested an experimental procedure based on both the elements mentioned above: stimulus rotation and extrapolated movement. The first element was derived from the mental rotation task that was initially developed for humans by Shepard and his colleagues (Shepard and Metzler, 1971; Shepard and Cooper, 1982). In a series of psychological experiments done in the 1970s, human subjects were asked to state whether an object that had been rotated from its standard form was the same as the standard form or was a mirror image of the standard form. Subjects' response time in tasks involving mental manipulation and examination of presented figures was found to vary in proportion to the spatial properties (size, orientation, etc.) of the figures presented. Reaction times were found to be a linear function of the angular rotation of the stimulus. In other words, reaction time indicated that humans took more time to respond when there was greater angular disparity between the standard and rotated forms. This isomorphism between reaction times and angular rotation suggested that subjects were rotating mental images to solve this problem.

Hollard and Delius (1982) were the first to conduct a mental rotation study simultaneously in humans and animals. Pigeons and humans chose which one of two alternative visual forms was identical to, or a mirror image of, a previously presented sample form. The two comparison forms were presented in various orientations with respect to the sample. The two species yielded similar accuracies, but although human reaction times depended linearly on the angular disparities, those of the pigeon did not. Humans appeared to apply a well-known, thought-like, mental rotation procedure to the problem, whereas pigeons seemed to rely on different recognition processes. Mirror image forms may be better discriminated by the pigeon's visual system than by the human one. In general, it was suggested that pigeons may even be superior to humans in their capacity to mentally rotate visual information, as they showed no increase in reaction time when judging the identity of stimuli presented at increasingly different angular orientations, whereas humans did (Cook, 2001).

Neiworth and Rilling (1987) elaborated tasks for pigeons and humans based on stimulus rotation on a video monitor. The important element of their procedure was based on extrapolation tests when subjects were required to extrapolate the location of a moving target that had disappeared. This approach was applied to testing the hypothesis that perceptual information is maintained by the subject and that mental transformations occur while the moving stimulus is absent. The experimental procedure thus includes two ingredients such as a perceptual trial, in which the stimulus and its movement are fully presented, and an 'imagery' trial, in which a portion of the stimulus transformation is invisible. By comparing the behaviour of subjects on perceptual and imagery trials, it is possible to test whether or not subjects can cognitively change a representation in the same way as they have learned to discriminate the perceptual change.

The task for pigeons involved observation of a clock hand stimulus that rotated from an initial location of 0° (12.00 position). On perceptual trials, the clock hand moved at constant speed and was always visible to the subject. Imagery trials were identical except the clock hand disappeared at the 90° position (3.00) for a specific delay and then reappeared at an appropriate location as if it had rotated with constant speed during the delay. On violation trials, the clock hand also disappeared at 90° but reappeared after a delay at a position inconsistent with constant velocity whether or not the hand was always visible. After discrimination training involving two locations (135° and 180°) pigeons showed immediate and positive transfer to a novel intermediate location (158°) and to a novel location outside the boundaries trained (202°).

The experimenters found an essential parallel between imagery in pigeons and humans based on evidence of an interaction between the perceptual illusion of momentum and visual memory. The fact is that humans do not accurately represent the final location of the moving object; instead, visual memory of the final location shifts forward by some small amount, a distortion similar to the real momentum that occurs while moving objects stop. This phenomenon, called *representational momentum* (Freyd, 1983), is assumed to be fundamental and automatic, and it turned out that pigeons made the same kinds of errors as humans when representing movement. In sum, the results of experiments supported structural, functional and interactive descriptions of an imagery process in pigeons.

Wasserman *et al.* (1996) examined pigeon's capacity to recognise depth-rotated stimuli. They used a discrimination learning paradigm to explore pigeon's recognition of line drawings of four objects (an aeroplane, a chair, a desk lamp and a flashlight) that were rotated in the depth dimension. The pigeons readily generalised discriminative responding to pictorial stimuli over all untrained depth rotations.

Monkeys also demonstrated a clear evidence of stimulus generalisation with drawings of depth-rotated objects (Logothetis *et al.*, 1994). After training monkeys for 8 months on successive same/different discriminations, the researchers tested their subjects with drawings of either the target shape at its given orientation or the target shape at a different depth-rotation orientation. 'Same' responses were found to be highest when the test drawing depicted the target shape at its given orientation, these reports systematically falling as the test drawings depicted greater rotation away from this orientation. It is interesting to note that stronger stimulus generalisation was obtained with rendered drawings of realistic objects than with computer-generalised wireframe and blob-like shapes and with exposure to multiple views of the target shape than with single views of it.

Cook and Katz (1999) have found some similarities between humans and pigeons in their mental representation of three-dimensional objects. In these experiment pigeons were taught to discriminate between computer-generated three-dimensional projections of cubes and pyramids. These object stimuli were then presented on each trial either dynamically rotating around one or more of their axes or in a static position at a randomised viewing angle. Pigeons were rewarded for pecking in the presence of only one of the objects. Tests with different rotational transformations of the stimuli suggested that the pigeons may have been using a three-dimensional perception of these objects as the basis for their discrimination. The pigeons' performance was consistently better with the dynamic presentations than with randomly oriented static views. Further, their performance was relatively unaffected by transformations in object size; the rate, direction and combination of motions; and changes in the surface colour of the stimuli. Three of the four pigeons showed some evidence of recovering the structure or shape of these objects from just the pattern of their motion on the display. When all contour and surface information was removed in test conditions leaving only the rigid projective geometry of the moving objects to guide their performance, these birds were again better at discriminating the dynamic stimuli. Cook (2001) suggested that pigeons may have some higher-level similarities with humans, as the experiments showed some capacity for recognising objects across different transformations and deriving structural information from the pattern of an object's motion.

There is some clear experimental evidence that complex visual information processing can be implemented in even smaller but also highly mobile animals such as bees, although it is still under discussion whether these results can be explained with the use of a mental representation paradigm. The ability to generalise a familiar image to visual transformations like a mirror image or a left–right transformation allows the recognition of familiar images from a different viewpoints for these small creatures. The fact is that bees learn and memorise not only the location of food sources but also specific floral features that allow them to discriminate between rewarded and non-rewarded flowers, and visual cues are of fundamental importance in this context.

The question of whether bees are able to transfer acquired information about a previously rewarded pattern to its mirror image and its left–right transformation was first raised by Gould and Gould (1988) in honeybees. Later this problem was studied on bumble-bees (Plowright, 1997). In Gould's study each stimulus consisted of four circles of the same diameter and different colours. The transformations were obtained by changing the positions of the circles. The same technique was used in the experiments with bumble-bees. However, in these experiments the possibility of the use of colour preference by the insects was not excluded.

Giurfa and co-authors (Giurfa *et al.*, 1999; Giurfa and Lehrer, 2001) in their experiments with honeybees used achromatic (black and white) stimuli whose transformations resulted in different alternatives for mirror image and left–right permutations. Individually marked honeybees were trained to collect syrup on vertically presented stimuli lying flat on the back walls of a wooden Y-maze that was placed close to a large open window in the laboratory, through which bees could enter the maze. Bees were trained either with a single pair of patterns or with six different pairs of patterns presented in a random succession. Within each pair one pattern was rewarded and the other was not. All patterns had four quadrants, each displaying a different stripe orientation. In multiple-pattern training the six rewarded patterns shared a common configuration different from that of the six non-rewarded ones. After both kinds of training, the bees preferred the mirror image and the left–right transformation of the rewarded pattern (or rewarded configuration) to a novel pattern. They also preferred the left–right transformation to the mirror image. The researchers explain this performance by: (1) matching with a retinotopic template of the trained patterns after training with a single pair of patterns; and (2) matching with a generalised pattern configuration after training with randomised sets of patterns. Although both strategies are based on comparison of an image currently perceived with one that has to be accessed from memory, they constitute different options as the former is less flexible while the latter allows categorisation of novel patterns.

In experiments of Plowright and co-authors (2001), bumble-bees were trained to discriminate between a reinforcing pattern (S+) and a non-reinforcing one (S−) which differed only in the configuration of four artificial petals. They were subsequently tested for recognition of the S+ rotated by 90° (S + 90). Experiment 1 used petals of four colours, and the other experiments used four symbols. The symbols either remained unchanged when the whole pattern was rotated (e.g. + in Experiment 2) or changed appearance (e.g. < in Experiment 3). The bumble-bees failed to recognize the S + 90 in the first two experiments, but in Experiment 3, the choice proportion for S + 90 in the presence of a new pattern was significantly higher than chance. The researchers concluded that bumble-bees can recognize a rotated pattern, possibly by using mental rotation, provided that a cue as to the extent of the pattern transformation is given.

The explanation of the results obtained on bees possibly does not invoke mental rotation processes. As Giurfa (2003) proposes, the processes involved in left–right transformation and mirror image generalisation seem to be based on 'lower' processes such as integration of pattern features into a simplified configuration and comparison of the resulting representation with the currently perceived image. In any case, bees can generalise the excitatory strength of a trained pattern to the mirror image and the left–right transformations. These insects are capable of the flexible use of the resulting representation to categorise and thus respond appropriately to novel patterns.

Chapter 16

Conceptual behaviour based on relations

Here we will consider experimental studies aimed at the problem of whether animals possess a type of reasoning that allows them to combine knowledge about specific relationships in order to infer another relationship. As we saw in the previous chapter, some animals can possibly operate with perceptual concepts based on categories such as 'people', or 'trees', in which application of the concept leads to the conclusion that novel exemplars belong to a given category. *Relation concepts* are based on common abstract relationships shared by sets of stimuli. For example, a generalised identity concept is shown by a subject that discriminates among novel sets of stimuli based on the perceptual *sameness* of identical stimuli, whereas a generalised *oddity* concept is shown by a subject that discriminates among novel sets of stimuli based on their perceptual difference. Finally, *associative concepts* are those in which stimuli are categorised on the basis of common associations with other stimuli, responses or outcomes. For example, disparate arbitrary stimuli, such as A and C, may become related to one another through their common relationships with a given mediated stimulus, B. Although the stimuli in such associative categories are not perceptually similar, they are treated in a similar fashion, as demonstrated by a transfer of function among the stimuli in a given category.

16.1 Cross-modal transfer in discrimination tasks

Inference of modality plays a key role in many situations when it is necessary to match relation properties of different stimuli. The ability to recognise by touch those things that we have only perceived, say, visually is a part of what is termed *cross-modal inference*. To do this, we need to compare signals received through different sensory channels and bring them into conformity with each other. The ability to solve problems based on cross-modal inference has been intensively studied as an essential part of human maturation. For example, in the experiments of Krekling and co-authors (1989), the ability to solve tactual oddity problems, and the transfer of oddity learning across the visual and the tactual modalities, was studied in 3- to 8-year-old children. Oddity tasks consisting of one odd and two equal objects were made from stimuli that were easily discriminated visually and tactually. The results showed that tactual oddity learning increased gradually with age. The growth in tactual performance begins later than visual, suggesting that children are more adept at encoding visual stimulus invariances or relational properties than tactual ones. Bidirectional cross-modal transfer of oddity learning was found, supporting the suggestion that such transfer occurs when training and transfer oddity tasks share a common vehicle dimension. The cross-modal effect also shows that oddity learning is independent of a specific modality-labelled perceptual context. These results are consistent with the view that the development of oddity learning depends on a single rather than a dual process, and that the oddity relation may be treated as an amodal stimulus feature.

Ladygina Koths (1923) was first to demonstrate that a non-human animal can possess the ability of cross-modal inference. In her experiments with chimpanzee Ioni, she placed into a bag several plane and volume figures, such as a prism, a cylinder, a plane circle, a plane square and a plane triangle. Being presented with one of those stimuli visually, the chimp accurately selected the same thing in the dark inside the bag by touch. Joni recognised a figure that had been perceived visually before, now on the basis of its tactile properties only. These results were later confirmed in experiments on chimpanzees that manifested the ability to recognise in photographs not only those things that they had seen before but as well those things that they had only perceived by touch (Davenport and Rogers, 1968).

In a recent study by Hashiya and Kojima (2001), a chimpanzee solved an auditory–visual intermodal matching-to-sample (AVMTS) task, in which, following the presentation of a sample sound, the subject had to select from two photographs that corresponded to the sample. The authors describe through a series of experiments the features of the chimpanzee AVMTS performance in comparison with results obtained in a visual intramodal matching task, in which a visual stimulus alone served as the sample. The results show that the acquisition of AVMTS was facilitated by the alternation of auditory presentation and audiovisual presentation (i.e. the sample sound together with a visual presentation of the object producing the particular sample sound). Once AVMTS performance was established for the limited number of stimulus sets, the subject showed rapid transfer of the performance to novel sets. However, the subject showed a steep decay of matching performance as a function of the delay interval between the sample and the alternative choice presentations when the sound alone, but not the visual stimulus alone, served as the sample. This might suggest a cognitive limitation for the chimpanzee in auditory-related tasks. In dolphins a similar cross-modal transfer between vision and echolocation has been demonstrated with the use of a matching-to-sample procedure (Pack and Herman, 1995).

16.2 Ordering and serial learning

The ability to put objects in order and to memorise them as a whole sequence is crucial for an intelligent action. *Serial memory* is the ability to encode and retrieve a list of items in their correct temporal order. The nature of the mental representation that allows to retrieve a list of items is still unclear. As many other approaches for studying animal intelligence, the experimental paradigm of what is also called *list learning* is derived from psychological studies, namely, from the classic experiments of Ebbinghaus (1885) (see Chapter 3) and Ebenholtz (1963), who first investigated the organisation of such sequences in experiments on the memorisation of nonsense syllables. Terrace (2001) suggests considering this classic paradigm together with the mastery of various types of mazes (Small, 1900) (see Chapter 1) because the results of both types of experiments gave rise to the classic theory which states that serially organised behaviour can be represented as a linear sequence of associations. Ebbinghaus explained list learning by reference to associations between successive items and between a particular item and its list position. Hull offered a similar explanation of maze learning by rats (Hull, 1943). Thus, associative principles that were used to explain how a human adult memorises a list of arbitrary items were used to explain how an experimentally naive rat learns a sequence of arbitrary responses, and vice versa (Osgood, 1953; Terrace, 2001).

Lashley (1951) rejected linear models of serially organised behaviour because they could not explain the knowledge of relationships between non-adjacent items. Recently cognitive ethologists and neurophysiologists have raised a question of serially organised association by augmenting linear structures of stimuli with hierarchical structures (Orlov *et al.*, 2002). This problem is based on the concept of *chunking* which was first introduced by Miller (1956) in his paper 'On the magical number 7'. Miller argued that a chunk was the basic unit for measuring the capacity of immediate memory (in current terminology, short-term or working memory; see Part III). The idea was that subjects could retain a large number of discrete items of information if they were encoded as chunks before they were transferred to long-term memory (LTM). For example, the 12 digits 1–4–9–2–1–7–7–6–1–8–1–2 could be encoded as three historical dates. In contrast to the enormous capacity of long-term memory, Miller estimated the capacity of short-term memory (STM) to be 7 ± 2 chunks and argued that the amount of information that is retained in STM is independent of the amount of information contained in each chunk.

One of the most relevant methods that allows studying serial learning in non-verbal subjects is the so-called *simultaneous chaining procedure*, which differs from those used in previous studies of serial learning in animals (Straub and Terrace, 1981; Terrace, 1987; Chen *et al.*, 1997; Terrace *et al.*, 2003). The main idea of this method is that a subject has not to simply recognise learned stimuli (multiple matching-to-samples), but to choose the stimuli in a definite sequential order. Unlike the successive chaining paradigm, a simultaneous chaining paradigm presents all list items throughout each trial (e.g. the numbers on the face of a telephone). In a successive chain, the subject encounters each cue individually (e.g. the choice points in a maze). A second difference is the variation of the physical configuration of list items from trial to trial. This prevents subjects from using a particular physical sequence of responses to produce the required list (for example, when making a telephone call with a sequence of learned movements on a number pad). To execute a simultaneous chain correctly, the subject has to respond to each item in a particular order, regardless of its spatial position.

Experiments on monkeys and pigeons showed that they learn sequences of stimuli much more readily in a case where the stimuli could be clustered into groups. Straub *et al.* (1979) trained pigeons to respond to randomly configured arrays of four colours in a particular sequence: red–green–yellow–blue. Pigeons learned the four-item list of colours. The extensive training time that pigeons need to master a four-item list (more than 3 months) suggests that four items may approach the limit of their memory span. For human subjects the classic remedy for overcoming limitations of memory span is to reorganise unrelated list items into chunks (Miller, 1956). The efficacy of that approach was evaluated with pigeons that were trained to learn five-item lists composed of colours and achromatic geometric forms. When the number of items increased, pigeons responded at chance levels of accuracy to subsets drawn from lists on which items were not clustered. Pigeons showed no signs of improvement on successive three- or four-item lists, each composed of novel items (colour photographs of natural scenes).

D'Amato and Colombo (1988) used the simultaneous chaining procedure to train capuchin monkeys to produce arbitrary four-item lists. Monkeys were trained on four lists, each containing four novel photographs of natural objects (flowers, fruits, animals). The task was to touch the simultaneously presented images in the correct order (A1–A2–A3–A4, B1–B2–B3–B4, C1–C2–C3–C4, D1–D2–D3–D4). When the monkeys had mastered this task, the items were shuffled, taking one item from each list, so that in two derived lists the ordinal numbers of the items were maintained (e.g. A1–D2–C3–B4) while in two others they were not (e.g. B3–A1–D4–C2). Lists with maintained ordinal position were acquired rapidly and virtually without error, while derived lists in which the ordinal position was changed were as difficult to learn as novel lists. This pattern of transfer to derived lists implies that the monkeys originally acquired some knowledge about each item's ordinal position, rather than only generating a chain of serial pair associations for each list of items. Monkeys acquired five-item lists more rapidly than pigeons. Pigeons showed no signs of improvement on successive three- or four-item lists, each composed of novel items (colour photographs of natural scenes). Another important difference between the serial skills of monkeys and pigeons was the ease of acquiring new lists. Monkeys trained to learn successive four- and six-item lists of different photographs became progressively more efficient at mastering each list (Swartz *et al.*, 1991; Chen *et al.*, 1997). More recently Terrace (2001) showed that rhesus monkeys can learn seven-item lists. Monkeys were first trained on three- and four-item lists. The subjects were then trained in the same manner on four seven-item lists. The monkeys not only mastered each list but they did so with progressively fewer trials on each new list. To place this achievement in a perspective, the probability of guessing correctly the ordinal position of each item at the start of training on a seven-item list is $1/7! < 0.0002$.

Another line of research addresses what is called *serial recognition*, that is, the ability of monkeys to recognise stimuli that are presented sequentially. In the first experiment, subjects were shown a sequence of novel photographs at the beginning of each trial. They were then required to select these photographs from a simultaneous display that included 'distractors'. Subjects were allowed to respond to the items in the display in any order they chose. To obtain a reward they had to select all of the items shown at the beginning of the trial and avoid responding to any of the distractors. Subjects were able to recognise all items of four-item sequences at high levels of accuracy even when those items were embedded in displays containing five distractors. Because there are no constraints on the order in which items could be selected, this paradigm is methodologically similar to free-recall studies with human subjects. One empirical similarity between the recall of sequentially presented items by monkeys and humans is the lengthening of the response time for each successive item that is reported. The monotonically increasing function relating response time to the order of reporting appears to reflect the difficulty of searching the contents of STM as more items are recalled rather than the time needed for a visual search of the actual display (Terrace, 2001).

16.3 Transitive inference

Such a type of reasoning as transitive inference (TI) allows us easily to reach the conclusion that, for example, A is bigger than C if A is bigger than B and B is bigger than C without having to see A and C side by side. The case of transitive inference is an example of a more general dispute between proponents of associative accounts and advocates of more cognitive account of animal behaviour (Allen, 2004).

Experimental procedure for studying transitivity is based on the technique of matching-to-sample that was described in Chapter 14. When a subject is taught to select stimulus B in the presence of stimulus A and to select stimulus C in the present of B, it is likely that the subject also will select A in the presence of B (*symmetry*), B in the presence of C (symmetry), C in the presence of A (*transitivity*) and A in the presence of C (combined symmetry and transitivity, or *equivalence*) without further training. When this occurs, the stimuli are said to participate (Roche and Barnes, 1997). Animal versions of the experimental scheme are derived from the original Piagetian version of the five-element task (Piaget, 1952) designed to test transitive inference in young human children. In that version, symbols or colours were paired with different length rods.

In experiments of Gillan (1981), chimpanzees first were trained to find food in five containers, A, B, C, D and E, each of which had a different colour. Containers were presented to the animals in pairs in accordance with the following schedule: A+B−, B+C−, C+D−, D+E−, where + denotes the container with food. The containers changed places irregularly so that discrimination could be achieved only on the basis of colour. After a number of trials the subjects persistently preferred A to B, B to C, C to D and D to E. To examine whether these relationships could be combined to lead to a novel inference, Gillan then gave test sessions that included, for the first time, the pair B and D. If apes are capable of transitive inference, then combining the knowledge that B is preferred to C with the knowledge that C is preferred to D should lead to the conclusion that B is preferred to D. One chimpanzee, Sadie, performed perfectly on this test choosing B in preference to D on all 12 tests trials. Although the results for the other two chimpanzees were not so good, the success of Sadie shows that at least one chimpanzee can solve the transitive inference problem.

Using the arbitrary matching-to-sample procedure, Schusterman and Kastak (1993) established that the California sea lion (a female named Rio) could relate stimulus pairs AB and BC with specific reinforcement contingencies. To obtain reinforcement (a piece of fish), Rio made a choice by sticking her nose into one of the two boxes containing discriminative stimuli. Rio was trained to select B conditionally upon presentation of A samples and to select C conditionally upon presentation of B samples, for each of 30 potential classes. Rio could immediately extend these relations without further training to stimulus pairs BA and CB (symmetry), AC (transitivity) and CA (equivalence). Ultimately, Rio had formed 30 three-member sets of stimuli, which were mutually substitutable within a matching-to-sample procedure. On another set of experiments with Rio, Schusterman and Kastak (1998) tested whether the discriminative function acquired in the matching-to-sample procedure by one member of an equivalence class transfers immediately and completely to the two other members of that class in a novel context. That is, if one member (A) of an equivalence class (ABC) becomes discriminative for a response, do the other members (B and C) also become discriminative for that same response? For instance, in one experiment two classes (numbered 1 and 2) each consisting of three members (A, B and C) were tested in a simple discrimination. The subject initially receives one simple discrimination pitting the B member of class 1 (B1) against the B member of class 2 (B2). If she selects B2 rather than B1, establishing B2 as the S+ and B1 as the S−, will she then select A2 rather than A1, when these stimuli are first presented as alternative choices? Will she also select C2 rather than C1 when they are also presented as alternative choices? In 28 of 30 tests, the sea lion immediately transferred the discriminative function acquired by one member of an equivalence class to the remaining members of that class. The elaborated relational concept, that of generalised identity matching, was re-tested after 10 years. Rio immediately and reliably applied the previously established identity concept to familiar and novel sets of matching problems (Reichmuth Kastak and Schusterman, 2002).

The use of a similar method has allowed other workers to suggest that transitive inference occurs in pigeons (Fersen and Lea, 1990), squirrel monkeys (McGonigle and Chalmers, 1992), hooded crows (Zorina, 1997; Lazareva *et al.*, 2004) and several other species (for detailed analysis see Pearce, 2000).

16.4 Relational matching-to-sample

The relational matching task is the most crucial one in testing animal knowledge about relationships between stimuli because it requires the most abstract reasoning abilities. If a non-human animal were able to learn this so-called relational matching-to-sample task, then this learning would constitute what some theorists consider to be the strongest evidence of abstract conceptual behaviour (Thompson, 1995). Until recently, the relational matching task had only been mastered by children of 5 years of age, or by chimpanzees first trained to use symbols to communicate (Thompson *et al.*, 1997).

In experiments by Fagot *et al.* (2001), baboons (*Papio papio*) successfully learned relational matching-to-sample. They picked the choice display that involved the same relation among 16 pictures (same or different) as the sample display, although the sample display shared no pictures with the choice displays. The baboons generalised relational matching behaviour to sample displays created from novel pictures.

Let us consider this experimental procedure as an illustrative example. The baboons were studied inside an experimental enclosure facing an analogue joystick, a metal touch-pad and a colour monitor that was driven by a computer. On the front of the enclosure were a view port, a hand port and a food dispenser that delivered food pellets into the enclosure in accordance with the prevailing contingencies of reinforcement. As stimuli, 72 computer icons were chosen and sorted into three sets. Two types of stimulus arrays were created from each of these icon sets: 'same' arrays and 'different' arrays. The stimulus set that was used to create the sample stimuli differed from the stimulus set that was used to create the choice stimuli. This difference meant that correct choice responses could not be based on the identity of the individual icons in the sample and the choice arrays; such identity was impossible. Correct responses could only be based on the relation of the icons in the sample and the choice arrays.

The baboons were individually placed into the test apparatus. The experiment involved multiple sessions per day of many trials that involved a two-alternative forced-choice matching-to-sample procedure. By manipulating a joystick, the baboons were required to place the cursor in such a way as to initiate the presentation. Then a same or a different sample array appeared for 500 milliseconds in either the left or the right half of the monitor. Immediately after sample stimulus offset, two choice arrays appeared on the vertical axis of the screen. One of these two choice stimuli was a 'same' array; it involved a single icon that was repeated 16 times. The other choice stimulus was a 'different' array; it involved 16 different icons. The location of the same and different choice arrays, either at the top or bottom of the screen, randomly varied across trials.

The choice stimuli remained on the screen for a maximum of 10 seconds, during which the baboons could make a single choice response by moving the cursor into contact with one or the other choice array. Failures to make a choice response in the allotted time were rare during training and never occurred during testing. Correct responses were followed by a high tone and one food pellet; incorrect responses were followed by a low tone and a short time-out. After food delivery or time-out, the next trial could be initiated by contact with the touch-pad.

The acquisition of relational matching-to-sample proceeded gradually in baboons. The results were clear: animals successfully acquired relational matching-to-sample to levels in excess of 80 per cent correct. The task was not easy for them; it took one animal 4992 trials to respond consistently in excess of 80 per cent correct, and it took a second one 7104 trials to do so. Earlier, the animals had learned to report 16-icon same versus different displays by making one choice response (e.g., 'up') for same displays and another choice response (e.g., 'down') for different displays (Wasserman *et al.*, 2001); it took only 600 and 700 trials to learn these. Relational matching-to-sample is obviously much more difficult for baboons to acquire than is same/different discrimination learning.

The experimenters were also interested in whether the baboons could generalise their relational matching-to-sample behaviour to novel sample displays. Here, too, the results were clear. On the very first session of testing, the two baboons discriminated the displays of novel sample items at a mean level of 70 per cent correct, whereas they discriminated the displays of familiar sample items at a mean level of 83 per cent correct. Two subsequent sessions of testing yielded very similar levels of discriminative performance. Such robust generalisation to the novel testing displays supports the hypothesis that these monkeys had indeed learned an abstract and general concept.

Recent studies have demonstrated that baboons surpass 3-year old children in their ability to solve the relational matching task (Bovet *et al.*, 2005). Whereas the baboons were able, after considerable training, to master this task, the 3-year-old children could not solve the relational matching task without any verbal guidance; they were only able to perform this task when the rationale of all of the steps was explained to them. Besides the children's linguistic capabilities, this disparity can be explained by analogical reasoning abilities that seem to be more developed in adult baboons than in young children.

CONCLUDING COMMENTS

Both animals and researchers have achieved remarkable results in displaying cognitive skills from one end and measuring them from the other end. Mainly on the basis of variants of discrimination tests, students of animal cognition have revealed abilities for categorisation, abstraction, mental representation, serial learning, transitive inference and relational matching in non-human animals such as apes, monkeys, horses, pigeons, crows, rats and some others.

Some uncertainty still surrounds the interpretation of the extraordinary capacities for categorisation, abstraction and imagery in pigeons and bees. It is still an open question as to whether these animals possess advanced cognitive abilities or whether they rather exploit much more simple discrimination based on very subtle perceptual variation in the properties of natural objects. However, it is indicative that the highest forms of categorisation are currently being intensively investigated in many species. We should further examine the limitations in animals' thought processes in parallel with their high levels of intelligence. As we will see in Part VII, members of different species are highly selective in their responses to stimuli and in forming associations. Members of some species display very fast and advanced learning within specific domains, closely connected with their ecological traits and evolutionary history. This enables us to claim that at least some species possess high cognitive skills. The most advanced of them will be considered in Part VI.

Part VI

Advanced intelligence in animals: rule extraction, tool-using and number-related skills

'Owl,' said Rabbit shortly, 'you and I have brains. The others have fluff. If there is any thinking to be done in this Forest – and when I say thinking I mean **thinking** – you and I must do it.'

A. A. Milne, *The House at Pooh Corner*

This Part is devoted to the most advanced forms of animal intelligence, and it starts from what was earlier called 'insight' and recently, in terms of cognitive ethology, is considered a component of rule extraction. Together with another important component of rule extraction, learning set formation ('learning to learn'), insightful behaviour is based on the animals' capacities for formation of 'what-leads-to what' expectancies which, in turn, are closely connected with their exploratory activity and 'latent learning' (Tolman, 1932). The theory of learning to learn (Harlow, 1949) describes the ability of animals to learn a general rule which can then be applied to rapidly solve new problem sets. Learning set formation still serves as a relevant measure of intelligence. Both classic and modern results in this field will be analysed in Chapter 17.

Working with tool-using animals in laboratory, experimenters have the opportunity to investigate what can be called 'distillation' of rule extraction, that is, to explore how a subject understands the laws of the physical world. As we will see from a brief review of field animal studies in Chapter 18, habits of tool use are distributed oddly in the animal kingdom. The crucial question is whether animals can abstract general principles of relations between objects regardless of the exact circumstances, or whether they develop specific associations between perceptible things and the concrete situations they are involved in.

Many intricate experimental paradigms have been elaborated to explore this problem.

Finally, basic number meaning, that is, knowledge of quantities and their relations, will be considered one of the highest cognitive properties in animals. This problem will be considered in Chapter 19.

Chapter 17

Insightful behaviour

17.1 What is insight?

Let us start this chapter with a quotation from Wolfgang Köhler's Address of the President to the 67th Annual Convention of the American Psychological Association (1959):

> What is insight? In its strict sense, the term refers to the fact that, when we are aware of a relation, of any relation, this relation is not experienced as a fact by itself, but rather as something that follows from the characteristics of the objects under consideration. Now, when primates try to solve a problem, their behavior often shows that they are aware of a certain important relation. But when they now make use of this "insight," and thus solve their problem, should this achievement be called a solution by insight? No – it is by no means clear that it was also insight which made that particular relation emerge. In a given situation, we or a monkey may become aware of a great many relations. If, at a certain moment, we or a monkey attend to the right one, this may happen for several reasons, some entirely unrelated to insight. Consequently, it is misleading to call the whole process a "solution by insight."

We see then that even the 'father' of the theory of insight in animals, in his late publications, considered a notion of insight rather fuzzy. As has been described in Part I, Köhler (1918, 1925) first revealed experimental evidence of insight learning in chimpanzees based on the ideas of Gestalt psychology. His series of experiments showed chimpanzees as using planning and foresight, that is, cognitive reasoning to solve a problem. Köhler devised an arrangement in which all of the elements necessary for the solution of the problem were in full view of the animal. In the 1910s Köhler conducted thousands of experiments with chimpanzees and summarised them in his book *The Mentality of Apes* (Köhler, 1925). His observations have been the subject of controversy ever since. The period of quiescence that sometimes preceded the solution by animals, its sudden onset, and its smooth, continuous emergence were proffered as evidence that (1) contrary to suggestions of learning theorists of the day, problem-solving was not necessarily a trial-and-error process, and (2) constructs such as 'insight' were necessary for an adequate account.

One of the most often referred-to examples of problem-solving by insight concerns Sultan, a chimpanzee, whom Köhler regarded as the brightest of a number of apes he worked with (and whom we have already met in Part I). Sultan sat in his cage, in which there was also a short stick. Outside the cage there was a longer stick, which was beyond Sultan's reach, and even further away was a reward of fruit. Sultan first tried to reach the fruit with the smaller of the sticks. Not succeeding, he tried a piece of wire that projected from the netting in his cage, but that, too, was in vain. Then he gazed about him and after a long pause suddenly picked up the short stick once more, came to the bars directly opposite to the long stick, dragged it towards him with the auxiliary, seized it and went with it to the point opposite the objective which he secured. From the moment that his eyes fell upon the long stick, his procedure formed one consecutive whole.

Modern researchers in the field of animal learning and cognition find it difficult to interpret the majority of Köhler's experiments because his apes may be simultaneously engaged in the same problem so that it is virtually impossible to understand the problem-solving abilities of any individual. Furthermore, all of his subjects had played with boxes and sticks prior to the experimental trials in which they then reached bananas that were hung from the ceiling of their enclosure by pushing a box under the bananas and then climbing onto the box, or by using a stick to shake the banana down. The absence of trial-and-error responses may thus have been due to the prior experience of the animals. As we have seen from the quotation above, Köhler (1959) himself was frustrated about understanding the nature of subject's awareness about relations between things that he called insight. In particular, Köhler distinguished two kinds of mistakes related to problem-solving in chimpanzees, that is, 'good mistakes' and 'bad mistakes'. If the chimpanzee tried to affix a box to a wall in order to reach a banana from the top of the wall, it is a 'good mistake' which can serve as evidence of the animal's understanding about dimensions; the chimpanzee is simply unaware of the box's properties. But chimpanzees in Köhler's experiments also demonstrated a lot of 'bad mistakes'. For instance, in one set-up, chimpanzees could only get bananas by removing a box. Here was something, Köhler expected, that even his awkward chimpanzees could 'do at once'. And yet, to his astonishment, the chimpanzees had difficulties in solving such problems; they often drew into the situation the strangest and most distant tools, and adopted the most peculiar methods, rather than removing a simple obstacle which could be displaced with perfect ease.

More recently, insightful behaviour has been defined as 'the sudden production of a new adaptive response not arrived at by trial behaviour' and 'the solution of a problem by the sudden adaptive reorganisation of experience' (Thorpe, 1963). There are two key words here, 'experience' and 'sudden'.

In problem-solving, the experience need not be directly associated with the problem at hand. Sometimes, when presented with a new problem, an animal will solve the problem at the first attempt because of experience in dealing with the component parts of the problem, although they had never been met together in just such a way before.

The important role of past experience in problem-solving is illustrated by the experiments concerning insight in pigeons (Epstein *et al.*, 1984). In some sense, this can be called 'pseudo-insight'. Researchers replicated with pigeons a classic 'hanging banana problem' which Köhler suggested to chimpanzees, and copied the design of Köhler's experiment. Pigeons that had acquired relevant skills solved the problem in a remarkably chimpanzee-like fashion. They were trained with the use of instrumental conditional technique to perform two behaviours: to push a cardboard box toward a target (a green spot placed at various locations on the floor) and to climb onto a box and peck at a miniature model of a banana which was suspended overhead. Banana pecking was reinforced with the food. These two behaviours were always trained in separate training sessions. Once they had learned, the experimenters confronted the pigeon with a situation it had not encountered during training. The box was placed in one part of the chamber and the banana was in another. At first, the bird appeared to be confused. It stretched toward the banana, turned back and forth from the banana to the box, and so on. Then, rather suddenly, it began to push the box toward the banana, sighting the banana and readjusting the path of the box as it pushed. Finally, it stopped pushing when the box was near the banana, climbed onto the box, and pecked the banana. In the subsequent experiments, Epstein (1987) taught a pigeon to (a) peck at a model of a banana, (b) climb onto a box, (c) open a door and (d) push a box toward a target. During testing, the pigeon was confronted with a banana hanging over its head and a box behind a door. The bird successfully combined all four behaviours to solve the problem: it opened the door, pushed the box out and moved it under the banana, climbed the box and pecked at the banana.

The possible contributions of different experiences were determined by varying the training histories of different birds. The experimenters concluded that successful performance was dependent upon the birds learning the basic skills in separate contexts that must go together to solve the problem. For example, it was demonstrated that only if the birds had learned to push a box towards a goal would they move it under a model banana hanging from the ceiling of the cage. Epstein argued that it is not enough to teach a bird to push a box; it must have learned to push it toward a goal. Presumably, the chimpanzees' experience with moving boxes around and climbing on them was necessary for them to solve the hanging banana problem.

Recently students of animal intelligence have argued that describing such behaviour as 'insightful' does not help us to understand it. If we do not know what behaviours an animal has already learned, a novel and complex sequence of behaviours that solves a problem seems to come from nowhere (Skinner, 1985). Rather than attributing the successful pigeon's performance to insight, Epstein offered a moment-to-moment explanation of their actions, based on principles of behaviourism (see Parts I and II). This is no surprise because Epstein had started as a graduate student of Skinner. As Skinner himself noted (1987), 'We collaborated on a variety of research, including a 3-year project which we eventually called "Columban [Pigeon] Simulation." Through careful construction of complex contingencies of reinforcement, we were able to get pigeons to exhibit behavior said to show "symbolic communication," "spontaneous use of memoranda," "self-concept," "insight," and other so-called cognitive or creative processes.'

Pigeon simulation demonstrated that animals can effectively combine previously learned skills in order to reach a goal even if these skills were never trained together. But this project had little in common with Köhler's main ideas that were based on the principles of Gestalt psychology. Köhler argued that 'sudden' (the second key word in the definition of insight) solution of a problem by animals is based on their understanding of the whole situation. He believed that chimpanzees grasped general principles or relationships and 'saw' the solution before carrying out an action. Perceptual reorganisation of the elements that constitute a problem situation allows animals (and humans) to see at one moment all the parts in relation to each other, forming a meaningful whole. Unlike Epstein's simulation, Köhler's scenario often demonstrated animals as being able to find unexpected techniques to reach a goal by using unexpected accessory items.

For example, Sultan learned to use such improvised means as people in order to solve the hanging banana problem. He seized a trainer by the hand and pushed him in order to jump on his shoulders and reach the prize. Sultan was very angry at the trainer when he did not grant the ape's wish. Once Sultan climbed on the trainer in order to use the man as a step-ladder but the trainer stooped so that the chimpanzee was not able to grasp the banana. Then Sultan jumped off, caught hold of the trainer's belt and tried to lift the man up, with groans. It soon became a habit with chimpanzees to use each other as step-ladders so they were crowded under bananas suspended over their heads. Chica, the most athletic chimpanzee in the group, preferred to stand a long bamboo stick upright and quickly climb up the 7 metres and grab the bananas before the stick fell over.

One task for the apes was to reach the hanging banana in a situation when the only box was available filled with four heavy stones. Among nine chimps only Sultan was clever enough to pull out all stones, one by one. He came to this solution after many attempts. First, he pulled out only one stone and tried to push the box with all his might but failed. Two stones were enough to be pulled out in order to push the box with an effort. After four repetitions, Sultan came to the most effective way of solving the problem and after that he always pulled all the stones out of the box. Other apes did not grasp the relationship between the stones placed in the box and difficulties they met when trying to move it forward. For instance, Grande took one stone out, but not so as to make the box movable but to use the stone as a step stool. She then estimated a situation correctly and did not try to climb onto the stone because it was too low. Nevertheless, like other apes, she did not guess how to act with the stones in order to make the box lighter.

Indeed, it is a reasonable criticism that many of the 'sudden' solutions of the chimpanzees in these experiments were achieved rather slowly in reality. Critics argue that only after years of practice and repetition did the apes learn the behaviour needed to solve the problem. Another criticism is that chimps have been observed to commonly swing sticks at nothing. They have also been seen enjoying themselves by climbing up on boxes and stacking them without a prize in view, so that the prize did not trigger the behaviour. Moreover, many factors were not estimated properly. In particular, some unaccounted factors could fix sequences of behavioural acts of animals in an unknown way thus making them 'ritualised'. Perhaps this was just the case in the experiments of Köhler's opponent, Pavlov, who replicated Köhler's experiments in order to prove that chimpanzees' behaviour is governed by contingencies of sequences of stimuli and reactions rather than by 'insightful' behaviour. At the same time, the results of these experiments by Pavlov and his students can set Köhler's results off and thus help us to understand insightful behaviour more clear.

Pavlov purchased two chimpanzees, Rafael and Rose. He performed, from 1933 to 1936, a number of experiments, including a replication of Köhler's building experiment. Confirming Köhler's findings, Pavlov explained the problem-solving process in terms of unconditional reflexes and the establishment, by Pavlovian conditioning and the Thorndikean method of trial and error, of temporary neural connections identical, on the psychological level, to associations. According to Pavlov, insight is achieved progressively – as the result of the organism's problem-solving behaviour. This statement contradicts Köhler's theory of a sudden subjective reorganisation of the environmental situation.

The most complex scenario was performed by Pavlov's student (Vatzuro, 1941). Rafael, the chimpanzee, was made to join together many complex behavioural patterns which he had already learned in separate training sessions. The chimpanzee was requested to (1) stack several boxes on top of one another and get a hanging stick; (2) to open a box with the use of the obtained stick and thus to get a rope from inside the box; (3) to open the second box with the use of the obtained rope as mechanical traction and take a cup from there; (4) to fill the cup with water from a large mess-dish; (5) to put out a fire in an apparatus and get the prize. This scenario was implemented on two rafts floating in a lake. In order to fill the cup Rafael had to get over to the second raft with the use of a long thin stick as a bridge and thus to reach a cistern with water there. Then he returned to the first raft balancing on the thin bridge with the cup filled with water. After putting the flame out, the happy chimpanzee could take his prize from the apparatus. Why wasn't Rafael clever enough to bend over the side of the raft and scoop some water with a cup? It is likely that he learned the complex sequence of feats as a ritual. It is also possible that Rafael learned to configure a sequence of events into a unique stimulus, like pigeons in the experiment by Nakajima (2001) which was described in Chapter 14.

Later Voronin and Firsov (1967) obtained rather different results in their experiments with Lada, a chimpanzee. Confronted with the same situation, Lada easily came to the natural solution of scooping water from the lake. The fact is that experimenters did not dictate to Lada an obligate sequence of behaviours connected with the cup and the cistern. During the training sessions, Lada was free to get water from different containers such as a jar, an aquarium, a bucket and a pan. She thus acted with water itself rather than with a concrete container. Like Köhler's apes, Lada was likely to achieve a goal through understanding the relations between things in a context of a situation.

So what is insight in animals? Based on recent data on animals' capacities for analogical reasoning, we can attribute insightful behaviour to animals' ability for fast integration of behaviours gained from their past experience and effectively applying this experience for solving a problem as a meaningful whole, in the context of a situation. There is a distinction between insight and insight learning. Insight is a part of the mental process that relates to understanding relationships; insight learning is demonstrated in the behavioural solution of the problem

(Klopfer and Hailman, 1967). Insight could be considered a part of rule extraction. To 'attain' insight, another part of rule extraction is necessary to be obtained by a subject, namely, the ability to 'learn to learn'.

17.2 Learning how to learn: learning sets

The concept of a learning set that was suggested by Harlow (1949) more than half a century ago still bridges a gap between the behaviourists' theory of learning by trial and error and the Gestalt hypothesis that learning can be achieved suddenly, or insightfully. Harlow hypothesised that trial-and-error learning and insight learning are related and that insight develops out of well-established connections between stimuli and reponses (S–R connections). He suggested the concept of a learning set to refer to an intervening period between S–R and insight. In other words, as Harlow proposed, when an animal learns a new kind of problem, it solves it by slow painful plodding trial and error. However, if it has had experience with a large number of problems of a single type then problems of this type can be solved insightfully. These abilities enable animals to adapt quickly to new environments, given that the individual is already equipped with a set pattern of understanding situations and solving problems in principle. This theory, 'learning to learn', describes the ability of animals to slowly learn a general rule which can then be applied in order to rapidly solve new problem sets.

Harlow suggested that learning set formation could serve as a relevant measure of intelligence, that is, the individual's ability to solve complex problems by using representations of previously experienced events. He claimed that performance levels on *discriminative learning set* (DLS) tasks across species are generally in keeping with an ordering base on cortical complexity (Harlow, 1959). Modern studies in cognitive ethology have demonstrated that the use of a single measure such as learning set formation is unlikely to test adequately all aspects of animal intelligence (Pearce, 2000). Nevertheless, an experimental paradigm elaborated for exploring learning set formation is widely accepted in cognitive studies.

Harlow used the Wisconsin General Test Apparatus (WGTA) (see Part III) to establish an experimental procedure for discrimination. This allowed the question to be asked as to how quickly and efficiently animals can learn to evaluate hypotheses. Harlow conducted his experiment so that his animal subjects (typically monkeys, but the same results were obtained in a number of other species) had to learn a long series of simultaneous discrimination problems. The monkey was first presented with two stimuli (a red block and a thimble, for example); one was predetermined 'correct' and reinforced with food (red block) and the other was 'incorrect' and not reinforced with food (thimble). After each selection, the objects were replaced and the monkey again chose a stimulus. Each trial reinforced the same stimulus (red block). The monkey had a 50 per cent chance of being 'correct' on each trial; however, it could increase its chances by adopting the 'win–stay/lose–shift' strategy. That strategy is in some sense the backbone of a hypothesis-testing approach: as long as what you are doing is correct, stick with it (win–stay); when it leads to a wrong answer, change it (lose–shift). For example, if the monkey chose the thimble and was not reinforced, it should shift to the red block for the award. If, however, it correctly selected the red block and was reinforced, it should stay with the reinforced stimulus and choose the same stimulus next time. The monkey continued throughout a series of six trials with eight pairs of stimuli (learning sets). Harlow found the monkeys to be averaging approximately 75 per cent correct responses by the sixth trial of the eighth set. He then began to look at the animal's behaviour during the second trial with the new stimuli, say, a box of matches and a half tennis ball. He found that the monkeys implemented the stay or shift strategy on the second trial of the six-trial set, which meant the animals did not relearn the strategy with each new set of stimuli, instead they applied the rule they had already learned. After 250-plus trials, the monkeys were about 98 per cent correct on the second to the sixth trial each with a new set of stimuli.

Harlow also asked whether his animals could acquire a win-shift/lose-stay strategy. In this later series of studies, a problem is always followed by a reversal shift problem with the same stimuli, after which a new problem is given, and then followed by its reversal shift. So, once you have solved a problem with new objects, you will now be placed on the next problem in which you will need to choose the stimulus exactly opposite to the one you just picked. In a particularly devious version of this procedure, we get you to the point where there are just two trials on each set of stimuli, the second being the reversal shift. Here, to do well, you must immediately pick the other stimulus if you choose correctly on the first trial (win-shift), and continue to pick the same stimulus if you chose incorrectly (lose-stay), since that loser will be the winner on the next round. Studies based on reversal learning have shown that apes, macaques and capuchin monkeys can generate abstract rules of learning such as 'select the object that was previously incorrect' (Rumbaugh and Pate, 1984; Washburn and Rumbaugh, 1989; DeLillo and Visalberghi, 1994).

Examining animals' ability to establish learning sets can be considered an essential part of testing their ability to create abstract concepts and, in a general sense, to grasp regularities. Experimental evidence of these abilities in different species has been described in Chapter 14. For example, in the experiments of Sappington and Goldman (1994) on discrimination learning in Arabian horses) cited above, the first - very simple - discrimination took the longest to learn, and was followed by a general decrease in the time taken to learn successive problems. This suggests that the subjects were able to use previously learnt information to facilitate subsequent learning, and supports the conclusion that horses have the ability to 'learn to learn'. Horses do this in the domestic training environment when they master advanced manoeuvres faster after learning a series of preliminary tasks. There are some experimental evidences of learning to learn in horses tested on serial reversals of a positional discrimination (Warren, 1965). The subjects quickly exhibited a rapid decline in error rates on consecutive reversals.

It is of no surprise that practically all students of cognitive skills in animals note that their subjects display learning set formation within the course of multi-stage experiments. Even Skinner's followers regard the phenomenon of learning to learn when shaping animals' behaviour in accordance with principles of operant conditioning. For instance, in experiments on shaping new (unnatural, novel, untrained) behaviour it took only a few days of continuously repeating the first step in conditioning (reinforcing new or more specific behaviours) for dolphins to emit an unprecedented range of behaviours (Pryor *et al.*, 1969; Herman, 1980) (see details in Chapter 6). After a small number of sessions, it had become apparent that dolphins not only knew that only novel behaviour would be reinforced, but also that only one behaviour per session would be reinforced. Dolphins and other species easily applied the experience gained to grasp rules for other games with their trainers that included other requests for animals. These laboratory studies reflect many situations in nature where experiences gained facilitate the animal's ability to solve new problems.

17.3 Latent learning and exploration

It is a natural idea that latent learning often precedes insightful behaviour. The term *latent learning* originated from a series of classic experiments carried out by Tolman and his co-authors in the 1930s. This phenomenon seems to be closely connected with what Tolman called the formation of cognitive maps which, in turn, is based on the principle of 'what-leads-to-what' expectancy (see Chapter 12). Beyond the visible results of animals' learning in mazes, sometimes learning would occur but there would be no observable change in the animals' behaviour. This form of learning is called 'latent learning'.

One example of experiments showing latent learning concerns a widely known study by Tolman and Honzik (1930a, b) in which rats could learn a route in a maze without obtaining reinforcement. Experimenters tested three groups of food-deprived rats in a maze. The rats in the first group were allowed to wander in a maze once each day and obtained food reinforcement on reaching the end location. The rats in the second group were allowed to wander in a maze once each day, but on reaching the end location received no reinforcement until the 11th day. The rats in the third group served as a control group and were allowed to wander in a maze once each day, but on reaching the end location received no reinforcement. It turned out that the rats in the first group quickly learned the way through the maze, while the rats in the third group simply moved aimlessly around the maze. However, the rats in the second group moved about the maze randomly during the first 11 days, but when they received reinforcement they learned the maze faster than the rats in the first group. Therefore it appears that the rats in the second group had in fact learned the correct route in the maze before reinforcement was given because they were able to select the appropriate route much faster than the rats in the first group. Apparently, they were learning all along, but it was latent learning and not evident until it was activated by a patent reward. It seemed that learning the maze was a kind of reward in itself.

Tolman's suggestion that 'disinterested' learning can occur in animals, without a special motivation and in the absence of reinforcement, generated violent discussion for at least two decades (Munn, 1950). The role of cognition in determining animal behaviour was more than problematic for explanation according to the theories of behaviourist conditioning. At the same time, experimenters using Skinner boxes were familiar with such a situation when a monkey was ready to perform a task, such as pulling a lever, simply to be able to look out of the box in which it was kept, so that the only reward seemed to be the chance to look around.

Although the term 'latent learning' was born in a laboratory, this type of learning, together with exploratory activity, is very important during the lifetime of an animal. Emphasising these natural aspects of latent learning, Barnett (1958) and then Thorpe (1963) chose the term *exploratory learning* and defined it as an association of indifferent stimuli, or situations, without patent reward. Performing exploratory behaviour, an animal is searching for 'news' in its surrounding and actively explores new surroundings.

For wild animals the necessities of life undoubtedly include learning about their home area. It is worth illustrating with several informative laboratory and field experiments. For example, in a laboratory study, two groups of mice were turned loose in a room with an owl. One group of mice were given a few days to familiarise themselves with the room before the owl was introduced. The second group was put into the room at the same time as the owl. The mice that were familiar with the room felt much better and fewer of them were caught (Metzgar, 1967). The survival value of an exploratory drive is obvious for wild animals and it is extremely high for small rodents. One can see how a squirrel, being placed into an unfamiliar area, estimates remoteness of different things by their visual displacement along with changes of a viewpoint. The squirrel makes characteristic movements with its head standing on its tiptoe before each jump. It learns all the trees in its home range in the forest and quickly makes long sequences of jumps on its learned way, so that it will be exposed to predators for the shortest possible period of time. However it would be problematic for the squirrel to escape predators in an area that is not 'zeroed in'.

There are many studies on exploratory behaviour in mice and rats in specially equipped 'mice-houses' as well as in the labyrinths of a city (Crowcroft and Rowe, 1963; Crowcroft, 1966; Kotenkova and Bulatova, 1994). In stable familiar surrounding a small animal knows where food is likely to be found and where there are holes into which it can dart at the approach of danger. It is of particular interest that rodents sometimes undertake over-exploration of their familiar surroundings. After time of rest and foraging, time of exploration comes when an animal makes the round of its area, inspecting everything, smelling at them, and looking and creeping into them. Observers agree that the animals are likely to be looking for something new in their home range.

Is exploratory latent learning possible in invertebrates? It is a natural idea that cockroaches

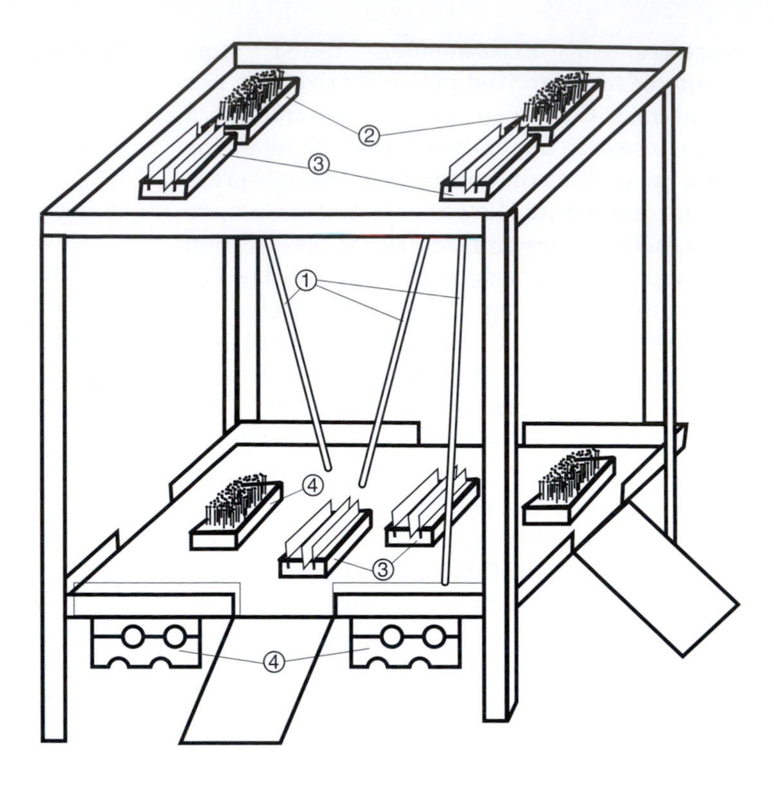

Fig. 17.1 Experimental set-up elaborated by Reznikova to study exploratory activity in ants. 1, cross-overs; 2, a maze simulating underground passages; 3, simulation of surface obstacles; 4, simulation of grass.

possibly have knowledge about the locations of shelters in their home range and thus know in which direction to run when the light is switched on in the kitchen where they happily live. Verron (1952) first studied latent learning in cockroaches in the laboratory. The group of insects that had an opportunity to familiarise themselves with a maze by walking on the transparent lid of the maze searched for an entrance much better than the control group which was not familiar with the maze. As we have already seen in Chapter 12 Thorpe (1950) revealed that digger wasps effectively find an optimal way to the vicinity of their nest from wherever they happen to capture their prey. He noted that these insects are in the habit of undertaking long 'aimless' trips along their home ranges which obviously help them to operate in familiar surroundings under changing circumstances.

Reznikova (1983, 2005) devised a method which can be used to compare levels of exploratory behaviour in ants and possibly in other small animals. The main idea was to estimate and compare the time duration which animals spend on exploration of models that simulate pieces of nature. Usually ants search for food in shelters, under stones, within grass columns, and so on. Each ant knows all the ins and outs in its territory, and the introduction of new things triggers its exploratory behaviour. The experimental models imitate natural situations such as underground passages, crevices between stones, irregularities in the ground, and columns of grass. The field experimental device includes double-decked arenas (0.5×0.5 m) equipped with models of several types: (1) mazes that imitate underground passages attached beneath the arena's surface, (2) parallel plates that imitate crevices, (3) 'brushes' made of vinyl bars that imitate grass stems. Ants can visit both levels of the device via bridges. Eight species were tested, possessing different ecological and ethological features such as foraging style (group or solitary), preferred tier of steppe ecosystem (ground surface, soil, grass), and some others (Fig. 17.1).

At the first stage ants were simply welcome to have some food on the open surface of arenas. There was no difference between species in their activity as freelance consumers. After three days the cafeteria closed and all models were placed onto arenas. Then two series of experiments were conducted with two groups of ants. The first group was presented with food placed into all types of models in order to examine whether ants of different species are able to penetrate all models. So it was the cafeteria experiment again. It turned out that ants of all species can effectively forage in all models. The second group was presented with models without food so that only 'disinterested' behaviour was displayed. All ants preferred to spend time on the open surface of the arenas. In the experiments levels of exploratory activity in different ant species were estimated by comparing the ratios of duration of time spent by ants on the surface of the arenas and in different models. A high correlation was demonstrated between the efficiency of problem-solving and levels of exploratory activity in ant species. Both characteristics were in turn connected with the ecological characteristics of the ants (Reznikova, 1982).

Together with other results, these data support a hypothesis that animals' ability to actively gain and retrieve information on a large number of locations, as well as the level of their general exploratory activity, are closely connected with ecologically and evolutionary traits of species and can be very specific in some species, based on their ability to learn news very quickly within species-specific domains. We will consider the problem of species specificity in connection with development of intelligence in Part VIII.

Chapter 18

Tool-using as a tool for experimental studies of animal intelligence

To what extent do animals' manipulations with physical objects express their intelligence? The main idea of this chapter is that, just as working with language-trained animals (about which we will learn in detail in Part VIII) enables experimenters to obtain unprecedented data on animals' cognitive skills, working with species that possess tool-using allows researchers to observe unparalleled behaviours based on problem-solving and rule extraction.

Tool-using behaviour is distributed rather oddly in different classes of animals and it is a relatively rare occurrence taxonomically, so it is not easy to choose appropriate species for studying instrumental problem-solving. Besides, sometimes species that have not been observed in the wild to be natural tool-users (for instance, bonobos, tamarins and blue jays) demonstrate wonders of engineering in laboratory studies. Anyway, a great progress has been made during the last two decades in our understanding of what animals think about the physical world, and the asset of tool-using animals should be appreciated.

18.1 Brief account of tool behaviour in animals

There are several detailed monographs about tool-using in animals. Among them, Beck (1980) systematises the great body of data on how tool use is distributed within a wide variety of species. Goodall (1970, 1986) was the first to summarise long-term observations on tool-using in chimpanzees, and McGrew in his books (1992, 2004) has given an encyclopaedic description of tool-using as a basic element of 'culture' in our closest relatives. Here I give a brief review of tool-using in wild and captured animals which act freely, i.e. not being forced by researchers to solve any special problems with the use of tools.

Concepts and definitions

When considering tool behaviour, *tools* must be clearly distinguished from *artefacts*. Artefacts are common in the animal kingdom: beehives and beaver lodges are among the finest examples – and they are made by some creatures quite low in the evolutionary scale (such as the caddis-fly larva that makes cases and nets). Tools are distinct from artefacts because they can be used to make other objects or to facilitate activities such as resource extraction. They are not by themselves of immediate and direct use. There are two distinctive tool behaviours in animals, namely, tool use and tool manufacture.

The generally accepted definition of animal *tool use* is that the animal is using an unattached environmental object to alter the form, position or condition of another object or organism, when the user holds or carries the tool during its use and is responsible for a proper and effective orientation of the tool (Beck, 1980). In terms of levels of sophistication, animals may or may not choose morphologically and functionally variable tools, showing selectivity amongst them so that a functionally appropriate tool is chosen for a task (Parker and Gibson, 1979; Tomasello and Call, 1997).

Tool manufacture includes four modes of tool behaviour. The simplest and most common is severing the fixed attachment between one environmental object and another (or the substrate) so that the first object can be used as a tool. This mode is termed *detaching*. Examples are breaking a branch from a tree for use as a club or uprooting a sapling which is then brandished. The second mode, *subtraction*, is removing an object or objects from another unattached object so that the latter can serve more usefully as a tool. For example, a chimpanzee is subtracting when it removes leaves from a twig when the twig is to be used for termite or ant fishing. If, however, the ape removes the leaves so that the leaves can be used for wiping, it is detaching. The third mode is *addition* or *combination* in which two or more objects are connected to produce an adequate tool. An example is a chimpanzee connecting two short sticks to produce one of sufficient length to reach food. The last mode of tool manufacture is *reshaping*. This involves fundamental restructuring of material to provide a functional tool. Examples include chimpanzees crumpling leaves to increase their absorbency as a sponge or unrolling a coil of wire to produce a straight reaching tool (Beck, 1980; Whiten *et al.*, 1999; McGrew, 2004). Recently came the first report of *customary use of tool-sets* – two or more different types of tool used in sequence to achieve a single goal – by a community of wild chimpanzees in the Congo Basin (Sanz *et al.*, 2004). Chimpanzees tap termites from their mounds firstly perforating with heavy sticks to punch holes in termite mounds. Having gained access, they use a lighter stick known as a fishing tool to extract termites. It is worth noting that chimpanzees select specific materials for their puncturing tools, often gathering the sticks far from the termite nests and transporting them to the nests, where the apes modify their selected sticks by stripping leaves and shortening them to a uniform length.

Let us survey how these modes are distributed in different species, from elephants to ants. As has been already noted, Beck (1980) collected in his book a huge amount of data about animals that have been observed to be tool users, among them such unexpected creatures as rodents, antelopes, ant-lions and sea anemones. Here we will consider only usual and frequently cited examples of animals' tool behaviour in order to create the necessary prerequisites for the analysis of instrumental problem-solving as an essential part of studying animal intelligence.

A club of tool users: description of phenomena and related problems

Asian elephants *Elephas maximus* and African elephants *Loxodonta africana* throw sticks, logs and stones at other animals and use sticks to scratch parts of the body; they wipe away blood with tufts of grass and cork up wounds with them (Douglas-Hamilton and Douglas-Hamilton, 1975; Chevalier-Skolnikoff and Liska, 1993). Elephants definitely express tool manufacture such as subtraction modifying branches for fly switching. The great storyteller Rudyard Kipling alluded to elephants fly-switching in the chapter 'The elephant's child' in his 1902 *Just So Stories*. The elephant's child, having acquired a new trunk (courtesy of the crocodile), starts for home and Kipling notes that, when flies bit him he broke off the branch of a tree and used it as a fly-whisk. Indeed, elephant's prehensile trunk equipped with the special 'finger' is capable of some of the same delicate manipulative movements performed by primates with their fingers. Use of branches for fly-switching is a common form of tool use in elephants. Hart and co-authors (2001) presented to 13 Asian elephants maintained under a naturalistic system branches that were too long or bushy to be effectively used as switches. Eight of these elephants modified the branches make them smaller and switched with the altered branch. There were different styles of modification of the branches, the most common of which was holding the main stem with the front foot and pulling off a side branch or the distal end with the trunk.

Sea otters provide one of the best-documented forms of tool use in wild animals. Fisher (1939) first described tool use in California sea otters *Enhydra lutris nereis* that hammered small molluscs on rocks placed on the animal's chests. Hall and Schaller (1964) made detailed observations of tool behaviour of these animals which have no match among other species in the sophisticated use of stones as tools. The otter surfaces with a single mussel and a stone, rolls so that it is floating on its dorsum with the stone balanced on its ventrum, and then immediately begins to pound the flat side of the mussel against the stone. The mussel is held with both forefeet. The otters pound the mussels in series of forceful blows at the rate of two per second. The longest feeding sequence by one otter involving tools lasted 86 minutes, during which 54 mussels were consumed and 2 237 blows were struck. Stones would sometimes be transported during subsequent dives and reused; one was used 12 consecutive times. There are many variations of tool behavioural patterns in sea otters. Sea urchins, crabs and lobsters are pounded on stones. Sea otters can also open clams by pounding them against another clam shell balanced on the ventrum. Houk and Geibel (1974) made observations of sea otters hammering an abalone from the substrate with a stone at a depth of 10 metres. An otter located an abalone in a rocky crevice, picked up a stone, and began to strike the abalone. The otter worked for 88 seconds and three times surfaced to breathe returning with the same rock. Indeed, sea otters offer an opportunity for further study of tool use.

The overwhelming majority of reports of tool behaviour in wild animals concern primates. Relatively simple behaviours such as dropping or throwing objects (branches, twigs, stones, excreta) towards intruders have been described by many authors in capuchin monkeys, howler monkeys, spider monkeys, squirrel monkeys, woolly monkeys and some other species (see Beck (1980) for a review). However, it takes many efforts to observe regular displays of complex tool behaviour in wild animals. Recently the regular cracking of palm-tree nuts with the aid of two stones ('hammer' and 'anvil') has been revealed in tufted capuchin monkeys in the Ecological Park of São Paolo (Ottoni and Mannu, 2001). A similar style of tool use by Japanese macaques has been monitored for 20 years (Huffman and Nishie, 2001). Vervet monkeys *Cercopithecus aethiops* use leaves to sponge up water in tree hollows (Hauser, 1988).

Chimpanzees are well established as the most proficient tool users and makers among non-human primates. At the beginning of the twentieth century Ladygina Koths (1923) and Köhler (1925) attracted the attention of scientists to skilful manipulations of objects by apes. The animals used sharp sticks and wires to reach insects in crevices, poles to vault for a suspended banana, twigs as levers to open lids. Apes used a great variety of objects for comforting themselves. They scratched their skin with twigs, stones and broken pieces of pottery, brushed their nails with sharp sticks and ears with folded paper, wrapped their hands in paper and leaves when they needed to work with spiky objects, and used leaves, sheets of paper and cloth to wipe themselves. Chimpanzees enjoyed the advantages of tool use for taunting other animals. For example, they called chickens by throwing pieces of bread and prodded them with sticks or wires.

The first scientific naturalistic study on tool use by wild chimpanzees was conducted by Nissen (1931). Since the 1950s, new observations have appeared in the primatological journals (see Goodall (1986) and McGrew (1992, 2004) for reviews). A systematic synthesis by Whiten and colleagues (Whiten *et al.*, 1999) describes various habits of chimpanzees including styles of the tool use at seven field sites in Africa. We will consider the 'cultural' aspects of these data in Part VIII. Here it is important to note that the authors described a complete set of tool behaviours such as pounding actions, ant and termite fishing, nut hammering, absorbing water and honey with chewed leaves, throwing twigs and stones, and so on.

Detailed observations of each particular type of tool use showed that chimpanzees use brains and not brawn to exploit resources ignored by their sympatric competitors such as, say, baboons and gorillas who lack instrumental technology (McGrew, 2001). For example, McGrew (1974) described in detail the so-called ant-dipping, that is, tool use by wild chimpanzees with driver ants. To reach the ants, a chimpanzee dips them with the use of shaped branches from rocks, fallen logs or other raised positions in order to avoid ants' attacks. To make the tool, a chimpanzee breaks off a branch of living woody vegetation, and then pulls off the leaves. If the tool becomes damaged or coated with soil during use, the ape may further modify it by biting off the worn segments. The chimpanzee is skilful enough to catch ants in a moment in a mass that is about the size of a hen's egg and contains about 300 ants which are then chewed and swallowed. Chimpanzees are the most intelligent insectivores (Fig. 18.1). Bonobos, gorillas and orang-utans may eat many insects, but without instrumental technology and thus with much more effort and downtime.

For brevity, we have just considered only one from the great number of examples of tool manufacture in anthropoids. It is worth noting that styles of life and technology vary greatly in different local populations. For instance, orang-utans in captivity often use tools, but no one before van Schaik and his colleagues (1999, 2003) had observed this behaviour in wild populations. Living high in the trees of the flooded forest in the north-west corner of Sumatra, members of a local population with a high density of orang-utans locate pulpy fruits, of which their favourite is locally called 'puwin'. Inside this fruit lies a rich source of protein and lipids; outside are tiny hairs that researchers describe from experience as feeling like 'plexiglas needles', capable of delivering a painful jab. By sliding a thin stick into a crack in the fruit, orang-utans can get the seeds out without having to handle the prickly husk.

Fig. 18.1 A chimpanzee is dipping termites. (Photograph by C. Tutin, Gombe National Park, Tanzania; courtesy of W. McGrew.)

In contrast to information from other great apes, which mostly show tool use in the context of food extraction, recent observations of Breuer *et al.* (2005) show that in gorillas other factors such as habitat type can stimulate the use of tools. These researchers first observed an adult female gorilla using a branch as a walking stick to test the depth of the water and to aid her in her attempt to cross a pool of water at Mbeli Bai, a swampy forest clearing in northern Congo. In the second case they saw another adult female using a detached trunk from a small shrub as a stabiliser during food processing. She then used the trunk as a self-made bridge to cross a deep patch of swamp. There is the intriguing possibility, as the authors propose, that using branches to test water depth or as bridges is a more common adaptation to this particular habitat. These observations show that functional demands other than obtaining food resources may also stimulate tool use, at least in apes.

Another group of organisms that seem to be obvious candidates for investigating the cognitive abilities related to tool use are birds. Although there are only a few bird species for which habitual use of tools is known, in at least two species – New Caledonian crows and woodpecker finches – skilfulness and flexibility in tool manufacturing can be considered close to that of primates.

Before considering the sophisticated tool usage in these two species, I give here several examples of tool behaviour in other birds.

It is known that Egyptian vultures *Neophron percnopterus* drop stones on ostrich eggs (Goodall and van Lawick, 1966). With other bird species that use stones for breaking eggs, black-breasted buzzard *Hamirostra melanosternon* (Aumann, 1990) and Egyptian vultures belong to a narrow section of creatures with claims to tool selectivity. Thouless *et al.* (1989) tested *N. percnopterus* by providing them with models of ostrich eggs and a range of stone sizes. The authors showed that vultures strongly preferred 46-g stones. In similar tests *H. melanosternon* preferred 40-g stones (Aumann, 1990).

Green-backed herons *Butorides striatus* obtain bait as diverse as live insects, berries, twigs and discarded crackers, and cast them on the waters. They then crouch and wait for the curious or hungry fish that comes to inspect the lure. The birds have even been observed carefully trimming oversized twigs to the proper dimensions (Walsh *et al.*, 1985). However, no studies have compared how well herons would do if they did not fish in this manner. Such an estimation recently has been made on 'land fishers', burrowing owls *Athene cunicularia* (Levey *et al.*, 2004). Their underground nests and surrounding areas are carpeted with stinky stuff. Observations showed that owls deliberately use mammal dung as a tool to reel in a meal and in the process substantially increase the number of dung beetles they eat. In 4-day tests with owls at ten burrows, researchers found that taking the dung away from the burrows' entrances left the owls with few beetles in their diet. Owls with a refurbished cow-dung decor, however, averaged ten times as many beetle meals during the test.

Recent studies on the cognitive abilities associated with tool use in birds revealed remarkable data on flexibility in tool manufacturing in New Caledonian crows and woodpecker finches. These are still the only two well-studied examples of tool modification found among birds. In contrast to context-specific stereotyped tool behaviour of many other bird species, New Caledonian crows and woodpecker finches are flexible in their use of tools.

Tool behaviour of New Caledonian crows *Corvus moneduliodes* has recently been used as the motif of a postage stamp from New Caledonia. These birds make and use at least three forms of tools to aid prey capture. Almost everything known about their tool use in the wild is from the work of Hunt (1996, 2000) and Hunt and Gray (2003, 2004). The crow's tools include hooks cut from pandanus leaves, stick-type tools made from tree twigs, fern stolons, bamboo stems, tree leaf midribs and thorny vines. To fashion a hook from tree twigs, crows detach secondary twigs from the primary one by nipping at the joint with their beaks, leaving a piece of the primary twig to form a hook. After a twig is detached, the crows remove leaves and bark, and have even been observed sculpting the shape of the hook with their beak. The crows carry tools with them as they fly to feeding areas, where they use hooked and narrow-tipped tools to fish out insects from hard-to-get spots.

The Galápagos finch *Cactospiza pallida* is known to use a cactus (prickly pear) spikes to poke out insects embedded in the branches or trunks of trees. Where there are no cacti, *Cactospiza* breaks off a short stiff twig from a tree. Woodpecker finches even modify their tools: they shorten twigs or break off side twigs that would prevent insertion into holes (Eibl-Eibesfeldt, 1961b; Millikan and Bowman, 1967). These birds spend more time using tools and acquire more food with them than do chimpanzees. Strong ecological correlates have been revealed related to their tool behaviours. In the arid zone during the dry season, when prey is scarce, finches spend half their foraging time using tools and obtain 50 per cent of their prey this way, whereas in the humid zone birds of the same species rarely use tools (Tebbich *et al.*, 2002). Experimental studies have demonstrated that trial-and-error learning is involved in the acquisition of tool use in young woodpecker finches (Tebbich *et al.*, 2001).

It is amusing that from many species only New Caledonian crows display such a high level of complexity of tool-related skills as manufacturing tools in a multi-step fashion, or by fine crafting (Kacelnik *et al.*, 2004). The crafting of tools in this species involves (1) selection of appropriate raw material; (2) preparatory trimming and (3) three-dimensional sculpturing (Weir *et al.*, 2002). Chimpanzees and orang-utans use crumpled leaves as sponges which could be regarded as 'reshaping' and all other known animal tool manufacture involves nothing more complex than detaching or subtracting objects from each other.

The second important thing is that the two most 'engineering gifted' species, namely, New Caledonian crows (Hunt and Gray, 2004) and chimpanzees (see McGrew *et al.*, 1996) are the only non-humans in which populations show routine tool use (Kacelnik *et al.*, 2004). As has been noted before, woodpecker finches (Grant and Grant, 2002; Tebbich *et al.*, 2002) and orang-utans (van Schaik *et al.*, 2003) show high frequencies of tool use in some populations, but there are others that never use tools. In all other animals – including sea otters operating with stones in very amazing style – there are either insufficient data to access tool-use frequency or tool use is known to be absent in many populations.

At the same time, even being the most prolific tool users, both species display a high degree of genetic adaptations involved in ontogenetic shaping of tool behaviour. Using hand-raised juvenile New Caledonian crows, Kenward *et al.* (2005) showed that chicks spontaneously manufacture and use tools, without any contact with adults of their species or any prior demonstration by humans. Similar results were obtained by Firsov (1977) in his experiments with artificially raised chimpanzees that manipulated with many things, used and manufactured tools and built nests for sleeping. Spontaneous tool use has also been recorded in woodpecker finches that were raised lacking contacts with tool-using adults (Tebbich *et al.*, 2001).

It is thus a very intriguing problem whether complex tool behaviour in animals is associated with a high degree of cognitive abilities or with strong genetic adaptation. If both domains are responsible for this type of behaviour, are they distinguishable? We will try to find an answer, at least partly, further in this chapter and in Part VII.

18.2 Experimental studies on tool use and cognitive abilities in animals

Comprehensive experimental investigation of solving problems using tools by animals may bring knowledge about the level of their apparent 'understanding' of the features that give tools their particular functions. Studies of tool use in animals include physical relations involved in animal's actions, categorisation of different sets of things by their functional properties, taking invisible forces into account by tool users, as well as animals' ability to anticipate their actions.

Experimental studies of relations between tool use and intelligence in animals have a dramatic history. For a long time primates have been in the lead, but recently new insights have come from birds. Recently discovered achievements of wild tool users are very impressive and may shift our imagination about animal intelligence into acknowledgement of the high level of their understanding of causal relationships. At the same time there is enough room both for enthusiasm and scepticism. For instance, just as Köhler referred to 'good and bad mistakes' in relation to tool use in his chimpanzees, Visalberghi (2002) has referred to a delicate balance between 'doing' and 'understanding' in capuchin monkeys. The fact that first ignited her interest in problem-solving by capuchins was the observation in the zoo of an adult male pounding an unshelled peanut with a boiled potato. Since that time, this researcher has conducted many experiments that have demonstrated, on the one hand, how successful capuchins are at solving problems, and, on the other, how relatively little they understand of what they do. So let us try to find together the balance between causality and idle enthusiasm in wild tool users, analysing the results of experiments aimed to studying instrumental problem-solving.

In fact, the first chimpanzees that appeared in laboratories at the beginning of the twentieth century received sticks and were faced with instrumental problems. More than a century ago Hobhouse (1901) suggested a problem that later would initiate many modifications and thus form a new experimental paradigm. The experimenter presented his chimpanzee Professor with a 'tube problem', asking it to push an item of food out of a short tube. Professor easily solved this problem. Yerkes (1916a) posed a variant of this problem to two apes. The task was complicated by the fact that the tube was too long to be taken in their hands (170 cm in length) so it was placed on the ground. One chimpanzee solved this problem after 12 days of attempts. An animal appeared to understand that to be grasped a food item has to be pushed away. A gorilla mastered this task only after being shown by the experimenter how to do this.

Roginsky (1948) and Ladygina Koths (1959) investigated primates' tool behaviour working mainly with Paris, the most curious and enthusiastic adult male from a group of chimpanzees in Moscow Zoo.

Together with those of Köhler and Yerkes, these experiments were the first to thoroughly study animals' problem-solving by means of tool use. They became prototypical for many modified versions that would be elaborated, so we consider results of the prototypical and modern experiments together here.

Means–end relationships

Understanding means–end relationships is one of the key elements of cognition in animals and this is also an important step in human cognitive development: a significant transition occurs at around the age of 8 months, when infants move beyond a reliance on what Piaget (1952, 1954) called 'circular reactions' (in effect, operant conditioning) to an understanding of means–end relationships (Willatts, 1999).

In order to study the understanding of means–end relationships, it is necessary to use problem-solving tasks in which the solution to the problem can in principle be perceived directly, without trial and error or previous experience of similar tasks. Means–end understanding is most clearly demonstrated if the subject shows an 'insightful' solution to the problem on the first trial, since correct performance at the end of a period of training may well represent the effects of operant conditioning. The usual way of studying means–end understanding in animals, introduced by Köhler (1925), is by offering a possible physical connection to an out-of-reach object of desire, for example a string that is attached to a piece of food (it is also called the 'string-drawing problem'). The food itself is out of reach of the animal, but the near end of the string is accessible. If the animal understands the physical properties of the string it uses it as a means to an end, i.e. pulls the food into reach with the string. It has been revealed in early studies that chimpanzees and monkeys readily cope with this task (Harlow and Settlage, 1934).

Recently Nissani (2004, 2006) has applied the string-drawing paradigm to Asian elephants. Having a great deal of experience with studying insects' behaviour (Nissani, 1977), he has compared design of this problem with a natural situation described as early as by Fabre (1879) from the life history of a digger wasp. The wasp seemed to lack means–end understanding. For instance, according to Fabre's observations, the wasp did not understand that she could grab her paralysed prey by a leg instead of an antenna. If the head appendages of her particular prey species suddenly ceased to exist, 'her race would perish, for lack of the capacity to solve this trivial problem' (cf. Griffin, 2001; Nissani, 1977). In Nissani's (2004, 2006) experiments elephants did not exceed insects in causal reasoning. In fact, the understanding of means–end relationships, as many other displays of problem-solving that have already been mentioned, give a very mixed picture for the comparative analyst. Let us consider different variants of this experimental paradigm.

One variant of problems to be solved by an organism in order to demonstrate understanding means–end relations is a *support problem*.

In this problem a reward is placed on a cloth. The reward itself is outside the subject's reach but one of the ends of the cloth is within reach. The solution of a problem consists of pulling in the cloth to bring the reward within reach. Roginsky (1948) suggested this problem to the chimpanzee Paris, and Piaget (1952) studied this problem with human infants. One-year-old children not only readily pull in the cloth, but, more importantly, they withhold pulling when the reward is not in contact with the cloth. This indicates that children at this age understand that spatial contact is necessary for the tool to act on the reward. The chimpanzee Paris displayed the same level of competence in solving the support problem. Spinozzi and Potì (1989) administrated more detailed support experiments with infants of apes and monkeys. In one condition the reward was placed on the cloth to the side. All these primates (one Japanese macaque, two capuchin monkeys, two long-tailed macaques and one gorilla) responded appropriately by pulling in the cloth when the reward was on the cloth, and withheld pulling when the reward was off the cloth. In a second experiment Spinozzi and Potì (1989) tested the generality of these findings by modifying the conditions of the off-cloth condition by placing the reward near the end of the cloth rather than to the side of it. The authors reasoned that if subjects had simply learned to respond appropriately to a specific configuration of the cloth and the reward rather than more general relations between them, they would respond inappropriately to this novel configuration. Under these circumstances, all subjects pulled in the on-cloth condition but not in the off-cloth condition. Later Spinozzi and Potì (1993) put this problem to two infant chimpanzees and one of them succeeded (a pessimist would rather say 'only one of them' succeeded). This illustrates the range of individual differences in problem-solving in animals which will be discussed in Part VII.

Hauser and colleagues (1999) turned next to a modified version of this task in which tamarins were asked to choose one of two pieces of cloth to obtain a food reward. Monkeys focused on changes to the cloths that affected its affordances. For instance, they rejected pulling cloths made of material which did not afford pulling such as pieces of cloth connected with chipped wood, sand or a broken rope. At the same time, they chose cloths of radically different shapes (i.e. triangles, circles, teeth-shaped) that functionally supported the food reward. Further they distinguished between cloths that supported the food reward and cloths that were merely in contact with the food and thus functionally inappropriate. Thus, tamarins distinguished the features that were relevant for the tool's function from those that were not (Santos *et al.*, 2003).

Some variants of the support problem turned out to be a challenging task for primates. This concerns the *rope problem*, or *string-pulling*: the subject is required to pull a cup containing a food reward with the use of a rope. The simplest variant is that in which the rope is visibly attached to the handle of the cup. This task has been easily solved by apes and monkeys in many studies. Roginsky (1948) conducted many experiments with two species of baboons (*Papio porcarius* and *P. hamadryas*), rhesus macaques, pig-tailed macaques and a mandrill. Monkeys were tested with 38 variants of the rope problem in which some ropes were connected with the reward whereas other were not, or were broken so as not to support the cup. All monkeys displayed good results in choosing relevant ropes, and one pig-tailed macaque conquered all tasks with the exception of one variant in which a very long (2.5 m) rope was connected to the cup by many zigzags. The macaque first chose a short straight rope that was not connected with the reward and learned to choose the long crooked rope after many trials (Fig. 18.2).

In another series of Roginsky's (1948) experiments the chimpanzee Paris was presented with

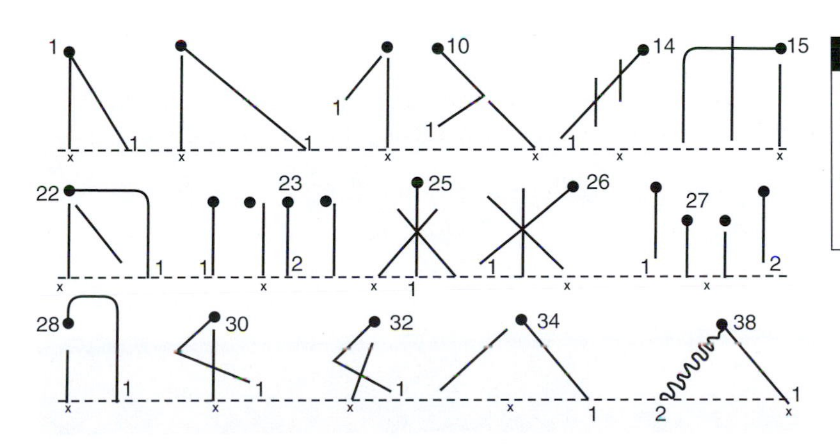

Fig. 18.2 The scheme of placing the ropes in Roginsky's experiments with apes. The digits correspond to the number of the task, X is the position of the animal and the circle marks the location of the bait. (Adapted from Roginsky, 1948.)

a piece of food placed into a cup out of reach and was expected to bring the cup within reach by pulling a rope that had to be passed through a handle on the cup so that two ends of the rope were brought together in the animal's hand. Paris did it right the first time but this was probably by chance because he was wrong the second time, and learnt to perform this task perfectly and systematically only after 30 additional trials. He made mistakes again when the problem specification was slightly changed, and thus failed to demonstrate full understanding of interrelations between the cup and the rope.

String-pulling behaviour has been studied in many species, and, as has already been mentioned, the picture is mixed. In particular, the string-pulling paradigm has been tested in a variety of birds. The crucial step in solving the problem for birds is a combination of several behavioural steps that must be completed in the right sequence. This includes reaching down, pulling up the string with the bill, placing the string on the perch, stepping on it with a foot, letting go with the bill, reaching down again, and repeating this cycle for at least five times. Most studies showed that the birds either had a hereditary coordination of movements that was adapted to normal feeding habits or they simply would learn the task through trial and error. Great individual variance has been observed in greylag geese *Anser anser* (Fritz *et al.*, 2000) and songbirds (Vince, 1956, 1958, 1961; Blagosklonov, 1974). Studies on ravens (Heinrich, 1995, 1999, 2000) and psittacids (Funk, 2002) have shown that at least some individuals solve the task without any former training or trial-and-error learning thus implying that the mechanism used was insight. Keas *Nestor notabilis* showed an extraordinary success rate in this study (Fig. 18.3). Except for a fledging, all individuals tested spontaneously pulled the reward in their first trial implying that they understood the means–end connection and the underlying physical properties of the task. When keas had to choose between two strings in eight different tasks they significantly selected the correct string in all tasks even when the strings were crossed. Therefore keas clearly comprehended the functional connection between the string and the reward. Great individual differences both in the performance and the methods used to obtain the reward exclude innate behavioural patterns (Huber, 2002; Werdenich and Huber, 2006).

Different species of Old World and New World monkeys and lemurs solved the problem successfully (for a review see Hauser *et al.*, 2002b).

Fig. 18.3 Solving a means–end problem by keas. (Photograph by I. Federspiel; courtesy of L. Huber.)

Fig. 18.3 (cont.)

There are long-term controversies concerning whether string-pulling behaviour is 'insightful' or is conditioned by trial and error in cats (Adams, 1929), rats (Tolman, 1937) and dogs (Shepherd, 1915; Köhler, 1925; Scott and Fuller, 1965; for a detailed review see Osthaus *et al.*, 2005). Recent experiments with the use of combined experimental techniques and a large sample of dogs showed that although dogs can learn to pull on a string to obtain food, they do not spontaneously understand means–end connections involving strings (Osthaus *et al.*, 2005). Instead of grasping regularities in means–end relationships, dogs rather demonstrate a high level of social intelligence relying on their owner's experience (Hare *et al.*, 2002). These results are in conformity with Pepperberg's (2004) data obtained with grey parrots. Pepperberg found that two language-trained parrots demonstrated no means–end understanding, but simply asked their human trainers to give them the treat, whereas parrots that had had no language training solved the problem easily. It appears that the availability of human-aided solutions to problems can sometimes inhibit the expression of animals' cognitive capacities. It is likely that the very modest feats of intelligence displayed by elephants in solving the means–end task in Nissani's (2004, 2006) experiments cited above were caused by the same effect, but this is not, of course, the only possible explanation of these results.

The stick problem

The stick problem consists of using a tool to bring in a reward that is not in direct contact with the tool. This situation entails putting the tool into contact with the reward and then sweeping the reward within reach. Solving this task demonstrates an ability to understand complex causal relations such as that the stick must be of appropriate size and material (long and rigid) and that only certain kinds of contacts (with the certain force and directionality) will be successful.

Roginsky (1948) put to the chimpanzee Paris a classic Köhler problem, that is, to bring a food reward placed out of reach within reach by pulling it with a stick. Getting used to the problem, Paris then selected appropriate tools and successfully adapted objects that did not satisfy his requirements; for example, he selected a stick of appropriate length, and splintered thick sticks into flimsy ones in order to act effectively. Later primates including baboons, macaques, orangutans, chimpanzees and tamarins were shown to be capable of solving the stick problems. Some of them used multiple objects in a combinatorial manner, placing one object inside another (for reviews see Tomasello and Call, 1997; Call, 2000).

Jones and Kamil (1973) presented the first detailed description of spontaneous solving of the stick problem in laboratory-raised blue jays *Cyanocitta cristata* that is likely to be based on understanding of causal relations by birds. The pioneering female jay was seen engaged in the following behavioural sequence. She ripped a piece of newspaper from the pages kept beneath her cage, folded the piece of paper, and then proceeded to thrust it back and forth between the wires of the cage, raking in food pellets too distant to be picked up directly with her beak. Experimenters were very surprised because there were no reports about tool-using in wild blue jays. It seemed that sophisticated tool-manufacturing displayed by that bird had been acquired by food deprivation that was aimed at maintaining the bird's health. After the authors observed the jay manufacturing tools, they isolated her from the other jays in the colony in a cage equipped for filming and presented her with pieces of papers. The bird displayed many times how coordinated beak and feet manipulations resulted in paper shreds. She then positioned a shred on one side of the pellets and by repositioning her grip on the paper made successive sweeping movements of the paper on the opposite side of the pellets, thus sweeping the pellets in an arc nearer a point where they could be reached with her beak from between the wires. On different occasions the experimenters presented the jay with a feather, a thistle, a piece of straw, a paper clip, and a plastic bag tie. In all cases the bird thrust the object between the wires and, except when using the thistle, was successful in raking in the pellets. The authors then tested a number of blue jays in their colony. Of eight hand-raised birds tested, five showed definite tool use, two displayed some components of the behaviour, and only one showed no sign of tool use. Even this single jay showed a high level of manipulation with the paper. Jones and Kamil (1973) consider tool-making in their subjects as indicative of the particular potential for behavioural adaptations typical for some species with highly generalised feeding behaviour, such as the northen blue jay.

The tube problem

As was noted earlier, the tube problem has served for more than a century as a relevant procedure for studying animals' understanding of the causal relations between the elements of the task. The simplest variant of this laboratory device is a transparent tube with food visible inside. To acquire the food, the subject must poke it out at the far end with a stick.

Ladygina Koths (1959) presented the chimpanzee Paris with many variants of the problem involving different types of tools that required different solutions. Tools were presented in groups of three in order to allow the chimpanzee to make a choice between more and less appropriate tools. He was also required to take some tools for completion by reshaping or detaching, or in other cases to combine several things together (Figs. 18.4 and 18.5). Three main variants of these tasks (the bundle task, the short-sticks task and the H-tool task) were later presented by Visalberghi and Trinca (1989) to capuchin monkeys and became popular tasks for studying possible mental manipulations of the elements of tasks in tool-using animals (Limongelli *et al.*, 1995; Povinelli, 2000; Tebbich *et al.*, 2001). In the *bundle task*, subjects were given a bundle of sticks taped together which as a whole was too wide to fit in the tube; the solution consisted of breaking the sticks apart. In the *short-sticks task*, subjects were given three short sticks which together added up to the length required; the solution consisted of putting them all in the same end of the tube to displace the food out of the other end. In the *H-tool task*, subjects were given a stick with transverse pieces at either end which prevented its

Fig. 18.4 The chimpanzee Paris solving the tube problem. (a) Paris manipulating with a coiled wire; (b) he takes the bait out of a tube using splinter chipped off a board; (c) he tries to put tied sticks into a tube and then unties them using his hands and teeth. (Adapted from N. Ladygina Koths, 1959.)

Fig. 18.4 (cont.) (b)

insertion into the tube; the solution consisted of removing the blocking piece from the tool.

As an honour to a galaxy of chimpanzees participating in cognitive experiments, Paris coped with the majority of tasks displaying remarkable quick-wittedness and enthusiasm. He straightened the wire, wound off the rope, took off cross-pieces, picked off widening with his teeth; he effectively reshaped, detached or combined tools, compared elements of tools at different stages of manipulations with them, and periodically checked whether the tool was yet suitable for inserting into the tube. At that, like all chimpanzees, Paris chronically lacked precision. For instance, he straightened the wire to the extent that only allowed inserting it into the tube with effort. One more turn would allow him to cope with the task easily but the

(c)

Fig. 18.4 (cont.)

chimpanzee preferred to puff over this work for a long time trying to insert the under-straightened wire into the tube until a goal was attained.

The tube problem has been presented to chimpanzees, bonobos and orang-utans. These results, together with those of Ladygina Koths, suggest that apes operate with foresight; at least, they obtained rapid success in the basic task, that is, in selecting sticks of the appropriate diameter to fit the tube. It does not mean that apes are foresighted without limit. They proved less successful in the H-tool task, and some chimpanzees failed this variant of the tube problem.

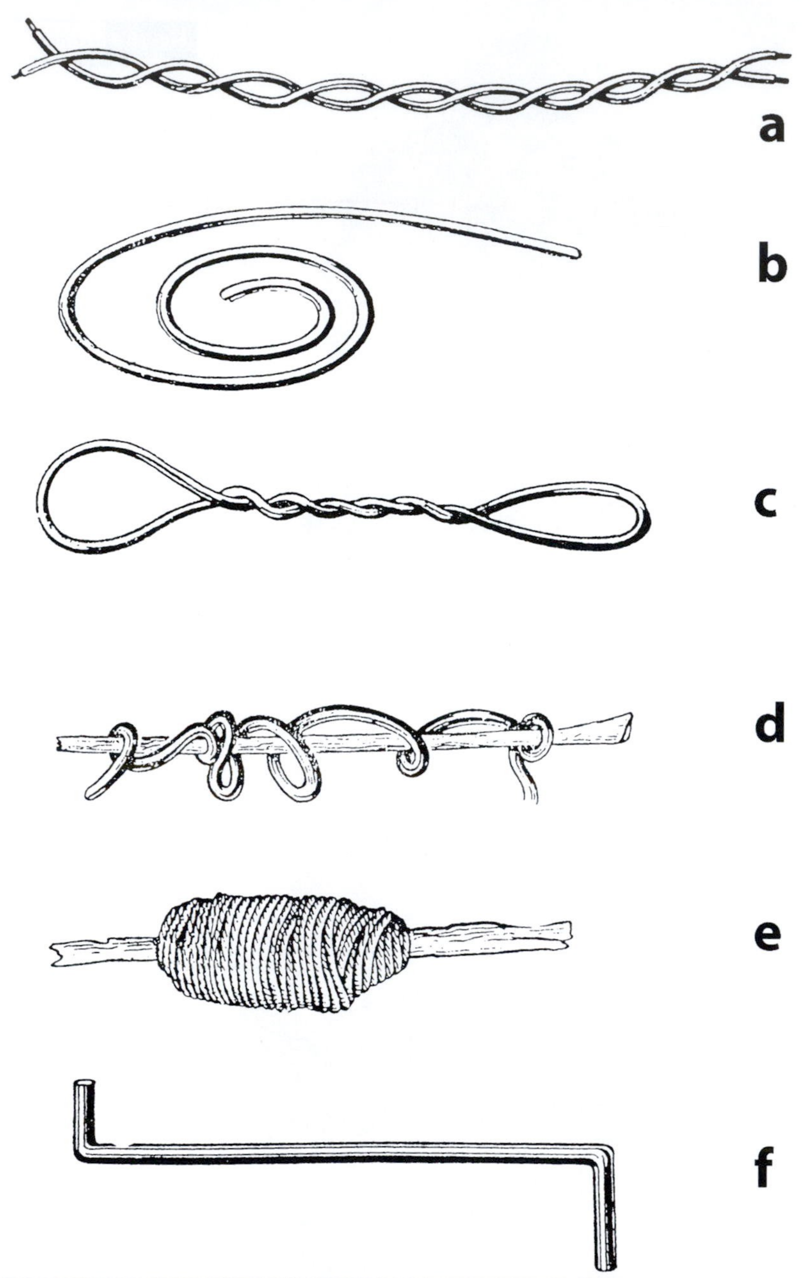

Fig. 18.5 Examples of billets (out of 155 options) given to the chimpanzee Paris which require modifications to be used to get the bait out of the tube (40 cm long, 3.5 cm in diameter). (a) Thin coiled wire, too wide to get the bait out with. Comments: Paris immediately uncoils the wire and uses it to get the bait out. (b) A piece of spiral-coiled wire. Comments: having taken the tube with the bait and the coiled wire in his hands, Paris hurls back the tool and for a long time tries unsuccessfully to take out the bait with his fingers. Then he turns back to the wire, bends its free end even further, but does not make enough effort to straighten the wire completely, and drops it. When the experimenter makes the task easier by suggesting a piece of wire coiled into a circle, Paris unbends it easily making it fairly straight. However, the resulting tool does not go into the tube. Paris repeatedly tries to wind it in from either end which finally makes the wire somewhat more straight. The chimp takes it out, unbends it and pushes the bait out. (c) Coiled piece of wire with loops of different width. Comments: the tool for pushing out the bait is a coiled piece of wire with loops of different width on either end. These loops can become narrower if the tool is

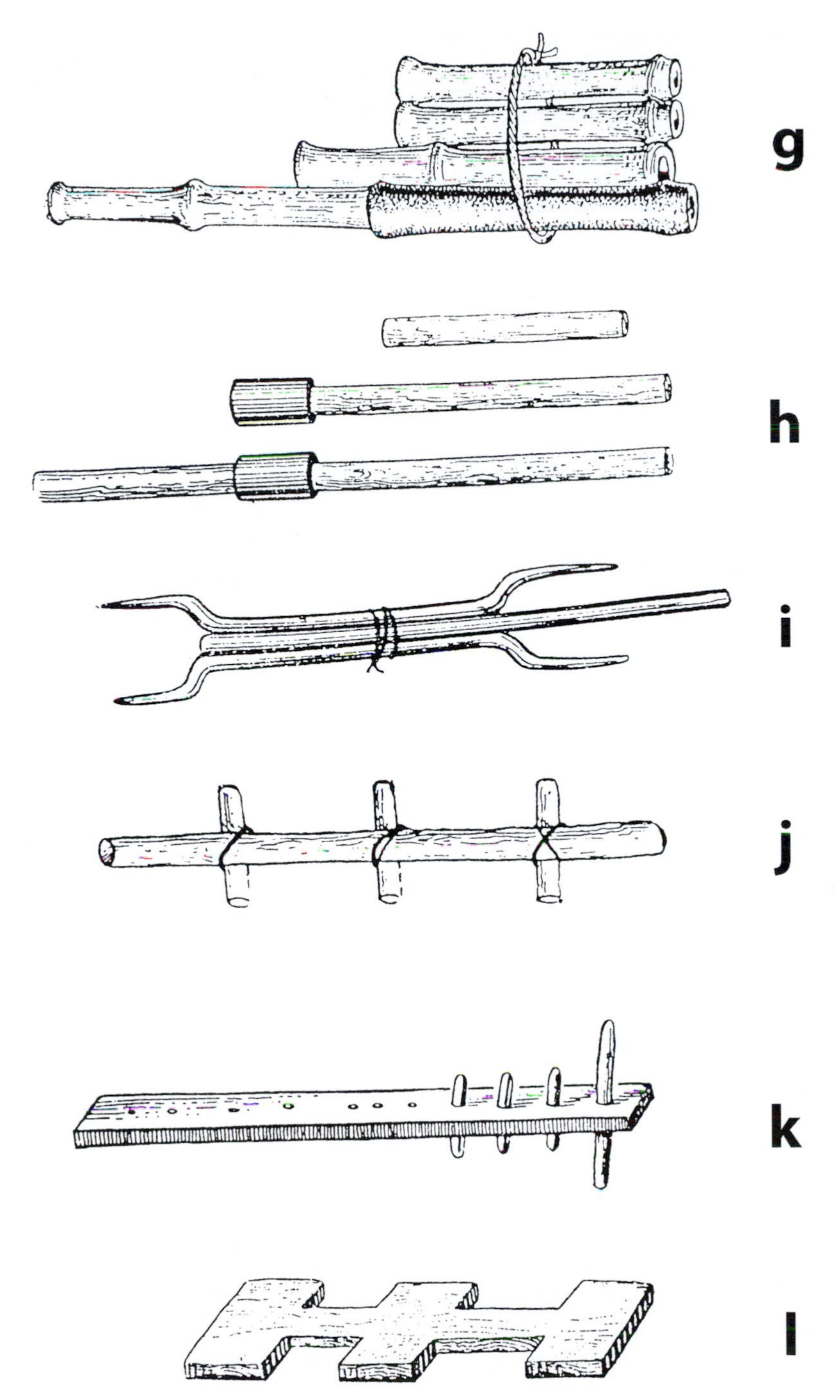

Fig. 18.5 (cont.)
pushed into the tube. At one end the tool is too wide and must be narrowed; at the other end it is narrow enough. The tool is 30 cm long. Paris looks at the tool, makes quick attempts at uncoiling the wire, then stops and pushes the narrow end of the tool in to the tube, getting the desired reward. (d) A wooden stick wrapped in thick wire. Comments: having taken the stick in his hands, Paris immediately starts on the hard task of taking off the wire. He holds the stick with his leg and pulls on the wire with his hand repeatedly from either end. After taking the wire off he pushes the stick in to the tube and gets the food. (e) A stick, wrapped around with rope at the centre: neither of its free ends can be used to push out the bait; the rope has to be taken off. Paris takes off the rope using his left hand, sometimes straightening the rope when it gets stuck, sometimes

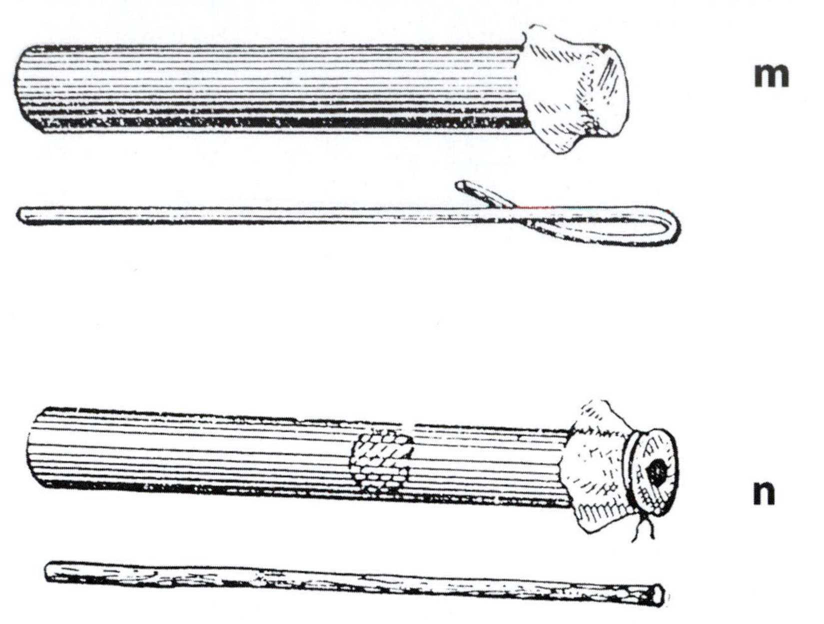

Fig. 18.5 (cont.)

tearing it off. Although after he has torn off the first few coils of rope the tool is narrow enough, Paris does not attempt to get the bait until all the rope is taken off. The unwrapped stick is immediately used to push out the bait. (f) A piece of wire with two bends of different lengths facing different directions. Comments: one end of the wire is bent 2 cm to the side and can be pushed into the tube; the other end is also bent at a right angle but cannot be pushed into the tube without unbending the wire. Paris does not distinguish the right end from the beginning: at first he tries to push the wrong end into the tube. Then he takes the wire out and pushes the other end in, hooks the bait and pulls it out. (g) A rubber tube with a reed in it and three more reeds, all tied together with a rope. Comments: Paris is given the four items tied together, a two-piece tool made of a rubber tube with a reed stuck in it (altogether 45 cm long) and three separate shorter reeds. Paris takes all four items in his teeth by the knot and goes into the corner. There, he unties the tools with an effort and takes the longest reed, which is stuck in the rubber tube. He pulls it out, pushes it back in and out again. Scratches his head with it, picks in his teeth, then breaks it both lengthways and across. Then he sits down and tries to use the remaining short sticks to push the bait out. He also puts some waste paper into the tube. After repeated unsuccessful attempts he finally pushes out the bait putting two reeds into the tube and topping them with the rubber tube. (h) Two sticks, one of them with a retainer into which the other stick can be put. Comments: on one end of the stick there is a metal retainer which can be used to attach another stick. Paris repeatedly tries to take the retainer off, then tries to split the stick without the retainer, and in the end the task is left unsolved. (i) A piece of wire with a fork at one end and a trident at the other. Comments: first the chimp tries to unbend one of the bent pieces on the end of the wire which is split in two, but this turns out to be hard. Then he unties the longest large straight piece, untangling the staples. Having taking the whole thing in pieces, Paris pushes the long piece into the tube and obtains the bait. (j) A stick with three pegs roped to it. Comments: Paris takes the tool and immediately starts untying the ropes which tie the pegs. With the peg-less stick he easily pushes the bait out of the tube. (k) A slat (38 cm long) with four pegs at the end (5.5, 1, 1 and 1 cm long). Comments: Paris pushes into the tube the end of the slat without the pegs. When the longest peg gets stuck he takes it out, and takes out one more peg, leaving the other two in (they do not obstruct pushing out the bait; neither does the short one which he already took out) and gets the bait. (l) A flat bar widened in three places, in the middle and at the ends. Comments: Paris nibbles the wider parts off and pushes the bait out. (m) A tube closed at one end with a film and a piece of wire with a loop at the end. Comments: the tube with the bait is wrapped at one end with a film. The other end is open. A tool is given, consisting of a piece of wire with a loop at one end. Having taken the tube and the wire, Paris pushes the looped end of the wire in the open end of the tube. Then he tries to get off the film using his teeth and hands. Finally he takes it off. Then using the looped end of the wire he pulls the bait towards himself and out of the tube. (n) A tube closed at one end with a pierced film. Comments: Paris is given a tube (15 cm long, 4 cm in diameter) one end of which is clothed with a film which, in turn, is pierced. There is bait packed inside the tube. A thin stick is also provided. Paris takes the stick and pushes it into the tube, but then stops it and starts to tear the film at the sides, using his fingers and teeth. Then he looks into the tube, tears off part of the film but not all of it, and makes a wider whole in the centre. Then he pushes the stick into the hole in the film and gets the bait out. (Adapted from Ladygina Koths, 1959.)

Visalberghi and co-authors (Visalberghi and Trinka, 1989; Fragaszy *et al.*, 2004) presented the tube problem to capuchin monkeys. The subjects were given three variants of the problem, namely, the bundle task, the short-sticks task and the H-tool task. Although all capuchins eventually solved these variations of the task, they made a number of errors such as attempting to insert the whole bundle and inserting one short stick in one end of the tube, and another short stick in the other end. Moreover, these errors did not decrease significantly over trials, suggesting that the capuchins understood little about the causal relations between the elements of the task.

Firsov (1977) suggested a variant of the tube problem that can be named the *reverse tube problem* (Fig. 18.6). Firsov conducted many experiments aimed at studying tool use with a group of chimpanzees that were freely housed on a small island in a lake for several summer months over several years. The reverse tube problem worked

Fig. 18.6 The reverse tube problem in Firsov's (1977) field experiments. (a) Subjects were required to extract a small food item from a hole in the ground; (b) baboons tried to solve the problem but failed. (Photographs by L. Firsov; courtesy of L. Firsov.)

as follows. Four apes (Silva, Gamma, Taras, Boy) were required to extract a small food item from a hole in the ground. The hole was 80 cm deep, that is, about 10 cm more than the length of the chimpanzees' arm. First reactions were similarly simple in all animals, that is, ineffective attempts to reach down the hole with the use of each of the four limbs in turn. After a short rest, an ape repeated attempts, now with less enthusiasm, and then started to look farther ahead. Of the four apes, only Silva manufactured tools to fit the hole. She prepared up to four tools and then selected the most appropriate, that is, long and elastic enough to pull the food item up by clasping it to one side of the hole. Other apes did not make tools from tree branches; instead, they picked sticks up, brought them to the hole and detached them with the help of their teeth and fingers.

A new problem presented by Visalberghi and Limongelli (1996) allows further examination of the limits of understanding of causal relations in animals. This problem, named the *tube-trap problem*, 'punishes' subjects who do not foresee the consequences of their behaviour. The apparatus consists of a horizontal tube with a closed 'trap' at its centre. Food is placed in the small metal cup inside the tube, and a stick is provided. The subject has to remove the food from the tube without it falling into the trap. Because the subjects are able to pull objects towards them as well as push them away, they have to either pull the cup when it is between them and the trap, or push the cup if it is on the far side of the trap (Fig. 18.7). Visalberghi and Limongelli (1996) found that one of their capuchins systematically solved the task pushing the reward away from the trap. Although this female seemed to be planning her moves in advance, the authors noted that in half of the trials she inserted the tool in the wrong end of the tube and upon seeing that the reward was moving towards the trap, she withdraw the tool, reinserted it at the other end and pushed the reward out. The experimenters probed further her understanding of the relation between the trap and the reward by inverting the tube by 180° so that the trap was on the top of the tube where it did not 'rob' the user of the reward. The monkey,

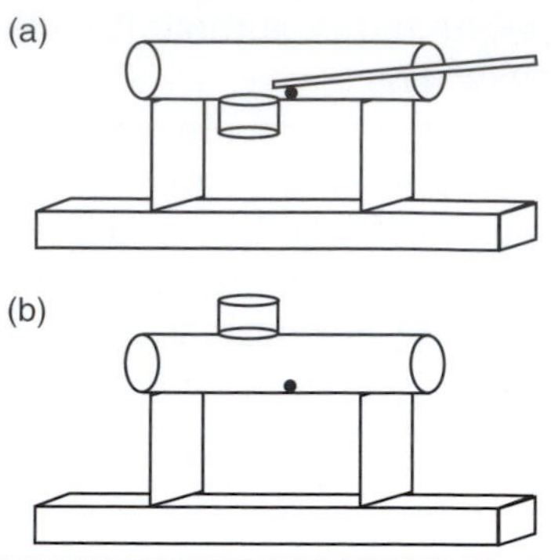

Fig. 18.7 Apparatus used in the tube-trap task consisting of a plexiglas tube with a hole lateral to the centre and a trap underneath (a). In the control condition (b) the tube was rotated, so that the trap was ineffective. (Adapted from Tebbich and Bshary, 2004.)

however, persisted with her strategy of pushing the cup away from the trap which indicated that she had only learned the simple rules without a deep understanding of the problem.

Limongelli *et al.* (1995) presented the tube-trap task to five chimpanzees who behaved at chance level for the first 70 trials, although two of them learned to avoid the trap during 70 additional trials. The experimenters then varied the location of the trap in the tube in order to learn whether the apes understood the relation between the position of the reward with respect to the trap, or whether they were simply using the simple rule of pushing the reward out of the side to which it is closest, thus avoiding the trap. In some variants of the new set of tasks the trap was located very close to one end with the food just beyond it, so that subjects actually had to push the food out the end from which it was farthest. In other cases, the opposite arrangement was used. The chimpanzees easily solved the problem. It should be noted, however, that the variations used in this experiment could still be solved by the rule 'push the food away from the trap', which could have been learned during the previous trials (Call, 2000). Unfortunately, the authors did not use the variant with the inverted (ineffective) trap. However, Reaux *et al.* (1999) used the inverted trap condition with one chimpanzee that was the only one from the group who was successful in the regular tube-trap task. Despite her mastery of the basic task, she continued to avoid the inverted trap – in the same manner as had Visalberghi and Limongelli's (1996) capuchin.

Although apes and monkeys, or at least some of them, have demonstrated a high level of selectivity in choosing relevant tools, and have even manufactured tools for solving some variants of the tube problem, they have not displayed a full measure of understanding the relations between the elements of the tube task.

Recent studies portray birds as being no less adept in passing these tests. Kacelnik *et al.* (2004) conducted series of experiments with New Caledonian crows applying a number of different techniques, including the tube-trap experiment used by Limongelli *et al.* (1995) and Povinelli (2000) to test primates. After about 100 trials with the apparatus, the crow Betty reached criterion (trap avoided on eight out of ten trials or more in three consecutive blocks of ten trials). This performance was comparable with that of apes and capuchins. When the trap was inverted during the testing phase, the response did not return to random; instead, Betty continued to avoid the now irrelevant trap. This is just the same result that had been observed in apes and monkeys. However, in similar experiments with woodpecker finches, one finch, Rosa, easily reached high criteria in solving the basic tube-trap problem, including reshaping tools when solving the H-tool problem (Tebbich and Bshary, 2004). To the credit of primates be it said that the other five woodpecker finches in the group displayed the same low understanding of the physical problem as apes and monkeys did. Besides, as Tebbich and Bshary (2004) reason, the conditions of the experiment did not confirm whether Rosa understood the function of the trap.

Do animals apply causal reasoning to tool-use tasks and, if so, how often?

Although animals often demonstrate incompetence in solving complex instrumental problems, there are many examples of how individuals effectively use their brains to exploit resources by means of tools (see Reznikova (2006) for a review). One of the most intriguing questions related to tool-using in animals is whether subjects are able to take invisible forces into account to guide their actions with different elements of a problem. Let us start with two controversial examples, both coming from early studies.

The first example concerns Rafael, a chimpanzee who participated in experiments with putting out a fire and was required to get a cup filled with water and move from one raft floating on a lake to another one balancing on a thin bridge with the filled cup in his hands (see Chapter 17). In a special series of trials Schtodin (1947) and Vatzuro (1948) investigated whether the chimpanzee was able to grasp the relation between the loss of water and a hole in the cup. A new Archimedes did not come in the person of Rafael. When he was given the cup with a hole in it, Rafael tried to fill it with water 43 times in vain. He did not notice the fact that accidentally closing the hole with his palm retained the water in the cup, so he did not use such a simple way to keep the water in. Being shown how to plug the hole with a short stick, Rafael took the stick out and continued to fill with water the cup with the hole. Finally, Rafael was presented with a small metal ball and was lucky to plug the hole the first time with the use of the ball. It was completely accidental. Playing with the ball, Rafael took it in his mouth, then took some water in his mouth and spat out the water together with the ball, into the cup; the ball hit the hole and thus stopped the gap. So the problem was solved. The chimpanzee fixed the connection between his actions and stopping the leak. But it is even more amazing that he later repeated this whole sequence of acts without any changes. Contrary to any sense, he always placed the ball in his mouth, and spat it out in the cup. Being presented with an intact cup, Rafael nevertheless dropped the ball into it, and when he received two cups, an intact one and the cup with a hole, he preferred to have the broken cup so as to perform his ritual with the ball and the hole again and again.

Perhaps 'ritual' is the key word here. Not only chimpanzees but members of many species persist on the behaviours that have brought them luck at least once. As was mentioned in Part I,

Tony, the dog belonging to the famous psychologist Morgan, refused all ways to lift the latch of a gate that his owner ever tried to teach him; instead, Tony repeated the behaviour that brought about such a fortuitous outcome. Since that, experimenters have described many similar cases. As Ladygina Koths (1935) reasoned, all chimpanzees are enslaved to their past experiences and hardly improve their methods of work. At the same time in the books of Goodall (1986), McGrew (1992, 2004), de Waal (2001) and others one can find many examples of how chimpanzees cheerfully cope with new problems in their natural life and, vice versa, how often they come to a standstill with repetitions of previously obtained solutions becoming senseless in changeable situations.

The second example concerning chimpanzees' reasoning about physical forces came from Firsov's (1977) experiments with his island colony of chimps (see above in this section). In this group, a young male, Taras, distinguished himself by very high frequency of spontaneous 'insightful' solutions of vital and artificial problems. For instance, he used a long stick in order to lift a sunken rope up and thus to pull a boat to him without getting his feet wet. The experiment was the following. Chimpanzees were presented with a food reward placed in an experimental apparatus equipped with a mechanical traction. A door of the apparatus was equipped with a latch spring and opened behind a subject's back when he pulled a handle that hung down from the opposite side of the apparatus. A rod was too long to reach the door with one hand while holding the handle with another hand or with a foot (Fig. 18.8). The problem turned out to be too complex for chimpanzees and gave rise to confusion. One male, Boy, being faced with this task, wallowed in the sorrow. A jar filled with syrup was visible through the transparent walls of the apparatus and attracted the chimpanzees. Taras, after several unsuccessful efforts, turned round sharply and walked to the edge of the forest. He broke off a long dry stick and returned to the apparatus with it. After several manipulations with the stick and the door Taras was lucky in fixing the door and reaching the reward through the jammed door. The next year, when meeting with the same problem, Taras made no unnecessary actions and obtained the reward the very first time using a long stick.

In fact, the distinction between performance and competence in wild users in the real world is a critical one for understanding whether they really take invisible forces into account. Some evidence that came from recent studies supports the claim that animals are able to get to the bottom of a solved problem. Among them, the experiments of Kacelnik and colleagues (Chappell and Kacelnik, 2004; Kacelnik *et al.*,

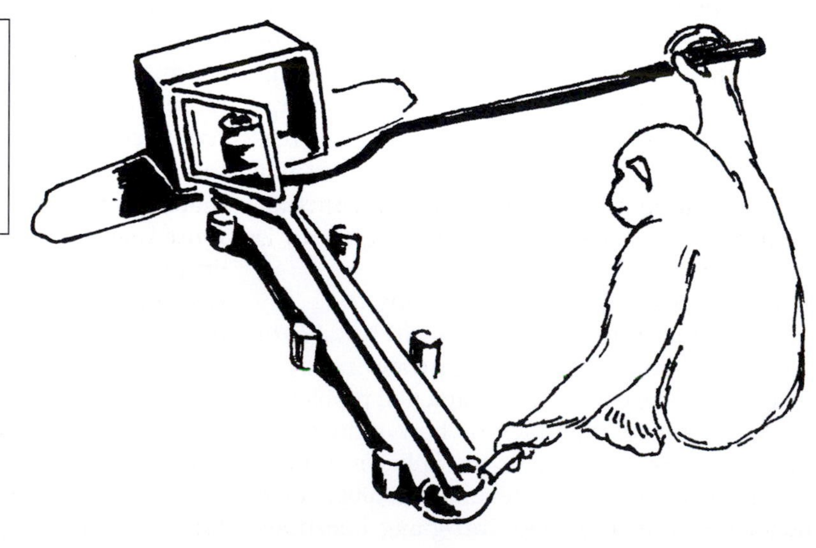

Fig. 18.8 The chimp Taras with one hand wedges the door of the apparatus open using a bent stick, and with the other hand pulls at the lever to get the can with syrup. (Drawing by I. Yakovlev after the photograph from Firsov's (1977) book.)

2004) with New Caledonian crows were directly inspired by similar experiments with chimpanzees described in Povinelli's book. The results obtained have enabled the authors to suggest that crows operate with some general principles rather than with specific associations.

In a 'rigidity' experiment a crow Betty was pre-exposed to two rake-like objects with different levels of rigidity in a non-functional context, and she was tested in a situation where only one of the tools would serve. The idea was to examine whether, when she needed to pick a tool among a set of objects that were familiar to her, but had not been used before as tools, she would choose according to the suitability afforded by the object's properties. The rakes differed at their wide ends. One had a solid end made of wood, whereas the other had a flexible head made of thin plastic. Betty was allowed to manipulate the tools freely for several days. The rakes were then placed in a box with a transparent lid that was internally divided into two compartments each containing a food-filled cup placed in front of the head of each rake. The cup could be retrieved from the box by pulling the rake with the rigid, but not the flexible head. The results suggest that Betty had learned the properties of the rakes and used this knowledge when choosing the tool. She was 100 per cent accurate on the first trial on each day (although she seemed to lose motivation quickly). Betty's success on the first trial of each session contrasts with the results of Povinelli's chimpanzees where six out of seven subjects performed at chance level throughout the experiment, and the only successful subject reverted to chance when a different experimental design was used. These results can also be considered to be related to the phenomenon of latent learning, whereas the results of the next, 'innovation' experiment, indicate that a crow is able to shape tools innovatively in anticipation of specific needs.

Researchers had an insight into the level of individual creativity of New Caledonian crows through a serendipitous observation that was made during the course of an experiment on tool selectivity. The question was whether the crows would choose a hooked piece of wire over a straight piece, where the task was to lift a bucket containing food (using the handle) from a vertical tube. On one trial Abel, a male crow, took the suitable (hooked) wire away, leaving Betty with an unsuitable straight wire. After attempting unsuccessfully to extract the bucket with the unsuitable straight wire, Betty spontaneously secured the distal end in a crevice and made a hook by pulling perpendicularly on the proximate end. With the hook thus made, she proceeded to retrieve the food. To explore this phenomenon further, experimenters repeated the task, but offered only the straight wire. Now, Betty bent the piece of wire and used it successfully at virtually every opportunity. She used at least two techniques to bend the wire and she often corrected the shape of the tool several times before attempting to use it. Betty did not have any experience with flexible wire or similarly pliant material prior to this episode nor was the technique she used possible with natural materials (Fig. 18.9).

These results again contrast with the achievements of chimpanzees in similar and even simpler tasks (Povinelli, 2000). For instance, apes did not think of straightening a hooked wire in order to put it in a hole and thus to pin a piece of an apple down and fish it out. This difference is amazing, the more so that the chimpanzees had previous experience of manipulation with both straight and hooked wires. These results do not establish a gap between the levels of intelligence of clever birds and silly apes; instead, they make it clear that our knowledge about animals' competence in the laws of nature is far from absolute. Some recent data are evidence of a relatively high level of understanding of tools' functional features in primates.

Santos *et al.* (2003) examined whether primates understand which features are causally most relevant to an artefact's function in the absence of any direct physical experience with that type of artefact. To this end, the experimenters used an *expectancy violation paradigm*. The logic behind the expectancy violation paradigm is that subjects will look longer at events that violate their expectations about the physical world than at events that are consistent with expectations. Researchers habituated captive cotton-top tamarins *Saguinus oedipus* to an event in which a

Fig. 18.9 Betty, the New Caledonian crow, solving tool-using tasks. (Photograph by A. Weir; courtesy of A. Weir.)

novel object – a purple L-shaped tool – pushed a grape down and onto a lower platform. The subjects were then presented with two trials. In one test trial, tamarins saw a tool of a different colour but a similar shape push the grape down the ramp. In the other test trial, subjects saw a tool of the same colour but a different shape (an I-shaped tool) push the grape down the ramp. The flat base of the new tool was too short to effectively push the grape. Results showed that subjects looked longer at the new shape test trial than the new colour test trial, suggesting that a change in the tool's shape was more important to its functioning than a change in the tool's colour. The experimenters extended their work to free-ranging rhesus macaques. They conducted the same expectancy violation experiments and obtained similar results.

How then can we reconcile these evidences of animals' competence with the many 'bad mistakes' (in Köhler's sense) introduced by members of different species in different experiments?

Perhaps an insight might come from studying ontogenetic development of tool use in animals. Recent studies have revealed a surprisingly high level of predisposition for manipulation with tools and - what is more important - for understanding of their functional features in members of species that are not natural tool-users. For example, in the absence of experience, infant cotton-top tamarins recognise that when an object is used as a tool, changing its colour or texture is functionally irrelevant, whereas changing its shape, size and orientation is functionally relevant (Kralik and Hauser, 2002). It is doubly astonishing, firstly, because this happens in infants, and secondly, because this species does not naturally use tools. As the authors suggest, tamarins possess innate recognition of functionally relevant features.

Experiments with the main technical bird species, woodpecker finches (Tebbich *et al.*, 2001, 2002) and New Caledonian crows (Kenward *et al.*, 2005), have revealed a very high degree of innate predisposition for manipulation with tools together with high selectivity of tools which is, as in the previous study on tamarins, based on innate recognition of functionally relevant features. In Part VII we will consider in detail problems concerning the predisposition of some species to develop complex behaviours including tool-using and tool manufacture.

Chapter 19

Numerical competence in animals

19.1 Criteria of numerical competence for comparative studies

Basic number-related skills, that is, knowledge of quantities and their relations, is one of the highest properties of cognition. A fundamental question in cognitive science is whether the sense of numbers is unique to humans or whether we share this capacity with other species. Of course, people who lead lives almost completely devoid of numbers remain in the ranks of human beings. In his recent book devoted to human 'mathematical brain', the neuropsychologist Butterworth (1999) describes people who have little or no sense of numbers. The clinical terms are *acalculia* for people who have lost their sense of numbers after a stroke and *dyscalculia* for people who were born without numbers. One of concrete cases described in this book concerns a woman who was blind to numbers greater than four. She could readily perform addition and subtraction, she could name numbers in sequence – so long as all the digits involved were less than or equal to four. In general, the condition called acalculia is evidence that in humans the brain may be biologically 'wired' for mathematics.

Cognitive ethologists use the term *numerosity* as a property of a stimulus that is defined by the number of discriminable elements it contains. It seems that our brain, as well as those of some other species, is equipped from birth with a number sense. This does not mean that animals have an abstract number conception. At the same time, being able to perceive numbers is helpful in many natural situations, for example, in tracking predators or selecting the best foraging grounds. How far does animals' numerical competence reach?

Perhaps no field of cognitive science is based on comparison between animal and human abilities to such a great extent as the field of studying animals' numerical competence. However, we are still lacking an adequate 'language' for comparative analysis. Practically all criteria for the comparison of number-related skills in animals and humans are derived from developmental psychology. The main difficulty in comparing numerical abilities in humans and other species is that our numerical competence is closely connected with abilities for language and for symbolic representation. It is likely that some species can judge about numbers of things and sounds, and maybe smells. A good example comes from the field experiments of McComb *et al.* (1994) with lions in the Serengeti National Park in Tanzania. The lioness leader identifies roaring that comes from individuals who are not members of the pride; she can also represent the defenders as known individuals. She is able to count distinguished roarers and the number of her sisters and compares the two numbers within the limit of four.

Recent approaches to studying numerical competence in animals are mainly based on criteria suggested by Gelman and Gallistel (1978) for children and then adopted for animal studies by Davis and Pérusse (1988). These authors distinguish several types of numerical competence and suggest that different cognitive or perceptual processes underlie each type. They divide numerical competence into the categories of relative numerousness judgements, subitising, estimation and counting.

Relative numerousness judgements involve the simplest decision processes since no knowledge of an absolute number is required. Instead an animal compares 'more' versus 'less'. Animal researchers often use the term *numerosity discrimination* instead of relative numerousness judgements.

Subitising is a form of pattern recognition that is used to assess rapidly small quantities of simultaneously presented items. This term was invented by Kaufman *et al.* (1949) who observed that when adult humans were asked to tell the number of items in an array there was discontinuity in their responses: if the number of items was six or fewer the subjects performed the task very quickly, but beyond this number their response time increased dramatically with the quantity of items. They suggested using the term 'subitising' in order to emphasise the difference of perception of small and large quantities by humans. Subitising is a sophisticated process that is defined by other researchers as 'magnitudes through an accumulator mechanism' (Meck and Church, 1983), or 'prototype matching' (Thomas *et al.*, 1999).

Estimation refers to the ability to assign a numerical label to an array of large numbers of items with poor precision. When we judge at a glance that there are about 50 ducks on a lake we are 'estimating'. Animals' judgements about large arrays have not been systematically studied yet; nevertheless some exciting results will be reviewed in this section.

Counting is the ability to discriminate the absolute number in a set by a process of enumeration. This involves tagging each item in a set, and applying a series of ordered labels as these items are 'counted off'. To count the number of peanuts in a packet of mixed nuts, for instance, we might put each peanut to one side, at the same time labelling them '1', '2', '3', etc. The numerical label we apply to the last peanut we find is the absolute or cardinal number of peanuts in the packet. Davis and Pérusse (1988) regard counting as a more sophisticated process than those involved in relative numerousness judgements or subitising (and presumably estimation too). They also discuss the concept or sense of number as an attribute of counting. This term implies an ability to transfer numerical discriminations across the sensory modalities (e.g. 5 sound pulses are equivalent to 5 light flashes) or across the modes of presentation (e.g. 4 red squares shown simultaneously on a computer screen are equivalent to 4 red squares presented one after the other).

Gelman and Gallistel (Gelman and Gallistel, 1978; Gallistel and Gelman, 1992) list four criteria that formally define the process of counting and have been widely accepted in comparative studies. They are:

(1) The one-to-one principle. Each item in a set (or event in a sequence) is given a unique tag, code or label so that there is a one-to-one correspondence between items and tags.

(2) The stable-order principle (ordinality). The tags or labels must always be applied in the same order (e.g. 1, 2, 3, 4 and not 3, 2, 1, 4). This principle underlies the idea of ordinality: the label '3' stands for a numerosity greater than the quantity called '2' and less than the amount called '4'.

(3) The cardinal principle (cardinality). The label that is applied to the final item represents the absolute quantity of the set. In children, it seems likely that the cardinal principle presupposes the one-to-one principle and the stable-order principle and therefore should develop after the child has some experience in selecting distinct tags and applying those tags in a set.

(4) The abstraction principle (property indifference). As it has been noted before, the realisation of what is counted is reflected in this principle. In experiments with children, a child should realise that counting can be applied to heterogeneous items like toys of different kinds, colour or shape and demonstrate skills of counting even actions or sounds. There are indications that many 2- or 3-year-olds can count mixed sets of objects.

It is important to note that Gallistel and Gelman (1992) do not consider counting to be a process dependent on language and so it can be presented within behavioural repertoire of non-human animals. They consider symbols that are needed to meet any of the criteria described above to be non-linguistic mental symbols ('numerons'), or internal tags the mind makes use of to enumerate a set of objects.

Piaget suggested that infants are born with no understanding of numerosity. His early experiments (Piaget, 1942) described infants' lack of numerosity as a poor perception of quantity conservation. Piaget argued that our very idea of numbers is constructed out of previously developed logical abilities. One of these was transitive reasoning (see Chapter 16): if A is bigger than B, and C is smaller than B, then A is bigger than C. If you figure that out correctly, then you are able to put the numbers in order. These logical abilities don't develop until at least 4 years of age, and are not functioning in their most abstract form until the teens. Further experiments have shown that infants possess some numerical competence at an early age, and this enables researchers to expand the correspondence between the abilities of humans and non-human animals.

19.2 Experimental approaches to studying numerical competence in animals

The earliest attempts to find experimental evidence of numerical abilities in animals were made by Kinnaman (1902) and Porter (1904). They used similar experimental procedures. Porter (1904) tested the response of house sparrows to numbers by hiding food under, say, the third container in a row, then recording which container a bird flew to first on repeated trials. Then he changed the number of the baited pot and tested the birds' ability to redirect its choices. After extensive testing, Porter concluded that the sparrows based their choice on the relative distance of the baited container from the end of the row. Kinnaman (1902) examined two rhesus monkeys. He aligned 21 boxes and trained monkeys to choose the boxes in the requested order. One of Kinnaman's subjects successfully mastered this task and searched for bait in six positions, whereas the second subject learned to search a goal only within three last positions after many attempts. For comparison, Kinnaman trained two children to solve the same task using marbles as rewards. The eldest child of 5 years of age demonstrated the same results as the 'retarded' monkey; the younger child of 3 years of age was able to find the marbles only in the first two boxes. A similar procedure has since been applied for studying many species and has brought contradictory results. For example, chimpanzees have shown great individual differences in their ability to solve such a simple task as searching for a reward in a second box from a proximal point. Some of them never solved this problem whereas racoons, skunks, martens and pigs did not find any difficulties in mastering this task (for review see Boysen and Hallberg, 2000).

Another set of early studies of counting in animals was based on the matching-to-sample paradigm. The results obtained in different species were rather modest. For example, Ladygina's pupil Ioni, a chimpanzee, whom we have already met many times in this book, hardly distinguished '1' and '2' stimuli during the matching-to-sample procedure. It took hundreds and even thousands of trials to teach monkeys, rats and racoons to distinguish between a card with one circle and another one with two circles. Obviously measurement capabilities of the methods applied were limited rather than numerical competence of animals.

The first real progress was made in this field by Otto Köhler (1941, 1956). He established a number of experimental paradigms and applied them for studying counting, such as simultaneous or successive stimulus presentation, as well as matching-to-sample and oddity matching procedures. The experiments were performed on a variety of animals, including squirrels, pigeons, jackdaws, a raven, an African grey parrot and budgerigars. These experiments had been so much appreciated that historical development of this field was subdivided in pre- and post-Köhler developments.

Köhler concluded that animals have two basic numerical abilities. One, based on a visual–spatial sense, enables them to assess the number of items presented simultaneously in a group, while the other allows them to assess the number of events that occur successively, or spread out in time (see Emmerton (2001) for a detailed description). Below we will consider the main training procedures elaborated by Köhler for studying these abilities. Many of these procedures are included in modern experimental techniques.

In one series of Köhler's experiments pigeons were trained to approach a strip of cardboard on which there were two sets of grains that differed in number. A bird had to choose the set containing a particular amount (e.g. 4 grains) and was allowed to eat this set as a reward. To prevent it eating the other set (of say 3 grains), Köhler shooed the bird away if it reached towards the incorrect group. At different stages of training, the correct set sometimes contained the larger and sometimes the smaller number of grains. The experimenter hid behind a screen, out of sight of the bird. The punishment of shooing a bird away was delivered in a standardised fashion by a mechanical device. The reactions of the birds were filmed to provide an objective record of their behaviour.

Another series of experiments were based on a matching-to-sample task. After looking at an array of blobs on a 'sample' card, the subject had to remove the lid from one of two pots in order to find a hidden food reward. On each lid was a different array. A correct choice, which led to reward, was to remove the lid with the same number of marks on as the sample card. Jacob, a raven, was particularly successful on this task. He could match the numbers of items on the sample card and the comparison lid even when the configuration of blobs and their sizes differed (both between and within trials), so that the only common feature was their number. One of the jackdaws was correct in its matching behaviour when the patterns of the blobs differed between the card and the numerically matching lid, but its choices were not statistically reliable when the sizes of the blobs varied. In this case, as Köhler recognised, the bird's performance did not guarantee that it was discriminating solely on the basis of the equality of numbers. Instead, it could have been comparing the overall areas of the stimulus marks on the card with those on the lids.

Other studies were aimed at birds' ability to 'act on number', as Köhler put it, i.e. to respond sequentially until a specific number of items had been obtained or events had been completed. For instance, pigeons and budgerigars were trained to eat only a given number of seeds from a much larger number they saw. So if they were required to take exactly 4 seeds their behaviour was scored as correct if they walked away after eating the fourth food item, but they were automatically shooed away if they tried to eat a fifth item. The accuracy of their performance was tested on trials in which this mild punishment was withheld.

It is of particular interest how birds behave overstepping the limits. For instance, a pigeon ate the fifth – allowed – seed and then slowly and warily sneaked to steal the sixth, then quickly grasped it and rushed away flapping its wings. Such expressive behavioural cues were later collected from many studies on a wide variety of species as 'behavioural indicators' of counting in animals (Krachun, 2002).

In another experiment with pigeons, peas were delivered one at a time down a chute into a large dish. In this experiment, the time interval between deliveries was randomly varied to prevent the birds estimating the total time that had expired, rather than the number of items taken. Köhler also argued that it was the number of peas, rather than the number of pecks, or the pecking rhythm that was important since the pigeons often had to peck several times at a rolling pea before they could grasp it.

The following series of experiments expressively illustrate bird's capacity to 'act on number'. With jackdaws, the task consisted of taking the lids off boxes until a specific number of hidden food items had been retrieved. An important feature of this experiment was that on successive trials, the same number of food items was differently distributed. There were both filled and empty boxes in the row. The bird had to open boxes until it found the allowed number of food items. For instance, it had to stop after finding the fifth item irrespective of how many boxes had been opened. Some birds were able to manage up to four tasks simultaneously. One subject was taught to lift black lids until two items had been found, green lids until three items had been found, red ones to find four, and white ones to find five items. Expressive behaviour of one jackdaw was of significant interest. This bird was taught to open lids until five items had been found. When it had found four items, it returned to the first box (that was empty now) and made a low bow in front of it. Then the bird made two bows near the second box (in which two items were found) and one bow in front of the third; after that the jackdaw returned to searching in order to find the fifth item. This interesting case enabled Köhler to suggest that birds are able to 'act on six' rather than 'count to six'.

From these and other studies, Köhler concluded that birds have at least a limited ability to discriminate objects or events on the basis of their numerosity and inferred that animals have some way of internally tagging the items they have seen or responded to. Köhler was careful to say that animals do not seem to count in the way that an adult human might by precisely enumerating items with a fixed series of symbolic labels (e.g. 1, 2, 3, 4). Rather, he argued that animals learn what he called 'unnamed numbers', so that four items might be represented by a series of inner marks or tags. He also noted that different species showed remarkable similarities in the limits to their ability to discriminate numerosity. However, for most species, the accuracy of performance broke down when the number of items or events they had to respond to was between 5 and 6, or 6 and 7 (Köhler, 1956).

19.3 Numerosity discrimination and estimation in animals

Animals estimate numbers

Numerosity discrimination requires only relative or ordinal judgements. For instance, a bird must show that it perceives one stimulus as having 'more' items compared with the other that has 'fewer', or one consists of 'many' elements while the other has 'few'. However, with these relative judgements the animal does not necessarily have to recognise the precise number of elements in a stimulus.

Thomas and Chase (1980) claimed that they ruled out the hypothesis of subitising in squirrel monkeys based on the fact that the subjects accurately distinguished numerosities of 7 from 8. It was later proposed (Thomas *et al.*, 1999) that the monkeys may have performed some kind of prototype matching that involved building prototypes from experience with different stimuli and then identifying novel stimuli by how closely they correspond to those prototypes. In other words, the monkeys constructed specific 'number prototypes' based on their individual experience and used them to match with novel quantitative stimuli.

There is a growing body of evidence for effective functioning of the number estimation system in non-human animals. Let us consider here two examples concerning birds. Emmerton *et al.* (1997) used a conditional discrimination procedure to study pigeons' ability for estimation. First, a bird had to peck at a visual array that was shown on a centre key. If the array contained 'many' items (6 or 7) then the pigeon had to peck at one of the side keys (e.g. the right-hand red-lit one) to obtain a food reward. If instead the centre array contained 'few' items (1 or 2) then the correct response was to choose the other side key (e.g. the left-hand green-lit one). Incorrect choices led to a time-out period of waiting several seconds in the dark ('punishment'). As was elaborated in Köhler's procedures described above, several tests were conducted in this experiment to make sure that the birds really discriminated the numbers of items in the arrays, rather than detecting some other confounds. A variety of these arrays containing 1, 2, 6 or 7 items were shown until the birds had learned to discriminate accurately between 'many' and 'few' items. Then they were tested not only with new versions of the 'many' and 'few' stimuli, but also with arrays consisting of the intermediate numbers 3, 4 and 5. These numbers were completely novel to them. Most of the 'many' choices were made when the centre test array contained 6 or 7 elements, and the least choices when the array consisted of 1 or 2 items. In this respect the birds treated novel arrays as they had the familiar training stimuli. When the intermediate numbers were shown, the birds' choices were distributed in an orderly fashion. Compared to their responses with arrays of 6 or 7 items, they made slightly fewer choices of the 'many' key when the test array contained 5 items, fewer when it had four items in it, and fewer still when there were 3 items. This distribution of choices indicates that pigeons can serially order numerical quantities.

In Smirnova *et al.*'s (2000) experiments four crows were trained to choose the greater array of elements in the range 1–12. All birds demonstrated high accuracy of comparison. They chose the greater arrays in 75.3 + 2.4 per cent trials including under the minimum difference between the compared arrays (Fig. 19.1). These researchers conclude that the upper limit of the range within which the crows are able to compare numerical attributes of arrays is close to 20.

In experiments by Kilian *et al.* (2003) a bottlenose dolphin was required to estimate numbers of visual stimuli and to choose a set containing more items versus fewer items. The dolphin was asked to discriminate simultaneously presented numbers of visual stimuli, basing on two-choice discrimination. For each trial, two floating hoops with a defined number of objects were hung from the hooks at the side of the dolphin's tank. The dolphin Noah remained at the opposite wall, holding his head out of the water. A short whistle was the starting signal for the dolphin to swim toward the stimuli. Noah made his choice by touching one object of one of the arrays with the tip of his snout. During the first phase, only the numerosities 2 versus 5 were presented. With the first stimuli used, a number of parameters, such as type, size and configuration of stimulus elements, varied with the number feature. After responding correctly to stimuli consisting of three-dimensional objects, the dolphin transferred to two-dimensional stimuli. New number pairings were presented in catch (non-reinforced) trials. All possible pairings were

Fig. 19.1 A hooded crow solves matching-to-sample problems based on the number of objects (Smirnova *et al.*, 2000). (Photographs by A. Smirnova; courtesy of A. Smirnova.)

Fig. 19.1 (cont.)

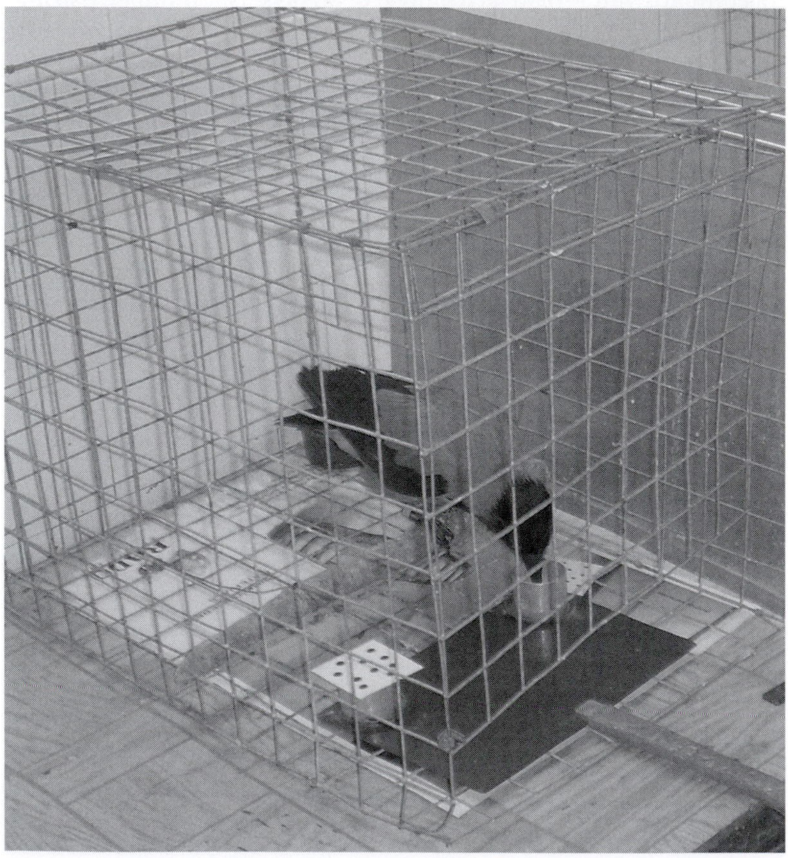

tested such as 3 versus 5, 2 versus 3, 4.versus 5, and so on. The high performance with numerosity pairings provides substantial evidence that the dolphin represented ordinal relations.

Animals judge proportions

It is a natural suggestion that animals have some ideas about the quantity of things independently of their dimensions. For instance, food quantities differ not just as a result of the number of edible items available but also because of their sizes, volumes, weight, etc. Woodruff and Premack (1981) examined how chimpanzees understand the relations between proportions and volumes based on one of Piaget's tests. Subjects were presented with beakers filled with water up to 25, 50, 75 and 100 per cent full, and with paper circles that were correspondingly shaded from 25 to 100 per cent. The task was to match proportions and volumes, that is, to cover a quarter-filled beaker with a quarter-shaded circle. In psychological experiments children master this task quite well at 4 years old. One adult and four juvenile chimpanzees were tested and the adult coped with the task quite well.

Pigeons were also shown to be able to estimate differences in proportions. Emmerton (2001) trained pigeons to discriminate differences in colour proportions within horizontal bars composed of continuous blocks of colour. They were then tested under a variety of conditions to see if they still responded accurately to differences in the relative quantity of colour when the stimulus displays were altered. Initially, the pigeons learned to discriminate between a red and a green horizontal bar when these two stimuli were presented simultaneously on a computer monitor. The 'correct' colour was counterbalanced across birds. Responses were sensed by a touch-screen, and a peck to the correct colour led to a food reward. Then the proportion of the two colours was varied in both stimuli. If a bird chose the bar with the greater proportion of the 'correct' colour (e.g. red in the following stimulus examples), it was rewarded. If it chose the bar with the complementary lesser proportion of 'correct' colour a time-out period ('punishment') followed. If the proportion of red to green was equal in both bars, one of the stimuli was arbitrarily programmed to be the correct one but the bird had to guess which one it was. The accuracy with which the birds discriminated the paired bars was correlated with the colour proportions in the bars. When there was no difference in proportion, their performance was at chance level. The new experiment used stimuli consisting of continuous areas of colour. In this case, proportion was no longer equated with the relative numerosity of component items but the results of both experiments were very similar: as the difference in the proportion of coloured areas or discrete items decreased so did the accuracy of the birds' discrimination. In one part of the study, the birds were presented not with paired horizontal bars but with arrays of small red and green rectangles.

19.4 Counting animals

As has already been noted, to demonstrate proper counting ability, several criteria must be met (Gelman and Gallistel, 1978). These include tagging items one by one, irrespective of their type, so that the final tag represents the precise or cardinal number of items in a stimulus.

Proto-counting

Let us start this paragraph with two examples concerning mini-brains' feats of intelligence.

The experiments of Chittka and Geiger (1995) enable us to suggest that honeybees can count landmarks or at least use the number of landmarks as one of the criteria in searching for food sources. Researchers worked with honeybees in a large meadow which was practically devoid of any natural landmarks that could be used by bees. They then set up their own landmarks, which consisted of large yellow tents. The bees were trained to take syrup from a feeder that was placed between the third and fourth tents. In the tests, the number of landmarks between hive and feeder were altered. It is interesting to note that individual foragers in a hive used different cues in their searching. Many bees continued to rely only on flying distance between the hive and the feeder. Anyway, the distance estimation of the bees as a group depended notably on the number of landmarks. If some family members encountered more landmarks on their way from the hive to the feeder than they had during training, they landed at a shorter distance than during control tests with the training landmark set-up. If they encountered fewer landmarks, they flew significantly further. Discussing their results, the authors consider it is unlikely that their bees meet the abstraction principle of 'true counting'. As was noted above, this principle states that after having learn to perform a given behavioural unit assigned to a certain number of objects counted, the subject should be able to transfer this knowledge onto a set of objects of a different quality. Since a transfer of the counting performance on different objects is unlikely to occur in honeybees, the observed behaviour is referred by the authors to as *proto-counting* (Davis and Pérusse, 1988).

In experiments by Reznikova and Ryabko (1993, 1994, 2001, 2003) red wood ants had to solve a counting problem based on their ability, similar to that of honeybees, to remember, pass and accept complex messages by means of distant homing, that is, by messages outgoing from the scouting individual, without other cues such as a scent trail or direct guiding. The detailed description of sophisticated 'symbolic language' can be found in Part IX. Here it is important to note that ants' numerical competence was studied using their own communicative skills. The researchers asked a scouting ant to transfer information about a number of objects to its nestmates and used quantitative characteristics of ants' communication for investigating their ability to count. The main idea of the experimental paradigm is that experimenters can judge how ants represent numbers by estimating how much time individual ants spend on 'pronouncing' numbers, that is, on transferring information about the number of objects.

In these experiments ant scouts were requested to transfer to foragers in a laboratory nest the information about which branch of a special 'counting maze' they have to go to in order to obtain syrup for food. The counting maze is a collective name for several variants of set-ups. All of them serve to examine how ants

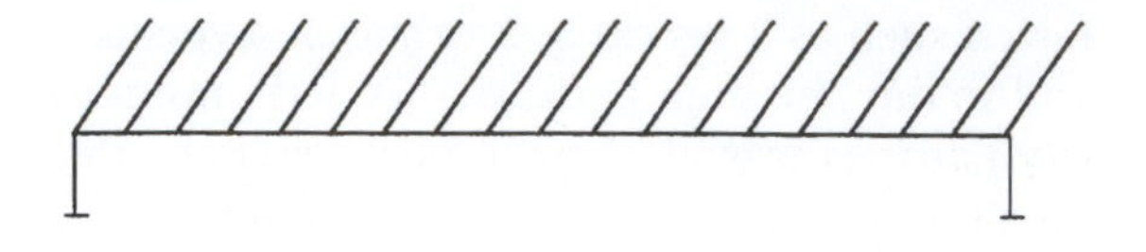

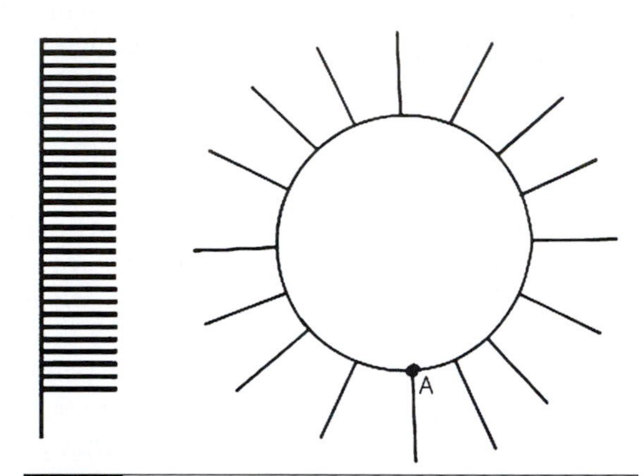

Fig. 19.2 The comb-like set-ups for studying numerical competence in ants: a horizontal trunk, a vertical trunk and a circle.

transfer information about numbers by means of distant homing. The first variant of the counting maze is a comb-like set-up consisting of a long horizontal plastic trunk with 25–60 equally spaced plain plastic branches, of 6 cm length, all horizontal (Fig. 19.2). Each branch ended in an empty trough, except for one (randomly chosen) filled with syrup. Ants came to the initial point of the trunk over a small bridge. The second variant is a set-up with vertically aligned branches. In order to test whether the time of transmission of information about the number of the branch depends on its length as well as on the distance between its branches, one set of experiments was carried out on a similar vertical trunk in which the distance between the branches was twice as great, and the branches themselves were three times and five times longer (for different series of trials). The third variant of the counting maze is circle-like set-up, that is, a horizontal trunk closed round a circle.

Ants were housed in a laboratory arena subdivided into two parts, one containing a plastic nest with a laboratory ant family (of about 2000

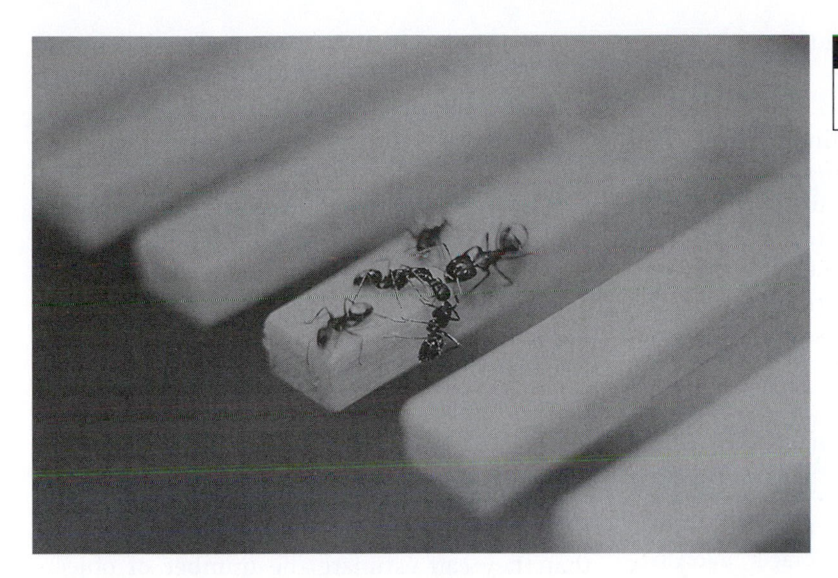

Fig. 19.3 An ant team on the counting maze. (Photograph by Zh. Reznikova.)

workers and one queen) and another containing one of the variants of the counting maze. In order to force a scout to transfer the information about food to its nestmates the researchers showed it the trough containing syrup (placing the scout directly on the trough) and then let it to return to the nest (see more detailed description of series of experiments aimed at studying ants' transfer of information in Part IX). After allowing it to contact foragers within the nest, the experimenters removed the scout and isolated it for a while, so that the foragers had to search for the food by themselves, without their guide. It is important to note that group-retrieving red wood ants organise their searching for food in such a way that the scout serves as a guide for a constant small team of three to ten foragers. So in at least several group-retrieving ant species large families, consisting of from hundreds of thousands to a million individuals, contain many small constant scout-foraging 'teams' or 'cliques' (see Reznikova and Ryabko, 1994; Anderson and Franks, 2001; Robson and Traniello, 2002; Reznikova, 2003).

The experiments were so devised as to eliminate all possible ways for the members of each foraging team to find a goal, except distant homing, i.e. an information contact with their scout. In order to avoid laying scent trails, the set-up was replaced with a fresh one during the time the scout spent within the nest. Besides, foragers were presented with a set-up with all troughs containing water instead of syrup in order to avoid the influence of the food smell. If the foraging team compactly reached the correct branch, then the trough filled with water was immediately replaced with another one filled with syrup in order to please the ants. In several series of experiments all branches were empty (Fig. 19.3).

The findings concerning number-related skills in ants are based on comparisons of duration of scout–foragers information contacts which preceded successful trips of foraging teams. The researchers measured the duration of contacts of the scout with its team when the scout returned from the experimental set-up being loaded both with syrup and information.

In total, 32 scout–foragers teams worked in three kinds of set-ups. The teams left the nests after they were contacted by scouts and moved towards the trough without their guides 152 times (remember that the scouts were deliberately removed). In 117 cases the team immediately found the correct path to the trough, without making erroneous trips to empty troughs. In the remaining cases, ants went to the empty troughs, and began looking for food by checking neighbouring branches. In all experiments (31 in total), foragers failed to find the food-containing trough when 'incapable' scouts were working. Such scouts were experimentally removed from the working part of the arena.

Since all set-ups had no fewer than 25 branches, the probability of finding the correct trough at random is not more than 1/25. Thus, the success ratio that was obtained experimentally can only be explained by information transmission from the scouts. The probability of finding the food-containing trough randomly in 117 cases out of 152 is less than 10^{-10}. In addition, ants, including scouts, placed in the set-up, without having information on which trough contained food, usually failed to find the food, even though they actively searched for it.

Data obtained on the vertical trunk are shown in Table 19.1 as an example.

In all set-ups the relation between the number of the branch i and the duration of the contact between the scout and the foragers t was linear, and may be described by the equation $t = ai + b$. The coefficient of correlation between t and i was high for different kinds of counting mazes (Table 19.2).

The likely explanation of results concerning ants' ability to search the 'right' branch is that they can estimate the number of objects and transmit this information to each other. Presumably they can pass messages not about the number of the branch but about the distance to it or about the number of steps and so on. Even if this is so, this shows that ants are able to use quantitative values and pass on the information about them. But it is worth noting that the relation between the number of the branch and the duration of the contact between the scout and the foragers is well described by the equation $t = ai + b$ for different set-ups which are characterised by different shapes, distances between the branches and lengths of the branches. The values of parameters a and b are close and do not depend either on the lengths of the branches or on other parameters. All this enables the authors to suggest that ants transmit information only about the number of the branch.

Table 19.1. *Results of experiments in 'vertical trunk 1'*

Experiment number	Number of food-containing branch	Duration of scout–forager contact (seconds)	Working team number
1	10	42	I
2	10	40	II
3	10	45	III
4	40	300	II
5	40	280	IX
6	13	90	II
7	13	98	I
8	28	110	III
9	28	120	X
10	20	120	X
11	20	110	III
12	35	260	III
13	35	250	X
14	30	160	I
15	30	170	III

Source: Reznikova and Ryabko (2001).

Table 19.2. *Values of correlation coefficient* (r) *and regression* (a, b) *coefficients for vertical trunk, horizontal trunk and circle*

Type of set-up	Sample size	Number of branches	r	$a \pm \Delta a$	$b \pm \Delta b$
Vertical trunk 1	15	40	0.93	7.3 ± 4.1	−28.9 ± 0.51
Vertical trunk 2	16	60	0.99	5.88 ± 0.44	−17.11 ± 0.65
Horizontal trunk 1	30	25	0.91	8.54 ± 1.1	−22.2 ± 0.62
Horizontal trunk 2	21	25	0.88	4.92 ± 1.27	−18.94 ± 4.75
Circle	38	25	0.98	8.62 ± 0.52	−24.4 ± 0.61

Source: Reznikova and Ryabko (2001).

It is interesting that quantitative characteristics of the ant's 'number system' seem to be close, at least outwardly, to some archaic human language: the length of the code of a given number is proportional to its value. For example, the word 'finger' corresponds to 1, 'finger, finger' to the number 2, 'finger, finger, finger' to the number 3 and so on. In modern human languages the length of the code word of a number i is approximately proportional to $\log i$ (for large values of i), and the decimal numeration system is the result of a long complicated development. Note that when using the decimal numerical system, people have to make simple arithmetical operations: for example, $23 = 20 + 3$. It is particularly obvious in Roman numerals: for example, $\mathrm{VII} = \mathrm{V} + \mathrm{II}$. The second series of experiments enabled the researchers to suggest that ants are also capable of simple arithmetic. We will consider these results at the end of this chapter.

Let us now turn to experiments with primates. Michael and Mary Beran (2004) discussing their results obtained on chimpanzees consider two possible mechanisms of relations between subjects and quantities that can match with proto-counting. In their experiments four chimpanzees were highly accurate in selecting the larger of two concurrent accumulations of bananas in two opaque containers over a span of 20 minutes. Bananas were placed, one at a time, into one of the two opaque containers outside the chimpanzees' cages. The chimpanzees never saw more than one banana at a time, and there were no cues indicating the locations of the bananas after they were placed into the containers. In other words, the location of the last item or the first item presented did not indicate the larger set. Thus, the subjects responded on the basis of what they had viewed throughout the entire trial. The performance of animals matched that of human infants and children up to 4 years of age in similar tests. The chimpanzees were successful even when the sets to be compared were sufficiently large, namely, 5 versus 8, 5 versus 10, and 6 versus 10.

Whether the chimpanzees knew exactly how many bananas were in each container remains unclear. The authors illustrate possible mechanisms of number representation by the apes by analogy with counting in human infants. At least, these results demonstrate extended memory for accumulated quantity in chimpanzees.

In the next paragraphs we will consider experimental studies in which some important criteria of true counting have been met by animals.

The ordinality principle

This principle may be well illustrated by the results of Brannon and Terrace (1998) obtained in co-operation with two rhesus monkeys, Rosencrantz and Macduff. It was shown that rhesus monkeys represent the numerosity of visual stimuli and detect their ordinal disparity. The exemplars were constructed from various abstract elements (e.g. circles, squares, triangles, bananas, hearts, etc.). As a control for non-numerical cues, exemplars were varied with respect to size, shape and colour.

The monkeys were first trained to respond to exemplars of the numerosities 1 to 4 in ascending numerical order (1, 2, 3, 4). To reveal the subject's ability for ordering stimuli, four exemplars, one of each set, were displayed simultaneously on a touch-sensitive video monitor. The configuration of the exemplars was varied randomly between trials. The subjects' task was to touch each exemplar in the ascending numerical order. Subjects had to learn the required sequence by trial and error and by remembering the consequences of their responses to each stimulus. Any error ended the trial, correct responses produced brief auditory and visual feedback, and food reinforcement was given only after a correct response to the last stimulus. The same stimulus set was presented on each trial for at least 60 consecutive trials. During the initial phase of training, subjects were trained on 35 different stimulus sets of exemplars of the numerosities 1 to 4. The opportunity to learn the correct order in which to respond to a new set of stimuli was eliminated during test sessions in which 150 new stimulus sets were presented only once (30 sets per session for five consecutive sessions). Rosencrantz's and Macduff's performances on familiar–familiar, familiar–novel and novel–novel pairs were compared.

The monkeys were later tested, without reward, on their ability to order stimulus pairs composed of the novel numerosities 5 to 9. Both monkeys responded in the ascending order to the novel numerosities. These results clearly demonstrated that rhesus monkeys represent the numerosities 1 to 9 on the ordinal scale.

The cardinal and abstraction principles

As was mentioned above, counting involves the ability to judge absolute, or cardinal, numerical amounts or numbers of responses. The mental tags, or symbols that represent each tagged item, must be applied in a particular order. Two-year-old children being repeatedly asked about how many items are there on the table count them again and again. Older children answer with a final count term because they understand cardinality, the idea that the last count term signifies the total number counted.

Some recent experiments have shown that the cardinal principle can be implemented not only in human brains. At least some animals can apply the label to the final item which represents the absolute quantity of the set. Pepperberg's investigations of the abilities of Alex, an African grey parrot, can serve as an impressive example. Alex had had extensive training in vocalising English words as verbal labels for different objects, shapes and colours (see details in Part VIII). In addition, he learned to respond to the verbal question 'How many?', spoken by his trainer, with a numerical label for the quantity of items he was shown. Alex was initially trained to use numerical labels to describe classes of shapes as 'three-corner' or 'four-corner' (Pepperberg, 1983). Then, over a period of several years, intermixed with training on other tasks, he was gradually taught the labels for between two and six objects (Pepperberg, 1987). In the first series of tests Alex could give the correct answer to questions such as 'How many X?' where X could be a cork, a plastic geometric figure, a piece of corn or a toy truck. For example, when presented with five corks and asked, 'How many?' Alex answered, 'Five'. This suggests that he understood the one-to-one mapping of symbol to object, property indifference, and cardinality. He also learned to tell the number of objects in a subset within a heterogeneous array. For instance, in one experiment, Alex was presented with a tray consisting of several different objects such as four corks and five keys. He was then asked about the number of corks and answered: 'Four'. The next test was even harder, because it required Alex to attend to several conceptual distinctions at the same time. Alex saw a tray with one grey cork, two white corks, three grey keys and four white keys. He was asked, 'How many white corks?' and answered 'Two'. Here, as in trials with other objects and properties, Alex had to attend to both colour and kind terms, and to use both pieces of information to determine the correct number of items. So Alex used 'spoken' numbers as abstract labels. Interestingly, when he was questioned about the number of unfamiliar objects in mixed arrays of known and novel items, his initial tendency was to respond with the total number in the array. Alex was equally accurate at judging subsets in the range of one to six items. He also had to process items defined by the conjunction of their properties (e.g. green and truck) when these items were scattered randomly amongst other similarly complex objects (e.g. blue trucks, green keys, blue keys) that would have acted as perceptual distractors. Although all the principles of counting were not directly tested in this experiment, this parrot obviously demonstrated an ability to respond in a flexible way to the absolute or cardinal number of things that make up a group, or part of a group (Pepperberg and Gordon, 2005).

In the experiments of Smirnova *et al.* (2000), hooded crows solved matching-to-sample problems based on the number of objects regardless of such attributes as colour, size and shape. The birds could select a card with an array that consisted of a blue rectangle and black dot if the sample was a card with an array that consisted of a red square and a green triangle, and vice versa. To solve this task, a crow had to determine the number of elements on the sample and compare it with the number of elements on the comparison cards. In the novel test the experimenter used heterogeneous graphic arrays consisting of five to eight elements of different shape, colour and disposition. Crows demonstrated the ability to recognise arrays by the number of elements itself and to apply the matching concept to the novel stimuli of numerical category.

19.5 Animals use symbolic representation of numbers

The ability of non-human animals to use symbolic representation of numbers looks quite sophisticated, although they of course do not use the logarithmic scale of numerals as we do. For example, to find out whether pigeons would be able to produce specific numbers of pecks in response to given symbols, Xia *et al.* (2000) used six symbols (A, N, T, 4, U and 5, some of them rotated) that were presented on one of the pecking keys (the 'symbol' key) in a conditioning chamber. Each symbol was associated with a certain number of pecks required. A second key (the 'enter' key) had to be pecked to indicate that the response requirement on the first key had been fulfilled. If this second key was pecked before the response requirement on the first key was completed, or if too many pecks had been delivered on the first key, a time-out followed. Pigeons only received a food reward for the exact production of the required number of pecks on the first key and a final single peck on the second key. After prolonged training, six out of nine pigeons reached a choice performance well above chance level for the first four symbols presented in a random order. Six of these birds managed to deal with five, and four birds also dealt reliably with all six symbols.

Matsuzawa (1985) trained Ai, a chimpanzee, to select an appropriately numbered response key from 1, 2, 3, 4, 5 or 6 whenever she was shown an array of a given object. Ai knew the count sequence up to 10, including the symbol for zero. When presented with a string of three different numbers on a monitor, Ai pointed out the correct sequence, from lowest to highest, independently of the intervals between numbers (e.g. 1–2–9, 0–4–5). This showed that Ai understood the relationships between numbers (Tomonaga and Matsuzawa, 2002): apparently she understood that each integer holds a particular relationship to other numbers in the numerical sequence. By carefully ruling out several alternative explanations for Ai's performance, the experimenters demonstrated that the chimpanzee could acquire some aspects of the number concept.

Beran and Rumbaugh (2001) studied representations of numbers in two chimpanzees, Lana and Mercury. The chimpanzees used a joystick to collect dots, one at a time, on a computer monitor, and then ended a trial when the number of dots collected was equal to the arabic numeral presented for a trail (Fig. 19.4). The chimpanzees were presented with the task again after an interval of 6 months and then again after an additional interval of 3.25 years. During each interval, the chimpanzees were not presented with the task, and this allowed the researchers to assess to what extent both animals retained the value of each numeral. It has been revealed that, despite performing at levels significantly better than chance, the chimpanzees' performance decreased across the two retention intervals. This was particularly true for the larger target numerals (6 and 7). The authors' explanation is based on the fact that the chimpanzees seemed to represent numerosity in a manner consistent with analogue magnitude estimation. By this is meant that the animals showed increasing variability in responding as a function of the size of the set represented by a numeral. As the target numerals increased, the variability in the number of dots selected to match those numerals also increased. The identical pattern was found at the 3.25-year retention test. This indicates that the chimpanzees can have approximate representations of numbers (Beran *et al.*, 2005).

19.6 Wild arithmetic: an insight from comparative studies

In the early 1990s developmental psychologists revealed that very young children can represent numerical quantities without the use of language. Even more importantly, they can

Fig. 19.4 The chimpanzee Lana in the experiment by Beran and Rumbaugh (2001) working on her computerised counting task. She controls the joystick with her left hand to move the cursor (+) on the screen. She has to move the cursor into contact with individual dots one at a time, until she collects a set equal in number to the target (in this case, the target is 7). (Photograph by M. Beran; courtesy of M. Beran.)

understand that addition increases the numerosity of the set of items, while subtraction does the opposite. Wynn (1992) explored whether 5-month-old infants can solve addition and subtraction problems taking advantage of the human infant's capacity for understanding object permanence (see details in Chapter 13). As many developmental psychologists do who work with pre-linguistic infants, she used 'looking time' as a relevant measure to judge the subject's understanding of a problem. Wynn showed an infant one, two or three identical dolls (Mickey Mouse) on a stage, as well as a screen moving up and down. The test trial started when an infant was bored, looking away from the stage. In the 'expected' test ($1+1=2$), an infant watches as an experimenter lowers one Mickey Mouse doll onto an empty stage. A screen is then placed in front of Mickey. The experimenter then produces a second doll and places it behind the screen. With the screen removed, the infant sees two Mickeys on the stage, an outcome that should be expected. In the 'unexpected' test, the infant watches the same sequence of actions, involving the same two Mickey Mouse dolls, but with one crucial change. When the experimenter removes the screen, the infant sees either one Mickey (i.e. $1+1=1$) or three (i.e. $1+1=3$). Infants consistently look longer when the outcome is one or three Mickeys than when the outcome is two. The same kind of result emerges from an experiment involving subtraction instead of addition. Wynn concluded that infants have an innate capacity to do simple arithmetic.

Starkey (1992) found even more advanced arithmetic in young children. She used a box in which a child could search for tennis balls without being able to look inside. Children were shown a small set of balls being put into the box, and then were asked to retrieve that set of balls. An assistant, in anticipation of each of the child's retrievals, secretly put another ball inside the search box so that their number remained constant. Children at 36 to 42 months old were able to perceive numerosities of up to 4. When children were shown an additional placement or removal of 1 to 3 balls and then asked to search for the set of balls, the question was whether they would retrieve the number of balls placed originally, or whether they would search for the number of balls after the addition or subtraction. Nearly all 18–24-month-old children searched for the set of balls after the addition or removal, signifying that they could understand the result of a simple addition or subtraction of up to 4. The above experimental results not only contradict Piaget's experiments that suggest poor number sense and arithmetical efficiency of children up to the age of 7, they also demonstrate an ability for more complex addition and subtraction problems, at the preverbal stage of a child's life.

Hauser *et al.* (2000) set up Wynn's $1+1=2$ task for the rhesus monkeys living on the Puerto Rican island of Cayo Santiago. In the first series of experiments subjects watched as an experimenter placed two eggplants behind a screen and then removed the screen. Subjects looked longer when the test outcome was one or three eggplants than when it was the expected two. Like human infants, rhesus monkeys appear to understand that $1+1=2$. Rhesus monkeys also appear to understand that $2+1=3$, $2-1=1$, and $3-1=2$. They seem, however, not to understand that 2 plus 2 equals 4. Comparable results have been obtained with adult cotton-top tamarins.

In the second series of trials the subjects observed experimenters place pieces of fruits, one at a time, into each of two opaque containers. The experimenters then walked away so that monkeys could approach. The monkeys chose the container with the greater number of apple slices when the comparison were 1 versus 2, 2 versus 3, 3 versus 4 and 3 versus 5 slices. It is worth noting that this technique differs from many others in that it reveals what animals think, spontaneously, in the absence of training. With respect to numerical abilities, adult rhesus monkeys and tamarins are comparable to 1-year-old human infants when it comes to summing objects. Spontaneous representation of number in these primate species is limited to a small number of objects. With training, animals surpass this limitation sometimes to a great extent (Uller *et al.*, 2001).

Boysen and co-authors conducted elegant experiments with chimpanzees' arithmetic. One of their pupils, Sheba, was at first shown a tray beside three placards. The tray contained up to three objects and the placards each portrayed either one, two or three discs. The task was to select the placard with the same number of discs as objects in the tray. In the next stage of the experiment, discs were replaced by the appropriate numerals. At the end of this training, she was able to select correctly the numerals 0, 1, 2, 3 and 4. In the next experiment Sheba was allowed into the room where oranges were hidden in as many as three different places. To gain a reward, Sheba was expected to inspect three locations, and then select the numeral that corresponded to the total number of oranges that she had seen. Finally, numerals were placed in two of the locations. Sheba responded correctly by selecting the numeral that corresponded to the sum of the two numerals that she just inspected (Boysen, 1993; Boysen *et al.*, 1996).

The game played between Sheba and the second chimpanzee, Sarah, displayed how hedonistic interests overshadow arithmetic skills, but their use of numerals restored to chimps their prestige as mathematicians that they had nearly lost. In this game (Boysen and Hallberg, 2000) Sheba played the role of selector and the other, Sarah, the role of receiver. Sheba was presented with a tray containing two containers filled with candies; one container always held more candies than the other. Sheba could point to one container. The amount of candies in this container goes to Sarah, and Sheba is left with candies in the other container. Logically, Sheba, the selector, should always point to the container with fewer treats. In so doing, she guarantees receiving the container with *more* treats. In reality, the chimpanzees could not inhibit their tendency to select the larger of two candy piles, even though a reversed reinforcement contingency dictated that choosing the larger pile resulted in gaining access to the smaller pile and vice versa. Sheba displayed signs of excitement such as touching, moving and rearranging the items in the array. Fortunately, Sheba knew numerals. In the key experiment, instead of presenting Sheba with a tray of candies, researchers presented a tray where each set of candies was covered by a card representing a distinctive numeral in dependence of how many items the set contained. Under these conditions, Sheba picked the card with the number '1'. Consequently, she received six treats while Sarah received one treat.

Results of the game experiment, although very expressive, can be considered an evidence of chimpanzee's awareness of ordinality rather than of their arithmetic skills. In experiments by

Beran (2001, 2004) chimpanzees successfully coped with tasks based on addition and removal of items within sets of 10 items. Each of two subjects watched how experimenters place food items into opaque cups. The quantities in each cup were presented as two (2 + 4; 3 + 2) or three (2 + 3 + 2; 3 + 1 + 4) sequentially presented sets. The task was to choose a desirable box in conformity with the chimpanzee's hedonistic interest. They performed at high levels in selecting the largest of two and even of three sets. Subtraction seems to be more complex for chimpanzees: they performed at above chance level for the task of removal of one, but not more than one item.

Reznikova and Ryabko (2000) elaborated a new experimental paradigm of studying ants 'arithmetic' skills that in principle can be extended to other social species possessing flexible behaviour, individual recognition and the need to pass and memorise complex 'messages'. The paradigm is based on a fundamental idea of information theory which proposes that in a 'reasonable' communication system the frequency of the use of a certain message and the length of that message must correlate. This correlation is described by the equation $\ell = -\log p$, were ℓ is the length of a message and p is its frequency of occurrence. The informal pattern is quite simple: the more frequently a message is used in the language, the shorter the word or the phrase coding it. For example, even in official documents, the words 'White House' are used instead of 'The Executive Branch of the Government of the United States of America'. Professional jargon, abbreviations, etc. serve the same purpose. This phenomenon is manifested in all known human languages. The second idea is that when using a complicated numerical system, one has to add and subtract small numbers. As was noted in Section 19.4, when using Roman numerals, VII = V + II, IX = X − I etc.

The main experimental procedure was described in Section 19.4. The scheme of the new set of experiments and the ants' main achievements were the following. Ants were offered a horizontal trunk with 40 branches. The trough with syrup was placed on different branches with different frequencies. At first there were two 'special' branches, say, number 10 and number 20; the food was placed much more frequently than on the others (i.e. in 2 cases out of 3). When the ants had learnt this, they changed their way of transmitting the information about the branch containing food. The time required for transmitting the message 'The trough with food is on the branch number 10' or '... number 20' by the ants was considerably reduced. This enabled the researchers to suggest that group-retrieving ants have a communication system with a great degree of flexibility. Furthermore, in those cases when the trough with food was placed on branches close to the 'special' ones (in the example described above – numbers 11, 12, 9, 8, 21, 22, 19, 18), the time required for transmitting the information about them by the ants also decreased considerably. Analysis of the length of time the ants spent for transmitting information about different branches suggested that the ants were using a mode of representing numbers similar to the Roman numerals in which the 'special' numbers (10 and 20 in this case) play the same role as the 'special' Roman figures V, X, L, etc. Thus, the ants were shown to be able to add and subtract small numbers.

Now let us consider plasticity of the ants' 'number system' in more detail. In fact, in the cited experiments the researchers examined the ants' aptitude for changing the length of messages in accordance with their frequency in the ants' communication. The experiments were divided into three stages, and at each of them the regularity of placing the trough on branches with different numbers was changed. At the first stages, in selecting the choice of the number of the branch containing the trough, a table of random numbers was used. So the probability of the trough being on a particular branch was 1/30 because only branches 1–30 were used. At the second stage the experimenters chose two 'special' branches A and B (N 7 and N 14, N 10 and N 20, and N 10 and N 19 in different years) on which the trough with syrup occurred during the experiments much more frequently than on the rest – with a probability of 1/3 for 'A' and 'B', and 1/84 for each of the other 28 branches. In this

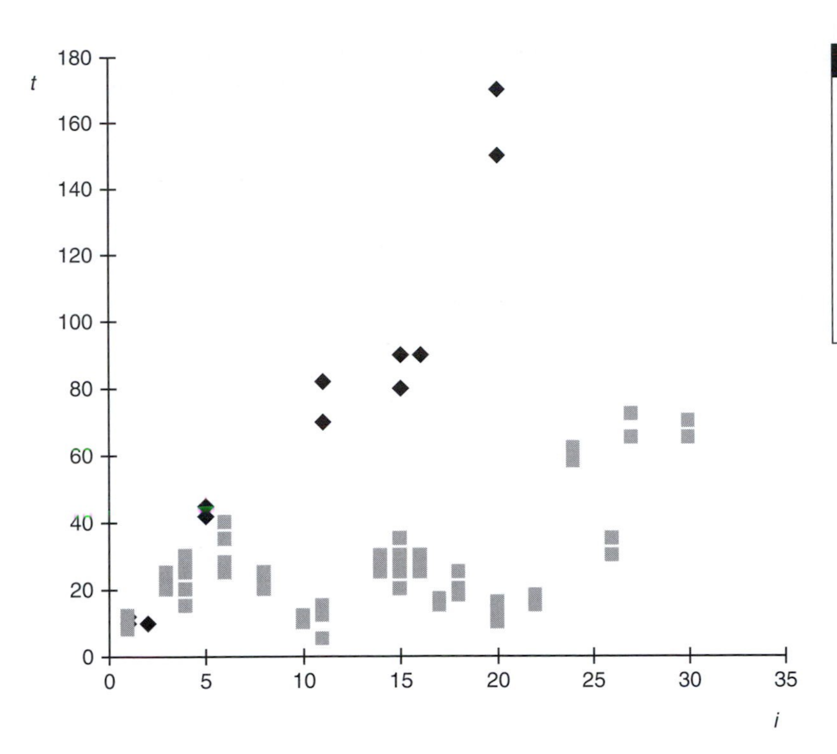

Fig. 19.5 Dependence of the time (in seconds) of transmission of information about the number of the branch having food (t) on its ordinal number (i) in the first and the third series of experiments in the ant *Formica polyctena*. Diamonds, the time taken for transmission of information at the first stage; Squares, the same at the third stage.

way, two 'messages' of the ants – 'The trough is on branch A' and 'The trough is on branch B' – had a much higher probability than the remaining 28 messages. In one of series of trials we used only one 'special' point A (the branch N 15). On this branch the food appeared with the probability of 1/2, and 1/58 for the other 29 branches. At the third stage of the experiment, the number of the branch with the trough was chosen at random again.

Now let us consider the relationship between the time which the ants took to transmit the information about the branch containing food, and its number. The data obtained at the first and third stages of the experiments are shown in the graphs (Fig. 19.5) in which the time of the scout's contact with the foragers (t) is plotted against the number (i) of the branch with the trough. At the first stage the dependence is close to linear. At the third stage, the picture was different: first, the information transmission time was very much reduced, and, second, the dependence of the information transmission time on the branch number is obviously non-linear: depression can be seen in the vicinities of the 'special' points. So the data obtained demonstrate that the patterns of dependence of the information transmission time on the number of the food-containing branch at the first and third stages of the experiment are considerably different. It means that the ants have changed the mode of presenting the data about the number of the branch containing food and rearranged their communication system. Moreover, in the vicinities of the 'special' branches, the time taken for transmission of the information about the number of the branch with the trough is, on the average, shorter. For example, in the first series, at the first stage of the experiments the ants took 70–82 seconds to transmit the information about the fact that the trough with syrup was on the branch N 11, and 8–12 seconds to transmit the information about the branch N 1. At the third stage it took 5–15 seconds to transmit the information about the branch N 11.

What about the ants' ability to add and subtract small numbers? Analysis of the time duration of information transmission by ants raises the possibility that at the third stage of the experiment the messages of the scouts consisted of two parts: the information about which of the 'special' branches was the nearest to the branch with the trough, and the information about the distance from this branch with the trough to this definite 'special' branch. In other words, the ants, presumably, passed a 'name' of the 'special' branch nearest to the branch with the trough, and then the number which had to be added or subtracted in order to find the branch with the trough.

In order to verify this statistically, the experimenters calculated the coefficient of correlation between the time of transmission of information about the trough being on the branch i and the distance from i to the nearest 'special' branch. The results confirmed the hypothesis that the time of transmission of a message about the number of the branch is shorter when this branch is closer to any of the 'special' ones. The high values of correlation coefficients showed that the dependence is close to linear. This, in turn, suggested that at the third stage of the experiment the ants used a 'number system' similar to Roman numerals, and the numbers 10 and 20, 10 and 19 in different series of the experiments, played a role similar to that of the Roman numerals V and X. All this allows the suggestion that ants of highly social group-retrieving species are able to add and subtract small numbers (Reznikova and Ryabko, 1996, 2000, 2001).

CONCLUDING COMMENTS

In this Part we have considered many elegant classic and modern experiments aimed at investigating to what extent animals are able to get to the bottom of a solved problem. In order to examine the limits of animal intelligence, researchers have confronted members of different species with problems that demand, somehow or other, mental reorganisation of cumulative experience based on rule extraction. The ability to form learning sets, that is, to grasp a general rule of solving a concrete problem and then apply this rule to solve new sets of similar problems, is one of the necessary prerequisites for insightful behaviour. Another is exploratory activity which extends an area for gaining information and premises to hidden learning in many species.

There is much work to be done to extend our understanding of whether at least some species share advanced characteristics of intelligence with human beings, or whether all animals think about the world in a way radically different from our own. Working with tool-using animals as well as with counting beasts serves as a good background for development of comparative studies of cognition.

It is still under discussion whether prolific tool-users such as chimpanzees and New Caledonian crows can take into account imperceptible physical forces or whether they are only capable of reasoning about perceptible things. Some experiments have revealed a great deal of flexibility and rationality, whereas others have established a distinction between performance and competence in wild users. To clear up this problem to a lesser or greater extent, developmental studies are needed that will allow a distinction to be made between inherited and acquired behaviour. This approach and the data obtained will be considered in Part VII.

Although our knowledge about advanced forms of animal intelligence has been much enlarged by applying new experimental paradigms including studies on counting in animals, we are still far from having an integrative measure of animal intelligence, such as when a teacher asks a little child to what limit can he or she count. Maybe ants are more competent than chimpanzees in numerosities; pigeons and crows are equal to our close relatives in applying number-related skills; maybe Darwin finches are not less and New Caledonian crows are even more advanced than chimpanzees in tool manufacture maybe not. We will see in Part VII that some species are predisposed to intellectual feats within narrow limits of solving life-or-death problems. It is possible in principle that group-retrieving ant species together with honeybees share with humans (within modest limits of course) such fundamental and interrelated properties of intelligence as predisposition to mathematics and symbolic language. What allows humans to have their heads in the clouds also allows social insects to search for food in complex situations and to treat their collective reason well, from small groups of individuals, to huge intercommunities. We will consider an intriguing problem concerning interrelations between intelligence, language and sociality in Part VIII.

In the course of the book, we have considered dozens of experiments conducted in boxes, which are more or less comfortable for the tested subjects. In this Part at last several outdoor experiments have been considered. The next Part, Part VII, will be devoted to the nature and specificity of intelligence in different species. We will examine members of species in the context of their natural life.

Part VII

Knowledge is power but not for all: species-specific intelligence

All animals are equal but some animals are more equal than others.

George Orwell, *Animal Farm*

As has already been mentioned in the Preface to this book, the lives of many creatures, from large to very small ones, are full of vital situations that demand minute-to-minute decisions from them: to fight or to flee, to court or to ignore, to dominate or to obey, to take from or to share with, to eat or to spit out. We can hardly presume that living things can be provided with inherited 'recipes' for all cases. One of the most interesting problems in studying animal intelligence is how inherited traits interact with individual and social learning during the development of complex behaviour, and, more concretely, how 'wired' stereotypes together with possible innate predisposition for selecting stimuli in environment and responding to them in specific ways influence the development of cognitive abilities.

Lorenz and Tinbergen elaborated the paradigm of classical ethology that was aimed at understanding the ultimate evolutionary reasons for behaviour. The basis of the general research programme for ethologists is frequently referred to as 'Tinbergen's four questions':

(1) How does the behaviour develop during the animal's lifetime, or what is the ontogeny of the behaviour?
(2) How does the animal's immediate environment activate the physiological mechanisms that generate the behaviour, or what are the proximal mechanisms of the behaviour?
(3) How does the behaviour contribute to survival and reproductive success, or what is the adaptive significance of the behaviour?
(4) What is the evolutionary history or phylogeny of the behaviour?

Now students of animal behaviour agree that many displays of behaviour have both genetically pre-programmed and learned aspects. So, at first sight, it seems trivial to discuss combinations of innate and acquired components of behaviour. However, concrete knowledge fills the empty honeycomb of the logical schema with the honey of sense, and can provide new ideas.

In this Part we will consider specific forms of interrelations between learning and instinct in animals, starting with the problem of individual diversity of members of species in relation to their intellectual potentials. When considering relations between innate and learned behaviour, a key question arises about what is *instinct*. In this Part I only touch on the aspects of the concept of instinct which deal with intellectual abilities of animals. I would like to refer the reader who is interested in classical concepts of instinct and its modern development to specialised books and reviews in this field (see, for example, Bateson, 1991, 2001, 2003, 2005; Heyes and Huber, 2000; Bekoff, 2002; Alcock, 2005).

The recent development of comparative cognitive studies in close relation with ethology has led researchers to a concept of guided learning in animals, that is, learning which is controlled by instincts. Different aspects of this phenomenon concern the predisposition of organisms to form some associations more readily than others and this includes displays of guided learning of different complexity, from relatively simple, such as 'innate' fear of enemies and preference for certain kinds of food, to very complex, such as tool manufacture and food storing. Starting with a classic problem of 'misbehaviour of organisms' in Chapter 20, we then will analyse different aspects of guided learning in Chapters 22 and 23.

In Chapter 23 seemingly paradoxical questions will be discussed about how an instinct is learned, how intelligence is wired, and whether 'innate learning' is possible, as a form of imprinting. Analysis of these intricate details of the whole picture of animal behaviour is based mainly on developmental studies. Recent data have laid the foundation of the concept of ecological intelligence, that is, the development of amazing cognitive skills in some species, within the frames of specific domains, aimed at solving some vital but restricted problems.

Chapter 20

Is finding a common metric of intelligence possible in real animal life?

In this chapter we will discuss the question of the possible possession of equal intelligence in members of species. In this book we have already met many examples of animals' limited capabilities to learn. Students of animal intelligence no longer share the belief of the early behaviourists that animals could learn all information with equal ease. We know now that animals represent the world in species-specific ways. Species differ in their abilities to perceive stimuli and to establish connections between them, and it is possible that sometimes species-specific behavioural patterns govern education in animals. The questions we will discuss in this chapter are the following:

Can we consider members of species equal?
Can we rank species within a universal intelligence scale?
Why do 'clever' animals make 'bad mistakes'?

20.1 Never laugh at fishes: some species and members of species are more intelligent than others

The eminent American neurophysiologist Donald Hebb was not lucky with his puppy Henry. Henry was part of Hebb's project aimed at comparing intellectual abilities in animals raised in an enriched environment and in deprivation. Hebb (1949) had previously reported that rats raised as pets performed better in mazes than normally reared laboratory rats. So, a litter of thoroughbred scotch terriers was divided into two parts where half the puppies were deprived in cages and the others were adopted by collaborators as family pets. In general, Hebb's prediction was true, that is, those dogs that were raised as pets in attentive families appeared to be faster learners than the poor dogs which, when infants, had seen nothing more than the wires of their cages. The one exception was Henry. He was extremely dull and was not even able to remember his way home, so the family members had to rescue him from the animal shelter where he had been caught as a stray. It was of no surprise that he proved himself to be the most retrograde individual when all the siblings were tested in labyrinths.

This example supports the hypothesis that individuals show a great deal of variability in their capacity of learning and memory. In this book we have already met many examples of how levels of individual variability exceed levels of species-specific ability to learn and solve complex problems. For example, some individual monkeys outperformed apes in solving complex instrumental problems that require tool-using, whereas one woodpecker finch and one New Caledonian crow outperformed them all (see Chapter 18 for details). Many papers devoted to experimental studies of animal intelligence end with the conclusion that at least one member among the specimens being confronted with a definite problem coped with it, and this gives the possibility of judging about intellectual potential of the species. For instance, in Gillan's (1981) study three female chimpanzees were asked to solve a complex transitive inference problem (see Chapter 16), and only one chimpanzee, Sadie, performed perfectly on the test in all 12 test trials.

Examining members of species in their performance on a series of learning problems, we have the possibility, at least illusory, of finding a level of complexity of a task at which all animals are equal and no one is 'more equal than others'. We also have the possibility of identifying the most 'advanced' individuals and thus estimating the limits of cognitive abilities within that species. For instance, in the study by Sappington and Goldman (1994) cited in section 15.3, four horses were presented with a series of problems, from simple to more and more complex, in correspondence to the accepted scale of different levels of learning abilities from simple discrimination to concept formation (Thomas, 1986). All the horses in the study coped easily with the first problem, a simple black versus white discrimination task. Three of the four subjects went on to learn at least one of the pattern discrimination tasks, and only one horse completed both the problems involving the concept of triangularity. As was also mentioned in the same chapter, very similar results were obtained in the honeybee studies of Mazokhin-Porshnyakov (1989). All members of the hive successfully learned simple problems of discrimination. A task that required generalisation on the basis of relative size, namely, to choose the biggest, or in other case the smallest, from a set of figures that varied in shape and colour, was solved by about half the bees in the hive. Finally, the tasks that required concept formations, i.e. concepts of triangulatity, two-ness and symmetry were solved by individual 'gifted' bees only.

When confronting members of closely related species with a similar experimental task, we gain the chance to reveal how the intellectual abilities of animals are tuned to the circumstances in which they live. Of particular interest are situations in which the intellectual potential of a species can be estimated by comparing problem-solving in different specimens.

For example, field studies of ants' ability to navigate 'round mazes' of different levels of complexity have shown a close connection between the percentage of individuals that solve problems and species-specific foraging systems in studied ants (Reznikova, 1975, 1982, 2005). Ants were presented with mazes containing food pellets. The maze of the first level of complexity contained one circle with one entrance. This could hardly be called a 'maze' because all the ants had to do was to enter the round box and get food. The maze of the second level consisted of two concentric circles placed one inside the other in such a way that their entrances were placed on two opposite sides. The most complex maze consisted of four circles (Fig. 20.1). In different series of field experiments, 100 mazes of the same level of complexity were placed in the ants' feeding territory. Two groups of ant species were compared, one group with a sole-foraging system (subgenus *Serviformica*) and other with group-retrieving (subgenus *Formica* s. str.). In sole-foraging species each forager searches for food by itself and thus undertakes long trips and makes its own decisions whether to catch a prey and to take it home or not. In group-retrieving species professional specialisation takes place in which a scout searches for food and a group of foragers share efforts to transport it. The feeding territories of these species are penetrated with foraging routes; systems of information transfer and food transportation allow them to exploit large food sources in optimal ways (Reznikova and Ryabko, 1994; Reznikova and Novgorodova, 1998; Robson and Traniello, 2002). In field experiments with mazes it turned out that in sole-foraging species nearly all active foragers could successfully obtain food pellets from the most complex mazes. In group-foraging species not all of the ants coped with the maze problem. Instead, there were 'top ten' individuals which successfully navigated mazes consisting of two circles, that is, about 10 per cent of the foragers.

Similar results have been obtained by researchers comparing the searching tactics and learning achievements of different rat species in radial mazes (Timberlake and Hoffman, 2002). Norway rats *Rattus norvegicus* and laboratory-reared desert kangaroo rats *Dipodomys deserti* displayed clearly distinguishable tactics of

Fig. 20.1 An ant in a round maze. (Photograph by I. Yakovlev.)

searching for food, closely related with their different foraging patterns in nature. In the wild, desert kangaroo rats use solitary foraging in a patchy environment; these animals do not use odour trails, but the use of landmarks for minimising travel distances is important for their foraging success. In experiments they applied a specific tactic in a maze dividing food sources into 'patches'. Norway rats are highly social animals that spend the greater part of their time above ground using trails established by the colony. They employed a tactic of trail-following or, being deprived of the use of smell, strongly preferred the tactic of 'central-place foraging'.

Besides these two studies on ants and rats, there are many other investigations which have shown that results achieved by members of different species in different laboratory environments depend on differences in their field environments such as distribution of food, sociality of the lifestyle, predator pressure, and many others.

The thorny path of searching for a common metric for measuring intelligence across a broad range of species has attracted comparative psychologists. Researchers have tried to classify the intellectual abilities of different animals and to rank them within a universal intelligence scale. For example, Thomas (1986) compiled data on concurrent discrimination learning for various fishes, reptiles, birds and mammals, including mice, rats, zebras, donkeys, horses and elephants. Of the species tested, only elephants and horses were able to discriminate the correct stimulus in 20 pairs of visual patterns concurrently. These were the only two species that were able to successfully complete so many concurrent discriminations. Krushinsky (1977) compared species on the basis of their performance on detour tasks that required extrapolation of trajectories of moving bowls with food (see Chapter 12 for details). He subdivided species into six groups basing on their ability to cope with these tests. Among them primates, dolphins, brown bears and crows comprised the top group, foxes, wolves, dogs and birds of prey belonged to the second group, whereas the 'lowest' group included voles. Some results matched intuitive concepts about 'animal intelligence' whereas others were difficult to explain. For instance, reptiles displayed outstanding capacities of extrapolation comparable with those of kites and falcons.

While a variety of laboratory tasks have been used, that of learning set formation has been widely explored since Harlow (1949) concluded that the results reflected evolutionary relationships. As was considered in Chapter 17, the learning set formation involves inter-problem improvement in performance seen in subjects which are given a series of discriminations involving different pairs of stimuli. Rumbaugh (1968) showed that in squirrel monkeys individuals differ to a great extent in their ability to pass the test. Some monkeys exceeded apes in their marks whereas others were not able to pass the test at all. Subsequent work showed that closely related species may have widely divergent performances, and that some 'lower' species may equal or exceed 'higher' species. Further, the ordering of species does not agree with that predicted from relative brain size. Macphail (1982) compared data from various sources concerning learning set formation in different species such as langurs, rhesus monkeys, bottlenose dolphins, cats, rats, squirrels, ferrets and minks, and concluded that it is not clear that any of the differences observed in performance in learning set formation (or any of the other types of behavioural studies considered) are due to differences in intellectual capacity. He cites a number of studies which demonstrate, as might be expected, that relative species performance is very dependent on the details of experimental technique.

The growing body of data indicates that cognitive abilities in different species are hardly comparable. Members of species achieve different results according to the context of the experimental situation. For instance, primates master visual discriminations faster when the task is presented in the context of a foraging situation compared to a traditional testing situation such as the Wisconsin General Test Apparatus (WGTA) (Menzel, 1996). The second example concerns dolphins: when they were tested on auditory rather than visual discrimination, their performance improved significantly (Herman, 1986).

Call (2002) starts his comments on the article by Bshary and co-authors (2002) on fish cognition with a note that when he told a colleague that he was reading an article on fish cognition, her first reaction was to laugh. However, research on fish promises to broaden our understanding of animal cognition in various ways: as fish are highly variable in their ecology, they can be used to determine the specific ecological factors that direct the evolution of specific cognitive abilities. Call (2002) concludes that standardisation does not have much room for the special cognitive abilities of some species, precisely because they are special and therefore not suitable for comparison across a number of species. This raises a crucial question about principal possibility of cognitive specialisation that will be considered further in this Part.

20.2 'Misbehaviour of organisms': learned behaviour drifts toward instinctive behaviour

From the great number of commands one can teach a pet dog the easiest is the command 'Give me your paw!' Your dog will readily give you its front paw; in some cases this action even does not demand any reward. However, if you try to teach the dog to give you its hind leg, it takes much time and effort. This is an example of the use of an innate behavioural pattern in training. The fact is that characteristic movements with the front paws are involved in the early behaviour of infant dogs: when sucking their mother puppies perform characteristic massaging movements of the teats with their front paws. Dog breeders call this behaviour by the German word *Milchtritt*. When puppies are growing up, a mother dog often feeds them standing up, posing as the legendary Capitol Mother Wolf does; then the sucking puppies have to crane their necks and massage the teats with one front paw leaning their weight on the other one. This behavioural pattern will manifest itself in adults in different contexts. A dog extends its front paw as a gesture of obedience. If you sit in front of a timid dog and begin to talk to it angrily, it is most likely that the dog will give you its paw as a gesture of apology.

Circus trainers have for ages built on the innate gifts of animals. Cats of different sizes from lions and tigers to pussy cats hitting the mark by exact jumping, sea lions balancing with balls, and many other animals that extend their innate repertoires under their trainers' tutoring and thus amaze the audience are amongst these examples.

As we have seen from data in previous chapters, it is obvious that the abilities of different animals to solve problems are tuned to the circumstances in which they live. This view dramatically contradicts with a fundamental concept of the Skinnerian branch of behaviourism that any organism can be taught to accomplish practically any behaviour if it possesses the relevant organs to perform and a patient trainer to shape its behavioural sequence. It was Skinner's former students, Keller and Marian Breland, who undermined his concept when they published in 1961 their paper 'The misbehavior of organisms' (the title parodies the title of Skinner's (1938) book *The Behavior of Organisms: An Experimental Analysis*). In it, they summarised 14 years' experience in the engineering control of animal behaviour by operant conditioning techniques. By the 1950s, conditioned behaviour had been exhibited in various zoos, fairs, tourist attractions and television shows. However, thoughtful animal trainers had run into a persistent pattern of discomforting failures. In their paper the Brelands gave several concrete examples of how their intentions of being good behaviourists were baffled by the misbehaviour of animals they worked with. Among these examples, the first concerned 'dancing chickens'. Well, they readily 'danced', i.e. scratched vigorously, round and round, over the disc for 15 seconds, at the rate of about two scratches per second until the automatic feeder fired in the retaining compartment. The chicken goes into the compartment to eat, thereby automatically shutting the door. The popular interpretation of this behaviour pattern is that the chicken has turned on the 'juke box' and 'dances'. The development of this behavioural exhibit was wholly unplanned. In an attempt to create quite another type of demonstration which required a chicken simply to stand still on a platform for 12–15 seconds, the authors found that over 50 per cent of the chickens developed a very strong and pronounced scratch pattern, which tended to increase in persistence as the time interval was lengthened (another 25 per cent or so developed other behaviours – pecking at spots, etc.). So it is hardly a practicable problem to teach a chick to stand still.

The last example – after, say, an example with a racoon which spent more and more time rubbing coins together (in a most miserly fashion) instead of dipping them into container, a pattern that was reinforced – was a pig that was conditioned to pick up large wooden coins and deposit them in a large 'piggy bank', a funny show if it were practicable. At first the pig would eagerly pick up one 'dollar', carry it to the bank, run back, get another; carry it rapidly and neatly, and so on, until the ration of food rewards was finished. Thereafter, over a period of weeks the behaviour would become slower and slower. The pig might run over eagerly for each dollar, but on the way back, instead of carrying the dollar and depositing it simply and cleanly, it would repeatedly drop it, root it, drop it again, root it along the way, pick it up, toss it up in the air, drop it, root it some more, and so on. The behaviour persisted and gained in strength in spite of a severely increased drive – it finally went through the rations so slowly that it did not get enough to eat in the course of the day. This problem behaviour developed repeatedly in successive pigs.

In some animals of different species, after its behaviour has been conditioned to a specific learned response the animal gradually drifts into behaviours that were entirely different from those that were conditioned. It can easily be seen that the particular behaviours to which the animals drifted were clear-cut examples of instinctive behaviours having to do with the natural food-getting behaviours of the particular species. 'Washing behaviour' in racoons, rooting or shaking behaviours in pigs, scratching and pecking in chicks are strongly built into the species and are connected with the food-getting repertoire.

The Brelands termed this phenomenon *instinctive drift*: wherever an animal has strong instinctive behaviours in the area of a conditioned response, after continued running the organism will drift towards the instinctive behaviour to the detriment of the conditioned behaviour and even to the delay or preclusion of the reinforcement. In a simplified form, it might be described as 'learned behaviour drifts toward instinctive behaviour.'

It was the Brelands' conclusion that the behaviour of any species cannot be adequately understood, predicted or controlled without knowledge of its instinctive patterns, evolutionary history and ecological niche. The most important behaviouristic assumptions turned out to be wrong: that an animal comes to the laboratory as a virtual *tabula rasa*, that species differences are insignificant, and that all responses are conditionable to all stimuli.

These ideas have helped to clarify the interpretation of many results produced by behaviourists. In fact, the Brelands had been the first to become Skinner's satellites, and had worked with his trained pigeons, and especially those who played bowling. Later a natural explanation of pigeons' skilfulness was suggested: it was likely that that pigeons' trainers built on pigeons' innate foraging behaviour, that is, the manner in which birds throw soil off in order to get seeds. This suggestion was supported by observations of how pigeons pecked a key in Skinner's chamber when they were expected to be rewarded by food grains or by water. Students who were asked to watch filmed trials and to distinguish for which kind of award were the pigeons waiting - grain or water - never made mistakes. Pigeons pecking for food did this with sharp pecks with open beaks, whereas birds pecking for water did so with closed beaks and touched the key for much longer (Moore, 1973). Remember, for instance, how a dog runs into a kitchen licking its lips in anticipation when it hears the call 'Doggy, take this!' though it neither smells nor sees the food yet.

Much evidence has been obtained of the specific relations between animals' actions and the kind of rewards expected. For instance, Shettleworth (1972, 1975) revealed 24 different behaviours in hamsters, some of which excluded other ones. Thus, those behavioural acts that were connected with locomotion increased in expectation of feeding time, whereas comforting and scenting behaviours were depressed. It was easy to teach hamsters to do some actions - such as digging, running and scratching - much more frequently during the period preceding feeding. However, it was practically impossible to teach them to perform other behaviours on the threshold of feeding time, such as washing and scenting territory.

In Chapter 1 I cited the example from the book by Morgan (1900), one of the founders of objective studies of animal intelligence. Morgan described the skilfulness of sheepdogs as an illustration of what he considered animal intelligence. As Morgan had it, an intelligent animal is what it is trained to be, its natural powers are under complete control of his master, with whom the whole plan of action lies. In the light of ideas of modern ethology, we know that the real picture is not so simple. Untrained sheepdogs have an innate tendency to run around a group of sheep and herd them back towards the shepherd. Possibly the dog regards the human as a participant in a hunt for food. A master builds on this natural herding tendency and trains the dog to drive sheep in response to shouts and whistles. Apparently it is difficult to train a sheepdog to drive sheep away from a human, and to leave one group of sheep to go and gather more. This difficulty is caused by a contradiction between the trainer's desire and the animal's innate predisposition for definite behaviours in the context of this situation.

There are many illustrative examples in this field. For instance, in the experiments of Sevenster (1968) male sticklebacks were trained to perform two tasks in order to gain an access to a female ready to court. One thing was easy to train: males quickly learned to swim through a ring in order to reach a female. However, they learned with difficulty to bait a stick in order to get access to the desirable object, a female. Instead of baiting it, they addressed courtship dances to the stick.

By 1970 more disturbing challenges to the behaviouristic view had appeared in many laboratory studies. Pavlov's idea that any perceptible cue could be taught, by classical conditioning, as a conditioned stimulus was dealt a severe blow by Garcia *et al.* (1972). They showed that rats can not associate visual and auditory cues with food that will make them ill, even though they can associate olfactory cues with such food. On the other hand, experimenters found that quails can associate not auditory or olfactory cues but only visual ones - colours - with dangerous foods. Later work by other investigators extended these results, showing, for example, that pigeons readily learn to associate sounds but not colours with danger and colours but not sounds with food. The obvious conclusion was that these animals are predisposed to make certain associations more easily in some situations than in others. The same kind of pattern was discovered in experiments in operant conditioning. Rats readily learn to press a bar for food, but they cannot learn to press a bar in order to avoid an electric shock. Similarly, pigeons easily learn to peck at a spot for a food reward but have great difficulty learning to hop on a treadle for food; they learn to avoid shock by hopping on a treadle but not by pecking. Once again it seems that in certain situations animals are innately prepared to learn some things more readily than others. The associations that are most easily learned have an adaptive logic. In the natural world odour is a more reliable indicator than colour for rats (which are notoriously nocturnal) in trying to identify dangerous food; the colour of a seed is a more useful cue for a pigeon to remember than any sounds the seed makes. Similarly, a pigeon is more likely to learn how to eat novel seeds if it experiments with food with its beak rather than with its feet.

In sum, animals that have innate biases concerning which cues they rely on and which procedures they attempt are more likely to ignore spurious cues, and they will learn faster than animals without inherent biases. The idea that animals are innately programmed to attend to specific cues in specific behavioural contexts suggests a mutually reinforcing relation between learning and instinct (Gould and Marler, 1987). This relation helps to reconcile the approaches of behaviourists and ethologists and to establish a link between ecology and cognitive ethology.

Chapter 21

An outline of instinctive behaviour

21.1 Displays of complex instinctive behaviour

When arguing about drifts towards instinctive behaviour as a part of learning process, we should discuss the question as to what is 'instinct' itself. The concept of instinct is one of the oldest in the behavioural sciences and therefore this concept has fuzziest boundaries. Early authors wrote about the 'imperial instinct' in 'queens' in social insects. As was described in Chapter 1, one of the first articles in this field was Ray's scientific text on the study of 'instinctive behaviour' in birds published in 1676. Darwin helped to prepare the ground for the study of the interplay of instinct and early learning, soon to be taken up by Spalding (1873), and later by the founders of ethology, Lorenz and Tinbergen.

Although the term 'instinct', through the years, has had many meanings, students of behaviour agree with a general concept that those behavioural patterns are considered to be 'instinctive' that are relatively constant throughout a species and are not acquired through previous experience or learning. Probably this idea is too simplistic, and the real picture of the intimate relationship between innate and learned behaviours is much more complex. For our consideration it is important to note that sophisticated behaviour in many species can be based mainly on inherited patterns.

There are many examples of species-specific behaviours in nature. If we recall just one kind of activity, such as 'dances', we then can list courtship dances in fishes (such as the zigzag dance in sticklebacks), birds and even insects (such as mating 'dances with balloons' in several fly species). Information-bearing 'dance language' in honeybees is much more flexible than courtship dances but the basic figures of the bees' dances are inherited. Recently the use of 'tactile dances' for ritualised interspecies communication has been described by Grutter (2004): in cleaning interactions between fish species, cleaners have evolved a specific behavioural strategy to avoid conflict or being eaten. Cleaners tactically stimulate clients while swimming in an oscillating dancing manner. This tactile dancing enables cleaning fishes to avoid conflicts with potentially dangerous clients.

Let us consider several examples of highly specific behaviour in animals based on innate stereotypes.

The first example concerns bowerbirds, a family of birds in which males attract mates by collecting brightly coloured objects which they display to females in 'avenues' or on 'maypoles'. From Darwin's time onwards, bowerbirds have been considered an example of a bird with a human-like aesthetic sense. What may be more important, these birds provide an outstanding model system for testing hypotheses about the evolution and functional significance of complex behavioural display. These hypotheses have been tested in long-term experimental works by Borgia and colleagues with all 19 species of bowerbirds living in Australia and New Guinea (Borgia and Presgraves, 1998; Patricelli *et al.*, 2002). An interesting relationship has been revealed between the colourfulness of the males' plumage, the visual properties of the objects they collect, and the complexity of the structures they build to house these objects. Some species of male bowerbird are relatively drab, but they make up for this by building elaborate structures decorated with brightly coloured objects. In other words, they fill a gap in their attractiveness with sophisticated behaviour. The latter aspect is of most interest here. Polygynous species of bowerbirds build stick bowers on the ground associated with a decorated display; females provide parental care at nests built in trees. Multifaceted male displays involve the decorated display court and the bower that act as a stage for an energetic vocal and dancing male display which is observed by the female from inside the bower. Courtship and mating occur in the bower. Bower quality, numbers of preferred decorations, vocal and dancing elements all contribute to male mating success. Male reproduction is skewed; one male may mate with 25 different females at his bower in one season. Males of different species choose colours of decorations in response to different light regimes. In *Amblyornis* bowerbirds, males in species using foggy ridge tops use predominantly black decorations while closely related species displaying on more sunlit slopes use a variety of decorations with a wide array of bright colours.

Cuvier (1825) considered that animals act as 'somnambulists', being moved by the power of their instincts. So the second Australian 'somnambular' bird that is probably fainting under the weight of its behavioural stereotypes is the so-called moundbird. The matter concerns various species of large-footed, ground-dwelling birds of the family Megapodiidae, found in Australia and the South Pacific islands. Their astonishing nesting habits were described in the early literature (Wallace, 1860; Banfield, 1908; Fleay, 1937). The three Australian megapode species – Australian brush turkey, orange-footed megapode and malleefowl – build mounds of leaf litter and other organic materials in which microbial decomposition produces an incubation temperature of around 34 °C. Direct solar radiation may also contribute significantly. These birds have to overcome hardship *ab ovo*. Megapode hatchlings are the most independent of all avian species. The large egg enables the hatching of a very advanced chick. One report of a chick unearthed from a mound states that it immediately flew about 10 metres. But what is even more amazing is that the chicks hatch deep in the soil, digging their way up to the surface, all by themselves, and then live without any parental care. A very complex behavioural scenario unwinds independently in each species. An especially hard lot falls upon males. First, a male builds a mound. A typical mound is 2–3 metres tall, up to 15 metres in diameter and can weigh over 50 tonnes, maintained by a pair throughout the year (Figs. 21.1 and 21.2).

Before a female lays an egg, the male digs a tunnel into the mound leading to an egg chamber. It takes him not less than an hour and after that the female walks out of the bushes and inspects the temperature in the chamber. If the female is not satisfied, she returns to the bush and the male has to dig another tunnel. He makes up to four before the female lays an egg. She lays several eggs, one every 3–5 days because each egg weighs about 10 per cent of her body weight. The female covers each egg with soil and walks away, and now the male takes up his duty. He defends the mound vigorously from snakes and other predators, and, what is the hardest duty, he attends to the mound by adding and removing litter to regulate the internal heat while the eggs hatch (Frith, 1956a). Early naturalists noted the melancholic and solemn style of movements of megapode males which lack the nice vivacity which is so typical of many birds. Even in midday sun when other creatures conceal themselves being languid with the heat, megapode males continue to dig in the soil, gloomy and lonely.

Megapodes can serve as a good example of intricate combination of morphological adaptations (such as notably large eggs, very advanced and well-equipped chicks, specific palatine thermo-sensors that allows birds to control the temperature), strong behavioural stereotypes and essential elements of flexible behaviour. It is a question of great importance as to how many degrees of freedom a brush turkey has.

Figs. 21.1 An adult brush turkey male walking down the slope of his incubation mound, which measures 3 metres in height. (Photograph by G. Ross; courtesy of A. Göth.)

Fig. 21.2 Two white brush turkey eggs in an incubation mound, opened for egg collection. Chicks usually hatch in the soil and then have to dig themselves to the surface. (Photograph by V. Chambers; courtesy of A. Göth.)

Let us compare the brush turkey's reaction to attempts to disturb its work with the corresponding behaviour of another fanatic digger, the digger wasp *Ammophyla*, whom we have already met in Chapters 12 and 17. Let the digger wasp dig her hole, bring several caterpillars as the food for future larvae and lay her eggs. Now she is covering up the hole with sand. If we disturb her and take all the contents out of the hole in view of its owner, the wasp will quickly return and close the hole up as if nothing was wrong.

The brush turkey male responds quite differently. When researchers took his nest material away, or packed it in different ways, or added some, he responded immediately by determining the change in temperature with his thermosensor and working nearly to the point of exhaustion in order to keep the temperature inside the mound at about 34 °C. In a special series of experiments, investigators built electrical heaters into the mound and frequently switched them on and off thus keeping the male alarmed. The bird responded to these distractions in a very effective way and soon celebrated victory (Frith, 1956b). So this bird appears not to be a whole 'somnambulist'.

Does this mean that the brush turkey behaves in a flexible and creative manner whereas the behaviour of a 'low' insect is governed by 'blind instinct'? Again, the picture is not so simple. As we have seen in Chapter 12, the digger wasp displays wonders of intelligence when she remembers where each of her several holes is situated and how many caterpillars should be killed and placed into each of them. Being disturbed at any other stage of her behavioural cycle than closing holes, she responds adequately and effectively. When researchers took from one to three prey items out of her holes, the wasp refilled each hole properly (Tinbergen, 1951).

It seems that species have different spaces for flexibility at different stages of their general and particular behavioural cycles. It is an essential part of the complexity of life that in some species nearly all behaviours include deeply intertwined innate and flexible components whereas in others these components can be distinguished with relative ease. In order to make this picture clearer, some specific features of instinctive behaviour will be analysed later in this chapter.

Returning to the special case of megapodes, one comparison can be of help. One can say that the flexibility of their constructive behaviour is at about the same (relatively high) level as that of beavers building their dams and lodges. Both species erect their imposing constructions under the guidance of inherited predisposition. Cuvier (1825) first raised two young beavers in isolation from adults and thus revealed that their engineering talents are highly predisposed. At the same time, both species flexibly adjust their efforts to the requirements of their environment. There is much evidence concerning flexibility of dam constructing in beavers (Wright *et al.*, 2002). As far as megapodes are concerned, one early observer of scrub fowls' behaviour described mounds that were constructed by birds living in a particularly rugged part of Dunk Island. They were made from boulders, and their inclination towards each other provided safe protection from rain. The author afforded this as illustration of 'purposive conscious action' (Banfield, 1908).

One can say that both species demonstrate a sophisticated combination of behaviours; some of them are based on innate stereotypes whereas others amaze observers with their flexibility and rationality. We will return to some, perhaps unexpected, aspects of behaviour of the same organisms further in this Part.

21.2 Room for intelligence in the context of selective perception and specific responses

It is a natural idea to study learning and intelligence in animals in the context of their natural environment and to investigate how species perceives stimuli before studying how they solve problems.

Selective perception

There are many interesting reviews of selective perception in animals (for example, Hinde, 1970; Kaufman, 1974; Freeman, 1991, 1995; Zentall and Riley, 2000; Dukas, 2004). It is known that many animals live in a 'smelly' world, and others live in a world penetrated with ultrasounds. Probably, rats live in a world dominated by sensations transmitted to their brains through their whiskers.

A huge body of recent investigations have derived from Tinbergen's (1951) ideas about the basic process of distinguishing between perceived and effective impulses in animals; this process is called *stimulus filtering*. Members of different species selectively perceive sounds, smells, as well as tactile and visual objects because they somehow filter incoming stimuli. A frog sitting in a marsh full of the sounds of the evening chorus of all sorts of frog species hears only the call of its own species. Besides, it is well known that some species can catch signals beyond the reach of other species. Bats, moths and dolphins use sonar, and this is only one small group of examples from a great array.

A pet subject for illustrating stimulus filtering in many books and reviews is a description of how a toad perceives prey units. Toads belong to the so-called 'sit-and-wait' predators and they remain motionless for long periods of time. When a prey item moves in their receptive field, the predator lunges with great speed (relative to the prey) and snaps it up. For example, when a toad spies a fly walking along a wall, it begins to orient itself in preparation for the tongue flick that will capture the insect; but if the insect stops, the toad shows no further reaction. The fly is still clearly visible, but now it elicits nothing in the toad's central nervous system that would compel the toad to continue its orientation. Somewhere between the eye of the toad and its muscles, the sight of the immobile fly is filtered out (Wallace, 1979). The toad sees the fly, certainly, but not as an orienting stimulus. The key to the toad's motion-based prey detector is the *receptive field*, the fundamental unit of its perception machinery. Each of the thousands of receptive fields in the toad eye consists of the following components: (1) a single ganglion cell that integrates information from the receptive field and relays a response back through the optic nerve, (2) bipolar cells that are all connected to the single ganglion cell on one synapse and connected on the other side to one or more receptor cells, (3) a circular cluster of receptor cells, the receptive field, that consist of (4) central excitatory photoreceptors that are loosely tethered together through bipolar cells, and (5) peripheral inhibitory photoreceptors that are connected to a single bipolar cell. This is the smallest *neural unit* of stimulus filtering found in the visual system.

Different stimulus filtering is found at the level of specific photoreceptors in that individual neurons pay attention to certain signals (e.g. rods versus cones, and if cones, then the kind of photopigment found in the cone). The stimulus filtering found in receptor sensitivity is hard-wired by evolution. However, the receptive field has special cellular interactions built into it that result in certain information being ignored and other information being acted upon. The receptive field is also the smallest unit of the toad's motion detector. The excitatory and inhibitory cells act in unison to either filter or detect objects from higher centres such as the optic tectum. Stimulus filtering in the receptive field is also capable of being modified by the animal's internal state – food-satiated or hungry. If a large object casts an image over the visual field, the light intensity changes on the photoreceptors. Both excitatory and inhibitory cells from many receptive fields are triggered. Because the ganglion receives impulses from both the excitatory and inhibitory cells (through their respective bipolar cells) the effect of the inhibitory cells cancels out the effect of excitatory cells. No impulse is sent from the ganglion cell to the optic tectum. However, if a small image passes over the visual field, the small image tends to trigger fewer receptive fields. The small image will also tend to excite some of these fields because the image hits many of the central excitatory cells, but only a few peripheral inhibitory cells. The ganglion cell receives impulses from the expiatory cells (through their bipolar cells), but with little inhibitory feedback, the action potential is relayed on to the optic tectum for further integration. The optic tectum receives inputs from all ganglion cells. Several clusters of ganglion cells form a higher-order receptive field at the level of the optic tectum that integrates information form the clusters of receptive fields. Consider objects of different shapes that might strike receptive fields. Receptive fields come in a variety of 'flavours'. Some are used for detecting long thin objects, others large objects, etc. One of the toad's favourite foods consists of worms – long thin objects. There are receptive fields that are tuned to fire when long thin objects pass across them. When several 'long-thin' receptive field detectors have the image of a bar pass over the receptors, their ganglion cells will relay the information to the optic tectum. The grasp reflex then takes hold, and the toad orients with both eyes. Once both eyes are locked on, other motor neurons cause the toad to lean forward, open its mouth and eat the worm.

It is important to note that if the coding (filtering) is done at the level of the receptor, it is called peripheral coding; if it takes place in the central nervous system, it is called central coding. An advantage of peripheral coding is that useless or irrelevant information is discarded early so that no neural activity is wasted on its transmission. On the other hand, peripheral coding is disadvantageous in that there are fewer neurons operating at the peripheral level than in the central nervous system. Therefore, any peripheral coding, or filtering, is likely to be less specific than central coding. The complexity and coordinating of any resulting behaviour will therefore be reduced. Central coding might be expected in animals living in rich or varied environments with high levels of potential stimuli that could be sorted and generalised in the brain. Animals, such as the frog, that rely more on genetically based and rather inflexible patterns tend toward dependence on peripheral coding. Mammals such as the cat and the monkey, on the other hand, show comparatively little peripheral coding. There are however notable exceptions, and the real picture is more various and complex (Munz, 1971).

It is worth noting that, in studying stimulus filtering in animals, researchers are not always able to remove their professional filters which prevent them from understanding how an 'outsider' species reacts to its environment. An amusing example comes from Vallortigara's (2000) paper about animals' 'left and right perceptual worlds'. The matter concerns visual lateralisation in many species. The first evidence of visual lateralisation in the intact animal was obtained in the domestic chick by temporary occlusion of either the left or right eye (Rogers and Anson, 1979; Rogers, 2002). This procedure revealed that the right eye is better at discriminating visual stimuli, such as grains from pebbles, and that the left is more reactive to emotionally charged stimuli. These results met with initial incredulity; a comment by an eminent animal lateralisation researcher, involved in studies of split-brain monkeys, clearly expressed this concern: 'These results lead to the plausible but revolutionary inference that a bird more effectively searches for food with its right eye while it watches for danger with its left!' (Hamilton, 1988). However, what at first seemed unbelievable turned out to be absolutely correct and was confirmed in a variety of other species of birds (Vallortigara, 2000).

We know now from a huge body of data that organisms perceive and react selectively in their species-specific ways. Studies of this kind pave the way for establishing a link between learning abilities and ecological traits of species.

Fixed action patterns: small room for learning

Members of many species display similar sequences of similar behaviours over and over again (Sisyphus is not a bad example here). As was noted in Part I, Lorenz (1937, 1950) and Tinbergen (1942, 1951) elaborated the paradigm of classical ethology based on some important ideas and definitions. One of the main ideas in their theory is that every species has a repertoire of stereotyped behaviours called *fixed action patterns* (FAP). Lorenz suggested that the fixed action patterns are (1) innate (and used the German word *Erbkoordination* which translates literally as 'inherited co-ordination' to describe them); (2) common to all members of the species (species-typical), and therefore they are as characteristic of the species as shared structural features; (3) once triggered by *sign stimuli* (see below), fixed action patterns proceed in the absence of the triggering stimulus. Fixed action patterns represent a motor program provided by a specific neural instruction, a component of a subject's nervous system, called the *innate releasing mechanism* (IRM). The innate releasing mechanism is the sensory mechanism that detects the signal, and the fixed action pattern is a fixed sequence of stereotyped acts.

The natural world is full of displays of fixed action patterns, and among them courtship is the most specialised. Characteristic sequences of behaviours in courting ceremonies are highly stereotypic in many species. Long before the concept of fixed action patterns had been accepted, Julian Huxley, Lorenz's teacher, in his study of the behaviour of great crested grebes, found that, in course of evolution, certain behavioural patterns lose their original functions (such as specific motions that are used for feeding or fighting) and become purely symbolic, useful only in communication. He called this process *ritualisation* (Huxley, 1942). In great crested grebes, for instance, a couple, facing each other, adopt an upright posture and shake their heads from side to side. Fixed action patterns in general, and particularly those appearing in courtship, can be used to follow the evolution of behaviour because they vary between related species of animals. Such analysis was one of the main interests of early ethologists. A stereotyped pattern of courtship behaviour in male ducks is one of the classic examples dating from Heinroth's (1911) observations.

In modern studies of evolutionary ethology courting rituals are now known to be one of the major causes of speciation in animals. There is an interesting example of ritualisation among the insects. Several thousand species of empid flies are known as 'balloon makers'. That is, the male flies capture an insect and enclose it in a frothy bag. The male carries this package around with him as a lure to entice a female to mate. Some species dispense with the froth. *Empis barbatoides* males capture little insects, usually weaker flies like bibionids (March flies), and dangle them in front of females as a preface to courtship. Members of other species present 'symbolic' gifts such as silken balloons (sometimes containing prey and sometimes empty). Other species show signs of sex reversal where females compete for males, and females have secondary sexual traits such as swollen abdomens aimed at fooling males into thinking they are full of ripe eggs. In other species, males attract females by displaying fake dead prey which is, in reality, merely a growth on the male's legs. The evolution of nuptial gifts is thought to be a classic case of ritualisation in courtship behaviour (Lorenz, 1965). Recently LeBas *et al.* (2004) have examined the DNA of several species, recording their behavioural data and investigating which factors different mating techniques are associated with. They suggest that the central problem of evolution of nuptial gifts in flies is connected with sexual conflict between males and females and variations in the effort that males and females put into courtship behaviour. This, in turn, is closely connected with cheating and counter-strategies in courtship.

As has already been mentioned, fixed behavioural patterns can be mixed with flexible behaviours. For instance, hunting in some birds involves instinctive behaviour. As a hunting falcon begins searching for prey, its behaviour is highly variable. At such times, the bird is hungry and might be equally pleased at the sight of a flying bird or a scampering mouse, so the falcon can catch its victim in the air or suddenly drop towards the ground. So the first stage of the falcon's hunt may be flexible, becoming focused only later when prey is spotted. When the falcon spots its prey, there is a point at which its behaviour can no longer be altered. It is committed to a specific pattern. Its actions now are performed in very much the same way as they have been at the same stages in previous hunts. This is the brief moment in which the falcon's feet are tightly clenched into a fist as it swoops at very high speed. A victim will be knocked from the sky and the falcon's hunt will be successful. So at this stage a falcon shows increasingly stereotyped behaviour as it ends each instinctive sequence with a fixed action pattern. A very similar pattern is characteristic for big dragonflies of the genus *Aechna*. When it sees a victim, a dragonfly clenches its six legs into a 'basket' and grasps a flying insect very effectively.

In general, the final strike of any specialised hunter, such as a bird of prey, a cat or a dragonfly, being a fixed action pattern, calls a series of muscles into play that contract for the same duration and in the same sequences under all conditions. Lorenz (1952) vividly describes a specific hunting blow in a pussy cat that is very much like that in a lion that hunts a zebra but now is aimed at the owner's foot.

Humans also exhibit fixed action patterns. Eibl-Eibesfeldt and Hass created a Film Archive of Human Ethology (see Eibl-Eibesfeldt, 1989) of un-staged and minimally disturbed social behaviour. They filmed people across a wide range of cultures with a right-angle reflex lens camera, i.e. the subjects did not realise that they were being filmed because the camera lens did not appear to be pointing at them. Eibl-Eibesfeldt had identified and recorded on a film several human fixed action patterns or human 'universals', e.g. smiling and the 'eyebrow flash'. Humans show a rapid brow-rising which coincides with raising the eyelids. Because all the cultures he examined showed this behaviour, Eibl-Eibesfeldt concluded that it was a human universal or fixed action pattern.

Nowadays the term fixed action pattern has been dropped from ethology and substituted by the phrases *behavioural patterns* or *behavioural acts* because behaviour is not as fixed as implied by the term fixed action pattern. There are subtle variations between and within animals in, for example, the duration of individual components. And, what is more important, fixed action patterns are not simply innate; they can be subtly modified by experience.

For example, long-term monitoring of satin bowerbird bowers shows that their highly specialised courtship behaviour includes elements of learning. Recent studies suggest that the attractiveness of a male display comes from the intensity of display and from the male's ability to modulate his display intensity in relation to female signals of comfort. Thus, in terms of sexual selection, in bowerbirds, experience and learning ability may ensure the quality of the male's signals (Borgia, 1995; Borgia and Coleman, 2000).

Although some fragments of behavioural patterns can be 'polished' by individual experience, it is very important to note that when animals perform a complex of behavioural patterns they lack the ability to recombine various segments of behaviours in their repertoire in new ways so as to achieve new goals.

Selecting parts of the environment: sign stimuli and releasers

As a rule, fixed action patterns in animals are triggered by specific *sign stimuli* (also called *key stimuli*) in the environment. After the stimulation, there is no need for any more stimuli for the continuation of the event. This was demonstrated by Lorenz's work with the greylag goose. When a greylag's egg is removed from its nest and placed in sight of the goose, the goose will extend its neck and roll the egg back in with its beak. If then the egg is removed out of sight when the neck extension and rolling movement has already started, the goose will continue as if the egg were still there.

Tinbergen (1942, 1951) described a variety of models that would release displays of behavioural patterns in animals. A classic and widely cited example concerns the reactions of sticklebacks to various models imitating specific features of males and females. In spring male sticklebacks change colour, establish a territory and build a nest. They attack male sticklebacks that enter their territory, but court females and entice them to enter the nest to lay eggs. Tinbergen used several models to investigate which features of male and female sticklebacks elicit the attack and courtship behaviour from male sticklebacks. A realistic model of a male stickleback but with a white (instead of the natural red) belly elicits little response from males in reproductive condition. Crude models having the red bellies were much more effective. Tinbergen's findings demonstrate that a model with a red belly is attacked, while a model with a swollen belly is courted by male sticklebacks. These two simple features: the red belly and the swollen belly serve as releasers that trigger two different behaviour patterns.

The terms *sign stimulus* and *releaser* are sometimes used interchangeably. However the term releaser stands for stimuli that have evolved to facilitate communication between animals of the same species. Sign stimuli are features of an animal's environment to which it reacts in a particular way. For example, a fly orchid looks like an insect and thus is a releaser which attracts pollinators.

Tinbergen (1951, 1976) conducted many experiments studying how sign stimuli and releasers work. In one of them, he investigated the stimulus responsible for releasing the gaping response in young thrushes. He discovered that the birds would gape at a protuberance (head) on the side of a body model. One might think that the birds would gape at the bigger head, but in fact Tinbergen found that the relative size of the head was more important than its absolute size. Presumably the birds preferred the 'head' that was the proper size for the model's 'body'.

It is important to understand that each instinctive act is not necessarily summoned by a single releaser. Instead, a particular response may be elicited by any one out of a number of releasers. For example, it was found that fighting behaviour in the male cichlid fish is elicited by five stimuli: (1) silvery blueness, (2) dark margin, (3) highness and broadness, (4) parallel orientation to opponent, and (5) tail beating (Seitz, 1940). It was found that any one of these stimuli would elicit hostile behaviour, and that any two would elicit about twice the reaction as one (although the relative strengths of reactions is very hard to measure precisely). Therefore, sometimes responses to releasers appear to be additive, so that the whole is equal to the sum of its parts. This additive effect is called the *law of heterogeneous summation* (Tinbergen, 1951).

Many examples illustrate how this rule works in different species. For example, in our experiments we applied Tinbergen's methods of presenting multiple models for animals in order to investigate whether red wood ants recognise images of their competitors and potential prey and, if so, what features are key for them (Reznikova and Dorosheva, 2004). Firstly, ants appear to react to models of enemies and potential prey on the basis of key signs which include darkness, bilateral symmetry and the presence of protuberances (legs and antennae). Additional features valuable for the ants were the size of a model and the speed of its movement. Moreover, the ants' reactions to models of carabid beetles satisfied the Tinbergen's principle of heterogeneous summation.

The whole picture of interaction between a subject and stimuli in its natural environment is clarified when we consider a phenomenon of *supernormal stimuli*. This is the phrase used to describe hyperbolic (usually artificial) stimuli that are more effective than the real thing in eliciting a behavioural response. It seems that the releasing value of any sign stimulus is not fixed. If blue is good, bluer is better (Wallace, 1979). Since the releasing value of each sign stimulus is modifiable, then a question arises regarding the extent to which a releaser may increase its effect.

In experiments by Tinbergen and Perdeck (1950) herring gull chicks pecked at a red spot on their parent's bill to induce their parents to regurgitate food. Chicks will also peck at a model consisting of a red spot against a yellow background. However it is possible to construct a model that is even more effective than a real head by using a red pencil with three white bars at the end. This is an example of a supernormal stimulus. In experimental situations animals look insane when they aim their parental care to supernormal stimuli and ignore their own offspring. For instance, birds such as oystercatchers, herring gulls and geese showed a preference for giant eggs that were products of experimenters' imitation rather than normal eggs that were bigger than their own eggs.

However, hyperbolic (and thus supernormal) stimuli may work in the natural world, not only in experiments. Certain parasitic birds, such as some cuckoos and cowbirds, have young that are larger and more 'babyish' than the host's. This may be the reason why the host attends to the young parasite in preference to its own young. Small passerine birds respond fanatically to the giant red gape of a cuckoo's chick. Undoubtedly this is a supernormal stimulus for parents. A possible evolutionary mechanism of some of such reactions has become more clear recently basing on investigations of relations between parents and chicks in barn swallows (Sacchi *et al.*, 2002; Saino *et al.*, 2003). Barn swallow nestlings beg vigorously for food from their parents by producing loud calls and displaying their bright orange gapes. The gape display is a reliable signal of health status since a challenge of the immune system with sheep red blood cells reduces the level of coloration, while artificial provisioning with lutein (the carotenoid causing the coloration) increases the level of coloration. Parent barn swallows respond to experimental manipulation of gape colour by changing their allocation of food. Gape coloration reflects the viability of offspring. Nestling begging calls also reliably reflect health status. Both current hunger status and long-term condition of nestlings affect their begging rate and the response of parents to the displays. This is a case when 'If orange is good, the most orange is the best', or, as the authors put it, 'Better red than dead'. A new series of questions is raised about a race between more and more 'supernormality' of stimuli and balanced interrelations between parents and offspring in some situations, and hosts and parasites in other situations.

We can see, then, that a knowledge of the perceptive capacities of an animal will not tell us what part of its environment is actually triggering a behaviour, since, as we know now, a certain behaviour may be stimulated by cues from only a small part of the environment (Wallace, 1979). In our effort to 'prepare' behaviour with the help of knowledge about key stimuli, we should note that some stimuli may work differently depending on the behavioural and environmental contexts.

In some cases, a set of releasers is necessary to elicit an entire adaptive response in animals. Honeybees may initially be attracted to a coloured paper flower, but they will not land unless the flower scent is also there, so in this case, the visual releaser alone will not initiate the entire behavioural pattern. This example also raises the question of what senses are brought into play in specific behavioural contexts.

We should also take into account that whereas some types of patterns such as capturing behaviour can be elicited at any time, other types of behaviour can only be released at special time. In other words, animals respond to certain environmental cues depending on of their physiological status at this moment. For example, in many species, none of the courtship signals has any effect outside the breeding season. Should a misguided male perform his mating patterns (such as the zigzag dance in sticklebacks) at the wrong time, a female may see him, but since the seasonal daylength has not caused her to produce reproductive hormones, she will remain totally unimpressed by his efforts. In many animals, in particular, in some species of fishes, lizards, social insects, birds, rodents, ungulates and monkeys, increasing levels of density of populations may switch patterns of aggressive territorial behaviour that are otherwise kept in 'doze mode'. In these cases concentration of smell and (or) number of encounters with conspecifics serve as sign stimuli.

It is important to note that a simple key stimulus may in some cases trigger an amusingly complex behavioural response. For example, it is likely that beavers react with dam-building to so simple an acoustic stimulus as the sound of flowing water. Building sequences seem to commence even when the animal is for away from the dam. It is possible that the sound of water running through the dam or over obstacles stimulates beavers to build even at a distance far from the dam (Wilsson, 1971; Richard, 1983; Zurowski, 1992). In hand-reared beavers building behaviour commences in the fourth month of life, so this behaviour is highly innate (Lancia and Hodgdon, 1983).

It would appear, then, that the beavers are governed by very simple rules in their business. But they are not. Beaver engineering also includes two other activities: lodge making and canal digging. Canals can extend hundreds of metres into the forest, and beavers can float branches from trees they have cut and thus move branches to safer feeding locations. The so-called beaver's lodge is a unique structure, that is, an oven-shaped house of plant material, woven together and plastered with mud, increasing gradually in size with year upon year of repair and elaboration (France, 1997; Collen and Gibson, 2001). To build and maintain these constructions, beavers behave not only on the basis of their innate patterns but also include a great deal of individual experience and social interactions.

In sum, however strongly may animals be equipped with innate behavioural patterns triggered by key stimuli in their environment, we cannot conclude that most species are equipped by behavioural formulae in all living instances. Neither can we believe that animals have time and abilities enough to learn what decisions are most effective in many vital situations. Instead, many species may be equipped with innate predisposition for conditioning specific behaviours in their lives. Let us explore this idea further.

Chapter 22

Guided learning and cognitive specialisation

There is now increasing evidence that animals may learn selectively only about a subset of the events they face. Studies of learning in ecologically relevant contexts have shown that animals more readily learn about complex functionally critical stimuli such as the appearance of a predator, features of their parents, colouration of venomous insects, or characteristics of their natal songs. The term 'guided learning' introduced by Gould and Marler (1987) acknowledges the adaptive nature of such phenomena. They argue that work done in the past few decades has shown that there is no sharp distinction between instinct and learning. The process of learning is often innately guided, that is, guided by information inherent in the genetic make-up of the animal. In other words, the process of learning itself is often controlled by instinct. It now seems that many, if not most, animals are 'pre-programmed' to learn particular things in particular ways. In evolutionary terms innately guided learning makes sense: very often it is easy to specify in advance the general characteristics of the things an animal should be able to learn, even when the details cannot be specified. For example, bees should be inherently suited to learning the shapes of various flowers, but it would be impossible to equip each bee at birth with a field guide to all the flowers it might visit.

It is important to note that predisposition of species to build up one set of associations more readily than another may create both a channel for very limited learning and prerequisites for outstanding cognitive development within a specific domain. In this chapter we will analyse different aspects of this phenomenon.

22.1 Learning preparedness: some associations can be built more readily than others

There are many examples in classic and modern ethological literature that some species demonstrate the so-called innate recognition of certain stimuli whereas others do not. In human studies a concept of *preparedness* was introduced by Seligman (1970, 1971) to explain why fears and phobias are so much more likely with certain stimuli (e.g. snakes, spiders) than with others (see Davey (1995) for a review). Seligman (1970) summarised evidence from animal instrumental learning paradigms to suggest that for a particular organism, certain behaviours differ in their potential to be successfully conditioned. He also analysed in detail the protocols of the 'Little Albert' study by Watson and Rayner (1920) and concentrated on the ease with which Albert was conditioned, the durability of his reactions, and on several other details of the experiment (see the description of the experiment in Chapter 2). In particular, Seligman (1971) noted that the experimenter did not get fear conditioning to a wooden duck, even after many pairings with a startling noise. Although Seligman seemed to misinterpret some details of Watson and Rayner's experiment (see Harris, 1979), he was, however, able to create a viable concept. In general, the concept of preparedness includes many aspects of readiness of organisms to form certain associations more easily than others.

There are many evidences of the fact that what animals can learn is often biologically influenced, that is, in many (maybe in most) cases learning is possible within the boundaries set by instinct. Several examples have been described in previous chapters. It is now clear that some animals have an innate predisposition towards forming certain associations. For example, if a rat is offered a food pellet and at the same time is exposed to X-rays (which later produce nausea), the rat will remember the taste of the food pellet but not its size. Conversely, if a rat is given a food pellet at the same time as an electric shock is delivered (which immediately causes pain), the rat will remember the size of this pellet but not its taste. Similarly, pigeons can learn to associate food with colours but not with sounds; on the other hand they can associate danger with sounds but not with colours. These examples of learning preparedness demonstrate that what an animal can learn has adaptive significance. The seed a pigeon eats may have a distinctive colour that the pigeon can see, but it makes no sound the pigeon can hear.

Let us consider in detail one of the most significant and representative sets of examples concerning predator recognition in animals. The idea that some birds possess an innate tendency to escape when they perceive a configuration of a bird of prey goes back to Spalding (1873). Up to now, some studies have suggested that predator recognition is innate in animals whereas others have stressed the importance of learning. Several different non-exclusive possibilities of how animals acquire predator recognition have been suggested: animals may have a genetically programmed ability to recognise predators without earlier experience with that very predator, or they may be sensitised to learn about particular stimuli very quickly. There may also be age-related maturation of the ability to express a recognition response. Furthermore, learning from conspecifics through social learning has been shown to exist in many animals (for review see Kullberg and Lind, 2002). Here we are interested in the first two sources of animal knowledge about possible danger, namely, innate predator recognition and predisposition for quick learning.

Tinbergen (1951) conducted an elegant experiment examining the reaction of naive chicks to a mounted silhouette model of a flying bird. The wings were symmetrical; at one end there was a short protuberance, at the other end a longer one. When a model was pulled along a guide-wire in one direction, it appeared to have a long neck and a short tail, like a goose. Flown the other way, it resembled a short-necked bird of prey. In the first case, it was ignored by young chickens, turkeys, ducks and geese on the ground. When they viewed it being pulled along the other way, however, they tried to escape. The results of this experiment are still discussed; moreover, the matter concerns such intricate problems as the 'innateness' of complex releasers. The interpretation of Tinbergen's results had been questioned since the birds under study were reared in a natural environment in which they had been more habituated to geese than to hawks.

Subsequent studies on several species of fish, birds, mammals and spiders have shown clear evidence for an innate component of predator recognition (for reviews see: Kullberg and Lind, 2002; Caro, 2005).

For example, Veen *et al.* (2000) compared predator recognition responses in two isolated but genetically similar Seychelles warbler (*Acrocephalus sechellensis*) populations, only one of which had experience with the egg-predating Seychelles fody (*Foudia sechellarum*). Individuals in the predator-free population significantly reduced nest-guarding compared to individuals in the population with the predator, which indicates that this behaviour was adjusted to the presence of nest predators. However, recognition responses (measured as both alarm calls and attack rates) towards a mounted model of the fody were equally strong in both populations and significantly higher than the responses towards either a familiar mounted non-predator or a novel mounted non-predator bird species. Responses did not differ with a warbler's age and experience with the egg predator, indicating that predator recognition is innate in this species.

Another type of reaction of naive birds to predators has been revealed in great tits by Kullberg and Lind (2002). Experiments in an aviary showed that 30-day-old naive great tit fledglings (*Parus major*) do not respond differently to a model of a perched predator than to a similarly sized model of a non-predator. Although chicks showed distress responses such as warning calls and freezing behaviour, they did not differentiate between the stimuli. In contrast, wild-caught first-year birds and adults responded differently to the two stimuli. Lack of recognition of a perched predator might be one explanation for the high mortality rate found in newly fledged great tits. Researchers suggest that the presence of parents emitting alarm calls is of importance to reduce predation risk during the time when fledglings are most vulnerable.

In principle, there are different scenarios for acquiring predator recognition, from cultural transmission in one species to innate fear based on characteristic features of the predator's image in others. In many species juveniles come into the world for the second time after their birth when they first emerge from a quiet, dark, natal burrow, or a hollow, a lodge, or a nest. A trade-off must be reached by juveniles between exploring and vigilance behaviours. Innate predisposition to distinguish between really dangerous and neutral objects can be adaptive.

In order to illustrate extreme 'innateness', let us recall brush turkey chicks (the megapodes) and consider their behavioural adaptations to the full measure of their suffering during childhood. As was described in Chapter 21, after a long incubation in a mound of leaf litter, the hatchlings dig their way out of the mound and strive to survive without any parental care (Figs. 22.1 and 22.2). Göth (2001) tested the response of 2-day-old chicks to predators in the semi-natural setting of a large aviary. She exposed chicks to a range of potentially scary predators or stimuli: a real cat, a dog, a rubber snake and a mounted silhouette model of a hawk. Naive chicks responded by crouching and freezing to flying predators. Any aerial stimulus, within a certain size and speed, probably causes chicks to crouch. It has also turned out that alarm calls of other rainforest birds may replace those of absent parents. Alarm calls of conspecifics elicit vigilant behaviour in chicks. It is likely that megapode chicks are equipped with innate generalised images of objects to fear based on visual and acoustic stimuli and among them there are several specific features for which the

Fig. 22.1 A megapode hatchling digging its way out of the mound. (Photograph by A. Göth; courtesy of A. Göth.)

Fig. 22.2 An apparatus for studying incubation and development of early behaviour in megapodes. (Photograph by A. Göth; courtesy of A. Göth.)

hatchlings are tuned and this allows them to distinguish between dogs and cats, and between alarm calls of their own and alien species.

Let us now consider another scenario that includes sensitivity to quick learning. A good example here is the experimental study by Mineka, Cook and colleagues on the shaping of fear in rhesus monkeys, a problem closely related to development of phobias in humans. Researchers have exploited the fear of snakes as a common behavioural characteristic among primates including humans (Öhman and Mineka, 2003). Agras *et al.* (1969) interviewed a sample of New Englanders about fears, and found that snakes were clearly the most prevalent object of intensive fear, reported by 38 per cent of females and 12 per cent of males. According to a review of field data (King, 1997), 11 genera of primates showed fear-related response (alarm calls, avoidance, mobbing) in virtually all instances in which they were observed confronting large snakes. The hypothesis that this fear is adaptive in the wild has been supported by field reports of large snakes attacking primates (Goodall, 1986; Mineka and Cook, 1988). Experiments showed that rhesus and squirrel monkeys reared in the wild were far more likely than laboratory-reared monkeys to show strong fear responses to snakes (Mineka *et al.*, 1980).

Further, a series of experiments revealed strong preparedness in rhesus monkeys for the development of fear of snakes as 'fear-relevant' (FR) objects. Firstly, important evidence has been obtained that laboratory-reared monkeys are not afraid of snakes, that is, they appeared not to be equipped with innate recognition of these fear-relevant objects. It turned out, then, that laboratory-reared rhesus monkeys can acquire a fear of snakes vicariously, that is, by observing other monkeys expressing fear of snakes. When non-fearful laboratory-reared monkeys were given the opportunity to observe a wild-reared demonstrator displaying fear of live and toy snakes, they were rapidly conditioned to fear snakes, and this conditioning was strong and persistent. The fear response was also learned even in those cases when the fear-demonstrating monkey was shown on a videotape (Cook and Mineka, 1990). Then the most intriguing stage of experiments followed. Videos were edited so that identical displays of fear in the demonstrating monkey were modelled in response to toy snakes and flowers, or to toy crocodiles and toy rabbits. The laboratory-reared monkeys showed substantial conditioning to toy snakes and crocodiles, but not to flowers and toy rabbits. At the same time, both toy snakes and flowers served equally well for a group of naive monkeys as signals for food rewards. These results provide a strong support for selective (guided) learning which is probably based on a specialised behavioural module. This study also illustrates how sophisticated the interaction between components of behaviour may be and how difficult it is to completely rule out 'non-obvious' environmental contributions.

Another example of guided learning comes from the experiments of Griffin *et al.* (2002) with the tammar wallaby (*Macropus eugenii*). Up to now tammars have only survived on predator-free Australian islands, but they evolved with a range of now extinct marsupial predators, such as the Tasmanian wolf (*Thylacinus cynocephalus*). In their experiments researchers used taxidermically prepared models of predators like foxes and cats, and a model of a size-matched non-predator (a juvenile goat) as a control. Foxes and cats are likely to share convergent vertebrate morphological features (such as frontally placed eyes) with historically important predators, but tammars have no evolutionary experience with these introduced species, or with goats. In training procedures, models were presented to tammars paired with an aversive event, that is, a human simulating a capture procedure with a net. Humans with net reliably evoke alarm responses including fleeing and foot-thumping in tammars. Researchers used simulated capture attempts as a standard fear-evoking stimulus. It was found that training with a fox model together with a fear-evoking stimulus enhanced adequate reactions in wallabies. It was then sufficient to show a fox alone in order to observe fear reactions in wallabies. Animals were then tested with an array of other unfamiliar stimuli to determine the specificity of this change in behaviour. Training with a fox produced increased responses to another model predator, a cat, but not to a size-matched non-predator (a goat). This suggests that wallabies may not acquire a fear response to any stimulus that is associated with an aversive event but rather may be predisposed to learn quite specifically about predators; at least, they have a bias to associate predators with frightening events.

Besides different variants of learning about a predator itself, there is also a gradient of preparedness in acquiring reactions to the species-specific alarm calls of other animals. How and when alarm call responses develop varies between species. It is known, for example, that some young birds and primates can recognise and respond to conspecific calls upon first exposure whereas other naive juveniles display poor initial discrimination between alarm calls and other auditory stimuli (Sherman, 1977; Krebs and Dawkins, 1984). Concrete intermediary variants revealed in playback studies shed light on the adaptive value of preparedness of these reactions in the context of the animals' life. For example, Mateo (1996) has demonstrated that ground squirrels *Spermophilus beldingi* develop responsivity to whistle alarm calls indicative of fast-moving predators earlier and more readily than to trill alarm calls, associated with slow-moving predators. It is worth noting that in some species animals, having assimilated alarm calls of their own species, can then learn and 'translate' heterospecific calls (see Chapter 31 for details).

22.2 Is it easy to distinguish between instinctive and learned behaviour?

Hopes for an easy distinction between instinctive and learned behaviour were abandoned when ethology entered the scene and when it was synthesised with the other sciences of animal behaviour and psychology. As Verplanck (1955) expressed it in his *Psychological Review*, 'Since learned behaviour is innate, and vice versa, what now?' Early ethologists believed that behaviour is largely instinctive, or innate, as the product of natural selection. Soon it became clear that individual behaviour develops as a jigsaw puzzle of environmental and genetic components, and very often it is difficult to find reliable signs by which to discriminate innate and learned behaviour. This problem has many aspects, from which we consider two in this section: how innate behaviour can prevent organisms from learning even though it is demanded by changes in the environment; and how conservative learned behaviour can prevent animals from learning something new.

Animals are pressed with innate rules

Let us consider several examples that demonstrate the genetic basis of behaviour and provide evidence that in some cases innate rules can block new learning.

The most reliable methods that allow discrimination between innate and flexible components of behaviour are based on a genetic approach. It is possible to select inbred lines of animals possessing certain specialised behaviours. In his famous experiment Tryon (1940) studied the ability of rats to find a way through a complex maze in order to obtain a food reward. Since some rats appeared to be fast learners, Tryon bred them with one another to establish a 'maze-bright' colony, and he similarly bred the slow learners with one another to establish a 'maze-dull' colony. The offspring of maze-bright rats learned even more quickly than their parents had, while the offspring of maze-dull parents were very poor at maze learning. After repeating this procedure over several generations, Tryon was able to produce two behaviourally distinct types of rats with very different maze-learning abilities. Clearly the ability to learn the maze was to some degree hereditary. It is important to note that genes that governed maze-learning abilities were specific to this behaviour, as the two groups of rats did not differ in their ability to perform other tasks, such as running a completely different kind of maze. This study supports a hypothesis about the genetic basis of specific domains responsible for certain learning abilities.

Although recent research has provided much greater details of the genetic basis of behaviour, let us go to a classic example that brilliantly illustrates a battle between inherited behavioural pattern and learning. Dilger (1962) examined two species of parrots, which differ in the way they carry twigs, paper and other materials used to build a nest. Fischer's lovebirds (*Agapornis personata fischeri*) carry single strips of nest material in their beaks, while peach-faced lovebirds (*Agapornis roseicollis*) carry material tucked under their flank feathers. The tucking behaviour was a leftover from ancestor species that lined their nests with small chips, which more easily stay put under the feathers. These two species can interbreed and produce sterile offspring. The hybrids showed a poorly organised mixture of the two strategies: they tucked nest material between their feathers but failed to go home, pulled it out, and started again. After two years they became partly successful, managing to transport some material back to the next site, but not in a manner that resembled either parent species. Sometimes they would just turn their heads toward their rumps without tucking, and would then fly off with the material. In later mating seasons poor hybrids transported the material in their beaks but they performed head-turning behaviour each time they took an item to carry it to the nest. Parrots are known as fast learners but in this case innate predisposition to a certain behaviour blocked the birds' ability to learn a relevant way of nest building, and their attempts to improve practical skills were mostly unsuccessful.

There is some more experimental evidence that innate rules may block learning. For instance, in experiments of Mazokhin-Porshnyakov and Kartsev (1984, 2000) honeybees and *Polistes* wasps were presented with the following task. Four small troughs with syrup were situated on a table covered with glass (see Fig. 15.3). Troughs contained a small amount of food so that foragers had to visit all of them to be sated. Each trough was placed on an icon of a geometric shape, and all of them were situated on the table in different orders, that is, diagonally, in the corners of a square and so on. The troughs containing syrup alternated with the troughs containing salt, and this was a penalty for an insect to find its proboscis in salt. Wasps quickly learned to escape the penalty and choose the correct pictures. Bees insisted on choosing the nearest trough each time and thus alternated syrup with salt. It seems, as the authors explained it, that innate rules of searching are more flexible in wasps than in honeybees. In honeybees learning was blocked by their strong innate rule for searching, that is, 'Remember what a bee plant looks like, take nectar and fly to the nearest one.' Since the task of flying around troughs simulates a natural situation, honeybees switched their behaviour to following one of their main searching rules. As was described in Chapter 15, honeybees are able to solve very complex problems which demand capacity for abstraction and categorisation. Those tasks are probably so different with natural problems that the innate rules did not prevent honeybees from learning.

Animals are bound by learned rules

In his book *King Solomon's Ring*, Lorenz (1952) describes how water shrews learn their paths in their home range. He considers this striking example of how learned novel behaviour may become highly routine and stereotyped. The water shrew is aquatic, but it also spends some time on land. Here it is nearly blind, and finds its way around mainly through its sense of touch and long whiskers. Once the shrew has learned a path it is bound to it, as Lorenz writes, as a railway engine is to its tracks and is as unable to deviate from them by even a few centimetres. In order to examine what happens if there is an obstruction on the path, Lorenz experimented by moving a stone which had been on one of the shrew's paths. This is what he found:

> The shrews would jump right up into the air in the place where the stone should have been; they came down with a jarring bump, were obviously disconcerted and started whiskering cautiously right and left, just as they behaved in an unknown environment. And then they did a most interesting thing: they went back the way they had come, carefully feeling their way until they had again got their bearings. Then, facing around again, they tried a second time with a rush and jumped and crashed down exactly as they had done a few seconds before. Only then did they seem to realise that the first fall had not been their own fault but was due to a change in the

wonted pathway, and now they proceeded to explore the alteration, cautiously sniffing the place where the stone ought to have been. This method of going back to the start and trying again always reminds me of a small boy who, on reciting a poem, gets stuck and begins again at an earlier place.

This clear example demonstrates that not only in laboratories but in nature learned behaviours may become 'automatic', and animals may follow these learned stereotypes for a long time even if such fidelity becomes completely useless. Chains of acts become 'ritualistic' as we have already seen when considering the behaviour of Pavlov's chimpanzees who performed complex and long chains of behaviours in order to gain access to a cistern and scoop water from there in order to put out a fire. The apes did not think of scooping water from the surrounding lake and thus solve a problem by a single operation. The learned stereotype became a ritual just as in shrews jumping over a virtual stone on their way home.

Dog trainers know that it is necessary to vary a sequence of commands given to a dog during each training sessions. Should the trainer several times repeat two commands in the same sequence ('sit–lie', and 'sit–lie' again), the dog will always perform these two commands one after another. The trouble is that it is much easier to teach a dog to perform something new than to teach it again and to force it to perform each command separately. The trainer now feels sorry that his dog is so fast a learner. Not being loaded with stereotypes so quickly wired, the dog would be more susceptible to a trainer's request.

22.3 A harsh environment for pluralism in animal societies: behavioural specialisation within populations

Two extreme approaches to considering species-specific behaviour exist in ecological and ethological studies: those that distinguish unique individualities of members of species and those that consider a population as a whole treating conspecific individuals as ecologically equivalent. Applying the ideas of evolutionary ecology helps to find a middle course and to reveal relatively stable fractions of populations that differ by sets of behavioural characteristics, a differentiation that covers routine differences of individuals by sex and age.

There are at least two levels of behavioural specialisation within populations. In some species members of a population comprise distinct groups that behave differently according to their evolutionarily stable strategies. In some cases members of these groups can be easily distinguished by certain morphological markers. Besides, more flexible individual specialisation can be expressed in differences in diets, techniques of getting food, forming searching images, escaping predators, nesting and so on. Relatively stable groups can exist in populations that differ by complexes of behavioural characteristics.

Evolutionarily stable strategies: a battle of behavioural phenotypes

The theory of the *evolutionarily stable strategy* (ESS) introduced by Maynard Smith and Price in 1973 is based on the concept of a population of organisms divided into several groups which use different strategies. A group is in a stable state if it is disadvantageous for any individual to change its strategy. In other terms the proportion of individuals using each strategy is optimal; natural selection suppresses any deviation from the current proportion.

Maynard Smith's best-known work incorporated game theory into the study of how natural selection acts on different kinds of behaviour. He developed the idea of an evolutionarily stable strategy as a behavioural phenotype that cannot be invaded by a mutant strategy. A classic example is a balance between hawk-like (aggressive) and dove-like (non-aggressive) individuals in natural populations. Maynard Smith and Price (Maynard Smith and Price, 1973; Maynard Smith, 1974, 1982) demonstrated that carriers of both aggressive and non-aggressive behavioural strategies can coexist comfortably and stably in populations for a long time, and neither aggressors nor non-aggressors can invade the population (see also McFarland (1985) for a detailed description).

Males of many species are characterised by alternative mating strategies and thus compose a representative set of examples concerning distinct behavioural strategies of carriers of different evolutionarily stable strategies. These strategies are based on complex behaviour sequences and thus may give to observers the impression of a deliberate choice of variants.

For instance, Sinervo and Lively (1996) revealed impressive mating strategies within populations of the side-blotched lizard (*Uta stansburiana*) native to California. These lizards have three mating strategies, distinct types of behaviour that constantly compete with one another in a perpetual cycle of dominance. Carriers of different behavioural strategies are marked by morphological signs. These researchers described the cycle of dominance in lizards in terms of evolutionarily stable strategies as the 'rock–paper–scissors' game.

In side-blotched lizards males have one of three throat colours, each one declaring a particular strategy. Dominant, orange-throated males establish large territories within which live several females. Orange males are ultra-dominant and very aggressive owing to high levels of testosterone, and attack intruding blue-throated males which typically have more modest levels of testosterone. Blue males defend territories large enough to hold just one female. These males spend a lot of time challenging and displaying, presumably allowing males to assess one another. Territories of both orange and blue males are vulnerable to infiltration by males with yellow-striped throats – known as sneakers. Sneakers have no territory of their own to defend, and they mimic the throat colour of receptive females. It is interesting that yellows also mimic female behaviour. When a yellow male meets a dominant male, he pretends he is a female – a female that is not interested in the act. In many cases, females will nip at the male and drive him off. By co-opting the female rejection display, yellow males use a dishonest signal to fool some territory-holding males. The ruse of yellows works only on orange males. Blues are not fooled by yellows. Blue males root out yellow males that enter their territories. Blue males are a little more circumspect when they engage another blue male during territory contests. Attack may or may not follow as blue males very often back down against other blue males. Indeed, neighbouring males use a series of bobs to communicate their identity, and the neighbours usually part without a fight.

Thus, each strategy has strengths and weaknesses and there are strong asymmetries in contests between morphs. Trespassing yellows, with their female mimicry, can fool oranges. However, trespassing yellows are hunted down by blue males and attacked. While oranges can easily defeat blues, they are susceptible to the charms of yellows. In contrast, contests between like morphs (e.g. blue versus blue, orange versus orange or yellow versus yellow) are usually more symmetric. Field data show that the populations of each of these three types, or morphs, of male lizard oscillate over a 6-year period. When a morph population hits a low, this particular type of lizard produces the most offspring in the following year, helping to perpetuate the cycle. This arrangement somehow succeeds in maintaining substantial genetic diversity while keeping the overall population reasonably stable. This is a good example of genetically based control over morphotype and behavioural type development (Sinervo and Clobert, 2003).

Another good example of male's alternative behavioural strategies came from the bluegill sunfish *Lepomis macrochirus*. Gross (1979, 1984) discovered that in this species males come in three different-sized morphs: (1) a large territorial parental male that courts females, and then defends a nest in which he rears eggs that the female oviposits; (2) a medium-sized satellite or sneaker male that mimics females, interrupts a courting territorial male and attempts to fertilise the female, and (3) a very small satellite male that dives in between a mating territorial male and female and squirts ejaculate in an attempt to fertilise the female. Gross argues that such morphs are condition-dependent tactics in which all males have the capability to become any of the alternative types. As a result, 'developmental decisions' are made, based up the condition of the male. Males that are in poorer condition become the smaller male types and the males that are in the best condition adopt the territorial-holder strategy. This study is considered an example of condition-dependent strategies contrasting with genetically based control over the development of a distinct strategy.

Individual behavioural specialisation and adaptation learning

Another form of learning is singled out here, based on the idea that animals often do not learn to do something really new in order to gain advantage from their environment; instead, they learn to select quickly and to manipulate readily with innate behavioural patterns, and this can be considered a separate form of learning, which can be called *adaptation learning*. This form of learning is closely connected with behavioural specialisation in populations: some specimens are predisposed to accomplish a certain part of a whole, species-specific, repertoire and consequently they learn more readily within this specific domain.

An impressive example of behavioural specialisation came from the study on how insects of different sizes and levels of intelligence catch jumping springtails (Collembola), small inoffensive creatures that nevertheless are equipped with a jumping fork appendage (furcula) attached at the end of the abdomen. The furcula is a jumping apparatus enabling the animal to catapult itself (hence the common name springtail), thereby changing sharply the direction of movement and escaping the attacks of predators. Reznikova and Panteleeva (2001, 2005) found springtail hunters in beetles of the family Staphylinidae as well as in several species of ants. Although beetles are taxonomically far removed from ants, there are three similar groups both in the beetles and ants: (1) good hunters that catch a jumping victim from the first spurt; (2) poor hunters that perform several wrong spurts until they catch a springtail; and (3) non-hunters that do not even show any interest in the victims. Behavioural stereotypes were similar in ants and beetles, with one great difference: ants were able to raise their hunting technique to the standard of the next level up whereas beetles were not. It turned out later that hunting behaviour in ants incorporated several variants of development; one of them is based on maturation rather than learning, while others include elements of social learning and different levels of flexibility. We will return to this sophisticated development of behaviour in Chapter 23. As far as beetles are concerned, there are three distinct types of behaviour relative to jumping victims in populations, and this is one of examples of individual behavioural specialisation.

Bolnick *et al.* (2003) present a huge collection of examples of individual behavioural and ecological specialisation for 93 species distributed across a broad range of taxonomic groups. In many species some specimens in populations are more risk-averse than others, possibly reflecting different optimisation rules (in terms of optimal foraging theory; see Charnov, 1976). Besides, individuals vary in their prey-specific efficiency because of search image formation. Individuals also vary in social status, mating strategy, microhabitat preferences and so on. In some species individuals constitute groups on the basis of relatively stable features. The bluegill sunfish serves as a good example of differentiation of individuals relatively to their foraging strategy (see review in Bolnick *et al.*, 2003). When a population of bluegills was experimentally introduced into a pond, individuals quickly divided into benthic and limnetic specialists. The remaining generalists constituted 10–30 per cent of the population and appeared to have a lower intake rate of food.

There is an example of more complex individual specialisation in the oystercatcher *Haematopus ostralegus*. In this species individual birds specialise both on prey species and in particular prey capture techniques such as probing mud for worms or hammering bivalves. Individuals that use bivalves tend to specialise in different hammering or stabbing techniques that reflect intraspecific variation in prey shell morphology. Individuals are limited to learning a small repertoire of handling behaviours, while additional trade-offs are introduced by functional variation in bill morphology. Subdominant and juvenile birds are often restricted to sub-optimal diets rather than those they would choose in the absence of interference competition (Goss-Custard *et al.*, 1984; Sutherland *et al.*, 1996).

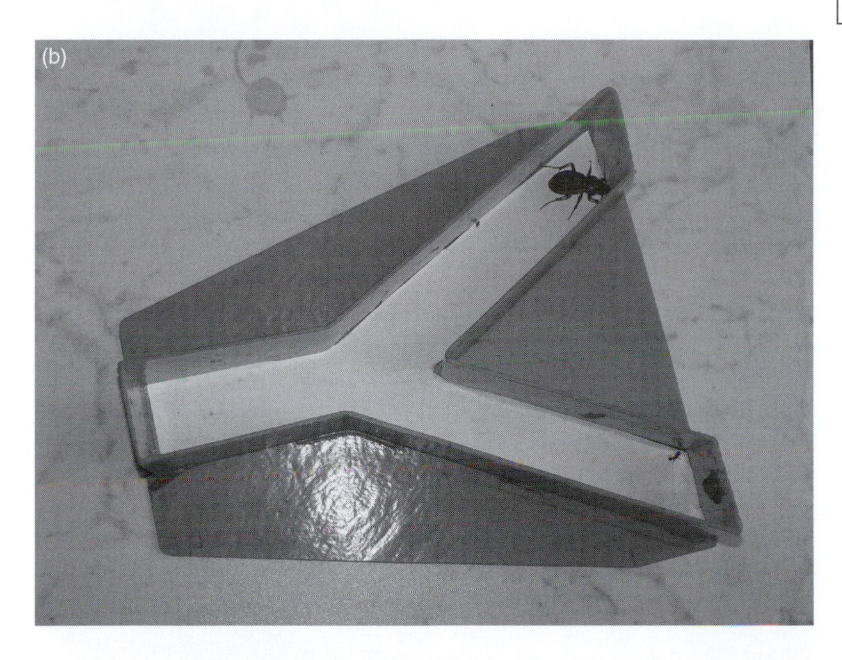

Fig. 22.3 Collisions between ants and beetles in the experiments of Reznikova and Dorosheva (2004). (a) Confrontation between an ant and a beetle; (b) a maze for studying carabids' ability to avoid clashes with ants; (c) an active ant tied up with a thread in one section of the maze; (d) the ant is biting the beetle; (e) the ant has just killed the beetle. (Photographs by E. Dorosheva and S. Panteleeva.)

In all cases described above behavioural specialisation within populations is based on an intricate composition of innate predisposition of individuals to choose a way of prey handling, to avoid risk or not, to dominate over conspecifics or to avoid conflicts, and so on. Some specimens can possess complex behavioural patterns that allow them to learn readily within a specific domain. This ability can be called *cognitive individual specialisation.*

The idea of adaptation learning is based on the experimental study on interrelations between ants and carabid beetles by Reznikova and Dorosheva (2004, 2005). To examine the ability of four carabid species to avoid collisions with *Formica polyctena* red wood ants, the researchers used two experimental techniques. In the laboratory, they tested carabids' ability to avoid a clash in a Y-shaped maze containing an active ant tied up with a thread in one section (Fig. 22.3). Four carabid species tested in the maze displayed a clear tendency to learn, that is, to modify their behaviour in order to avoid encounters with ants. In all four species individuals comprised three groups that differed by their ability to learn to averse clashes: 45–76 per cent made fewer then 35 per cent of errors in the maze ('good learners'), and 17–39 per cent made more than 65 per cent of errors ('bad learners'). The remaining intermediate group was relatively small in all species (from 10 to 20 per cent). This can be considered individual specialisation relative to the ability to avoid danger by means of learning. 'Good learners' successfully avoided conflict with ants, each applying one out of a set of stereotyped behavioural tactics: (1) to

Fig. 22.3 (cont.)

attempt to go round the ant; (2) to turn away after touching the ant with the antennae; (3) to turn away without a contact; (4) to avoid a dangerous section altogether; (5) to stop near the ant with legs and antennae hidden (Fig. 22.4). Members of different species appeared to have specific preferences for definite sets of tactics. For example, for *Pterostichus oblongopunctatus* tactics 1 and 2 are preferable, whereas *P. magus* prefers tactics 2 and 5. Complementary field experiments (Reznikova and Dorosheva, 2004) enabled the researchers to suggest that an effective combination of tactics allows carabids to penetrate ant foraging territory and to particularly avoid interference competition.

It is important to note that in the cited experiments animals did not learn something new such as remembering a new path in a maze or pressing a lever in response to a presenting stimulus. Instead, facilitation of manipulations with

Fig. 22.3 (cont.)

Fig. 22.4 Behavioural tactics used by carabids to avoid clashes with the ant in the maze. (a, b, c, d) Attempts to go round the ant; (e) turns away without contact; (f) avoids a dangerous section altogether; (g) stops near the ant with legs and antennae hidden. (Photographs by E. Dorosheva and S. Panteleeva.)

innate behavioural patterns by animals took place in these experiments. It is worth noting that animals very quickly learn to 'juggle' with behavioural patterns that are ready for operation. Perhaps, adaptation learning can be implemented on a wide variety of species.

'Ecological intelligence'

As we have already seen in this book, there are some impressive examples of complex behaviour in animals closely related to the ecological traits of the species. In those cases when these behaviours include sophisticated learning one can think about 'ecological intelligence'. Let us consider one of the most astonishing examples concerning the ability of food-hoarding animals to memorise hundreds of sites of food location.

Observing birds that are attracted to a winter feeder in a garden, one can see, among others,

Fig. 22.4 (cont.)

great tits, blue tits and marsh tits. The great tits and the blue tits congregate at the feeder, eating as fast as they can. They interrupt their meal only to chase away their competitors. A marsh tit nonetheless darts in, grabs a peanut and flies off. It is back almost immediately to grab another. It stores the peanuts nearby, each in a different site, until the feeder is empty. Then it searches out its hidden food. The marsh tit is one of the food-storing species. Another example is the nutcracker. This pale-grey bird with black wings and a long beak flits through woodlands, collecting seeds during times of plenty and tucking them away for a hungry winter's day. During a year, each bird buries 22 000 to 33 000 seeds in up to 2500 locations, and researchers estimate that the bird recovers two-thirds of them up to 13 months later. Nutcrackers carry their seeds as far as 25 km to cache them (Balda, 1980). Detailed studies have demonstrated that food-storing

Fig. 22.4 (cont.)

birds really do remember large numbers of storage sites over long periods, and their memory could be an example of cognitive specialisation.

Food hoarding (or caching) is a fundamental adaptation of animals to variation in food supplies. A detailed systematically review of food hoarding in birds, mammals and arthropods is given in Vander Wall's (1990) book. Sherry (1989) cites one of the earliest descriptions and, what is even more important, the earliest experimental paradigm of the study of food-storing behaviour. This is attributed to Baron von Pernau (1660–1731) whose observations and methods were unknown until they were rediscovered by Stresemann (1947). Von Pernau provided a remarkably accurate description of the basic methods that are used today for observing food storing in birds in captivity. He allowed a tit to collect and hide seeds in a room, then rehid them and examined what the bird had remembered.

Fig. 22.4 (cont.)

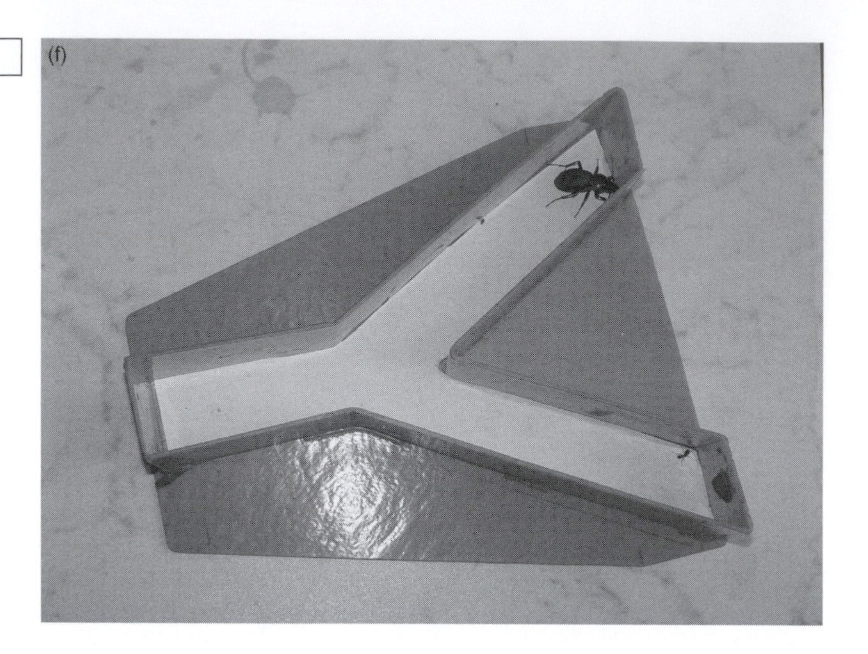

Based on a similar method, Brodbeck and Shettleworth (Brodbeck, 1994; Brodbeck and Shettleworth, 1995) compared spatial memory abilities in food-storing chickadees and non-food-storing juncos. Birds were allowed to return after a few minutes or hours to food which they had encountered briefly once before. The food was a peanut hidden in a brightly decorated block of wood on the wall of a large aviary. Four new feeders in four new locations were used on every trial. When the birds returned directly to the baited feeder on a high proportion of trials, they were tested to see what they remembered by swapping the formerly baited feeder with another one of the feeders on occasional unrewarded tests, thus dissociating location from pattern and colour cues. Chickadees nearly always went first to the former location of the baited feeder, even though it was now occupied by a feeder that looked entirely different. In contrast, juncos chose about 50 : 50 between that feeder and the formerly baited feeder in its new location. Clayton and Krebs (1994) obtained similar results comparing food-hoarding Clark's nutcrackers with two closely related but non-storing species.

Experiments by Balda and Kamil (1992) with Clark's nutcrackers challenged the idea that the birds were following some subtle sensory clue to the seeds themselves instead of relocating particular sites. The birds were able to find their caches in the sandy aviary floor irrespective of whether or not the researchers put landmarks such as cinder blocks, wall posters, and so on in the room. In a further refinement of the experiments, experimenters built a floor that was 30 feet by 50 feet (9×15 metres), with 330 holes drilled in it to hold either cups of sand or plugs. Still, the birds maintained their high success rate in relocating their hiding places. The researchers also tested one of their graduate students at caching. The student hid seeds and 30 days later found only about half as many caches as a bird typically did.

The very specific ability to memorise site locations in food-storing species seems to be closely related to some important brain adaptations. Just as in mammals, in birds the hippocampus is important for spatial memory. It is consistent with the idea that food storing involves the evolution of a modular cognitive capacity: in food-storing species of birds the hippocampus is larger relative to the body size and telencephalon volume (most of the rest of the brain) than in non-storing species. Krushinskaya (1966) was the first to show that lesions that include the avian hippocampus disrupt food cache recovery in food-storing birds. More recently many detailed findings have appeared in this field which include, in particular, that the hippocampal enlargement that occurs as animals store more food during the autumn is accompanied by increased neurogenesis. There are also studies in which males and females of the same species differ, perhaps only seasonally, in space use and concomitantly in hippocampal volume (Smulders *et al.*, 1995; Sherry and Duff, 1996). In principle, a relationship between the relative size of the hippocampal formation and the degree of specialisation for food hoarding has been shown in many avian species (Krebs *et al.*, 1989; Sherry *et al.*, 1989). However, recent results have revealed a surprising differences between species from continents, with North American species possessing significantly smaller hippocampi than Eurasian ones (Lucas *et al.*, 2004).

What is important to note is that in many tests of different design food-storing species excelled closely related but non-storing species at spatial tasks but they were no better than the others in operant tests or colour memory. This enables researchers to conclude that food-storing animals such as many species of parids (tits and chickadees), corvids (jays and nutcrackers) and rodents (kangaroo rats, squirrels and others) possess a relatively narrow cognitive specialisation based on their extraordinary spatial memory. This may be called a *species-specific endowment*.

It is a topic of controversy to what extent displays of sophisticated behaviour of 'species-specific geniuses' (or 'cognitive specialists') can be explained in terms of cognition rather than the accomplishment of fragments of hard-wired species-specific repertoire.

Let us consider the very complex behaviour of food-hoarding western scrub jays which seem to anticipate problems yet to come. Emery and Clayton (2001) looked at how the jays reacted to the possibility that some neighbour might steal a food cache. Some of the jays in the laboratory had criminal histories of snitching food that another bird had buried. The researchers allowed the criminal jays as well as ones with clean records to hide treats. When the birds stashed their seeds in private, they didn't take opportunities to move their treasure to a new hiding place. However, if the researchers let another bird get near enough to watch the caching, the criminal jays took the next opportunity to recover the treat and hide it in a different place. The birds with no experience of thievery didn't re-cache. The authors are careful not to say that animals mentally travel in time the same way people do. Still, they argue for continuing to push questions about animals' mental abilities.

Analysing different aspects of the findings described above we can conclude that food-hoarding behaviour develops on the basis of a specific behavioural domain seated in a particular brain structure. Within these frames animals demonstrate sophisticated mental abilities and high flexibility of memory. Seemingly, a compromise has just been found between innateness and flexibility.

However, the picture is not simple. The next example shows that the range of flexibility may be relatively narrow. The matter concerns the so-called 'pilfering avoidance hypothesis' which has been debated in behavioural ecology since the 1970s (MacDonald, 1976). Studies on rodents and birds have shown that members of the same species can use several constant scenarios of hoarding and several scenarios of pilferage. They shift their strategies, for instance, from larder to scatter hoarding, depending on many factors among which risk of pilfering plays a key role. Animals detect the risk of pilfering by encountering potential competitors by scent and visual cues or directly after the tragic experience of emptied caches (Hansson, 1986; Clarke and Kramer, 1994; Jenkins and Breck, 1998). There is evidence from many species that animals may alter cache behaviour in the presence of competitors and that a personal 'criminal past' facilitates the alteration of strategy. For example, in experiments by Preston and Jacobs (2001) kangaroo rats were forced by researchers to play the roles, in turns, of 'stealers' and 'victims'. Similarly with Clayton *et al.*'s scrub jays, rodents were likely to change their strategies more readily after experience as victims. Analysis of a huge body of literature has led Vander Wall and Jenkins (2003) to a model that is based on the imagination of the lives of many rodents and birds under circumstances of reciprocal pilfering. Levels of cache pilferage are often high, that is, from 2 to 30 per cent per day, but victims can return a favour to stealers. It is likely that processing of food-hoarding interacts with inter- and intraspecific competition in many species at an evolutionary level, and that certain sets of strategies have evolved as a basis for shifting from one strategy to another depending on the level of the risk.

All these results enable us to consider sophisticated 'anticipation' in food-storing animals within a paradigm of 'adaptation learning' (Reznikova and Dorosheva, 2004) rather than in terms of tactical deception and taking into account rich social context in birds (Emery and Clayton, 2001). In order to avoid the risk of pilfering from conspecifics, animals are likely to choose more and more quickly and effectively a relevant strategy from a set of strategies that already exists rather than to contrive something completely new as an answer for an intervention.

Chapter 23

Developmental studies of animal intelligence: role of innate and acquired behaviour

One of the most intriguing questions in ethology is to what extent complex behavioural patterns in animals, such as flight, nest building, hunting and tool-using, are determined by inherited programmes. What one may say definitely is that each species should be investigated experimentally, and perhaps theoretical constructions are of little help. Lorenz (1935) spoke of 'instinct–learning intercalation' describing situations when innate and acquired components become integrated into one behavioural sequence. It is interesting to separate these components experimentally and to investigate how early experience influences the behavioural scenario in different species. Development studies help researchers to unravel strapwork of innate and learned components in animals.

A good example here is food-storing behaviour that was considered above (Chapter 22). The way food-storing behaviour develops and can later be modified indicates that food-storing in birds is better described as a hard-wired behaviour which the birds perform relatively independently on current need and anticipation of future consequences (Shettleworth, 2001). Fledglings begin storing avidly, at around 40 days of age, even when they have ample food and efforts are made to deprive them of storable items and suitable substrates (Clayton, 1995). The items chosen and the expertise with which they are inserted into suitable sites change with experience, as does the selection of sites in adults (Hampton and Sherry, 1994). However, birds store food even under circumstances where they seem unlikely to be anticipating retrieving it. In the laboratory, some birds persisted indefinitely in storing peanuts in places where they dropped out of reach (Shettleworth, 1994).

In this chapter we will analyse experimental approaches for studying the development of complex behavioural patterns.

23.1 How an instinct is learned: early experience

Starting with developmental studies we should first of all take into account major differences in the ontogenetic scenarios of species. In mammals and birds two main types of ontogenetic scenarios are usually distinguished. Some animals are born in a relatively undeveloped, helpless state and are very dependent on their parents for survival. These animals are known *altricial* animals. Tree-nesting birds, rats and humans can serve as examples. Other animals are said to be *precocial* and are born almost fully developed and are able to move about. Examples include poultry (and of course wild ducks, galliforms and turkeys), hoofed animals and some others, for example, guinea pigs and hares.

It is important to note that taxonomically close species can possess different types of development. For instance, hares and guinea pigs are precocial whereas rabbits, rats, mice, hamsters and many other rodents are altricial. What may be even more important is that there are a great many specific behavioural types of development within each of these two major categories. For example, in the small rodent *Baiomys taylori* the young adhere by suction to the mother throughout the day liberating her only at night so she can collect some food. Young rats and mice enjoy their mother's good graces for very short periods round the clock. In wild rabbits suckling is possible for pups for a few minutes only once a day, and young tree shrews see their mother briefly once in two days (for a review see: Manning, 1979).

Different schemes of interaction between parents and offspring in combination with effects of early experience, and the effects of maturation of the nerve and motor system, result in a variety of ontogenetic scenarios. That is why it is difficult to squeeze the behaviour of species into the limits of general schemes. However, we can consider several sets of examples to illustrate sophisticated interrelations of different factors that shape animal behaviour. We will be focusing here on the impact of processes of maturation and early experience on the development of species-specific behaviour.

Determination of the provenance of a species-typical behaviour is possible by the use of the so-called *Kaspar Hauser* (deprivation) experiments. Kaspar Hauser monkeys, chicks and other subjects take their name from a real prototype: an early nineteenth-century 'wild child' found wandering the streets of Nuremberg who was thought to be the princely heir to the royal throne of Baden. Kaspar seemed to have been raised in isolation and could hardly behave like a human being. Adopted by a kind tutor, Hauser was taught the ways of civilised society; meanwhile, growing political machinations threatened to destroy him, and finally Hauser was found dead (probably killed) 5 years after he was first discovered.

In deprivation experiments with animals subjects are given no opportunity to learn the skill components under investigation. For nest-building, for example, young birds might be prevented from seeing a nest or nesting material until they arrive at breeding age. If these birds are then immediately able to build a nest successfully, nest-building does not depend on these sorts of experience. Spalding (1873) first applied this approach to reveal the innate components in behaviour of birds, and Cuvier used this method to observe dam-building in beavers. Ladygina Koths (1935) revealed innate features in chimpanzee's nest-building. Many authors since have used the deprivation procedure in order to separate innate and learned components in animal behaviour. This method is often criticised because the 'deprived' environment is still an environment so it is difficult to obtain clear results. Nevertheless, this approach sometimes allows the veil of mystery to be lifted from developmental problems.

Another experimental procedure that helps to separate components of behaviour is *cross-fostering*. Cross-fostering techniques, in which infants are taken from their biological parents and reared by unrelated parents from the same and even from alien species, is widely used in behavioural research. Many cross-fostering experiments have demonstrated that members of species retain at least main features of their species-specific behaviour and do not follow the habits of their foster parents. There are many stories about how people switch the eggs of a brooding duck with those of a brooding hen, later to enjoy the anxious antics of the hen when her brood takes to the water, and on the other hand the futile efforts of the duck to coax her hatchlings into the duck pond.

Let us consider examples illustrating how the development of behaviour in infant animals depends upon maturation, individual and social experience.

In many instances the developing behaviour patterns appear very clumsy. Only gradually does the initial lack of coordination disappear, which may be the result of maturation processes, learning processes or a combination of both. Maturation closely interacts with practice which corresponds rather to beating of nerve tracts than to learning. Actually, practice improves behaviour regardless of reinforcement.

The development of pecking in newly hatched chicks is an example of the interaction between maturation and practice in the development of behaviour. Newly hatched chicks have an inherited tendency to peck at objects that contrast with the background. Cruze (1935) studied how this improvement occurs. He measured pecking accuracy by testing chicks individually in a small arena with a black floor onto which he scattered several grains of millet. Each chick was allowed 25 pecks; each peck was scored as a hit or miss. Cruze used ten groups of chicks of different ages and different times of being kept in dark (from 24 to 120 hours) and in light (from zero to 12 hours) to practice pecking. He concluded that as the chicks mature there is a steady improvement in accuracy even if they have not had any opportunity to practice pecking. At any age, 12 hours of practice greatly improves accuracy. So, pecking improves as a consequence of both maturation and practice.

The title of this section, 'How an instinct is learned', is borrowed from Hailman's (1967) monograph about the development of 'instinctive' behaviour in herring gulls. Herring gull chicks peck at a red spot on their parents' bills to induce them to regurgitate food. Hailman (1967) tested Lorenz's claim that this behaviour is innate. It turned out that at birth herring gull chicks peck equally often at a model of their own species and at a model of a laughing gull. Moreover, they peck not only at their parents' beaks but at other parts of their head and body as well and thus have a hazy idea of what their real parents look like. After 6 days in the nest they show a preference for the model of their own species. In this case pecking improves as a consequence of maturation, practice and learning because the chicks are rewarded with food by their parents if they peck precisely at the relevant part of their beaks (which is marked by a red spot).

Eibl-Eibesfeldt's (1961a, 1970) classic experiments with squirrels *Sciurus vulgaris* demonstrate how learning interacts with maturation and how behaviour units integrate into one functional whole. Squirrels possess the movements of gnawing and prising, but they must learn how to employ these behaviour patterns effectively to open a nut. Experienced squirrels can do this with a minimum of wasted effort. They gnaw a furrow on the broad side of a nut from base to tip, possibly a second one, wedge their lower incisors into the crack and break the nut open into two halves. Inexperienced squirrels, on the other hand, gnaw without purpose, cutting random furrows until the nut breaks at one place or another. The first improvements in the technique can be seen when the furrows run parallel to the grain of the nut and are concentrated on the broad side of the nut. The squirrel follows the path of least resistance, and in this way the activity of the squirrel is guided in a specific direction by the structure of the nut itself. The squirrel continues with its attempts to prise open the nut, and it keeps repeating those actions that have led to success. In this way most squirrels acquire the most efficient prying technique. Adult squirrels seem to make short work of gnawing nuts in very similar ways but we know now that although their behaviour sequences contain fixed action patterns as elements, these elements are combined by learning into acquired coordinations (Eibl-Eibesfeldt, 1970). Promptov (1940) described many examples of similar development of behaviour in birds and concluded that such a similarity when most of members of species possess the same elements of innate behaviours and have to learn the same results in species-specific stereotype.

Cross-fostering experiments help us to estimate to what extent members of species develop their species-specific way of life. A common human practice with cross-fostering is switching the eggs in poultry, for instance, between hens and ducks. Slavery in ants is an experiment with switching eggs conducted by nature. As Hölldobler and Wilson (1995) have said, the terms 'slavery' and 'slave-making' are employed loosely; the activity is more akin to capture and domestication. In one way or another, members of one ant species kidnap the eggs of another species and raise them in their home. Well, cocoons, not eggs, and this amendment is important because this allows slave-makers to obtain nearly ready workers and thus save on raising their own brood at the stages of eggs and larvae. Being hatched in a foster family, 'slaves' feel themselves at home, and they seem not to take over the behaviours of their slave-makers. Ant species that serve as slaves belong to the subgenus *Serviformica* which is characterised by sole foraging and the high agility and enterprise of its members.

In our laboratory, we carried out several cross-fostering experiments in order to reveal details of behaviour of both slaves and slave-makers (Harkiv, 1997; Reznikova, 1996, 2004). We chose *Formica sanguinea* slave-makers because these ants are able to live without slaves and can completely look after themselves, in contrast to highly particularised amazon-ants of the genus *Polyergus* that can only fight but not work with their sickle-shaped mandibles and are fully dependent on their slaves. However, *F. sanguinea* as slave-makers differ essentially from *F. cunicularia* and other species of the subgenus *Serviformica* (their potentially slaves) in characteristics of movements, the contour of a searching trajectory (as the shuttle-movement of a pointer and yaw-movement of a foxhound) and the manner of foraging. To the eye of an observer who is not a specialist on ants, these members of the genus *Formica* are very similar (in their sizes, images and colour). In our experiments we forced not only *F. sanguinea* to adopt pupae of *F. cunicularia*, but vice versa as well, so several laboratory groups of potential slaves became kings for a few days. Being hatched in foster families, adopted ants behave as members of those families. It turned out that natural slaves (*F. cunicularia*) take on some intimate features of behaviours of their natural slave-makers, namely, several details of running movements and contours of searching trajectories. Several quantitative characteristics of behaviours in slaves' ethograms also became nearer to those of members of foster families. Experimenters then were able to distinguish between free-living *F. cunicularia* and adopted individuals. The slaves, however, kept all the characteristic features of their agile and sole-foraging manner which differentiate them essentially from group-foraging slave-makers. In contrast, ants of slave-making species now adopted as 'slaves' did not imitate any details of the behaviours of members of their forced 'slave makers' and kept all details of their innate behavioural features including movements, trajectories and foraging manner. Moreover, they occupied high positions in the family hierarchy. These results suggest that only minor shifts in ants' behaviour can be caused by fostering, and that subdominant species have more tendencies to change their behaviour than dominant ones.

It is interesting that results similar to these ant studies were obtained in cross-fostering experiments with canine species. Mainardi (1976) asked his fox terrier Blue to adopt a 10-day-old fox Kochis. The experimenter was lucky to observe many interesting details of the partnership between Kochis and dogs. Firstly, this concerned playing behaviour. Both dogs and foxes usually use highly stereotyped behavioural patterns in their games. Like an enslaved ant in our experiments, Kochis behaved as a member of his own species but some details of his behavioural repertoire shifted to the side of his social environment, that is, to the dog's side. Again as in ants, the difference was quantitative and concerned the frequency of display of several behaviours shared by dogs and foxes in nature. Being rare as elements of playing behaviour in a 'normal' fox, these elements seemed to be distinctly more frequent in Kochis' playing behaviour. Like members of our research group who easily distinguished between enslaved ants and free-living members of the same species, the dogs easily recognised the 'hybrid' behaviour, and they behaved rather differently towards a 'normal' fox and the fox raised by the dog as a foster mother.

There are many other examples obtained from cross-fostering experiments that illustrate how conservatively members of different species keep details of their species-specific behaviour. For instance, Hinde and Tinbergen (1958) described how young tits learn to hold a large piece of food with their legs in order to make pecking it to bits more convenient. Young chaffinches are not able to learn this behaviour even after being raised by tit foster parents.

The development of birdsong illustrates how genetic and environmental factors interact during the development of behaviour in different species (Slater, 2003). There is a wide variety of variants of song development in birds, from fully originate (and innate) to fully mocking (and obtained by imitation). In the majority of bird species all elements of their vocal repertoire are innate as well as the structure of the species-specific song (Nottebohm, 1970). That is why if you buy a yellow domestic chick and ask your home canary to raise it, you will nevertheless enjoy cock-crows in the near future. However, under the same circumstances, a young bullfinch will please you with a depleted variant of a canary's song.

Let us try to clarify this question and return to the chaffinch first. In contrast to, say, young bullfinches and greenfinches which can learn to imitate the song of other species while infant chaffinches cannot do this. Deprived of hearing songs of their own species and being presented with songs of other species during playback experiments, they sing only their species-specific songs, and the same is true of Kaspar Hauser chaffinches that have heard no songs at an early age. The only species that chaffinches are able to follow is the tree pipit. The song of this species contains one element resembling an element of the chaffinch's song, although these songs differ a lot in their structure. The chaffinches are predisposed to distinguish and reproduce this element in their own song (Thorpe, 1961). It is somewhat different in the zebra finch *Taeniopygia guttata*, which learns its song from those who feed it. If a society finch feeds a young zebra finch, it will learn the society finch song, even though zebra finches are singing in the adjacent cage. However, if it is fed by both a society finch and a zebra finch, it will learn the song of the zebra finch. Thus a preference for the song of the species as the model becomes evident (Immelman, 1972).

Marler's (1970), now classic, experiments revealed one of the most complex types of interaction between innate and environmental factors in the development of birdsong. He studied white-crowned sparrows which appeared to have geographically stable dialects. Within the same species, there are regional variations in birdsong. Although these differences could be interpreted as evidence for a genetic basis for birdsong, research has shown that young birds learn the dialect from adults in their area. There were three variants of the development of birdsong in Marler's experiments: (1) development under normal conditions; (2) development in isolation from white-crowned sparrow song, and (3) development after deafening. Under normal conditions, from 10 to 50 days of age, the young male's template accepts the adult male white-crowned sparrow song as a model; for example, it rejects the swamp swallow song as a model. The improved template now specifies the dialect he has to learn. The young bird does not sing, but the model is remembered for 2 months or more. The maturing male begins singing its sub-song (similar to babbling in human children) at about 150 days of age. During this period vocal output is gradually matched to the dialect specified by the improved template. At about 200 days of age full song begins; it is a copy of the model he learned in his youth.

If infant birds are raised in soundproofed chambers in a laboratory, they emit a crude but recognisable song that contains elements from the normal song. As the bird cannot learn a dialect, it does not sing and retains its basic unimproved template for 2 months or more. The maturing male begins sub-song (at about 150 days). Vocal output develops to match the specifications of the unimproved template. There is no dialect, but some species qualities persist. Full song begins, based on the unimproved template, at about 200 days.

Marler explained this by postulating that young birds are born with a crude template of what their species song should sound like. They match this template to the song they hear around them during development, so that the template is sharpened. When the bird is isolated during this memorisation phase, all it can produce is the crude template. The song of a deafened bird is even cruder than that produced by the isolated bird because although the deafened bird may have an exact template, it cannot hear its own output, so it cannot compare the song that it sings with the internal template.

The white-crowned sparrow may be unusual in having a sensitive period for memorisation that is over before the bird itself begins to sing. Other birds may show a longer period during which they learn a song (Slater, 1983) Returning to song development in zebra finches on the base of recent studies, it is now more – but not completely – clear what choice young birds make of the tutors from which they copy. While timing is largely restricted to a sensitive phase there is some flexibility in this so that sounds heard earlier or later may sometimes be produced if experience is restricted. Young birds prefer tutors similar in song and appearance to those adult males of which they have earlier experience, but they also agree in their preference for some tutors over others for reasons yet to be determined (Riebel *et al.*, 2005). It now appears that most songbirds come equipped with innately encoded song types which are activated by auditory experience, stored in memory and then winnowed down to a set of songs as a function of social interaction with community members (Marler, 1997; Slater, 2003).

There are some, still surprisingly limited, data indicating that types of development of vocal repertoires are distributed oddly in primates. Firsov (Firsov, 1983; Firsov and Plotnikov, 1981) raised infant chimpanzees under artificial feeding in full isolation from their own species, and clearly demonstrated that they develop species-specific vocalisation containing all the same elements as normally raised animals. Hammerschmidt *et al.* (2001) studied six squirrel monkeys *Saimiri sciureus* for their vocal development over the first 20 months of their life, using a multi-parametric acoustic analysis. Four of the animals were normally raised, one animal (Kaspar Hauser) was deprived of adult species-specific calls, and one animal was congenitally deaf. Both acoustically deprived animals remained within the variability range of the normally raised animals, suggesting that the ontogenetic changes found were mainly maturational. At the same time, Hauser (2000) argues that vervet monkeys learn at least part of types of their calls. Vervet's acoustic signals are considered a part of their 'symbolic language', a topic to which we will return in Parts VIII and IX.

23.2 How intelligence is wired: innate complex patterns or acquired coordinations?

One of the main features of 'instinctive' behaviour is that fixed action patterns are common to all members of the species (species-typical) and therefore they serve as characteristics of the species. We thus may expect that all sticklebacks will perform their zigzag courtship dance in similar ways, all squirrels will gnaw nuts applying similar techniques, and all wolves will stop aggression aimed at them by high-ranking individuals by turning the unprotected neck as a sign of strict obedience. There are, however, individual deviations, and we can start from the three examples of animals behaviour listed above.

Mainardi (1976) observed how several wolves hunted one poor male wolf that definitely demonstrated signs of obedience turning its neck, bending down, wagging and low smiling. All these signals were of no help: his aggressive mates tore him to shreds. Something was wrong with his signs of obedience, and we will never know exactly what. At least, we no more believe in the absolute power of signs of obedience in animal behaviour, although they work in most cases.

In the previous section I cited the results of Eibl-Eibesfeldt's (1961a) experiment with squirrels gnawing nuts. Most of them acquired the most efficient technique. There were, however, displays of individual variance. Some squirrels learned to open nuts by gnawing a hole through a few closely spaced furrows instead of following the path of least resistance. One squirrel achieved almost instant success by gnawing a hole into the base of the nut, and continued to use this technique. This example shows how early experience can influence the display of species-specific behaviour.

Male sticklebacks have to seek a compromise between their aspiration to call to the female with their zigzag dances and the risk of attracting predators' attention. In open places they dance on the verge of danger, and the intensity of zigzags depends on the level of risk (Candolin and Voigt, 2003). Environmental influences thus shift displays of species-specific courtship behaviour, although it is rather conservative.

Of course, it is logically impossible to test whether behaviour would develop the same way in all environments. At the same time, the 'sameness' of behaviour between members of the same species does not exclude the possibility that all members of a particular species share common learning experiences. Eibl-Eibesfeldt (1970) calls learned behavioural sequences *acquired coordinations*, taking into account that they contain fixed action patterns as elements which are combined into the functional whole by learning. Variants of hunting behaviour in mammals and birds can serve as good examples of acquired coordinations.

It is tempting to include all complex species-specific stereotypes into this set. We do however stand a chance of finding a few individuals in a population who display very complex behavioural patterns 'at once and entirely'. Even a single individual of that sort displays clearly that we are dealing with an innate behavioural pattern. Insight thus can be obtained from studies at the population level. We have found it in our laboratory when we first applied the Kaspar Hauser method to studying development of two complex stereotypes in ants, namely, aphid-tending and hunting jumping victims.

Firstly, we examined the question of whether ants learn to milk aphids or whether this very complex stereotype is innate (Reznikova and Novgorodova, 1998). It is known that in many ant species aphids' sweet excretions are one of the main sources of carbohydrates for adult family members. Until recently, nothing was known about the roles of innate and acquired behaviour in ant–aphid interactions. We conducted experiments on three control colonies of red wood ants and three experimental Kaspar Hauser colonies of the same species composed of individuals raised from pupae in separate laboratory nests and deprived of the experience of communication with adult ants as well as with aphids. We timed the behaviour of 230 individually labelled Kaspar Hauser ants throughout their daily cycles of activity until the end of their seasonal activity.

Under natural conditions, the behaviour of an aphid-milker during direct contact with an aphid is stereotypical and specific. An ant strokes an aphid's abdomen with its antennae, which are folded so that their ends are close to the ants' trophi. In this way the ant is begging for a drop of the sweet excretion; the ant immediately catches the drop and puts it into its crop. It is important to note that during trophallaxis, i.e. the exchange of liquid food belched from the crop with other ants, the antennae are folded in similar manner. Trophallaxis is the basic process common not only to ant species but also to all social hymenopterans. Kloft (1960) compared the aphid's abdomen to the head of an ant offering liquid food. The behaviour of the aphids closely resembles that of ants during trophallaxis and apparently triggers the same behavioural stereotype in naive aphid-milkers. However, the behaviour of an ant eating carbohydrate liquid food from an open trough or encountering various objects is distinctly different. In this case the ant feels the object with extended, almost straight antennae, with the frequency of tapping reflecting the degree of the ant's interest in the object; however, the position of the antennae themselves does not change.

In the deprivation (Kaspar Hauser) experiment, when ants encountered aphids for the first time, they perceived the aphids as any other unknown object; the ants felt the aphids with extended antennae and did not remain near them for long. An ant behaved in this manner until it accidentally touched a drop of an aphid's excretion and had to test it when cleaning its antennae or legs. After this, the ant's behaviour substantially changed: instead of tapping at the aphids the ant began to stroke them with folded antennae, thus asking for the sweet excretion. This change was gradual. At first, the ant only slightly folded the antennae, so that they tapped the aphid's sides (in normal contact, the ants pats the aphid's back); the movements of the antennae were uncoordinated. We observed this stage in the behaviour of all 'beginners'. After successful contact with the first aphid, an ant began to perceive other aphids. The ant stopped them and tried to milk them, lengthening the contact until a drop of excretion emerged.

Fig. 23.1 Characteristic aphid-milking postures of red wood ants. (a) An experienced ant (bent antennae, ready to catch the dew drop); (b, c) a naive ant (awkward movements, widespread antennae). (Drawings by T. Novgorodova.)

At this stage, the movements of antennae were more coordinated; however, the ant was usually unable to catch the drop in time, and hence, was compelled to clean itself continually (Fig. 23.1). The development of milking behaviour, including the stage of asking for and waiting for the drop of excretion, was accomplish 60–90 minutes after the ants had faced aphids for the first time. The behaviour of naive ants during their subsequent contacts with aphids did not differ from the behaviour of control, 'wild' ants.

We think that, in this case, innate recognition of the objects of the species-specific instinctive behaviour occurred. When the ants perceived the stimuli that came from the aphids, innate recognition was complemented with acquired reactions, and all the behavioural elements formed an integrate behaviour. Apparently, social facilitation (see Part VIII) was also involved in this process. Those ants that were the first to appear in the aphid colony took considerably more time to develop the behaviour. In sum, direct interaction with aphids in ants consists of short innate behavioural sequences which are likely to be the same as those involved into the process of trophallaxis, one of the basic processes integrating a family of social hymenopterans into the whole. Whereas the process of aphid-milking seems to be completely hard-wired, the division of labour in groups of aphid-milkers appears to be determined by social experience.

It was difficult for us to believe that such a complex behavioural pattern as aphid-milking could be innate in ants. Further, it was very likely that another complex behaviour, hunting for jumping springtails, is learned in those ant species which encounter springtails very often (a good chance to eat them) but lack special morphological structures (snap-on mandibles) intended to facilitate catching springtails in highly specialised tropical dacetine ants. As was briefly noted in Chapter 22, we carried out many experiments with 'wild' and naive families of *Myrmica rubra*, abundant inhabitants of soil and litter in the forest and steppe-forest zones. To observe the interaction of the ants with active prey, we put live jumping springtails (*Tomocerus sibiricus*) into glass containers (6 cm in diameter and 12 cm in height, 30 animals per container). The containers had a gypsum bottom covered with splinters made out of plastic bottles (this substrate mimicked forest litter but was transparent). Ants were put into containers

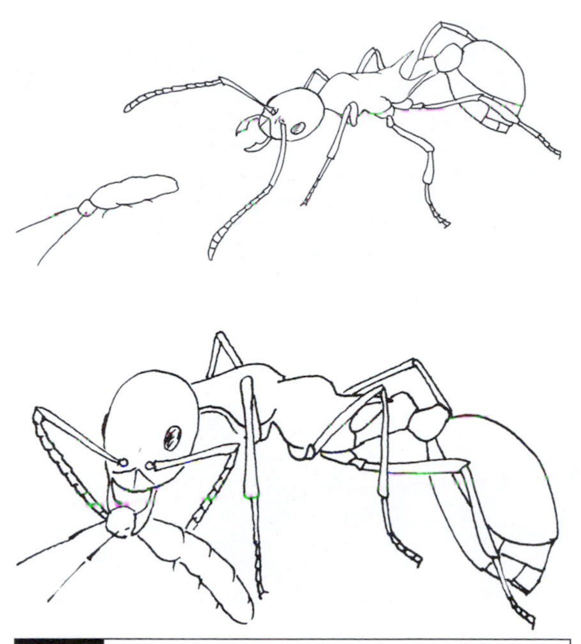

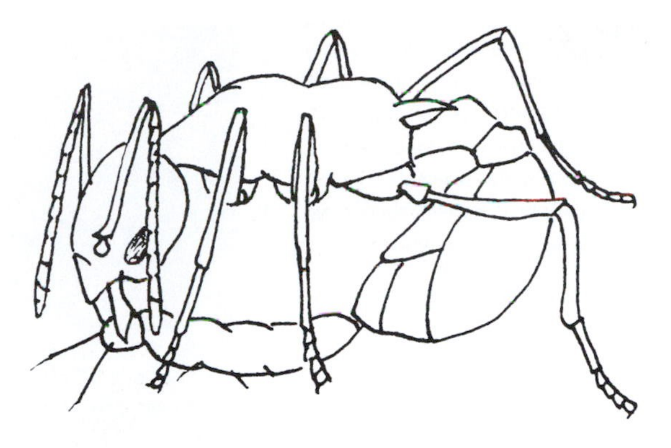

Fig. 23.2 The sequence of actions of a *Myrmica* ant when it hunts a jumping springtail. (Drawings by S. Panteleeva.)

individually, by transferring them with a small brush. Their responses to the prey were timed and video recorded; the records were then viewed in the slow-motion mode, so that individual frames could be fixed and drawn in detail (Reznikova and Panteleeva, 2001). The members of the control families caught the jumping springtails quite effectively. For instance, in one series of experiments, 116 of 214 tests ended in catching the prey; in the remaining tests, ants also responded to the springtails aggressively. Successful 'springtail hunters' demonstrated searching and sufficiently specific behaviour. These ants moved relatively fast and freely through the bulky artificial litter. Once the ant found itself in the immediate proximity to a springtail, it attacked the prey: it bent its abdomen and head to the thorax, jumped to the springtail and fell on it in the similar manner to the way a fox falls on a mouse (Fig. 23.2). At the final stage of this behaviour sequence the ant differs sufficiently from a fox by setting its sting in motion. However, naive individuals behaved rather differently. One Kaspar Hauser ant after another, being at the age of full maturity, behaved towards a potential victim in the same friendly manner as if it was its nestmate. We observed numerous antennal contacts of ants with springtails; i.e. ants responded to springtails as to conspecific animals rather than potential prey. If the springtail made occasional jerks, the ant jumped aside. As our laboratory and field observations have demonstrated, it took from several weeks to several months of practice and trial and error to build up the character of a successful springtail hunter. At this stage of our study we believed that individual and social experience strongly dominated the development of springtail hunting in *Myrmica* ants; moreover, this species lacks specialised morphological adaptations for catching jumping victims. This suggestion agrees with the fact that under natural condition the higher population density of potential victims was fixed in the ants' feeding territory, the greater numbers of successful hunters were encountered within the ants' families.

We nevertheless did not stop our efforts to examine naive *Myrmica* ants in order to reveal more details of the scenario of hunting behaviour (Reznikova and Panteleeva, 2005). At last, in one of experimental families, we found seven out of 123 naive ants that were able to catch the prey, and all of them exhibited an 'at once and entirely' fixed action pattern during the final act of the hunt and had no noticeable differences from the adults (Fig. 23.3). One of them caught the prey at an age of 7 days and the others at an age of 14 days. In contrast to the control ants, however, they stayed with their prey in the laboratory arena instead of transporting it to the nest. If we transferred them, together with their

Fig. 23.3 Experiments for studying the development of hunting behaviour in *Myrmica* ants. (a) an ant pursuing a springtail; (b) an ant seizing a springtail; (c, d) naive ants treating springtails in a peaceful manner, as nestmates. (Photographs by S. Panteleeva.)

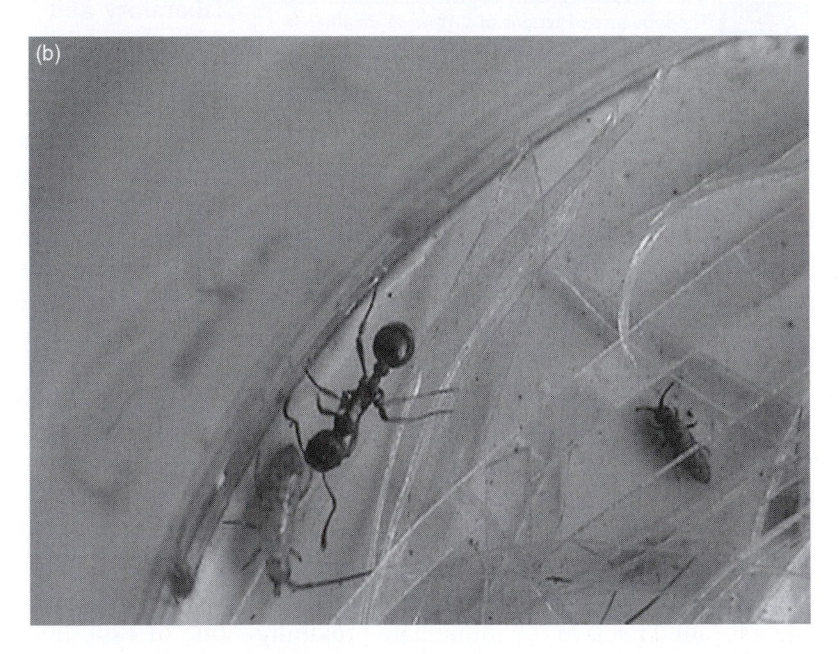

prey, to the nest by means of a brush, they left the springtail near the brood. Having found the killed prey, the members of the naive family that had stayed in the nest carried it to the remote part of the nest, far away from the brood, and did not use it to feed the larvae. Our observations showed that the broods of the naive families fed on fodder eggs laid by adult worker ants. Thus, the hunting in the young families 'ran idle', i.e. the prey was not used for its intended purpose.

The 'at once and entirely' fixed action pattern of ants hunting for a jumping springtail, which is so difficult to catch, indicates that the specific stereotype of hunting behaviour may be expressed as an integrated set of behavioural sequences. However, the expression of this stereotype is variable within a family. Only in a

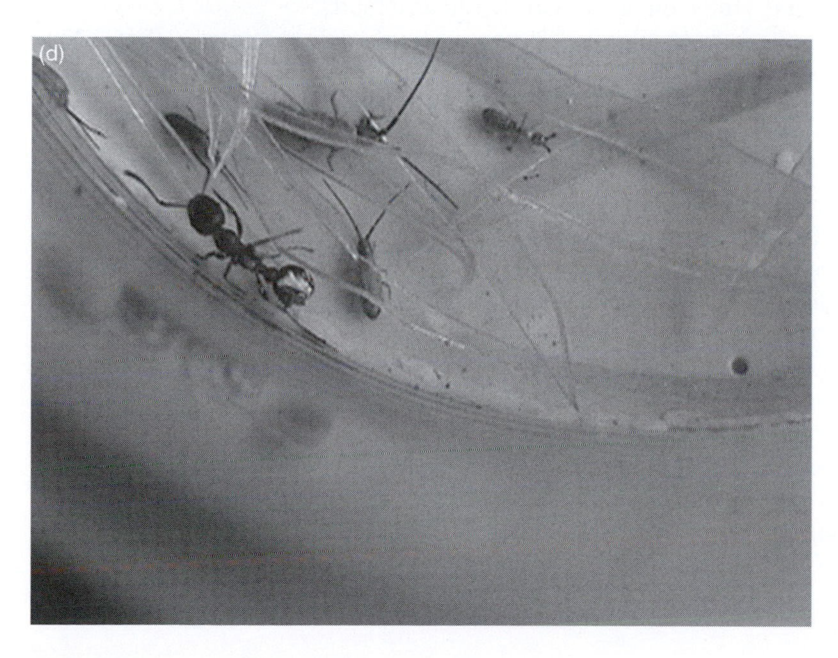

Fig. 23.3 (cont.)

small proportion of ants (fewer than 10 per cent) was hunting behaviour expressed at an early imaginal stage. For comparison, recall that the formation of such a complex behavioural pattern as begging for honeydew from symbiotic aphids was observed in all naive ants within 60–90 minutes after the first contact with a drop of honeydew. In contrast, the scenario of the formation of interaction with the difficult-to-handle prey requires the multi-stage maturation and completion of the species-specific fixed action pattern, which probably includes elements of social learning (see Part VIII). This example enables us to examine complex behavioural patterns in many members of species investigated because apparent learning by trial and error can hardly be distinguished from maturation, and a single exemplar possessing the 'at once and entirely' spontaneous behaviour sequence can argue for 'instinctive' behaviour.

Very similar results to ours on hunting *Myrmica* were obtained in the study on tool use in New Caledonian crows, indicating that one individual can change the experimenter's mind from recognising individual experience to maturation as the main factor guiding complex and apparently intelligent behaviour. We have already described New Caledonian crows in Part VI as the most prolific avian tool-users (see Hunt, 1996; Kenward *et al.*, 2005). Students of tool-using in this species have elaborated a hypothesis about cumulative cultural evolution (Hunt and Gray, 2003; Kacelnik *et al.*, 2004). However, a recent developmental study on a group of four hand-raised juvenile individuals (Kenward *et al.*, 2005) showed that one individual spontaneously manufactured and used tools in sophisticated manner, without any contact with adults of its species.

Experimenters raised chicks in artificial nests and subsequently transferred them to enriched aviaries that contained twigs of assorted shapes and sizes, and food items hidden in holes and crevices. None of the subjects was ever allowed to observe an adult crow. Two of them were housed together and were given regular demonstrations by their human foster parents of how to use twig tools to retrieve food. The other two were housed individually and never witnessed tool use. All four crows developed the ability to use twig tools. Although the tutored crows paid close attention to demonstrations, there was no qualitative difference between them and the untutored birds in their tool-oriented behaviour. Experimenters first observed successful food retrieval from a hole by the tutored birds when they were 68 and 72 days old and by the untutored birds at 63 and 79 days old.

The authors also tested the juvenile's response to leaves from trees from genus *Pandanus*, similar to those from which wild crows make tools. The leaves were mounted on wooden frames so that the birds could access them roughly as they would in the wild. On the first day that he was presented with *Pandanus*, the untutored individually housed male Corbeau (then aged 99 days) produced a straight tool from one side of the leaf by using a swift 'cut–tear–cut' action. Immediately after producing the tool, Corbeau carried it to a crevice where food was often hidden and used it as a probe, a sequence that he has since repeated several times to successfully retrieve food.

This finding demonstrates that New Caledonian crows have inherited characteristics that support tool-making and use and that possibly guide fast learning processes enabling birds to solve their living problems by applying tools. The fact that inherited predisposition can account for a complex behaviour such as tool manufacture highlights the need for controlled developmental investigations in those species that seemingly show intelligent and especially culturally transmitted behaviour. Social learning however may play a significant role in transmitting specific techniques within populations even if it only facilitates maturation of innate behavioural patterns. We will consider this aspect of learning in Part VIII.

Chapter 24

Imprinting

Imprinting, in general sense, is the phenomenon by which many animals form special attachments to objects to which they are exposed during specific, sensitive (and 'critical') periods of their life.

Within the whole spectrum of combinations between innate and learned behaviour imprinting provides an amazing and very curious example of genetic and environmental influence on animal behaviour. Lorenz (1935) used the noun *Prägung*, or *imprinting*, to refer to the process of rapid bond formation early in the life of precocial birds (ducks, geese and the like). Lorenz himself 'imprinted' this idea from his teacher Oskar Heinroth, a German zoologist and ethologist. In his paper (1911) Heinroth described the behaviour of incubator-hatched greylag goslings. These newly hatched birds showed no fear, and attached themselves readily to humans. Such human-attached goslings do not show any inclination to approach and stay with parent-geese; they behave as if people were their parents. This kind of attachment behaviour was described by Heinroth by the verb *einzuprägen*, corresponding to the English 'to stamp in' or 'to imprint'; and the phrase 'stamped in' had earlier been used by Spalding (1873), and by Thorndike (1898) in relation to firmly acquired modes of behaviour. Further imprinting has been found in many species, from ants to humans, and it appears to include reactions not only for visual but also for acoustic and olfactory stimuli.

Indeed, the first scientific description of the phenomenon of imprinting was done by Spalding (1873). He discovered that newly hatched chicks followed almost any moving figure. Spalding regarded such behaviour as 'unacquired', that is, instinctive rather than learned. He then concluded that early learning takes over at some stage from instinct; chicks which at first follow instinctively then learn who their mother or mother-substitute is, and are eventually able to discriminate between her and other figures. Spalding drew attention to a remarkable feature of such early behaviour, namely that a chick would follow its mother only if it had the opportunity to do so early enough in life. If faced with its mother for the very first time after the opportune or *sensitive period* has passed, the chick would fail to follow her and would show no affinity whatever to her; furthermore, it would not subsequently be able to develop any attachment to its mother. We now say that a chick becomes imprinted to the mother-figure when it learns its characteristics, and forms a tie to it. Spalding reported that this development was confined to a short period soon after hatching; and this has since become known, with some degree of justification, as the critical period for imprinting.

Lorenz went further than Spalding in that he specified the characteristics of imprinting that differed fundamentally from what he called 'ordinary learning', and thereby generated much interest in it, which, in turn, has resulted in further systematic observations and experimentation in this area of animal behaviour. Precocial chickens turned out to be ideal subject for studying impinting. Lorenz recalled in his Nobel Lecture:

> Selma Lagerlöf's *Nils Holgersson* was read to me – I could not yet read at that time. From then on, I yearned to

become a wild goose and, on realising that this was impossible, I desperately wanted to have one and, when this also proved impossible, I settled for having domestic ducks. In the process of getting some, I discovered imprinting and was imprinted myself. From a neighbour, I got a one day old duckling and found, to my intense joy, that it transferred its following response to my person. At the same time my interest became irreversibly fixated on water fowl, and I became an expert on their behaviour even as a child.

Lorenz started his systematic study on imprinting in the middle of the 1930s. From the 1950s to the 1970s many experiments were conducted on imprinting (Guiton, 1959; Gottlieb, 1961, 1971; Bateson, 1966, 1979; Landsberg and Weiss, 1976). Hess (1959) suggested an experimental set-up that allowed the observation and recording of behavioural patterns of ducklings and goslings following different things that imitated their mothers such as balls, boxes and dummy birds of different sizes and colours.

The main and most widely cited of Lorenz's experiments with gosling was the following. Lorenz divided a nest of goose eggs, leaving half with the mother and putting the rest in an incubator. The geese raised by their biological mother showed normal behaviour, following her around during their youth and growing up to interact with other geese. When the incubated eggs hatched, the goslings spent their first few hours with Lorenz instead of their mother. From then on, these goslings followed Lorenz around, showing no recognition of their own mother or even adults of their own species. As adults, these geese continued to prefer humans over members of their own species. Lorenz's experiment showed that geese have no instinct telling them who their mother is, or who is a member of their species. Instead, they respond to and identify with the first moving object they encounter.

Lorenz and his followers marked out several specific features of imprinting. Firstly, imprinting could take place only during a brief critical period in the individual's life; and secondly, once it had taken place, it could not be reversed. This feature dramatically differs from associative learning which is highly volatile, usually fading with time. Furthermore, imprinting was reported to occur very rapidly, without any trial and error; and, above all, imprinting would show itself, at maturity, in a courtship directed towards the original mother-figure or figures similar to her. One more amazing characteristic of imprinting is that stressful stimuli fortify the 'stamping in'. If there is an increased level of stress at the time of the original imprinting, the learning paradoxically becomes faster. Hess (1964) demonstrated that if in the laboratory set-up obstacles are placed in the runway between the duckling and the followed object then the response the duckling subsequently exhibits is more determined and energetic. It may be that this enhances an individual duck family's level of imprinting at times of greatest need, for instance when the threat of predators or the distraction of other broods is a particular problem.

It was later questioned whether these features would separate imprinting sharply from other forms of learning. Indeed, Lorenz himself, some 20 years after the appearance of his early papers, expressed the view that imprinting might be a type of conditioning. Whether it is, or is not, would depend partly on how narrowly or broadly conditioning is defined.

Several forms of imprinting are distinguished. *Filial/maternal imprinting* is a kind of imprinting in which the offspring will follow its mother, or respond to the mother's calls. The same term is also used with respect to mothers that give birth to offspring: maternal imprinting occurs during a brief period after parturition in which mothers learn to recognise the voice and/or odour of their young.

Filial and maternal imprinting is typical for precocial birds and mammals (that is, those born in a relatively mature state). It also may well be – but this is still somewhat controversial – that imprinting also occurs in altricial species (that is, those that are rather immature when born). Altricial neonates are unlikely or unable to stray from their home base in the first few days of life and therefore do not need the same response as precocial youngsters. They learn similar lessons rather later in life during what are called 'socialisation periods'. These apply when the animal's sensory, motor and thermoregulatory systems are fully functional and they learn to move away from their mother and to interact with others of the same and other species. The window of opportunity for learning varies in different species. In dogs it is from 3 to 10 weeks and in cats from 2 to 7 weeks, while in primates it is usually 6–12 months. Stimuli to which the youngsters of each species are exposed during these window periods will be accepted as 'normal' (Immelman, 1972).

Most of the work on imprinting and in particular on filial imprinting has been done with birds because they are convenient research animals. They will imprint on practically anything – balloons, boxes, or even the cover under which an experimenter is hiding so as not to influence the behaviour of his charges (see Wallace, 1979). Colour and shape do not seem to matter, but the imprinting takes place faster if the object contrasts with the background and if it makes some sort of noise. Klopfer (1959) found that young wood ducks actually imprint on sounds rather than on visual stimuli. In an experimental situation, a young wood duck will approach any sound, but later it will follow only familiar sounds. In nature these birds nest in dark holes, so that during the imprinting period they may never even see their mother.

In filial imprinting, once the young individual has formed an attachment to a particular object it avoids novel objects. There are conflicting pressures on the young to readily recognise and to follow its parent but also to recognise and to avoid other adults, as well as heterospecifics that are potential predators (Hinde, 1970). Such conflicting pressures in filial imprinting resemble those involved in sexual imprinting and mate selection, when it is advantageous to distinguish conspecifics from heterospecifics (Irwin and Price, 1999). We will see further that although filial imprinting is separable from sexual imprinting, the processes are similar in many ways.

It is possible that the human infants are tied to their mother by imprinting-like connections. Lorenz (1935) drew attention to certain analogies in human behaviour to the occurrence of imprinting to, or with, inappropriate objects; he had in mind human ways of acting which, as he put it, 'appear in the form of pathological fixations on the object of an instinct'. Based on the ethological concept of imprinting the British psychiatrist Bowlby (1982) and the American developmental psychologist Ainsworth (1982) elaborated the attachment theory of parent–child and other close relationships. According to this theory, an *attachment* is a strong affectional tie that binds a person to an intimate companion. Human researchers have found many analogies between behaviours in human infants and infants of other primate species at early stages of establishing intimate connection with their mothers.

The dramatic effect of early experience and consequences of disturbance of imprinting-like behaviour in primates were first revealed in the investigations of H. F. and M. K. Harlow (1962). They raised rhesus monkeys without their mothers. The animals had only surrogate mothers, which were either covered with towelling or consisted of bare wire mesh. Attached to these models were bottles with nipples, from which the baby monkeys could suck their milk. The monkeys that were raised in this way later proved to be poor mothers. They did not nurse their young or only did so after some time, and even mistreated them, being rejecting or aggressive. Here an early childhood experience led to substantial disruptions of later social behaviour. In human beings this phenomenon is called *hospitalism* (Spitz, 1945). Emphasising this dramatic role of early experience, Harlow (1971) called his book about these experiments *Learning to Love*. It turned out a little bit later that there was no need to separate infant monkeys from their mothers for so long in order to observe displays of hospitalism in them. Hinde (1974) carried out much less traumatic experiments with the same species. He allowed mothers and their infants to develop normal bonds and only removed a mother from a group for several days. The young was not left alone, instead, it could enjoy attention from other females in the group. Infants, however, survived with difficulty, and the more so the less harmonious were their previous relationships with their mothers, as if they possessed less 'margin of safety'. Hinde was able to distinguish between normal monkeys and those that had been separated from their mothers in their childhood even after several years.

Recent studies on wild elephants in Africa revealed displays of hospitalism in adolescents caused by seemingly weak changes in their social bonds. Being isolated from their supportive older caregivers ('allomothers') because of fragmentation of their social environment by humans, young elephants display symptoms associated with human post-traumatic stress syndrome: abnormal startle response, depression, unpredictable asocial behaviour and hyperagression (Bradshaw *et al.*, 2005).

Sexual/social imprinting is a kind of imprinting in which a young offspring imprints on members of its own sex, so that when the offspring mature, they prefer their own species. Sexual imprinting arises as a consequence of learning about individuals and can create mate preferences within species (Bateson, 1966). As we just have seen, sexual and filial forms of imprinting are closely connected and determine in many respects the future of adult's life. For instance, in duck species in which the sexes have different appearances only the male ducklings imprint sexually on the mother. In these species, male ducklings exposed to adult males during their imprinting period will later form homosexual bonds, even in the presence of females (Schutz, 1965).

Lorenz (1937) related a story about a male bittern which was raised by a zoo-keeper. Although the bittern was maintained with a female of its own species and eventually paired with it, the mis-imprinted male would drive the female away whenever the zoo-keeper approached and try to get the keeper to come into the nest to incubate the eggs. Thorpe (1956) reported the case of a gander's alleged seven-year fixation to an oil-drum. Subsequent controlled experiments have confirmed the power of sexual imprinting. Cross-fostering experiments with sexually imprinting species have succeeded in creating preferences of one species for another. For instance, pigeons raised by doves will later prefer to court other doves, rather than members of their own species (Sluckin, 1965).

It is worth noting that some traits are affected by sexual imprinting more than others. Lorenz noted with some amusement that jackdaws that had imprinted on him would court his favour by presenting him with juicy fresh earthworms and would even attempt to introduce these into his earholes. However, when not sexually aroused, these birds would happily join other jackdaws in flight. In sexually dimorphic species (in which the external appearance of males and females differ), sexual imprinting varies depending on whether the youngster is a male or a female. While a male mallard duckling will identify his future mate by relating it to the appearance of his mother (or attachment figure), the same does not apply to a female. For falcons imprinted on humans a combination of human and avian stimuli are required in order to elicit sexual responses.

Sexual imprinting is one of the several known non-genetic, yet social factors which influence mate preference and can facilitate the formation of new species (Bateson, 1978, 1983; Todd and Miller, 1991; ten Cate and Vos, 1999). Whether sexual imprinting can support the evolution of novel traits is still under debate (Hörster *et al.*, 2000). Laland (1994) designed a model which showed that when there is an asymmetrical mate preference sexual imprinting can support the evolution of novel traits. There is experimental evidence that sexual imprinting on both natural and artificial novel traits takes place. Ten Cate and Bateson (1989) showed that in Japanese quail *Coturnix coturnix japonica* offspring prefer mates which differ slightly in plumage characteristics from their parents. Experiments with artificially changing traits demonstrated that Javanese mannikin *Lonchura leucogastroides* males and females became sexually imprinted on a red feather on the forehead as a novel trait in parent birds. Those females imprinted on the red feather showed a strong preference for another novel red trait, red stripes at the tail, similar to a male with the red feather. Females transferred the preference for a red feather to the other novel red trait (Plenge *et al.*, 2000; Witte *et al.*, 2000). Further studies demonstrated that not all kinds of red ornaments are suitable for birds to be imprinted on, and that a red bill is too much. Males and females raised by a red-billed father showed even a strong rejection to conspecifics of the opposite sex with a red bill (Hörster *et al.*, 2000). Seemingly, in other experiments females showed no preference for males with novel blue traits (Plenge *et al.*, 2000). Thus, not all kinds of novel traits can be sexually imprinted on birds.

The formation of social attachment, that is, social imprinting, usually requires the presence of conspecifics at an early age. For instance, dogs pass through a critical period for the development of social relationships at age 4–6 weeks (Scott, 1992). During this time they form a close social bond with conspecifics or with a human as a substitute, regardless of whether they are punished, fed, or treated indifferently. Scott and Fuller (1965) have emphasised that the internal processes on which this readiness for contact is based seems to be more important than the external factors.

There is a growing body of evidence that domestic animals include humans into their social environment and thus treat them as conspecifics (Broom, 1981). I was impressed by one slide from Altbäcker (2004) presentation at the 2nd European Conference on Behavioural Biology which contained a notice 'Beware of rabbits' together with a photo of a furious male rabbit trying to drive a human away from its territory. Rabbits are highly territorial animals. In Altbäcker's experiments fear of humans has been eliminated in rabbits by exposing infant rabbits to the smell of humans at early critical periods. Early experience interfering with conspecific recognition resulted in rabbits fearless of humans. They gamely defended the boundaries of their territories from 'giant rabbits', as they considered human intruders.

Knowledge about imprinting has practical applications in farming, veterinary and breeding practice. Chinese peasants have for centuries capitalised on the tendency to imprint in making ducks more effective in controlling snails that otherwise damage rice crops. By imprinting ducklings onto a special stick, the peasants can not only take their brood out to the paddy fields as required but, by planting the stick sequentially in different parts of the plantation, they can ensure that molluscs in all areas are imprinted for predation. Hawks are also subjects of sexual imprinting: falconers take advantage of this by wearing special hats to collect the sperm of their amorous human-imprinted charges for use in breeding programmes. Some stud-farms are practising imprinting of foals at birth to acquaint them early with such things as halters, clippers and trailers. Exposing a foal to these objects and practices within the first 45 minutes of its life demonstrates to the foal that certain sights and sounds offer no danger. Here breeders are guided by the ethological principles of 'imprint training' suggested by Miller (2004). However, experimental investigations showed that foals hardly changed their behaviour against their current behavioural stereotypes. Early handling that began 2 hours (Williams *et al.*, 2002) and 5–7 hours (Lansade *et al.*, 2005) after birth appeared to have only a short-term effect on the foal's reactivity and manageability. It appeared to be easier to shape the foal's behaviour toward humans through the human's positive interactions with the dam. It is likely that observations of its mother by the foal during the early stages of its life facilitate the human–foal relationship (Henry *et al.*, 2005).

Filial and sexual imprinting forms a basis for bond formation as well as for species and sex recognition. It may be easier to appreciate the role of imprinting in animals' self-determination if we consider species that do not necessarily meet conspecifics in their development. Interspecific brood parasites (such as some cuckoos and brown-headed cowbirds), and megapodes are good examples. For brood parasites, several mechanisms of sexual and social self-determination have been hypothesised. It was suggested that the timing of imprinting is delayed until the young leave the nest and interact with conspecifics, that innate recognition of a conspecific's features is more important than imprinting, and that both genetic factors and learning are involved in the development of social attachments (for review see Göth and Jones, 2003). As is known, brood parasites are mostly altricial and are initially confined to the nest where they cannot actively seek contact with conspecifics. Conversely, a megapode chick, being a typical representative of a precocial bird, might possess imprinting. However, it seems to be too much advanced. As Göth and co-authors (Göth and Jones, 2003; Göth and Evans, 2004) have revealed by observations of incubated Australian brush turkeys *Alectura lathami*, these 'packing' chicks show 'social behaviour without social experience' and this requires no postnatal learning. As we have seen earlier (in Chapter 21), megapode chicks hatch underground in mounds of leaf litter, where external heat sources incubate the eggs, and then they dig their way to the surface to set themselves on the road to completely independent life. Chicks hatch asynchronously and receive no parental care, so imprinting cannot occur (Göth and Evans, 2004). Wong (1999) found that megapode chicks do not imprint on moving objects after hatching like chicks of other galliforms do. Nor do they have a tendency to aggregate with other chicks. They are capable of finding adequate food alone and of detecting predators innately, and these features should enable them to survive without assistance from others (Göth, 2001).

Whereas young megapodes display independence from an early age, young ants aspire to tight social bonding with their nestmates. As do many other social animals, ants possess social imprinting and this enables them to occupy an appropriate place in their society. The most striking example of imprinting in ants is the learning of colony odour (see Hölldobler and Wilson (1990) for a review). The period of greatest sensitivity is usually within first few days after eclosion of the young from the pupa, although in some species the learning can begin during the larval stage. There is some evidence that when no contact with nestmates is permitted during the sensitive period, later social behaviour can be seriously impaired.

All these examples show imprinting as a very specific innate form of learning which is likely to be under hard pressure of natural selection in many species. To put studying of imprinting in the context of animal intelligence, it should be noted that in some cases imprinting appears to be an initial stage of a long-lasting specific learning rather than a brief restricted event in an animal's life.

For example, Charrier *et al.*'s (2003) experimental field study of mutual recognition in mothers and pups of fur seals *Arctocephalus tropicalis* has revealed surprisingly long-term recognition. In pinnipeds, and especially in otariids, mothers and pups stay within close range during about a year and they develop a strong capacity to recognise each other's voices. Pups become able to discriminate their mother's voice a few days after birth. For females this discrimination seems to occur earlier, probably during the few hours after parturition. However, during lactation mothers are confronted with a major problem: the change of characteristics of their pup's call caused by the maturation of their brains and vocal tracts. During the lactation period mother seals have to alternate suckling periods with foraging trips at sea and periodically leave their pups alone in the colony for 2–3 weeks. The mother–pup vocal recognition system is particularly effective, since, in spite of the high risk of confusion due to the great density of individuals in the colony, mother–pup pairs have been shown to meet up in less than 11 minutes when the mothers return from a foraging trip at sea. Researchers carried out playback experiments at the end of the rearing periods to test whether mothers could recognise the changing voices of their pups. They presented females with recorded signals of their own pups at different ages, as well of other pups. The tests were carried out when mothers and pups were separated in the colony. Playback experiments demonstrated that females remembered and still recognised all the successive immature calls of their own pups and easily distinguished them from other pups' calls. Thus, mothers must be capable of permanently learning their pup's vocal characteristics. In the northern fur seal *Callorhinus ursinus*, another otariid species with a similar pattern of maternal attendance, mothers and pups are able to mutually recognise their calls for at least 4 years (Insley, 2000). This may be redundant regarding adaptive behaviour, and the initial impulse of long-lasting memorisation is likely to be originated by imprinting.

Recent studies have brought many new facts and ideas concerning imprinting, and at the same time they have made the boundaries of the concept of imprinting fuzzier and raised new questions and problems. There is compelling evidence that the social bond that develops through imprinting entails an 'addictive' process that is mediated by the release of endorphins, the brain's own opiates (Hoffman, 1996).

Bateson (2000; see also Bateson and Martin, 2000) has formulated a model in which imprinting is not an instantaneous and irreversible process but a much more flexible and less peculiar phenomenon. Imprinting does not necessarily occur immediately after birth but has a more flexible sensitive period affected by both experience and species-specific features. Imprinting is not a monolithic capability but is composed by several linked processes: (1) detection of a relevant stimulus guided by a predisposition to what the animal will find attractive; (2) recognition of what is familiar and what is novel in that stimulus, which involves a comparison between what has already been experienced and the current input; (3) control of the motor patterns involved in imprinting behaviour.

Although imprinting can be functionally distinguished from learning involving an external reward, both types of learning are deeply connected, as suggested by the possibility of transfer of training after imprinting.

In both filial and sexual imprinting animals can make associations between multiple traits that distinguish individuals or species. It is still an open question as to what traits are really learned and why some traits are not learned during imprinting (Witte *et al.*, 2000; Hörster *et al.*, 2000). It is possible that characters of species-specific predisposition to establish strong attachment links based on attractive traits act like 'filters' so that one trait loses attractiveness and others take on special significance during certain periods of the animals' life. Not all 'filters' are compatible with evolutionarily established tendencies so some traits have more power than others to be imprinted on, and some traits can not be imprinted at all.

CONCLUDING COMMENTS

Ecological and evolutionary aspects of animal intelligence are promising and fascinating topics of research. It is a challenging problem to understand the specificity and limits of wild minds. Classic ethologists considered 'nature versus nurture' actually a false dilemma: in almost all animals behaviours there is a mixture of both. Indeed, simple 'nature/nurture' or 'instinctive/learned' dichotomies have now been abandoned, and attention is now focused on experimental investigation of what does and what does not influence behavioural development. Modern researchers are interested in how genetic and environmental factors interact.

There are, however, new problems in this field raised by recent experimental studies. We have already found a great many examples where members of many different species develop extraordinary abilities to solve their problems on the basis of very specific stereotyped behaviour that can be shown to be innate. Does this mean that cognition in animals can be imagined as a thin layer of gloss on inherited species-specific stereotypes? Or do animals enjoy flexible and creative cognitive skills? Or should we seek a compromise between these alternatives?

Summarising the data concerning innate and learned rules that establish strict boundaries for flexibility in animal behaviour, we can conclude that even those signs of behaviour that are considered to be unique characteristics of instincts in classic ethology, may be attributable to learning on closer examination. Instinctive acts are often triggered by simple key stimuli but we know now that animals can learn to pick out some simple stimuli and ignore others in a changeable environment. Hard-wired stereotypic behavioural patterns similar in all members of a species can actually be learned stereotypes just because the animals pick out the same simple stimuli and learn the same. Indeed, this is not the whole truth about the relative importance of genetics versus environment in animal behaviour.

Here I am not trying to analyse the balance between nature and nurture in animal behaviour in general. Instead, I only concentrate on the problem of the development of complex behavioural patterns. Analysing different aspects of the findings described above we can conclude that complex behaviour can develop on the basis of specific behavioural domains. Within these frameworks animals demonstrate sophisticated mental abilities and high flexibility of memory. In some cases a compromise between advanced intelligence and innateness can be found if we consider behaviour within the framework of the paradigm of 'adaptation learning' (Reznikova and Dorosheva, 2005). That is, in order to solve vital problems animals are likely to choose more and more quickly and effectively a relevant strategy from a set of strategies that already exists rather than to contrive something completely new as an answer for a challenge from their environment.

Being an advocate of animal intelligence, I nevertheless would like to emphasise that inherited predisposition can account for different complex behaviours such as food-hoarding, hunting, tool manufacture, communication, and even Machiavellian interaction with conspecifics. It seems that there are only narrow paths for the display of flexibility in animal behaviour. To clarify the question of interactions between inherited traits and individual and social learning it is necessary to consider social interrelations and 'language' behaviour. The question of flexibility of communication systems seems to be crucial in consideration of animal intelligence. We will discuss these problems further in this book.

Part VIII

Wisdom through social learning

Several Horses and Mares of Quality in the Neighbourhood came often to our House upon the Report spread of a wonderful Yahoo that could speak like a Houyhnhnm, and seemed in his Words and Actions to discover some Glimmerings of Reason. These delighted to converse with me; they put many Questions, and received such Answers, as I was able to return. By all these Advantages, I made so great a Progress, that in five Months from my arrival, I understood whatever was spoken, and could express myself tolerably well.

Jonathan Swift, *Gulliver's Travels: The Wisdom of the Houyhnhnms*

Members of many species spend a great deal of their time in the company of conspecifics. Animals can assimilate essential information by observing their companions, that is, when, where and what to eat, with whom to mate, whom to fear, and how to spend spare time if there is some. In principle, as we have seen from Part VII, all information can be picked up from internal resources by the development of an inherited programme. However, social learning and communication give animals great possibilities to improve the adaptability and flexibility of their behaviour in conformity with concrete and changeable vital circumstances. In many natural situations the boundaries between flexibility and conservatism are rather fuzzy. Social learning can sometimes generate behavioural traditions, and some of these traditions can be paradoxically conservative and thus hardly distinguishable in appearance from innate forms of behaviour. If we want to know what part of a whole repertoire falls to the share of social learning, we definitely can not gain this knowledge in the mind's eye; instead, we must conduct developmental studies and carry out special experiments.

Previous parts of the book were devoted to investigations of the role of individual experience and inherited behavioural programmes in animal life. In this Part we turn to the role of social learning. In modern ethology and comparative psychology the study of social learning is a specific and rapidly developing direction with its own notions, definitions and hypotheses. We will consider different forms of social learning, from relatively simple ones such as social facilitation, to the most complex such as tutoring and maintaining traditions in animal societies. In general, we will develop a concept of how animals acquire information and skills from other individuals by means of observations on their behaviour.

Chapter 25

Ecological and cognitive aspects of social learning

Social learning is said to occur when direct or indirect social interaction facilitates the acquisition of a novel pattern of behaviour. It usually takes the form of an experienced animal (the demonstrator) performing behaviour such that a naive animal (the observer) subsequently expresses the same novel behaviour sooner or more completely that it would have done using individual learning. Social learning in modern reviews (see, for example, Zentall and Akins, 2001; Caldwell and Whiten, 2002) refers to any situation in which the behaviour, or presence, or the products of the behaviour of one individual influence the learning of another.

25.1 Different forms of social learning: brief description and definitions

Many authors have differentiated several possible types of social learning, in terms of the mechanisms potentially involved. One of the most complex manifestations of social learning is *imitation learning*, or *observational learning* (Hall, 1963). Thorndike (1898) defined imitation as 'learning to do an act by seeing it done'. Based on his studies of insightful behaviour of animals, Köhler (1925) suggested that imitation demands elements of consciousness and understanding of the fact that a modelling subject possesses similar features to the learner.

The ongoing nature–nurture debate has generated the discussion about whether animals can be credited with the ability to copy their companions' behaviour and whether this ability can be compared to the superior capacity for imitation in humans. Cognitive ethologists have focused on imitative learning because it seems likely to be implicated in the origin of human culture. In animals imitative learning can generate behavioural traditions or proto-culture. Imitation can be considered as the opposite to instinctive behaviour. However, seeming displays of imitation can be triggered by instinctive responses. As we have seen in Part VII, many species can perform complex sequences of behavioural acts being triggered by a simple definite releaser. In some situations, animals can perceive such a releaser when observing conspecifics. For instance, when a pigeon sees another pigeon pecking, it also begins to peck.

In order to distinguish between imitation and other forms of behaviour, Thorpe (1956) marked out *true imitation* as copying a novel behaviour for which there is no instinctive tendency or for which other parsimonious explanations (Morgan, 1884) can be ruled out. The development of experimental studies of social learning has led to the acceptance of somewhat broader definitions. Thus, according to Heyes (1993), imitation occurs when an observer learns about responses, actions or patterns of behaviour as a direct result of conspecific observation. To be more precise, the observer will learn to execute a behaviour that is topographically similar to that performed by the conspecific. Call and Tomasello (1995) insist that for 'true' imitation to occur the observer needs both to recognise the goal of the demonstrator and to realise that reaching this goal is only possible by copying the acts of the demonstrator.

There is the distinction between imitation and *mimicking*. Mimicking is said to occur when the response that is copied does not lead to an immediate, tangible reward. Imitation will be used to refer to responses that have been copied, and that lead to a reward. In mimicking, the observer copies the action precisely but does not understand the goal of the demonstrator. It is important to note that both mimicking and imitation demand that the learner has to interpret the acts and movements of the demonstrator in relation to its own body and movements.

A striking example came from Moore's (1992) experiment with the African grey parrot which displayed capacities to mimic humans. The experimenter spent a few minutes every day with the parrot in its room. During this period, the experimenter would utter a word or phrase, perform a stereotyped movement in front of the parrot, and then leave. A videotaped record of the parrot's behaviour revealed that during the periods when it was alone, it mimicked both the utterance and the action of the experimenter. For example, on leaving the room the experimenter waved goodbye and said 'Ciao'. After a year the bird was heard to say 'Ciao', and at the same time it waved a foot in the air. As this action was never performed in the presence of the human, its repeated occurrence was definitely not influenced by a possible reward. This mimicking behaviour is rather sophisticated, in particular, because the parrot's view of its own leg would be very different from the sight of the experimenter waving; before the waving action could be copied, some transformation of the visual memory of the experimenter's response would be needed, and how this transformation might take place is not known (Pearce, 2000).

Mimicking can help social animals to acquire skills involving tools, and it also serves social functions. Seyfarth and Cheney (2002) have described how monkeys use mimicking in their Machiavellian interrelations within groups. For instance baboons attract the attention of other group members by mimicking the fighting calls of high-ranked individuals. Mimicking also plays an important role in communication. Certain birds, as we have seen in Part VII, learn their songs by mimicking adult birds. The ability to mimic acoustic signals underlies vocal learning in birds and marine mammals. Recently an advanced ability to mimic acoustic signals has been discovered in elephants. One African zoo elephant reproduced the sounds of trucks, and another mimicked the acoustic call of the Asian elephant (Poole *et al.*, 2005). Researchers have appreciated the advanced abilities of apes for mimicking when teaching them elements of human gesture sign language (Wallmann, 1992) (see details in Part IX).

The power of forms of social learning simpler than 'true imitation' has been underestimated for a long time. 'Social learning' in its recent meaning includes a wide range of categories of different levels of complexity. Several phenomena that were once seen as clearly imitative have since been explained in terms of simpler mechanisms resembling imitation.

Contagious behaviour is exemplified by a rule such as 'if others are fleeing, flee also'. The idea is that the stimuli produced by the performance of a particular behaviour serve as triggers for others to behave in the same way. Possible examples of contagious behaviour include flight responses, movements in flocks or shoals, and chorusing by birds, frogs and dogs. Laughing and yawning are excellent examples of contagious behaviour in humans (Provine, 1996). Zentall (1996) argues that contagious behaviour must have a genetic basis, i.e. it must involve the triggering of an instinctive response.

Social facilitation, in its wide meaning, is defined as an enhancement of performance of definite behaviour when another person is present. Originally, the theory of social facilitation was intended by Zajonc (1965) to explain the effects of an audience on human performances. Zajonc (1965) argued that the presence or action of the demonstrator might affect the motivation state of the observer that eventually leads to better performance. Applying the logic of social facilitation to ethology, Clayton (1978) defined social facilitation as an increase in the frequency of a behavioural pattern in the presence of others displaying the same behavioural pattern at the same time. She suggested that social facilitation in animals could be due to the reduction of isolation-induced fear. Recently many authors have considered social facilitation a basic form of social learning that can explain by more mundane means some phenomena that had previously been treated in terms of 'animal culture', such as milk bottle opening by tits and potato washing by Japanese macaque monkeys. We will consider these and other examples further in this Part.

Response facilitation is a close but narrower notion than social facilitation. If observation enhances the relative frequency of an act that is already in the repertoire of an animal, it is referred to as response facilitation. This can be illustrated with results of experiments with naive animals. For instance, in ground squirrels the presence and behaviour of conspecifics affect juvenile alarm-call responses. Visually isolated juveniles were less likely to respond to playback alarm calls (Mateo, 1996). Naive ants reared in isolation both from socially experienced ants and from symbiotic aphids providing them with sweet excretions display species-specific patterns of aphid-tending earlier and more effectively if they have the opportunity to observe experienced aphid-tending nestmates (Reznikova and Novgorodova, 1998).

Stimulus enhancement (e.g. Spence, 1937; Galef, 1988) is said to occur when the presence of an individual draws an observer's attention to a particular object (the 'manipulandum'), thus enhancing the observer's opportunity to learn about the object. The result of this narrowing of behavioural focus is that the individual's subsequent behaviour becomes concentrated upon these key variables. Generalised stimulus enhancement focuses the observer's attention on an object as a whole while localised stimulus enhancement focuses the attention on a functional part of the object only. However, the observer does not copy actions of the demonstrator, and the actual actions of the observer are acquired on the basis of trial and error.

Social facilitation and stimulus enhancement from more experienced individuals can serve as proximal mechanisms fostering safe incorporation of novel foods, spread of knowledge about predators and other dangers, and even increasing effectiveness of mate choice.

Observational conditioning takes place when the demonstrator's actions provide the observer with the opportunities to learn that the appearance or movement of an object signals the occurrence of an appetitive or aversive event. The observer thus learns the relation between some part of the environment and the reinforcer, that is, a Pavlovian association may be established (Zentall and Levine, 1972; Heyes, 1993). Socially transmitted food preferences (Galef, 1988) represent a special case of observational conditioning. The mechanisms responsible for socially acquired food preferences appear to have strong simple associative learning components (e.g. learned safety or the habituation of neophobia to the novel taste), for which the presence of a conspecific may serve as a catalyst. Furthermore, these specialised mechanisms may be unique to foraging and feeding systems.

One of the best examples of observational conditioning is in the acquisition of fear of snakes by laboratory-reared monkeys exposed to a wild-born conspecific in the presence of a snake (Mineka and Cook, 1988). This example was discussed in Part VII. Presumably, a fearful conspecific serves as the unconditioned stimulus, and the snake serves as the conditioned stimulus. It appears that exposure to a fearful conspecific or to a snake alone is insufficient to produce fear of snakes in the observer.

When observation of a demonstrator allows an animal to learn how the environment works, a form of learning is involved which has been labelled *emulation* (Tomasello *et al.*, 1987). Whereas stimulus enhancement changes the salience of certain stimuli in the environment, emulation changes the salience of certain goals (Byrne, 2002). In emulation the learner gains information from observing a demonstration, but in achieving the same goal, may use a different method. The investigation which prompted the recognition of this process involved chimpanzees learning from a trained conspecific how to rake food items into a cage (Tomasello *et al.*, 1987). The data showed that chimps exposed to the skilled demonstrator learned how to use the rake, unlike controls, who were unsuccessful in the task, despite manipulating the tool just as often. Animals, however, did not copy the precise strategy employed by the trained conspecifics. Instead, the observers were learning from the demonstration the 'affordances' of the tool. The meaning of *emulation learning* (Call and Tomasello, 1994) has been expanded to incorporate observational learning about the properties of objects and potential relationships among them. Emulation can also account for some findings of observation learning that had previously been treated as imitation.

25.2 Ecological aspects of social learning

The opportunity for exchange of information among individuals is one of important benefits of living in groups. Species differ in their abilities to use socially acquired information and, in particular, in their abilities to learn through traditions (Fig. 25.1). Members of social groups often monitor the behaviour of their companions in an attempt to gain information about the location of foraging sites or approaching predators. Using social cues provided inadvertently by individuals engaged in efficient performance (*inadvertent social information*, ISI), animals rely on *public information* (PI) about the quality of the resource (Valone, 1989; Giraldeau *et al.*, 2002; Danchin *et al.*, 2004). In many cases, for group-living animals the only socially acquired information available to individuals is the behavioural actions of others that expose their decisions, rather than initial stimuli on which these decisions are based. So an individual has to make a choice between using socially generated cues or relying on a personal decision basing on the stimuli that are gained directly from its environment. A readiness to pay attention only to socially generated cues can reflect the level of conformity of an individual or of a whole group.

As has been already noted more than once, social learning rarely, if ever, appears as a single source of knowledge during normal ontogenesis; instead, social learning combines with the consequences of maturation and of individual experience. The question of how readily animals learn – in particular, by means of social cues – about predators, food, suitable partners and other vital things is important not only from the theoretical but also from the applied perspective. For example, in conservation management programmes of endangered species, reintroduced and translocated individuals as well as those reared in captivity are vulnerable

Fig. 25.1 Friendship and social learning in chimpanzees. (Photograph by M. Vančatová.)

to predation and starvation risks. It is practically important to estimate the contribution of each component into an integral picture of behaviour of a certain species, that is, the role of individual experience, inherited patterns and readiness for social learning. There are many experimental data demonstrating the role of social learning in concrete situations of foraging, mate choice and anti-predator behaviour in different species.

The role of social learning in foraging

The idea that animals may observe others to gain information about resource quality arose mostly in foraging context (for a review see Danchin *et al.*, 2004). Animals can use socially gained cues in the context of searching patchily distributed food or making decisions about food availability and appropriateness.

One of the central problems in behavioural ecology is whether individual group members can improve the estimation of quality of a food patch by combining their prior knowledge of the distribution of prey with the obtained public information. For example, European starlings *Sturnus vulgaris* and red crossbills *Loxia curvirostra* use a patch-foraging scenario (Charnov, 1976) and exploit prey hidden in the soil (starlings) or within a tree trunk (crossbills). These birds must probe repeatedly to estimate the current quality of a foraging patch. Both species observe their flockmates's probing success and use this as public information to decide when to leave a patch in search of another (Templeton and Giraldeau, 1995). Public information can also be obtained heterospecifically. For instance, nine-spined sticklebacks *Pungitius pungitius* use the feeding behaviour of three-spined sticklebacks *Gasterosteus aculeatus* to select the most profitable patch to exploit (Coolen *et al.*, 2003). In ant species communities, the dominant species uses behavioural cues from more agile subdominants to obtain hidden food items (Reznikova, 1982, 2003; see details further in this chapter).

Social facilitation of eating novel food has been found in many species. Nevertheless, there is a debate and equivocal evidence about the role of this source of knowledge in the formation of foraging habits in animals. It is likely that different ontogenetic scenarios imply different roles of social learning in a wide variety of species. For example, in mammals preference for a particular food is shaped under the influences of inherited predisposition, information gathered from the smell of mother's milk and faeces and supplemented with the smell and taste of food items (Galef, 1993; Altbäcker *et al.*, 1995). Infants manage to get novel food directly from the mother's mouth or by pushing her away from the food without ceremony. In some avian species the young also learn to choose appropriate food from parents and then from conspecifics whereas in others individual learning based on inherited predisposition dominates (Lefebvre and Palameta, 1988).

Animals living in groups monitor each other every moment of their periods of activity and react on specific motions which send messages that a food is available. For example, Brown and Laland (2002) have shown that a specific darting motion serves as the cue to naive fish to learn to forage on novel prey items. They found that 100 per cent of the individuals that paired with pre-trained fishes learned to accept the novel prey. Naive fishes paired with equally naive individuals actually performed even worse (50 per cent) than the individuals learning in isolation (73 per cent), due to 'social inhibition'.

Although social learning would be particularly advantageous when food can be poisonous, and some concrete data support this hypothesis (Mason and Reidinger, 1981; Fryday and Greig-Smith, 1994) some species do not rely on social information. For example, Japanese macaques and tufted capuchins rely on their own experience and not on what they see other group members doing in response to a decrease of food palatability (Visalberghi and Addessi, 2001). However, it is easier for many species to acquire food preference socially than to learn by themselves to avoid food that is poisonous. In Galef *et al.*'s (1985) experiments with rats some of the critical features of the social interactions preceding formation of food preference were revealed. The experimenters used a simple apparatus to allow one rat to smell food on an anaesthetised demonstrator rat (that one can call a 'sleeping beauty' rat). An observer rat was placed into the basket of the apparatus, and an anaesthetised (and thus unintentional) demonstrator was placed into a wire-mesh basket (Fig. 25.2). Some demonstrators had food dusted on their faces, and others had food placed directly into their stomachs through a tube. In both cases the observers subsequently showed a preference for the diet of the flavour that had just been fed to the demonstrator. However, if the rear end of the demonstrator was dusted with food and placed foremost in the basket, then only a slight preference for the food was demonstrated. Finally, if a wad of cotton wool, rather than a rat, was placed in the basket, then despite being dusted with food, there was no change in the attractiveness of the food. Thus the demonstrator does not need to be active to encourage the development of a food preference in another rat. But the demonstrator should be a rat, and the observer must be sure that the demonstrator touched the food with its face, not with its tail!

Fig. 25.2 The apparatus used to allow one rat to smell food on an anaesthetised demonstrator rat. (Adapted from Galef *et al.*, 1985.)

In experiments with domestic chickens Turner (1965) revealed that young chicks respond strongly to a pecking model of a hen. Sherwin *et al.* (2002) demonstrated that avian social learning might not be fundamentally different that of mammals, and that similar features of social interactions influence food preference in these groups of animals. In particular, it turned out that the more enthusiastically a demonstrator pecks novel food items, the more items observers consume. It is interesting to note that even in such socially independent chicks as brush turkeys (see Chapter 21 for details) young chicks react positively to a pecking conspecific. This has been demonstrated by Göth and Evans (2004) in experiments with the use of naive chicks and realistic pecking robots (Fig. 25.3). Megapode chicks do not form bonds with their parents. They hatch with a general tendency to

Fig. 25.3 A brush turkey chick robot, pecking at the ground. (Photograph by A. Göth; courtesy of A. Göth.)

respond to some common features of food objects, such as contrast, movement and reflective surfaces, and while trial and error is initially important, they successfully aim their pecks at edible items soon after hatching. Besides, megapode chicks show a predisposition to respond to conspecific chicks of similar age. A pecking conspecific indicating food might speed up the transition from trial-and-error searching to more selective pecking through social facilitation.

Mate choice copying

It is an intriguing question to what extent social factors can influence the choice of sexual partner. Female mating decisions are often influenced by exposure to the mating interaction of others. This style of mating behaviour is called *mate choice copying*, and is said to occur when the probability of an individual selecting another as a sexual partner increases because other individuals (of the same sex) have selected the same partner (Gibson and Hoglund, 1992). Mate choice copying has been reported in several species of birds and fishes. To estimate the role of social information in mate choice, it is necessary to separate the signals deliberately produced by displaying males from the cues that are inadvertently produced by females that make their choices.

Dugatkin (1992) has elaborated an experimental paradigm to investigate this problem. In his study on guppies *Poecilia reticulata* two males were secured at the ends of an aquarium, one with a demonstrator female nearby. The observer, another female, placed centrally, watched the other female interact with one of the males. When, after the demonstrator had been removed, the observer was allowed to choose between the two males, she consistently chose the male that the first female had chosen. Multiple comparisons with choices that were made by control females enabled researchers to suggest that females follow the rule: 'If this male is good for another female, he is good for me', that is, they utilise the presence of the female near a male as an indication of his quality. At the same time, females take into account key features of males themselves. For the guppy, tail coloration is significant: in a case where males differ slightly by this sign, a female will choose a more 'popular' male, but if the male has a pale-coloured tail, he has no chance of being selected by a female (Dugatkin and Godin, 1993). There are, however, many other mating patterns in fishes that do not include mate choice copying (for a review see Brown and Laland, 2003).

Galef and White (1998, 2000) have suggested an interesting experimental technique in order to explore social influences on reproductive behaviour. They used Japanese quail as a model species. Researchers changed the typical look of birds by adding to them a novel trait, namely, a white hat. Females that observed that males with novel traits mated successfully preferred males that possessed similar white hats. Applying the same methodical trick, Swaddle *et al.* (2005) have revealed mate choice copying in zebra finches. Females preferred males that were wearing the same leg band colour as the apparently 'chosen' males, that is, those that been demonstrated as being paired with other females. These studies show that mate preference can spread rapidly through a population by social mechanisms, affecting the strength of sexual selection even in a monogamous species.

Public information about danger

Acquiring information about danger such as predators by the use of social cues can substantially decrease the level of lethal risk for group-living animals. Utilising information gained from observing conspecifics is especially advantageous as it allows an animal to adopt appropriate behaviours without the need to independently verify the approach of a predator (Brown and Laland, 2003; Caro, 2005).

The general tendency to copy the flee responses of an entire group (flock, herd or shoal) is based on a simplest form of social learning, namely, contagion. A panic reaction of a single individual can trigger similar reactions of other members of the group. Individuals react to the flight response of neighbours rather than directly to the advancing predator itself. Synchronous predator responses seem to be cooperative at least in some species. For example, in herring schools, attacks from predatory fish and killer whales induce massive predator-response patterns at the school level, including bend, vacuole, hour-glass, pseudopodium, herd, split and 'tight-ball' formation within the shoal (Axelsen *et al.*, 2001). It is a question still under discussion whether the repertoire of predator responses observed in large groups of fishes and birds can be interpreted as a range of cooperative tactics to trick predators, or different individual tactics determine the outcome at group level (Couzin *et al.*, 2005).

Many researchers have reported social learning at a group level when, after observing predator responses of a neighbouring group, a school of fishes or a flock of birds reacts much more readily to the approach of a predator (for a review see Brown and Laland, 2003). For example, minnows showed a significant increase in the frequency of flight responses after observing the flight responses of minnows in a neighbouring tank that had been threatened by a predator (Magurran and Higham, 1988). Animals often use the predator responses of other species as social cues to learn to avoid danger, and this may include the formation of reactions to a novel predator. For example, Curio (1988) has shown that blackbirds learn to mob a species of bird to which they were initially indifferent (Australian honeyeater, *Philemon corniculatus*) once they have seen conspecifics apparently mobbing it.

25.3 Cognitive aspects of social learning

Thorndike's (1911) winged (but disputable) words 'Apes badly ape' generate a series of questions such as 'Do monkeys ape?' (Visalberghi and Fragaszy, 1990); 'Do rats ape?' (Byrne and Tomasello, 1995); followed by Tomasello's (1996) revision 'Do apes ape?' to which I have added 'Do ants ape?' All these questions are derived from a discussion about which, if any, form of social learning is more intelligent. In particular, Whiten (2000) asks a question: Which is more intelligent? Imitation or emulation?

There is a growing body of evidence in the literature that observational learning, irrespective of whether it includes imitation or 'only' emulation or stimulus enhancement has cognitive implications. It is generally assumed that imitation is a more sophisticated cognitive process. First of all it concerns an imitative translation process which includes cognitive implications of how organisms view the behaviour of others, relative to their own behaviour. It implies the ability to take the perspective of another. For this reason, researchers have tried to distinguish imitation from other kinds of social learning and influence. Recent reviews (e.g. Tomasello and Call, 1997) have concluded that only humans, or in some cases chimpanzees, can truly imitate. However, some researchers have claimed that some species of birds can imitate (Zentall, 1996). In fact, researchers are still not certain what mechanisms underlie this ability (Zentall, 2003).

Emulation also demands feats of intelligence as it implies that the learner can select from the model's performance just the new information it needs, and then efficiently combine this information with its own practical knowledge to deal with the task in its own way. I agree with Whiten (2000) in that it may be appropriate to think of imitation versus emulation as a useful but qualified distinction, which makes sense only in relative terms in a particular context, rather than in absolute terms. So in some situations it may be reasonable to consider 'observational learning' without clarifying the types of social learning.

There are several experimental paradigms for comparative studying cognitive aspects of social learning. A paradigm that is known as a 'do as I do' test allows the testing of imitation as *kinaesthetic visual matching* (Mitchell, 1993), or as a process that is an 'especially demanding variety of visual cross-modal performance' (Heyes, 1993). One can then ask if an animal can learn to match any behaviour of another 'on cue'; in other words, can the animal learn the general concept of imitation and then apply it when asked to do so in a 'do as I do' test. Virginia and Keith Hayes (1952) gave intensive training to Viki, their young chimpanzee. They taught her by using an imitation set: whenever Vicki responded to the order 'Do this, Vicki' by imitating the experimenter's actions, she was rewarded. Viki learned to respond correctly to the command 'Do this!' over a broad class of behaviour. More recently, Custance *et al.* (1995) have replicated this result under more highly controlled conditions. Using the 'do as I do' procedure, the experimenters found that actions on parts of the body that cannot be seen by the performer were just as readily copied as those that could be seen. Both chimpanzees and young children were able to imitate acts that involved parts of their body they simply could not see (facial expressions, head movements, touching parts of the body out of sight) just as well as those in sight. The importance of behaviour that cannot be seen by the performer (e.g. touching the back of one's head) is that it rules out the possibility that some form of visual stimulus matching might account for the behavioural match. The establishment of a 'do as I do' concept not only verifies that chimpanzees can imitate, but it also demonstrates that they are capable of forming a generalised behaviour-matching concept (i.e. the chimpanzees have acquired an imitation concept) (Zentall, 2003).

An instrumental method that gives wider possibilities for comparative studies of social learning is known as the *two-action method*, or *two-ways action/one outcome* as there are two possible actions which can be performed on one object. Imitation can therefore be tested by finding out whether subjects tend to perform whichever of the two actions they have seen. This can control for displays of other types of social learning such as stimulus enhancement and emulation. This method was first applied by Thorndike (1911) in his studies on chicks. Thorndike noted that those chicks which had had the opportunity to observe how their companions escaped from a puzzle-box coped with this task faster. He then divided the demonstrating chicks into two groups, and trained each of the groups to escape by two different ways. Both ways were available for the observers. They, however, chose the way that they had seen done by their demonstrator. This method has been developed in many studies. For example, Dawson and Foss (1965) trained budgerigars *Melopsittacus undulatus* to remove a lid from a cup using either their beak or their foot. When naive budgerigars were allowed to observe one of these techniques they showed a significant tendency to use the same method as their demonstrator.

More recently, the two-ways action/one outcome paradigm was successfully used to show evidence of observational learning in common marmosets (Voelkl and Huber, 2000), Norway rats, (Heyes *et al.*, 1994), ravens (Fritz and Kortschal, 1999) and some other species. The majority of studies have used very simple manipulations of simple objects. For example, European starlings were requested to use either pushing or pulling actions in manipulating a plug for access to a food reward (Fawcett *et al.*, 2002), whereas pigeons (Zentall, 1996), and Japanese quails (Akins and Zentall, 1998) either pecked or stepped on a treadle.

In order to examine cognitive aspects of more complex, sequence imitation, experimenters have combined the two-ways action/one outcome paradigm with the use of *artificial fruit*, that is, a device that must be opened for a food reward. To open the fruit, several defences have to be removed, as happens with many natural foods used by many species, especially, by primates and parrots. The artificial fruit may be of different levels of complexities, from a simple plastic container that can be easily opened, say, either by teeth or by extremities, to complex devices equipped with bolts, latches and so on. This combined method offers the possibility of a 'gold standard' in comparative imitation research.

Even the use of relatively simple 'fruits' have enabled researchers to pose the question of whether an all-or-none test for imitation is appropriate, as successful imitation may depend on the integration of many different sources of information. For instance, common marmosets displayed some fairly complex social learning, but did not imitate the entire structure of the observed action (Fig. 25.4). In the experiments of Caldwell *et al.* (1999) one animal was trained to open an artificial fruit to get a food reward. Another animal was allowed to lick a food reward from the outside of the fruit. Each of these models was observed by four subject

Fig. 25.4 A marmoset manipulating simple artificial fruits. (Photograph by B. Voelkl; courtesy of L. Huber.)

animals, all of which were subsequently given an opportunity to open the fruit. Although none of the subjects was able to open the fruit, marmosets appeared to be capable of some degree of imitative matching, and the movements made, though futile, were clearly directed by what they had seen from the demonstrator. Analysis of the subjects' actions towards the apparatus showed distinct differences between the groups: in their exploratory behaviours, those that had observed the manipulation model used their hands more, and those that had observed the licking model used their mouths more. The first group also devoted attention to the components manipulated, whereas the second focused on the parts they had seen licked.

The use of the artificial fruit paradigm across different species led to the conclusion that some primates are more skilled imitators than others. In experiments of Whiten *et al.* (1996) chimpanzees and young children were presented with adult human models opening an artificial fruit in one of two alternative ways. In one experiment the defence consisted of a pair of bolts that had to be either poked out through the back, or pulled out at the front with a twisting motion to open the lid and to gain the edible treat inside. In another experiment a pin was spun round and removed using one of two different methods, after which a handle could be disabled by either pulling it out or turning it to one side, allowing the lid to be opened. Chimpanzees were found to copy the method they witnessed being used to remove the bolts, as did children. However, while the children also imitated the method of removing the handle, the chimpanzees did not – all tended to use the same method of pulling out. Thus, in a situation when young children learned a technique with quite high fidelity, chimpanzees did not copy all they witnessed so faithfully. Further studies enabled these researchers to suggest that, as has already been noted, the capacity for true imitation is restricted to humans and apes only, or more precisely, to children and chimpanzees raised in a human environment.

It is still not completely clear to what extent our close relatives use their brains when copying the behaviour of companions. Several authors consider chimpanzees and gorillas as being able to copy hierarchically organised patterns of food processing in nature and in laboratory (Byrne and Byrne, 1993; Byrne and Russon, 1998) whereas others argue for a kind of sequential emulation in apes (Whiten, 1998; Povinelli, 1999), all demanding abilities to extract and remember the basic plan of the action sequence.

It has been shown recently that autistic children display widely ranging imitation deficit whereas they do not differ from normal children in performance in emulation tests (for a review see Heyes, 2001). This enables us to consider

Fig. 25.5 A kea watching how its mate is coping with the means–end task. (Photograph by I. Federspiel; courtesy of L. Huber.)

imitation a part of normal development in our species which includes predisposition for copying the actions of close company. It is interesting to note that cross-fostering experiments revealed shifts to foster parents' behaviour just to that extent to which members of the adopted species are predisposed for mimicking and imitation. As we have seen in Part VII, this has been clearly demonstrated on birds, and we can also recall an experiment in which a young fox performed species-specific behaviour of his own species with only minor shifts towards the behaviour of his foster mother, the dog (Mainardi, 1976). Similar results were obtained with ants raised by members of other species. It is, however, noteworthy that minor but distinct changes in motor patterns were revealed in adopted ants similar to those of their 'mentors' (Reznikova, 1996, 2003). As far as human infants are concerned, nobody could think about cross-fostering experiments but there are some documentary evidences from abandoned children who grew up together with animals (in most cases, with dogs). These infants copied a number of behavioural acts from their companions: eating by licking, walking on four feet and using sounds similar to those of the dogs (Miklósi, 1999). Meltzoff (1988) argues that humans are genetically predisposed to imitate others, and that this predisposition allows us to become imitative generalists.

Some insight that helps to clear the grey area between imitation and other types of social learning and to throw light on the cognitive aspects of these has come from studies of exploratory or curiosity behaviour, for which social learning may be particularly adaptive (Miklósi, 1999; Huber *et al.*, 2001). As we have seen in Chapter 17, there is a high level of specificity in displays of exploratory behaviour, and the tendency of animals to explore details of their environment correlates with coping with intelligence tests (Reznikova, 1982).

Huber *et al.* (2001) have investigated how social learning affects object exploration and manipulation in keas. This New Zeland parrot, as the authors note, has been used as an example of curiosity in birds for a century, and its natural habitat is thought to have led to the evolution of extreme behavioural flexibility (Fig. 25.5).

Researchers adapted the experimental paradigm used by Whiten *et al.* (1996) to investigate imitation of foraging technique in chimpanzees (see above in this section). Five young keas were allowed to observe a trained conspecific that iteratively demonstrated several techniques for opening a large steel box. The lid of

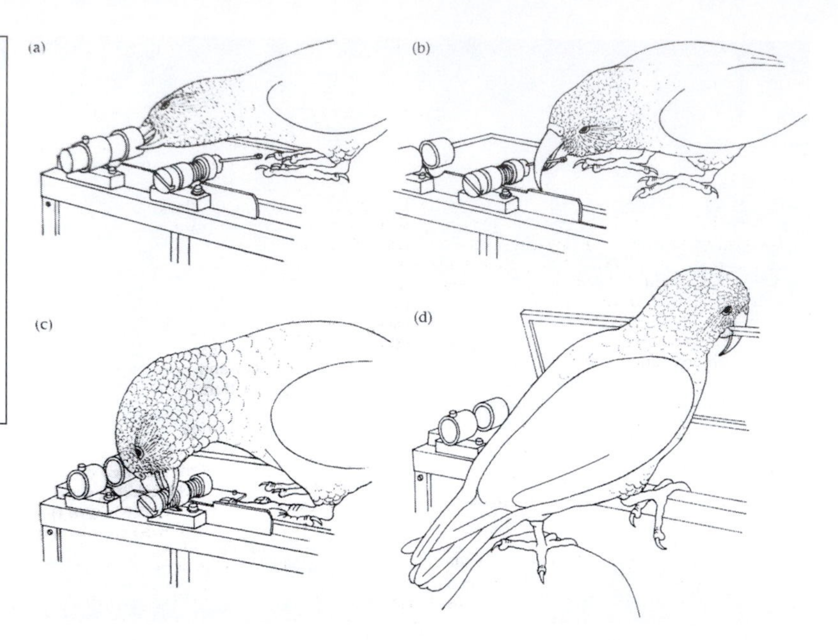

Fig. 25.6 Illustrations of the experiments of Huber *et al.* (2001) on social learning in keas. Above: schematic representations of the actions of a demonstrator opening the artificial fruit during observer training: (a) poking the bolt out of the rings; (b) pulling the metal split pin out of the screw; (c) rotating the screw; (d) raising the lid. Below: a kea in action. (Adapted from Huber *et al.* (2001); photograph by L. Huber; courtesy of L. Huber.)

the box could be opened only after several locking devices had been dismantled: a bolt had to be poked out, a split pin had to be pulled, and a screw had to be twisted out (Fig. 25.6). The observers' initial manipulative actions were compared with those of five naive control birds (non-observers). Although the kea observers failed to open the box completely in their first attempts, they explored more, approached the locking device sooner and were more successful at opening the boxes. These results provide evidence for effects of social facilitation and both generalised and local stimulus enhancement on object exploration in this species. The data obtained also suggest that the keas definitely learned something during observation. Although their initial attempts did not match the response topography or the sequence of model's actions, the birds' efficiency at unlocking the device seemed to reflect the acquisition of some functional understanding of the task through observation, that is, emulation learning.

There was no evidence of true imitation in keas, and a salient explanation given by the authors is that the keas' propensity for exploration, object play and demolition runs counter to the exact reproduction of movements demonstrated by others. Keas are justly described as 'chimpanzees among birds'. Their dynamic and playful style of life does not coincide with close watching and imitating actions of others. Being attracted by a conspecific to explore a novel object does not necessarily lead to slavish copying but may lead to learning which parts of the object are worth exploring (Huber, 1998). Together with data on anthropoids, the cited results on parrots enable us to regard emulation learning as being cognitively quite demanding.

It is important for comparative imitation studies that many factors should be taken into account in order not to place a species in a list of 'backward' species. Among these factors, motivation is of great importance as well as the ranking and 'self-confidence' of the individuals who play the role of demonstrator. For example, von Holst (1937) observed that in wrasses, fishes which live in shoals, it is enough to extract a forebrain in an individual and it will become a recognized leader in its shoal. The fact is that, if surgically deprived of its forebrain, a wrasse loses its shoal-forming reactions. The lobotomised fish dares to swim where it pleases and the shoal tags along behind it.

In wrasse shoals, as in many other fish species, no fish has individual knowledge of another. For members of personalised animal societies it is important to copy actions of highly ranked individuals. This was taken into account in Huber *et al.*'s work on keas cited above: highly ranked birds were appointed as demonstrators. Another example comes from Vančatová's (1984) study in which imitation behaviour in capuchins *Cebus apella* was clearly demonstrated while in other studies members of this species were unable to learn to use a tool efficiently that they had repeatedly observed being used by others. In the cited experiment the monkeys were highly motivated by the nature of the reward, that is, a little mouse to eat instead of items of vegetarian diet, and, what is not less important, it was the dominant individual who was used as a demonstrator.

In cases of interspecies social learning naive animals benefit from watching a skilled demonstrator mastering detour tasks in order to obtain a reward. Let us consider two examples.

The first example concerns Pongrácz *et al.*'s (2001) study of the role of social learning in interactions between pet dogs and humans. Dogs had to solve several variants of a detour problem. The ability of dogs to solve detour problems was described in Chapter 12. Here we are interested in the analysis of dogs' ability to improve their results by observing a human demonstrator. In series of tasks, an object or food was placed behind a transparent, V-shaped wire-mesh fence. In certain groups of subjects tested, two opened doors were offered as an alternative way to get inside the fence. In Experiment 1 the doors were opened only in trial 1, then they were closed for both trial 2 and 3. Other dogs (Experiment 2) were first taught to detour the fence with closed doors after having observed a detouring human demonstrator, then the doors were opened for three subsequent trials. In Experiment 1 all the dogs got the object through the doors in trial 1, but their detouring performance was very poor after the doors had been closed, if they had to solve the task on their own. However, in another experimental group dogs were allowed to watch a detouring human demonstrator after the doors had been closed. These dogs showed increased detouring ability, in comparison to the first group. In Experiment 2 the dogs tended to keep on detouring along the fence even if the doors were opened, giving them a chance to get behind the fence by a shorter route. These results show that dogs use information gained by observing a human demonstrator to overcome their own mistakenly preferred solution in a problem situation. In a reversed situation social learning can also contribute to the emergence of preference for a less adaptive behaviour. However, only repeated individual and social experience leads to a durable manifestation of maladaptive behaviour.

The second example concerns the use of public information heterospecifically by two ant species connected by interspecies kleptoparasitic relations. What is of special interest here is that a subdominant species serves as a scout for a

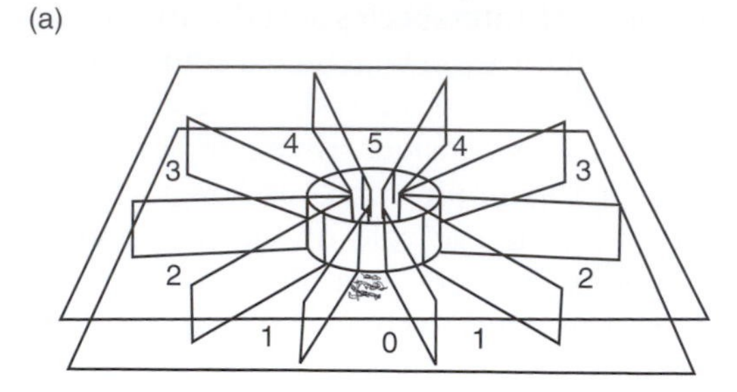

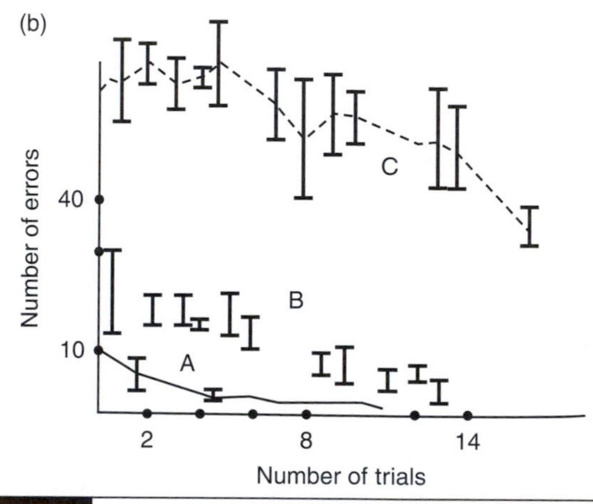

Fig. 25.7 Experiments on social learning in ants. (a) Circular maze with the bait in one of the sectors; 0–5, numbers of the sectors. (b) The change in the number of errors during the training: A, *Formica, cunicularia* (scouting species); B, *F. pratensis* (scroungers) in contact with *F. cunicularia*; C, *F. pratensis* without contact. (Adapted from Reznikova, 1982.)

dominant one when it is necessary to obtain food in complex situations (Reznikova, 1982). In the Siberian steppes *Formica pratensis* is a territorial dominant over *F. cunicularia*, driving its workers away from favourite nest sites and food finds. Occasionally, *F. cunicularia* foragers can steal food items that have been temporarily laid on the ground by *F. pratensis* foragers, but in general they have to rely on quickness and luck to gather food before *F. pratensis* arrive on the scene. For their part the *F. pratensis* use the *F. cunicularia* as scouts. Reznikova (1982, 2001) tested this form of interspecies transfer of information in field experiments presenting the ants with food baits hidden in a 'sectorial maze'. Meat baits were offered in a central area, the approach to which was divided into ten equal sectors (Fig. 25.7). This apparatus was presented in plots visited by only one of two species, and in plots visited by both. Improvements in the ease with which the bait was found over periods of exposure to the maze for 3 hours at each experimental plot were recorded. The improvements of ants' actions were assessed by assigning error points which were the sum of the sector numbers entered before the bait was reached. Individual marking showed that on each plot where the species were active a constant group of about ten *F. pratensis* and two to three *F. cunicularia* were working.

It turned out that within 1–15 minutes of the beginning of trials the bait was found by *F. cunicularia* ants and quickly dragged away. In the case when mistakes were made they ran around and entered the correct sector. This species needed not more than three or four visits to remember the right sector. *Formica pratensis*, on the contrary, roamed about the sectors for an hour, or even two, and found the bait only by chance. Later on their searching became orderly, but the sum of errors remained high. If *F. pratensis* had the opportunity to observe *F. cunicularia* then they did not touch the bait at all for the first 30–40 minutes, allowing the subdominantants to drag it away. During this time *F. pratensis* foragers stood near the maze watching the actions of the scout species. Then they drove *F. cunicularia* away and collected the food pellets themselves. Judging by analogous data obtained from the example of *F. uralensis* and *F. picea*, and also of *F. pratensis* and *F. rufibarbis* (Stebaev and Reznikova, 1972; Reznikova, 1994), it may be assumed that relations in ant communication of this type are in general characteristic of ecologically similar pairs of species playing the role of dominants and subdominants.

These examples of interspecies social learning can be attributed to the generalised notion of observational learning which includes elements of stimulus enhancement and emulation and can be considered cognitively demanding.

Chapter 26

The spread of innovation within populations

As we have seen from the previous chapter, in laboratory studies experimenters create 'innovators' by themselves. They choose active and exploratory animals that have high ranks in their groups, train them to solve a problem and after training let them 'inculcate' new knowledge in naive members of social groups. This mode of experimental investigation helps to enlighten the process of social learning and to estimate the learning potentials of different species. However, such an approach does not allow the investigator to learn how innovations spread within populations in the wild.

It is an intriguing question whether a single individual prodigy or maybe several advanced individuals could propagate a new tradition in an animal community. To catch sight of the transmission of novel behaviour in groups of animals, detailed observations in natural populations are needed, supplemented by experiments in captivity. Sometimes researchers are lucky to witness the gradual establishment of a new tradition. In the majority of cases described in literature new traditions concern vital situations such as feeding techniques or fear of predators. However, intricate patterns of social behaviour such as specific modes of grooming or mating rituals can also serve as subjects for discussion.

In this chapter we will consider possible ways of establishing new behavioural traditions within groups of animals as well as the characteristics of potential innovators.

26.1 The ways in which behavioural traditions spread

The phrase 'population-specific behavioural traditions' is used to describe behaviours that have the following properties (Nagell *et al.*, 1993):

(1) They are acquired through experience, rather than being innate.
(2) They are found throughout a well-defined population.
(3) They persist from one generation to the next.
(4) They are absent in other populations of the same species.

The spread of novel feeding methods through a population, as a particular case of fixing population-specific behavioural traditions, has been documented for a number of terrestrial and avian species (Roper, 1986). Two of the most famous cases are milk-bottle-top opening by birds in Britain (Fisher and Hinde, 1949), and washing sweet potatoes (yams) by Japanese macaques *Macaca fuscata* (Kawai, 1965). In both cases, the spread was initially thought to be due to imitation, but more recent work has cast doubt on this (Sherry and Galef, 1984; Whiten, 1989). Let us consider briefly these two examples together with some analogous studies which appreciate the role of simpler forms of learning than imitation.

Blue tits and great tits in Britain are notorious for their ability to break through the foil tops of milk bottles in order to drink the cream at the top. This skill is believed to have originated in a small group (Fisher and Hinde, 1949), and its spread to the rest of the population has been attributed to imitation. However, the results from a series of experiments by Sherry and Galef (1984, 1990) using black-capped chickadees, suggest that the spread of this habit was promoted by more simple means of social learning than imitation. If a bird happens to come across a bottle that has already been opened, it will drink the milk. Once it has drunk from the bottle, apparently the bird will be very much more likely to break through the foil tops in the future. Pavlovian conditioning provides one explanation for this outcome (see Pearce (2000) for details). Sherry and Galef (1990) report that their subjects were unlikely to open foil tops when they were tested in isolation. In an attempt to answer the question of how the birds came to open foil tops in the first place, experimenters examined the behaviour of a naive bird that had access to a foil-covered container of cream when it could see another naive bird in an adjacent cage. The mere presence of this second bird was sufficient to encourage the first bird to peck at the foil cap and eventually open it. Sherry and Galef (1990) consider social facilitation as the main mechanism responsible for the origins and perhaps spread of milk-bottle opening among certain birds. The reasons for social facilitation of pecking in this concrete situation are not fully understood, but the presence of the second bird may serve to reduce fear, or to encourage foraging responses, in the experimental subject.

An example of the fixing of a behavioural tradition in primates is provided by a group of Japanese macaque monkeys who wash sweet potatoes (yams) before eating them (Kawai, 1965; Itani and Nishimura, 1973). In 1952, on the island of Koshima, scientists were providing a group of 22 Japanese macaques with sweet potatoes dropped in the sand. An 18-month-old female named Imo found she could solve the problem by washing the potatoes in a nearby stream. Imo's name has become legendary as one of the first personalised innovative animals described in scientific literature. The researchers also scattered grains along the beach. The monkeys had to pick the grains from the sand, one grain at a time. Then Imo threw a handful of sandy grain in the water. The sand sank and the grain floated, making it easy to scoop up. Again, other members of Imo's troop eventually learned how to throw their grain in the water.

By March 1958, 15 of the 19 young monkeys (aged 2 to 7 years) and two of the 11 adults were washing sweet potatoes. Up to this time, the propagation of the innovative behaviour was on the individual basis, along family lines and playmate relationships. Most of the young monkeys began to wash the potatoes when they were 1 to 2.5 years old. Males older than 4 years, who had little contact with the young monkeys, did not acquire the behaviour. By 1959, the sweet-potato washing was no longer a new behaviour to the group. Monkeys that had acquired the behaviour as juveniles were growing up and having their own babies. This new generation of babies learned sweet-potato washing behaviour through the normal cultural pattern of the young imitating their mothers. By January 1962, almost all the monkeys in the Koshima troop, excepting those adults born before 1950, were observed to be washing their sweet potatoes. If an individual monkey had not started to wash sweet potatoes by the time he was an adult, he was unlikely to learn it later, regardless of how widespread it became among the younger members of the troop.

Nagell *et al.* (1993) have suggested that the spread of this habit is due in part to stimulus enhancement. The attention of a naive monkey can be drawn to a potato when it sees another monkey pick one up. The naive monkey may then pick up its own potato and for social reasons follow the experienced monkey into the river. At this point, the naive monkey may learn by accident the benefits that accrue from placing the potato in the water.

Indeed, stable embedding of a new feeding technique in wild populations does not necessarily mean that there is imitation underlying cultural transmission. Behavioural habits can be based on mechanisms of social learning that are simpler than imitation.

For instance, stimulus enhancement explains the acquisition of pine-cone stripping behaviour of black rats *Rattus rattus*. Terkel (1996) found that although naive rats never learned to strip cones unaided, the animals were capable of learning the trick if partially striped cones were provided, and especially so if they were exposed to cones with progressively fewer rows of scales removed. Young rats pay close attention to whatever their mother is eating, and often manage to steal partially eaten cones from her.

Combination of social facilitation, stimulus enhancement and individual learning are likely to underlie the formation of 'subcultures', or behaviourally specialised 'cultural clans' in animals. There are a lot of examples in the literature. For instance, populations of crows in Kamchatka specialise in different techniques of getting food from humans. Some flocks regularly steal alms at cemeteries whereas others track skiers in winter and gatherers of mushrooms in summer stealing food from them when they make stops. Dolphins in Shark Bay show a similar specific foraging specialisation – being fed by humans at Monkey Mia Beach – in which not all of the population takes part. This variation appears to be maintained by vertical cultural transmission, since most of the dolphins taking advantage of the feeding are the offspring of females who were themselves fed (Smolker *et al.*, 1997); hence the specialisation is probably learned while swimming with the mother.

Clans of Norway rats specialise in catching fishes or frogs, stealing fishes from fishing nets, harvesting molluscs, and stealing eggs and chicks from birds' nests. Clans dwelling on different sides of a lake display different techniques of catching frogs. Galef (1980) has conducted laboratory experiments simulating his own observations in nature on how Norway rats dive for molluscs. It turned out that young pups are able to adopt this technique from their mothers.

Having been lucky to catch the first manifestations of novel behaviour within a wild population and then monitoring it during long periods, several researchers have reported on very interesting cases of inculcation of new habits, mainly, feeding techniques. For instance, the regular cracking of palm-tree nuts with the aid of two stones ('hammer' and 'anvil') by Japanese macaques was fixed from the very first case and then monitored for 20 years. During this time, about 80 per cent of population adopted this method (Huffman and Nishie, 2001). The spread of a novel feeding technique was described in humpback whales *Megaptera novaeangliae* (see Rendell and Whitehead (2001) for a detailed review). In the southern Gulf of Maine, a novel complex feeding technique, 'lobtail feeding', was first observed in 1981, and by 1989 had been adopted by nearly 50 per cent of the population (Weinrich *et al.*, 1992).

Not only new feeding techniques can be transmitted socially. Social transmission can also apply to group-specific vocalisations, courtship displays, grooming postures and so on.

A good example of the transmission of a characteristic posture in social groups of apes has come from primatological studies. This concerns the so-called 'grooming hand-clasp', a unique grooming posture in which two individuals sit facing one another, simultaneously raise one arm overhead, and support the other by holding hands or wrists to create an 'A-frame' posture. The grooming hand-clasp was first observed by McGrew and Tutin (1978) in a chimpanzee community inhabiting the Mahale Mountains in Tanzania, whereas the same pattern has never been reported for the chimpanzees of Gombe Stream, who live only 170 km to the north. The authors believe it 'unlikely that the two populations had time to differentiate markedly through genetic drift' (McGrew and Tutin, 1978). De Waal and Seres (1997) suggested that learning is the most likely mechanism underlying differences in behaviour between populations of the same species that live within a distance at which genetic exchange is likely or possible. These authors found the grooming hand-clasp developed spontaneously in a captive group of approximately 20 chimpanzees in the Field Station of the Yerkes Regional Primate Research Center, Georgia. Researchers followed the development of this new social custom during a 5-year period. In the beginning, hand-clasps were always initiated by the same adult female named Georgia. She initiated the posture mainly with her adult female kin. In the second year Georgia initiated the pattern with several partners, mostly immediate adult kin. In later years, these relatives began initiating the pattern at rates similar to or higher than those of Georgia herself. Over the years the posture increased in frequency and duration and spread to the majority of adults and also to a few adolescents and older juveniles. When the apparent originator Georgia was removed from the colony, the pattern nevertheless persisted.

Another well-documented local tradition is the rhythmic hand- and foot-clapping during grooming in captive apes' community has been described in bonobos *Pan paniscus* at the San Diego Zoo. Over the years, transmission of this pattern took place: these group-specific gestures were adopted by several individuals introduced as juveniles into the San Diego colony (de Waal, 1989).

It is worth noting that, in contrast to the cases of spread of the novel feeding technique described before, transmission of the new grooming posture is based on a highly stereotyped behavioural sequence which apparently displays in the 'at once and entirely' manner in chimpanzee and bonobo originators. The role of social learning cannot be doubted in a process of infection of community members but the role of inherited predisposition probably is not less. We will consider this problem in the next section.

26.2 Possible mechanisms of establishing new customs in populations

All the examples of specific behavioural traditions fixed in populations that have been described above are based on the spread of innovations. It means that behavioural patterns obtained by innovative animals through their individual experience are then assimilated by observers. However, the inclusion of members of animal communities in new behavioural traditions may be based on initial performances by a few carriers of 'at once and entirely' complexes of actions. Let us consider from this standpoint the two examples which have been described in Chapter 23.

The first example concerns hunting for jumping springtails in *Myrmica* ants (Reznikova and Panteleeva, 2001, 2005). A strong correlation between dynamic density of victims in different plots and number of skilled hunters in ant families inhabiting these plots enabled us to suggest that encounters with lucky hunters promote awakening of predisposed hunting behaviour in those specimens that possess only a 'sketch' of this complex hunting stereotype. It turned out that about 10 per cent of family members display hunting for difficult-to-handle victims at early age, while within the rest of family members hunting behaviour is socially biased. In this part of the family, the more frequently ants observe collisions between victims and able hunters the more effectively their hunting patterns are shaped.

The second example concerns tool use in New Caledonian crows. Social learning in combination with predisposition for manipulation with tools has been considered main mechanisms underlying tool manufacture in this species. Preparedness for tool-using should be strong because the New Caledonian crow is a single bird species displaying total spread of similar tool use within populations (see Chapter 23 for details). Discovery of 'at once and entirely' tool manufacture in a single chick in Kenward *et al.*'s (2005) study enables us to suggest a similar mechanism of spread of complex behavioural patterns within birds' population to that of hunting in *Myrmica* ants. It is very likely that presence of individuals equipped with an inherited complete behavioural stereotype is necessary for triggering this stereotype in other members of population, and individual experience and maturation accomplish fixation of behavioural tradition in populations.

Such a strategy should be evolutionarily stable in populations. We can call this strategy *triggering dormant behavioural patterns*. From the point of view of the present discussion this possible mechanism of fixation of behavioural patterns is of interest because its widespread presence in populations can be attributed to social facilitation, where carriers of whole patterns to be spread serve as catalysts of social learning.

Let us return to the grooming hand-clasp referred in the previous section as a social custom which has been described as a newly implanted one for several communities of wild and captive chimpanzees (McGrew and Tutin, 1978; de Waal, 1989; de Waal and Seres, 1997). It may be that the process of infection of chimpanzee communities by the new social custom which is based on highly stereotypic behavioural sequence such as grooming hand-clasp has much in common with the two examples just described. One can suggest that if there is a carrier of a specific behavioural pattern within a social group, then performance of such a pattern can trigger dormant behavioural patterns in some members of the group by means of relatively simple forms of social learning.

Another example that could be revisited from the standpoint of the hypothesis of triggering dormant behavioural patterns is preparedness for development of fear of 'fear-relevant' objects in some species including humans. In Chapter 22 the experiments of Mineka and co-authors with rhesus monkeys were described in detail. At the first stage of those experiments the evidence has been obtained that laboratory-reared monkeys are not equipped with innate recognition of snakes. Nevertheless, the laboratory-reared monkeys can readily acquire a fear of snakes by observing wild conspecifics expressing fear of snakes. It may be conjectured that, when testing a sufficiently large group of naive monkeys in the laboratory, experimenters will find a few carriers of the whole fear pattern, that is, individuals equipped with innate recognition of snakes. Perhaps those few carriers of phobias of snakes and spiders within populations of our species mentioned in Chapter 22 reflect an evolutionarily stable strategy in hominids.

One of the most important factors that influences the effectiveness of transmission of new behavioural patterns within a community is the frequency of performance of this pattern which may be affected both by the frequency of displays by a single individual, and the number of skilled (or equipped from birth) demonstrators. Triggering of dormant behavioural patterns can be based on cumulative effect. There are several field observations and experiments that support this hypothesis. Beck and Galef (1989) found that multiple demonstrators allowed naive rats to learn more quickly about the proper diet. In experiments with pigeons Lefebvre and Giraldeau (1994) demonstrated that the rate of adoption of innovations increases with the number of knowledgeable 'tutors' and the number of uninformed observers. Experimenters presented pigeons with the problem of pecking open a stoppered inverted test tube to obtain seeds. In the first series of experiments they presented observers with one, three, six or nine demonstrators. In the second series of experiments they added two, five and eight naive observers to an initial group of observers. The observers learnt more slowly with increasing number of naive birds. At the same time, they learned more quickly with increasing tutor numbers. The rate of learning increased exponentially as a result of a dual effect of increased tutor numbers and reduced numbers of naive birds which became informed and switched to tutoring. It is worth noting that the authors call informed pigeons tutors without quotes. It seems however that more simple forms of social learning than tutoring have been involved such as social facilitation and stimulus enhancement.

When analysing situations of transmission of new social customs and feeding techniques within populations an interesting question arises about which features animals have to possess in order to be innovators, which is considered in the next section.

26.3 What it is to be an innovator

As we have already seen from the analysis of examples of innovative behaviour, the main body of data concerns the use of a new food or application of new feeding techniques. Some authors consider the frequency of innovative behaviours, for a given taxonomic group, a useful indicator of its behavioural plasticity and its tendency to use novel means to solve environmental problems (Wyles *et al.*, 1983; Lefebvre *et al.*, 1997). Lefebvre *et al.* (1997) collected 322 foraging innovations in avian species from nine British and North American ornithological journals and analysed them in connection with measures of relative forebrain size. Innovations were documented from field studies and included such examples as herring gulls catching small rabbits and killing them by dropping them on rocks (Young (1987), cited in Lefebvre *et al.*, 1997), or house sparrows systematically searching car radiator grilles for insects (Simmons (1984), cited in Lefebvre *et al.*, 1997). The authors found that relative forebrain size in different species was related to innovation frequency in the two zones, the British Isles and North America. It seems that at a taxonomic level of innovative behaviour demands at least relevant brains.

What about characteristic features of innovators at the individual level? What individual dispositions required for becoming innovative? Only a little is as yet known about the starting conditions of innovations. For example, in a study with guppies, in which the fish had to make a choice quickly between holes in a partition of an aquarium, Laland and Reader (1999) found that females were more likely to innovate than males, smaller fish more likely than larger fish and food-deprived fish more likely than non-deprived. However, apart from these differences related to sex, size and motivation they found that individuals who repeatedly innovated in the past did so again with a higher probability than past non-innovators. Thus some individuals in guppies seem to express 'personality' differences in their tendency to innovate.

The expression of individual behavioural and physiological phenotypes or *coping styles* is defined as the way to cope behaviourally and physiologically with environmental and social challenges, irrespective of life-history state, sex or motivational state. The existence of different coping styles has been shown for various animal species including humans (Broom, 1996, 2001; Koolhaas *et al.*, 1999). In mice and rats, for example, aggressive individuals ('proactive copers') entrained more rigid routines, spent less time exploring novel environments and were less alert to changing stimuli in known environments than less aggressive individuals ('reactive copers'). Similar patterns were found in great tits (Verbeek *et al.*, 1994), domestic pigs *Sus scrofa domestica* (Hessing *et al.*, 1993) and a cichlid fish, *Steatocranus casuarius* (Budaev *et al.*, 1999). In ants scouting individuals that can first solve complex searching problems and usually attract foragers to novel objects, are smaller, have a more diverse behavioural repertoire and are much more agile than other members of their colony (Reznikova and Ryabko, 1994; Reznikova and Novgorodova, 1998).

Experimenters at the Konrad-Lorenz Research Station in Grunau investigated the spread of the ability to trigger a food dispenser in a free-living, semi-tame flock of greylag geese for several years (Fritz *et al.*, 2000). These birds live in affiliative social units of pairs, families or sibling groups (Lorenz, 1979), and their members scavenge without interference. Usually one or two individual within such groups operated the food dispenser whereas the others exclusively scavenged. Pfeffer *et al.* (2002) investigated hormonal and behavioural correlates with the individual's ability to perform operant tasks in hand-raised greylag goslings. Results suggest that becoming an innovator may be contingent upon individual coping styles. A tendency was revealed that males are more successful in coping with new tasks whereas females are biased toward learning by means of stimulus enhancement. Individuals that displayed elements of innovative behaviour have a higher level of corticosterone than conservative geese.

These as yet limited data enable us to suggest that predisposition to innovative behaviour is based on some definite genetic features. In a changeable environment a wide spectrum of adaptations is tested for defects and this includes behavioural adaptations. As McGrew (1992) has noted, in many situations in which researchers have identified innovations within a population, they could be predicted based on some essential change in environment such as shortage in food, forced migrations and so on. Under such circumstances new customs 'invented' by a few innovators can be more useful than the species-specific stereotypes that had been valid before. However, this does not mean that members of a community will readily copy the novel life style. Usually animals observe the odd behaviour of their conspecific curiously but keep aloof.

Do innovators try to spread their new behavioural pattern? In other words, can animal teach each others?

26.4 Can animals teach?

In this section we will first analyse a set of examples concerning the role of parental 'teaching' in one of the most interesting and enigmatic process studied by behavioural ecologists and ethologists, namely, the process of building the character of a wild hunter. Then we proceed to general definitions and ideas concerning the possible role of teaching in spreading of innovations within populations.

Active tutoring ('teaching') can be considered the most complex form of sharing knowledge in animal communities. A working definition of teaching widely accepted among students of social learning was suggested by Caro and Hauser (1992):

> An individual actor (A) can be said to teach if it modifies its behaviour only in the presence of a naive observer, B, at some cost, or at least without obtaining an immediate benefit for itself. A's behaviour thereby encourages or punishes B's behaviour, or provides B with experience, or sets an example for B. As a result, B acquires knowledge or learns a skill earlier in life or more rapidly or efficiently than it might otherwise do, or that it would not learn at all.

There are two main processes of transferring information within populations in which tutoring can be involved: (1) polishing of species-specific behavioural patterns and (2) spreading of innovations. Both processes are parts of social learning.

Back-fitting of species-specific behaviour by efforts of tutoring parents is quite usual in animals, particularly in vertebrate predators. In many species polishing of searching and hunting behaviour makes up an integral part of their ontogenetic development.

However, tutoring as an instrument for spreading innovations is a rare phenomenon in the wild. One can say that this concerns a level of complexity which is over and above the 'general plan' of species-specific behaviour. The fact is that innovations can spread within populations by means that are simpler than tutoring. We have already seen this from previous sections and will consider it again in the last section of this chapter. Although the real role of tutoring is not great, the analysis of teaching in animals is very important for estimating the limits of their cognitive abilities. Even isolated observations on instances of teaching in the wild are valuable.

There is a large body of data about how predators teach their offspring to kill victims. This concerns predators of different sizes and styles of hunting, from giant whales to little jerboas. At the same time, it is known that members of many species grow up as self-made hunters. For instance, polecats *Putorius putorius* learn very quickly how a mouse must be grasped by the neck so that it cannot bite back. During normal ontogenesis complex hunting behaviour can mature with the assistance of individual experience. It can be considered a complex process because innate releasing mechanisms mature as well and they become increasingly more selective through individual learning. This scenario is widely distributed in animal species, which can be illustrated by Eibl-Eibesfeldt's (1970) examples of the prey-catching behaviour of toads, frogs and others. Even in such advanced hunters as the Mustelidae (to which the polecat belongs) parents' instructions have little part. If that is so, is it necessary for predators to teach the young? Is it possible that parents' instruction run idle in animals? Special investigation is needed in each case if we want to know whether parents take part in shaping of a character of a wild hunter. Let us consider several examples.

It has been known for a long time that in felids and other carnivores mothers modify their predatory behaviour in a series of stages. It was described by Leyhausen and Toinkin (1979) in domestic cats *Felis catus*. Adult females pursue, capture, kill and eat prey in a smooth sequence with little hesitation between acts. However, when cats become mothers and their kittens start walking out of the nest, mothers alter their behaviour and carry prey to their kittens to eat it in front of them. Next, they carry live prey directly to their offspring and allow them to play with it but recapture it if it escapes. Finally females take little part in prey-catching at all, merely moving toward prey initially while kittens chase, capture and dispatch it efficiently. Mothers give characteristic mewing calls to their kittens in all of these situations.

The sequence of the mothers' acts is so logical from a human point of view that it undoubtedly looks like successive shaping of the youngs' hunting behaviour. Under controlled laboratory conditions Caro (1981) tested the alternative hypothesis about the role of mothers' teaching in shaping hunting behaviour in kittens. In Caro's study kittens between the ages of 4 and 12 weeks were exposed to domestic mice as a live prey. In one series of trials mother cats were present, while in other trials the mothers were absent. Control kittens received identical exposures but without their mothers being present. Behaviour of mothers was also recorded. The results obtained suggest that maternal behaviour reduces the age at which kittens acquire predatory skills. For instance, when 6-month-old kittens were tested on their predatory abilities, experimental subjects delivered significantly more bites to the nape but not to other regions of the mouse's body than did control kittens, that is, they more easily applied the method by which adult cats dispatch rodent prey. However, Caro (1981) found it hard to say whether maternal behaviour is sensitive to developmental changes in kitten behaviour. These two processes seem to go in two parallel courses. Indeed, the timing of each step in the mothers' predatory sequence might not be contingent upon improvement in their cubs' predatory skills, but rather change according to their individual time-course. Certain mothers started to leave prey with their kittens at very early stages when the young were not able to react to it. Many aspects of mothers' predatory behaviour were significantly negatively correlated with increasing skills of their offspring. Caro (1987, 1994, 2005) has since continued by studying interactions between mothers, cubs and prey in wild cheetah *Acinonyx jubatus* in the Serengeti National Park, Tanzania. This researcher found very slow progress in young cheetahs' education seemingly as a gradual result of great maternal efforts. It turned out that the cubs' hunting skills remain poor up to and beyond independence from their mother, showing surprisingly little improvement in the 10 months after first being introduced to prey. In sum, Caro's data enable us to be careful with the conclusion that it is just maternal tutoring that make cats skilled hunters.

A good example that illustrates how difficult it is to judge about the roles of maturation and investments of parents in shaping hunting behaviour is the behaviour of ospreys *Pandion haliaetus*. Meinertzhagen (1954) provided rich description of adult ospreys encouraging their fledglings to catch fishes. At first, the adult perched away from the nest with fish in their talons but would not feed the young, despite their screaming for food, repeatedly flying away in an apparent attempt to encourage the young to follow. On the first day the fledglings did not leave the nest, but on the next two days when the young flew off the nest to a rock, they were fed. On the following day, the young followed the parents to hunt over a lake. Each parent caught a fish, carried it toward the young and then dropped it, but caught it again and secured it before it hit the water. After having repeated this many times, one of the young finally caught the fish in a stoop and carried it to the rock to eat it. The less successful sibling now flew to the rock to share the catch, but the parent arrived and literally pushed this offspring off the rock forcing it to take wing again. The process of dropping a fish was repeated until the second fledging finally caught it and went back to the rock to eat. On the fifth day the same procedure was observed with each fledging following a parent around and unsuccessfully attempting to catch fish that were dropped for it in mid air. When the fish reached the surface, the parents would retrieve it until eventually youngsters descended to the water and picked up the fish. On the seventh day, the adults drove the offspring away from the lake and they were not seen again. So, the period of education took six days.

Seemingly, this study provides a strong evidence of a definite role of parental teaching in the building of the hunting behaviour in osprey. Nevertheless, observations of hand-raised young ospreys showed that they successfully caught fishes within three days to three weeks of being released into the wild, in the absence of parental instruction (Schaadt and Rymon, 1982).

These data enable us to suggest that parental 'instructions' run in parallel with maturation of hunting behaviour of young, and that at least one aspect of the function of parental teaching in animals is to awake dormant behavioural patterns; repetition of instructions in mammals and numbers of encounters with successful hunters in ants possibly have a cumulative effect. This does not mean that teaching should be excluded from consideration of ontogenesis of hunting behaviour.

Transmission of innovations by teaching

Active teaching as a mean for sharing new experience would seem to be very rare in the animal kingdom, even in apes. The important role of imitation in the social life of anthropoids does not necessarily means that they can teach each others. Young chimps learn how to break twigs from trees, strip away the leaves, and insert them into termite holes by observing adults. The steps required to extract termites in this manner are lengthy and complex. Without the demonstrations of adults, many chimps would probably never become very successful termite fishers. However, part of the acquisition of this tool use appears to relate to innate characteristics of chimpanzee behaviour. All young chimpanzees amuse themselves by playing with sticks and poking them into holes. It seems as though the chimps are able to observe the more skilled adults and translate their juvenile play into a successful means of securing food. But this does not mean that young chimpanzees follow the instructions of adults (Goodall, 1986).

During 10 years of investigations Boesch (1991, 1995) observed interactions between mothers and their young among Tai chimpanzees in the context of nut-cracking. He divided his observations of mother–offspring interrelations into 'stimulation', 'facilitation' and 'active teaching'. Observations of stimulations and facilitation included such things as mothers leaving intact nuts for their infants to crack (which they never did for other individuals) or placing hammers and nuts in the right position near the anvil for their infants to use. Stimulations were observed on 387 occasions of interactions between mothers and their children. Stimulation differs from the common behavioural pattern for adult chimpanzees, when they carry their hammers during nut collection and consume the nuts that they have placed on an anvil. Mothers incurred a foraging cost by having to find more nuts and another tool for opening them.

Active teaching was observed only twice and involved direct intervention on the part of the mother in her offspring's attempt to crack open a nut. In one example, a 6-year-old male had taken a majority of his mother's nuts, as well as her hammer stone. After the young male placed a nut on the anvil, but prior to opening it, his mother approached, picked up the nut, cleaned the anvil, and put the nut back in a different position, more suitable for opening. The young male cracked the nut and ate the kernel. In the second example, another mother reoriented the hammer for her 5-year-old daughter who then succeeded in opening several nuts by maintaining the same grip on the hammer that her mother had used.

One more example of isolated cases of teaching in animals came from the experimental investigations on free-living scrub jays by Midford *et al.* (2000). Experimenters trained models (demonstrators) and then followed them as they modelled the task in the presence of naive (observer) animals. Jays had to learn that a class of objects (bright plastic rings) indicated the presence of buried food (peanut pieces) in a specific location, the centre of the ring. Birds were trained in their family groups to perform the task during the summer season, and were allowed to perform the task in the presence of juveniles in later years. Jays living in 18 control families received partial exposure to the training situation, but received no exposure to the ring before being presented with the task in the presence of their young. Juveniles in 16 families with trained jays were able to witness demonstrations and to scrounge peanut pieces from the models as they completed the task. These 41 juveniles learned much more of the task than the 33 juveniles in control families. What is important for our narrative is that the authors observed three cases of active teaching in two separate families. In each case, the highly ranked bird (the breeder of the family) dug in the centre of a ring until it uncovered the food. Then, rather than taking the pieces of nuts, the adult either departed or stood over the depression it dug, pointed its bill downwards, towards the pieces, until the juvenile took them. This differs markedly from the usual behaviour of jays after finding bits and, as the authors told it, fall within the definition of teaching.

Chapter 27

Culture in animal societies

In this chapter I briefly analyse the complex and fascinating problem of what is culture in animals. The relationship between cultural and genetic evolution was identified by Wilson (1998) as one of the 'great remaining problems of the natural sciences'. Several studies, adopting Dawkins's (1976) concept of the 'meme' as the unit of cultural evolution, have examined factors influencing the transmission and success of memes in animal cultures. The presence of cultural processes within animal societies is an area of some controversy (see: Whiten, 1989; McGrew, 1992, 2004; Wrangham *et al.*, 1994; Boesch, 1996, 2003; Tomasello, 1999, 2000; de Waal, 2001; Fragaszy and Perry, 2003; van Schaik, 2004). How to treat cultural behaviour in animals much depends on its definition. Many definitions in the literature attribute cultural traits only to humans. At the other end of the scale is considering culture as a 'meme pool' in populations which can include all cases of the regular use of public information in populations basing on very simple forms of social learning (Laland and Brown, 2002).

Arguing about animals' variants of culture, some authors concentrate on clearly delineated customs within social groups, considering them elementary units of culture, whereas others insist on taking into consideration 'ethnographic patterns' in animal societies and complexes of unique interrelations within groups ('pieces of unique atmosphere') transmitted by means of social learning. Complexes of characteristic behaviours in whales and dolphins (Rendell and Whitehead, 2001), as well as very specific interrelations in a troop of olive baboons (Sapolsky and Share, 2004), both transferred through social learning but not genetically, serve as good examples.

Many cognitive ethologists agree now that human beings are biologically adapted for culture in ways that other primates are not, as evidenced most clearly by the fact that human cultural traditions accumulate modifications over historical time (the *ratchet effect*: Tomasello, 1999). As we have already seen above and will meet further in this book, our species is likely to possess some uniquely powerful forms of cultural learning, enabling the acquisition of language, discourse skills, tool-use practices and other conventional activities.

Not arguing about restricted notions of 'culture' in animals, I would rather adhere to broad definitions based on social learning as the main mechanism of 'cultural transmission' of behavioural patterns in animal societies. I see here a fascinating perspective of estimating the limits of the power of social learning in non-humans that allows some species to improve adaptiveness of behaviour by non-genetic means. In this context, I consider broad definitions acceptable, such as 'culture is information or behaviour acquired from conspecifics through some form of social learning' (Boyd and Richerson, 2005), and 'animal tradition that rests either on tuition of one animal by another or on imitation by one animal of acts performed by another' (Galef, 1992). Imanishi (1952) defined culture as 'socially transmitted adjustable behaviour'. In this broad sense, cultural learning is widely accepted by students of animal behaviour. Defining culture as a package of behaviours, the working description given by Nishida (1987) is useful: 'Cultural behaviour is defined as behaviour that is (a) transmitted socially rather than genetically, (b) shared by many members within a group, (c) persistent over generations and (d) not simply the result of adaptation to different local conditions.'

27.1 Empirical approaches for studying animal culture

The empirical study of cultural processes in animals is generally approached in two major ways: controlled laboratory experiments on the mechanisms of social learning and field descriptions of behavioural variation (Lefebvre and Palameta, 1988). Both make important contributions to our understanding of culture.

The first approach focuses on experimental study of the cognitive processes underlying cultural transmission. In general, controlled laboratory experimentation is the preferred methodological tool; this gives the approach the advantage of controlled conditions and hence less chance of ambiguity in the interpretation of data. However, the studies do not necessarily relate to what occurs in the wild.

The second approach is field-based; here culture is deduced from patterns of behavioural variation in time and space, which cannot be explained by environmental or genetic factors (Nishida, 1987; Boesch, 1996; Whiten *et al.*, 1999). This approach has been likened to ethnography in the social sciences (Wrangham *et al.*, 1994), and thus is called the 'ethnographic approach' in recent ethological literature. The strength of this approach is that it is firmly rooted in what the animals actually do in the wild, with the unavoidable weakness that results can be more ambiguous than those derived from controlled experiments – such studies cannot usually tell us much about which specific social learning processes are involved in producing the observed behavioural variation (Rendell and Whitehead, 2001).

Practically, students of chimpanzees, the most 'cultural' after our own species, have elaborated the following steps to identify cultural variations:

(1) Show that behavioural differences between chimpanzee populations are not consistent with a genetic explanation – for example, where a boundary between different methods of tool-use occurs within the range of a subspecies (e.g. at a large river), rather than between subspecies.

(2) Check that the behavioural differences cannot be explained by ecological factors such as availability of suitable raw materials for making tools.

(3) Study the transmission processes used by animals in controlled experiments: can they learn by watching others? If so, what kind of things do they learn?

Well-designed experiments of this kind can guide researchers to the most likely learning mechanisms at work in the wild.

27.2 'Crucibles of culture' in animal societies

As has already been noted, we argue about cultural changes in animal societies in those cases when animals learn new habits of living and pass them along to the next generation. In such a situation the spread of a certain innovation results in stable conservation of a new custom that is further maintained and transmitted in a train of generations through social learning. Culture thus is displayed as the presence of geographically distinct variants of habits. Even in this limited sense, culture was long considered to be a uniquely human trait. Ethologists have investigated the problem of animal culture for decades but only in the last few years has a clear picture of cultural diversity in several 'elite' species begun to emerge. Insight into cultural evolution has come from a comparative geographic approach when researchers have thoroughly studied behavioural customs in different populations and thus revealed 'crucibles of culture' in animal societies.

The main methodological difficulty on the way of studying animal culture is to recognize innovations in the field. Even when the origin of a certain innovation had been observed, it is difficult to predict a living trajectory of this innovation. As has been noted earlier in this chapter, innovations can be spread by means of relatively simple forms of social learning and even low-end innovations can lead to extensive cultural change. Remember the Japanese macaques with their potato washing. By using water in connection with their food, the Koshima monkeys began to exploit the sea as a resource in their environment. Sweet potato washing led to wheat washing, and then to bathing behaviour and swimming, and the utilisation of sea plants and animals for food (Kawai, 1965).

At the same time, there are reasons to believe that new skills do not spread easily in animal populations. As Kummer and Goodall (1985) note, of the many innovative behaviours observed, only a few will be passed on to other individuals, and seldom will they spread through the whole troop. For example, Goodall (1986) observed two instances of the use of stones by adolescent chimpanzees to kill dangerous insects. She supposed that this usage of stones would become customary in that reference group. But this had not happen in the following 30 years, and the innovation faded away.

The chimpanzee is clearly the most interesting animal from a cultural point of view. Different populations of chimpanzees seem to have their own unique behavioural repertoires, including such things as food preferences, tool use, gesture signals and other behaviours, and these group differences often persist across generations. After collecting a great body of data in the wild (Goodall, 1964, 1968; McGrew, 1977; Boesch and Boesch, 1983; Ghiglieri, 1984; Nishida, 1990; and others), the first intimation that chimpanzee possess 'material culture' came with McGrew's (1992) book about chimpanzee's tool use. Since then, new observations have appeared and some researchers have argued that individual communities of chimpanzees have their own local traditions (Matsuzawa and Yamakoshi, 1996; Humle, 1999). The grand synthesis was done by a collective of primatologists published in *Nature* (Whiten *et al.*, 1999; see also a review by de Waal, 1999).

The researchers discovered the various habits of chimpanzees at seven field sites and clearly distinguished behavioural patterns. Some of them concern tool use, such as ant-dipping, termite-fishing, nut-cracking, honey-dipping, drinking water with leaves, and so on. Others concern characteristic behavioural habits such as rain-dances, hand-clasp grooming, details of courtship rituals, and so on.

For example, some populations fish for ants with short sticks, eating insects from the stick one by one. Only in one population have the apes developed the more efficient technique of accumulating many ants on a long rod, after which all insects are swept into the mouth with a single hand motion. Another impressive difference concerns the use of leaves for drinking water. In different communities chimpanzees use a 'leaf sponge' crumpling leaves in their mouth, soaking them in tree hollows with their hands and sucking the water from them. The other type technique is a 'leaf spoon' where apes use a leaf like a spoon, without crumpling them up, to scoop out the water.

After compiling a preliminary list, Whiten *et al.* (1999) rated behavioural patterns on a scale from customary to absent at each field site, and the ecology of each field site was taken into account. For instance, chimpanzees will not sleep in ground nests (as opposed to tree nests) at sites with high leopard or lion predation. Such ecologically explainable differences were excluded from the list leaving 39 behavioural patterns. The researchers found no evidence that habits vary more between, than within, the three existing subspecies of chimpanzees. So genetics cannot account for the observed variability.

Taking into account many results obtained in captivity, the authors suggest that mechanisms of social learning simpler than imitation could be involved in the processes of the formation of new customs in chimpanzee populations. The most commonly suggested mechanism is stimulus enhancement, in which the attention of an observer is merely drawn to a relevant item such as a stick. Socially learned experience is combined with self-practice in young and adults. Field and laboratory experiments have enabled primatologists to observe more flexible tool-using behaviour than in 'orthodox' communities with stable traditions. For example, in field experiments, presenting apes from the Bossou group with containers filled with water and juice, Tonooka (2001) revealed a new technique for the use of leaves for drinking, that is, 'leaf folding': the chimpanzees folded leaves in a manner resembling the side ribs of a bellows or Japanese origami (paper folding).

Until recently chimpanzees were considered the only species among the great apes to possess elements of 'material culture'. Nowadays researchers consider that chimpanzees display the highest level of manufacturing ability but are not the only species sharing with humans the membership in the club of animals with culture. Besides Africa's gorillas and chimpanzees the great apes include orang-utans, the fabled red apes of the forests of Indonesia. Orang-utans *Pongo pygmaeus* are less social than other primates, living a rather solitary life in the wild. They are slow in movement, not leaping vigorously from limb to limb like chimps or crashing through undergrowth like gorillas (Galdikas and Briggs, 1999). Thirty years of field observations of the shy South-east Asian orang-utans have enabled an international group of researchers to conclude that these apes definitely have the ability to adopt and pass on learned behaviours (Fox *et al.*, 1999; van Schaik *et al.*, 2003; Wich *et al.*, 2003).

Studying six populations of orang-utans in Borneo and Sumatra, Indonesia, researchers identified 24 examples of behaviours that have been defined as cultural variants. Many of the culturally transmitted behaviours involve tool use such as using sticks to dig seeds out of fruit, to poke into tree holes to obtain insects, or to scratch; using leaves as napkins or as gloves to protect against spiny fruit. Twelve other behaviours, such as making a pillow with twigs, were seen only rarely or were practised by only a single individual. Practices that are common in one group and absent in another are of great interest to researchers because variations in these behaviours found among the different populations seem to be cultural. For example, in a Sumatran swamp, one particular group of orang-utans like a fruit that was protected by needle-like spines, and to get to the edible seeds inside, the apes used a tool. With a sharp stick, they prised open the fruit to extract the seeds. Only a single group of the six observed has discovered how to use sticks to extract insects from tree holes or to wedge out seeds from fruits. Such tool use is common among chimpanzees, but the Sumatran orang-utan band puts a unique twist to the practice – they grip the stick with their teeth instead of their hands. On the far side of the river another group of orang-utans have plenty of sticks available, but they do not use them on fruit; most ignore the fruit, others smash it to get the seeds. The stick trick seemed to be an invention created by one group that was passed along. This is what researchers call a 'cultural boundary'.

A group of orang-utans in Sumatra has learned to use leaves as gloves when handling spiny fruits. The dainty use of a napkin has been discovered by one band. Apes in a Borneo band routinely wipe their faces with leaves and parents teach the social skill to their young. A second Sumatran band has learned the unique skill of getting a drink by dipping a leafy branch into a water-filled tree hole and then licking the moisture from the leaves.

Comparative data on the geography of tool use in northern Sumatra have shown that ecological and genetic factors are involved as necessary preconditions, but that the geographic variation in orang-utans is cultural, as in chimpanzees. Thus, the incidence of skilled behaviours such as tool use in a social unit must be explained with reference to the ontogenetic process of skill acquisition: invention, diffusion (importing skills invented elsewhere) and social transmission.

Recent data obtained by Krützen *et al.* (2005) allow marine mammals to be added to the catalogue of culturally transmitted forms of tool use in non-human populations. In Shark Bay, Western Australia, wild bottlenose dolphins (*Tursiops* sp.) apparently use marine sponges as foraging tools. Sponge-carrying came to the attention of scientists 20 years ago when a boater reported seeing a dolphin in Shark Bay with a 'tumor' on its beak. The tumour turned out to be a sponge, and in 1997 researchers proposed sponge-carrying as the first known example of tool use in dolphins (Smolker *et al.*, 1997). Dolphins have devised a way to break marine sponges off the sea floor and wear them over their snouts when foraging. Researchers believe that dolphins use sponges as a kind of glove to protect their sensitive rostrums when they probe for prey in the substrate. Unlike in apes, tool use in this population is almost exclusively limited to a single matriline which is part of a large albeit open social network of frequently interacting individuals. To discover whether tool use is a genetic trait, or one transmitted culturally, Krützen and colleagues analysed DNA from 13 of 15 spongers, only one of which was male, and 172 non-spongers. They found that most spongers were maternally related – sharing the same mitochondrial DNA, which is only transmitted through the female line. A comparison of their nuclear DNA showed that the spongers were closely related, suggesting that spongers are descendants of a recent 'Sponging Eve'. However, the pattern of sponging among the dolphins could not be explained by a 'gene for sponging' – the trait's pattern of inheritance just did not fit. The researchers conclude that the behaviour is culturally transmitted, presumably by mothers teaching the skills to their sons and daughters, although they have not actually observed this feat taking place.

Tool use is the most amazing but not the only population-specific behavioural trait that enables cetacean biologists to claim that marine mammals possess culture (Whitehead, 1998; Deecke *et al.*, 2000) or at least traditions. Mann and Sargeant (2003) have listed many population-specific patterns concerning foraging strategies, styles of diving and other behavioural traits, and many of them have been clearly demonstrated to be transmitted by means of social learning.

27.3 Dialects as cultural traits

The communication behaviour of animals can change because of genetic differences, maturational effects or learning. Communicative traditions in animals are expected to be socially learned. Recently many authors have considered *dialects* as a part of animals' culture. Dialects are specific communication traditions of sympatric living or neighbouring groups of subpopulations. Potentially, dialects could arise from communications whose functional structure is determined by genetic templates. Recently a growing body of literature has included examinations of dialects in many species. Some variations of communications are likely to be based on genetic and not social influences. Such are variations in dance dialects in honeybees (Rinderer and Beaman, 1995; Johnson *et al.*, 2002) and possibly in alarm calls in ground squirrels (Randall *et al.*, 2005). Even in chimpanzees, differences in the species-typical male call ('pant hoot') which serves for communication with conspecifics over long distances have been interpreted in terms of ecological rather than social factors (Mitani *et al.*, 1999). Nevertheless, many studies have concentrated on the cultural constituents of dialects. Among cultural communication systems in animals, song dialects in two very different groups of animals, birds and whales, are perhaps the most intensively studied.

Birds, especially songbirds and parrots, are well known for their ability to learn complex communication signals. Vocal dialects have been documented in a range of bird species (Thorpe, 1958; Baptista, 1975; Marler, 1984; Slater, 1989). Many birds, such as the white-crowned sparrow, chaffinch or parrot, can develop local song dialects. For example, the song of male white-crowned sparrow *Zonotrichia leucophrys pugetensis* forms about 12 dialects along the Pacific Northwest coast (Nelson *et al.*, 2004).

It was a long-standing hypothesis that avian dialects contribute to reproductive isolation between populations (Mayr, 1942, 2001; Marler and Tamura, 1962; Nottebohm, 1993). Behavioural studies have shown that in many bird species variation in songs results from learning, and that cultural transmission of song is the rule for most oscine birds – nearly half of the world's avian species (Kroodsma and Baylis, 1982; Baker and Thompson, 1985). Recent genetic examinations also revealed a low degree of correspondence between population genetic structure and dialect boundaries in many songbird species (for review see Wright and Wilkinson, 2001). Interestingly, studies in humans have repeatedly found a correspondence between geographical variations, genes and languages (Sokal *et al.*, 1988; Cavalli-Sforza, 1997). This enables researchers to speculate about 'real differences between humans and birds in the degree of co-evolution of genetic and cultural traits' (Wright and Wilkinson, 2001).

A good example of a recent examination of the relationship between genes and culture in birds is the study of Wright *et al.* (2005) on the yellow-naped Amazon *Amazona auropalliata*. In this species dialects are confined to communal roots of 50–200 parrots. Playback experiments have shown that most birds attending a roost within a dialect only use calls specific to that dialect; the rare exceptions are some birds at roosts bordering two dialects that produce the calls of both neighbouring dialects. To test the correspondence between dialects and genetic population structures, Wright *et al.* (2005) have compared geographic variations in microsatellite allele frequencies at nine sites in Costa Rica. There was no relationship between the genetic distances between individuals and their dialect membership, and high rates of gene flow were estimated between vocal dialects based on genetic differentiation. The results suggest that the observed mosaic pattern of geographic variation in vocalisations is maintained by learning of local call types by immigrant birds after dispersal. The lack of concordance between vocal dialects and population genetic structure in the yellow-naped Amazon mirrors that found in a range of songbird species with vocal dialects (see Wright *et al.* (2005) for a review).

Now we pass on to dialects in cetaceans. Dialects have been documented in several species of whales and dolphins. Among them, the vocal dialects of killer whales *Orcinus orca* have been studied most extensively (Ford and Fisher, 1983; Yurk *et al.*, 2002; Tarasyan *et al.*, 2005). The north-eastern Pacific Ocean is home to two distinct forms of killer whales. Resident killer whales live in large stable groups and feed exclusively on fishes. Transient killer whales live in smaller social groups and prey only on marine mammals. The two different forms do not interbreed and rarely interact. They show striking differences in their vocal behaviour. Residents frequently emit a variety of vocalisations whereas transients are usually silent. Vocalisations of residents include echolocation clicks, tonal whistles and pulse calls. The most common pulsed vocalisations of resident killer whales are 'discrete calls', which can be divided into distinct call types. Long-term studies have shown that resident killer whales have a complex system of vocal dialects: different social groups have repertoires of between seven and 17 structurally distinct call types. Comparison of two call types made by two matrilineal social groups of resident killer whales have suggested that vocal learning is not limited to vertical transmission from mother to offspring, and that distinct acoustic repertoires persist in the northern resident community in the form of acoustic clans. Frequent prolonged acoustic contacts between members of different clans occur; however, clan boundaries rather than boundaries between matrilines are the barriers to vocal matching and horizontal transmission (Deecke *et al.*, 2000).

It is interesting that existence of a 'crucible of culture' reflected in dialect clans of killer whales possibly has generated a 'mirror culture' in populations of harbour seals that learned to distinguish between characteristics of communications of those acoustic clans of killer whales which are dangerous or harmless for marine mammals. Playback experiments have shown that harbour seals *Phoca vitulina* responded strongly by panic diving to the calls of mammal-eating killer whales (transient populations) and unfamiliar fish-eating killer whales that migrate with salmon shoals and use vocal dialects unfamiliar to local populations of seals. Harbour seals learn the vocal repertoire of resident fish-eating killer whales and do not react to their signals, thus saving much energy due to their capacity of complex acoustic discrimination (Deecke *et al.*, 2000).

To what extent dialects can be considered a part of language behaviour is one of intriguing questions of the large topic of intelligent communication in animals which will be considered in Part IX.

CONCLUDING COMMENTS

Social learning plays an important role in the processes of 'tuning' behaviour in group-living species and in those that live solitarily but at least have contacts with relatives at early stages of ontogenesis. Readiness to gain information from conspecifics reflects both the conformity prevailing in animals' society and the flexibility that enables animals to improve their individual behaviour in changeable environment.

The capacity to learn from others and about others allows members of a species to decrease the cost of being equipped with an inherited suite of a great number of behavioural characteristics. Being given extra guidance by means of social learning, animals can increase their fitness and make relationships with their environment more flexible and thus more efficient. It is possible that social learning has more fundamental importance as a part of evolutionary strategies of many species than we thought before. In principle, it could be more adaptive for populations to have dormant 'sketches' of complex behavioural patterns being implemented in several carriers and then spread by means of social learning under suitable circumstances. Integration of behaviour thus takes place not only at an individual level but at the population level as well. Behavioural strategies of carriers of such 'sketches' of behavioural patterns can be evolutionarily stable. Following these strategies does not require feats of intelligence from animals, being based on relatively simple forms of social learning. Moreover, in these cases social learning underlies the species' predisposition to learn certain behavioural patterns. Several examples can be adduced from experimental studies on ants, crows, apes and some others.

Animals' ability to develop completely new behaviour by observing innovations invented by a single or a few advanced individuals must be based on intelligence rather than automatic population processes. The effectiveness of new behaviours performed by 'wild prodigies' may be evident for conspecifics but this does not mean that many imitators will subscribe to the same activity. Usually animals observe innovators and try to stand aside. Innovations are most often extinguished within a viscous environment of wild minds. One can say that non-humans teach badly and learn poorly, and that preparedness is the best teacher for animals.

It is very likely that, as Premack and Premack (1996) have stated, humans possess a unique 'pedagogic disposition' to exploit the learners' 'predisposition to culture', for teachers to demonstrate correct performance for the benefit of the learner.

In general, social learning is based on differences existing between members of animal communities, that is, on behavioural specialisation in populations, and in some situations, on cognitive specialisation of individuals.

With the exception of rare cases of 'true teaching', social learning is based on 'public information' rather than on the deliberate exchange of messages. Animals definitely have something to communicate to each other. The question of whether and to what extent intelligence is involved in their communication will be considered further in Part IX. We also passed over such an original form of social learning as 'social learning from an information centre'. This feature is typical first of all for eusocial organisms and will also be considered in Part X.

Part IX

Intelligent communication

'Taffy dear, the next time you write a picture-letter, you'd better send a man who can talk our language with it, to explain what it means.'

Rudyard Kipling, 'How the first letter was written' (*Just So Stories*)

A hero of Carlo Collodi's *Adventures of Pinocchio*, the carpenter Geppetto, made his wooden marionette in silence in a room until he made its mouth, and no sooner was it finished than it began to laugh and poke fun at him. Starting from very early experiments on animal intelligence, we have considered them in silence and now it's time to repay a means of communication to our subjects and to try to understand what they can say to each other and perhaps to experimenters as well.

In this part we will consider a question of how addressed signalling works in animal societies. It is intuitively clear that many social species must possess complex communications. The question of the existence of a developed natural 'language' in non-humans remains so far obscure. Of course, just as we have seen the same problem when considering animal cultural behaviour, how to treat language behaviour in animals is heavily influenced by our decision on which definition of language to adopt. Many definitions in the literature make language exclusively a human trait. Even following broad definitions of language, animal experts are oscillating between questions 'Why animals don't have language' (Cheney and Seyfarth, 1998) and 'What's so special about speech?' (Hauser, 2001). The main difficulties in the analysis of animal 'languages' appear to be methodological. In this part of the book we will shortly consider the main experimental approaches for studying complex communication in animals. What is beyond doubt is that studying the communicative means of different species is a good tool for judging their cognitive abilities.

This part contains analysis of experimental approaches for studying animal language behaviour. The term 'language' being attributed to animals is a point at issue so I prefer to use the notion of 'language behaviour' rather than 'language' in those cases when descriptions of complex forms of communication do not require the use of definite terms. We will concentrate on methodological aspects of studying animal language behaviour and on the role of language behaviour in animal intelligence.

At least three main approaches to a problem of animal language behaviour have been applied recently. Firstly, there is direct dialogue with animals based on language-training experiments. Having been applied to apes and one grey parrot, this approach has revealed astonishing mental skills. It is important to note that this way to communicate with animals is based on adopted human languages. Surprisingly little is known yet about the natural communication systems of the species that were involved in language-training experiments. The second approach is aimed at direct decoding of animal signals. The third approach to the study of animal communication has been suggested based on the ideas of information theory (Ryabko, 1993; Ryabko and Reznikova, 1996). The main point is not to decipher signals but to investigate just the process of information transmission by measuring the time duration which animals spend on transmitting messages of definite length and encoding them. Application of this approach has already allowed demonstration of the fact that a few highly social ant species possess possibly one of the most intricate forms of animal communication known, and it is a great challenge to extend these experimental schemes and approaches to the study of other species.

Chapter 28

Can animals exchange meaningful messages?

Understanding the 'languages' of animals seems to be an attractive and hardly achievable skill for humans with which many legends are associated. The title of Lorenz's (1952) book *King Solomon's Ring* refers to a legend about King Solomon who possessed a magical ring which gave him the power of speaking with animals.

In 1661 Samuel Pepys wrote in his diary about what he called a 'baboon': 'I do believe it already understands much English; and I am of the mind it might be taught to speak or make signs' (as cited in Wallmann, 1992).

More than two hundreds years later, Garner (1892) experimentally tried to 'learn the monkey tongue very much in the same way men learn the language of a strange race of mankind'. In his book *The Speech of Monkeys* Garner (1892) argued that the reasoning of monkeys differs from that of humans in degree, but not in kind, and if it be true that humans cannot think without words, it must be true of monkeys. Garner spent much time in the zoological gardens at Washington and Cincinnati conducting experiments in which he showed or gave different things to monkeys and then took them away recording with a phonograph all sounds that the animals emitted during repetitive events. Garner examined several species including a chimpanzee, a spider monkey, several rhesus macaques and others. With the help of the phonograph Garner conducted first playback experiments. He artificially changed elements of the animals' 'speech' and then compared how animals reacted to variants of reproduced 'words'. In particular, Garner insisted that he succeeded in distinguishing a word that meant 'food' in the capuchins' language from other ones that meant 'bread' and 'vegetables/fruits' (Garner argued that capuchins sound similarly when they express a desire to get a carrot, an apple or a banana). The word 'food', according to Garner, was also used by capuchins as a friendly greeting.

Garner anticipated some findings of the present day. For instance, applying a similar experimental paradigm, Hauser and Marler (1992) found that rhesus macaques produce five acoustically distinctive vocalisations when they find food. Three vocalisations ('warble', 'harmonic arch' and 'chirp') are restricted to the discovery of high-quality, rare food items, whereas the other two ('grunt' and 'coo') are given while waiting for and eating lower-quality, common food items. An individual's hunger level is positively correlated with call rate, but not with the type of call produced. That is, the type of vocalisation produced is influenced by the type of food discovered, and not by the discoverer's hunger level (Hauser, 2000).

Attempts to identify categories in animals' communication are based on a natural idea that the complexity of language behaviour should be connected with high levels of sociality and cooperation in animals' societies. In the 1960s and 1970s elegant but ambiguous experiments were conducted with highly social intelligent animals such as chimpanzees and dolphins that were asked to pass some pieces of information to each other.

In Menzel's (1971, 1973a, b, 1974, 1975) classic experiments a group of chimpanzees living in an enclosure searched for hidden food (see also Chapter 12). The experimenter placed a piece of food hidden in a box in view of a leader. An informed animal succeeded in leading others to the reward by drawing attention to it through actions such as tapping others on the shoulders or repeatedly glancing at them while heading in the direction of the food. Eventually, those chimpanzees who were naive to the location of the bait seemed to have learned to recognise individuals most competent in finding food and followed them until rewarded. Menzel did not explicitly test whether the followers understood that the leader knew the location of the hidden food. However, he argued that chimpanzees somehow learn from the leader what kind of the reward was hidden. They more readily searched for fruits than vegetables, and displayed signs of exciting and fear in anticipation to find a scaring toy (a snake) hidden in the box. Menzel suggested that chimpanzees possess the means for transferring information about both location and properties of objects but it remained unclear how they do this.

The dolphin experiments carried out by Bastian (1967) and Evans and Bastian (1969) concerned cooperative behaviour of two young dolphins which could involve intelligent communication. Researchers tried to teach a male dolphin, Bazz, and a female, Doris, to communicate through an opaque barrier. First of all, while they were still together, the dolphins were taught to press paddles when they saw a light (session A). If the light was kept steady, they had to press the right-hand paddle first; if it flashed then the left-hand one. When they did this correctly they were rewarded with fish. At the second stage (session B) of the experiment, they were taught to follow a sequence: Doris had to wait for Buzz to press the signal first. When the light came on under those conditions and stayed on, Doris waited for Buzz to hit his signal. When he didn't, Doris eventually made a sound, and soon after Buzz pushed the correct signal. Doris followed suit, and they got their fish. As soon the dolphins had learned this manoeuvre, they were separated (session C). Now they were situated in adjacent pools that allowed them to hear but not see each other. The paddles and light were set up in the same way, except that the light which indicated which paddle to press first was seen only by Doris. In order to get fish both dolphins had to press the paddles in the correct order. Doris had to inform Buzz which this was, as only she could see the light. The dolphins demonstrated essentially perfect success over thousands of trials at this task. It seemed that Doris was transferring novel information to Bazz by means of acoustic signals. Indeed, the transmission of this type of information would imply that dolphins have the ability to create acoustic symbols to represent something as arbitrary as a flashing or steadily lit light.

Appreciation of Bastian's paradigm came from Markov and co-authors (Markov and Ostrovskaya, 1990; Zanin *et al.*, 1990) who used adult female dolphins in an experiments modelled closely but not exactly on Bastian's. They replaced light signals with balls of different size and incorporated role-reversals successfully into the same experiment. In Bastian's experiment, the role-reversal segment was not completed, but enough data were accumulated, before they shut down the experiment, to show that there was not an easy exchange of roles and thus probably no 'words' that were easily used by both dolphins. In experiments of Markov and co-authors the dolphins did cooperate in joint work with a high frequency (83 per cent) and in 90 per cent of all joint work pressed the same paddle. After the high percentage of session A, though, the response percentages began to decline until the dolphins refused to 'talk' to each other in the session C. This breakdown in communication was due to one of the dolphin's (Jenny) dominance over the other (Kora) and her actions stopped the role-reversal (in which Kora would 'tell' Jenny which ball to press). Nevertheless, the results obtained showed that dolphins can coordinate each other's behaviour. The relative ease with which the dolphins reversed roles enabled the researchers to suggest that the dolphins were using an existing equivalent of a 'word' from their own communication code.

Despite these supportive experiments with a role-reversal stage completed, the question about dolphins' natural 'language' still remains unclear. One can agree with Morton's (1971) opinion that if dolphins do have a language they are going to extraordinary lengths to conceal the fact from us.

Firmly resolved to discover what animals really conceal from us, we should first define the term 'language' and identify a field of comparative analysis of human and animal languages.

Chapter 29

Communication, speech and language: what falls to the share of non-humans?

In this chapter we will consider some notions and definitions that help us to specify limits for reasoning about animals' language behaviour as one the most intelligent forms of relationship.

29.1 Communication

The term 'communication' enjoys a wide variety of meanings, which is of no wonder because communication is a diverse and widespread phenomenon that serves as a substrate of any social behaviour. It is difficult to imagine either social behaviour of any level lacking a means of communication or any transfer of information that has nothing to do with social relations.

Some definitions include codes of specific signals which animals use to communicate with each other. For instance, Vauclair (1996) defines communication as follows: 'Communication consists of exchanges of information between a sender and a receiver using a code of specific signals that usually serve to meet common challenges (reproduction, feeding, protection) and, in group-living species, to promote cohesiveness of the group.'

In group-living animals communication serves many important functions such as (1) to advertise individual identity, presence and behavioural predispositions; (2) to establish social hierarchies; (3) to synchronise the physiological states of a group during breeding seasons; (4) to monitor the environment collectively for dangers and opportunities; (5) to synchronise organised activities (migration, foraging).

Many species have well-developed forms of inter-individual exchange of information based on specific social signals related to courtship, defence of territories, ranking, care of offspring and other forms of social interrelations. There is a huge body of examples concerning the use of expressive signals within a wide variety of species. For instance, lizards have gestures such as submissive and aggressive circular forelimb waving, tail lashing, head bobbing, back arching and so on (Sinervo and Lively, 1996).

To identify complex forms of animal communication that may be attributed to language behaviour, the important feature is purposiveness. It is often difficult to decide whether animals intend to share information with conspecifics or whether they use inadvertent signalling. We can consider language behaviour the most complex form of communication which take place when animals advisedly transfer the information to each other.

29.2 Speech

It is intuitively clear that communication is too broad a concept and speech is too narrow for consideration as the form of information transfer in animals. Speech is a specially designed system for language and serves as a very effective mean of communication. According to Kimura (1979) there are two major forms of language in humans: speech and the manual sign language of the deaf. Besides, many ways of signalling languages exist in human culture such as

Morse code, drum signalling, flag signalling, whistles, semaphore, smoke signalling and so on. Evidently, human beings universally and preferentially employ speech to communicate.

In the great majority of definitions in the scientific literature speech is stated to be a form of communication specific to humans. Phonetic assurance of speech in humans is supported by the coordinated actions of several articulators (jaw, lips, tongue, velum and larynx) and by specific brain structures. The task for a child learning to speak is to reproduce, or to imitate, the patterns of articulatory gesture specified by the acoustic structure of the word heard. A capacity to imitate vocalisations is confined to a few species of birds, certain marine mammals, certain monkeys and humans. Speech development has both universal and language-specific aspects. Perceptually, speech already has a unique status for the human infant within a few hours or days of birth. Neonates discriminate speech from non-speech, prefer speech to non-speech, and prefer their mother's voice to a stranger's.

Perceptual studies of infants from 1 to 6 months of age, based on the habituation–dishabituation experimental paradigm, have shown that infants can discriminate virtually any speech sound contrast on which they are tested, including contrasts not used in the surrounding languages (these experiments have been described in Part II as an expressive example of studying habituation: the amplitude of a baby sucking its dummy and several other behavioural cues were used as characteristics of infants' reactions to novel stimuli). Over the second 6 months of life infants gradually cease to discriminate non-native sound contrast. During the period in which an infant is perhaps first attending to the different contexts in which words are used, it also gradually sifts speech sound contrasts that are functional in the surrounding language from those that are not (Studdert-Kennedy, 1983; Eimas, 1985; Menyuk *et al.*, 1986). A real-world problem facing the human infant is how to segment the continuous acoustic stream of speech into functional units such as phonemes, words and phrases. It has been shown that infants are equipped with mechanisms that enable them to extract abstract rules which subsequently may form the foundation upon which grammars are constructed (for a review see Hauser *et al.*, 2002a).

These findings suggest that our species is highly predisposed for the development of speech as the specific form of communication, and they are in harmony with Chomsky's (1968) point of view that there is an innate apparatus in humans generating a universal grammar.

Studies by Santos *et al.* (1999) have shown that the same method of studying reactions to speech sounds can be used with human infants and non-human primates. Instead of the suckling response used as the behavioural cue in human infants, for monkeys, researchers recorded a head-orienting response in the direction of the concealed speaker. The habituation–dishabituation experimental paradigm provides a tool to explore similarities and differences in perceptual mechanisms.

Results of playback experiments have shown that such animals as cotton-top tamarins not only attend to isolated syllables but also attend to strings of continuous speech. Tamarins demonstrated the ability to discriminate sentences of Dutch from sentences of Japanese in the face of speaker variability which enables to suggest that they are able to extract acoustic equivalent classes. Specifically, having been exposed to continuous acoustic stream of syllables, tamarins, like human infants, were able to compute relevant statistics (Hauser and Fitch, 2003).

It is important to note that neuroanatomically, there is a direct connection from the cortical face area to the laryngeal motoneurons in humans but not in monkeys. Bilateral destruction of this area distorts human vocal utterances severely, while those of the monkey remain unaffected. These and other findings suggest that the cortical face area participates in voluntary articulation in humans. At the same time, discrimination experiments have shown that there is a left hemisphere (right ear) advantage for the recognition of species-specific (but not others) calls in monkeys. So the principal neural prerequisites for decoding speech sounds seem to be already present in the monkey (Weiss *et al.*, 2002).

As we have seen, speech is just one form of language, the most complex and likely to be confined to humans. Unlike the broad notion of communication, and the specific notion of speech, the concept of language should be more useful for reasoning about animal communication.

29.3 Language

The use of language is one of the most complex human skills and it is no wonder that many scientists define language in such a way that only human beings can be said to be capable of it. For instance, Hornby (1972) defines language as 'Human and non-instinctive method of communicating ideas, feelings and desires by means of a system of sounds and sound symbols'. Chomsky (1986) considers language the prerogative of humans that is aimed to facilitate free expression of thought and clarify one's ideas, as well as to help establish social relations and to communicate information. Behaviourally, language can be defined as a system of self-generated movements, composed of definable units, which can arbitrary represent some object, event or intention on the part of the mover (Kimura, 1979). Human language is the most sophisticated communicative system which includes thousands of spoken and written languages. As a reflective communication, language can be used to transfer ideas, to convey feelings, to lie and to exchange fantastic imaginations. It is a debated question whether any but human animals share these features. As with most questions about animal cognition, there is a problem of methodology and interpretation. As we have already seen in this book, many species share with humans such cognitive abilities as abstraction, categorisation and classification. Animals' ability to lie and sympathise with others will be discussed in the next Part. All these abilities are essential for the development of language but it is still difficult to decide whether they are sufficient. What can we interpret to be equivalent to human language depends on how we identify language and to what extent our concept of language allows comparisons between species.

It is now agreed that language cannot be described or defined by one single feature; it is a polymorphus concept. Hockett (1960, 1963) identified a range of characteristics that described essential features of language. Hockett's longest list includes 16 features. Some of them are only applicable to human spoken language and are not relevant to all forms of language in general, including human sign language. However, some of the design characteristics are seen as essential features of any form of language and are useful criteria to assess claims about animal language. Several authors have used Hockett's set of characteristics to complete a list of criteria that an act of communication must fulfil if it is to be regarded as language (Aitchison, 1983; Anderson, 1985; Pearce, 2000).The following list provides a useful framework for evaluating the linguistic skills of animals.

(1) *Interchangeability*. Language is a two-way process that involves both sending and receiving the same set of signals. This is distinctive from some animal communications such as that of the stickleback fish. The stickleback fish make auditory signals based on gender (basically, the males say 'I'm a boy' and the females say 'I'm a girl'). However, a male fish cannot say 'I'm a girl', although he can perceive it. Thus, stickleback fish signals are not interchangeable. In many social species any member of a group can produce and perceive universal signals, so their communication systems meet this criterion.

(2) *Specialisation*. Language is not a by-product of some other biological function; it has a special function for communication only. Humans enjoy the most specialised system of communication. Returning to the stickleback, the male reacts to biological aspects of a female's messages such as a swollen belly, and the female reacts to the change of the male's coloration, so the male's message is more specialised.

(3) *Discreteness*. The basic units of language (such as sounds) can be organised into discrete units and classified as belonging to distinct categories. There is no gradual, continuous shading from one sound to another in the linguistics system, although there may be a continuum in the real physical world. Thus speakers will perceive a sound as either a [p] *or* a [b], but not as a blend, even if physically it falls somewhere between the two sounds. It is important that discrete units can be broken apart to form new signals.

(4) *Arbitrariness of units.* Language is composed of discrete units, and the form of the signal does not depend on the referred thing. Messages consist of arbitrary units but not of icons describing an object. This important property returns us to the epigraph to this part of the book. Taffy, the girl in Kipling's story, sent her mother a letter consisting of icons, and a wrong interpretation of pictures resulted in a great deal of trouble.

(5) *Displacement.* Language can be used to refer to things that are not present in space (here) or time (now). These requirements are known as 'spatial displacement' and 'temporal displacement' respectively. Human language allows speakers to talk about past and future. Speakers can also talk about things that are physically distant (such as other countries and other planets). They can even refer to things and events that do not actually exist. Many researchers consider that the honeybees' dance language meets this criterion. Bees are able to refer to objects that are distant in space and even in time. For instance, Lindauer (1960) reported that in the night bees performed dances signalling feeder sites that they had visited the previous day. Apes easily meet this criterion with the use of intermediary artificial language but we know little about the potential of their natural communication.

(6) *Semanticity.* Communication meets this criterion if signals (or words) convey specific meanings. We presume that for an animal a signal has meaning if it can activate a representation of the event to which it relates. Thus vervet monkeys appreciate the meanings of alarm calls they hear (Cheney and Seyfarth, 1980, 1990). It is likely that honeybees may appreciate the meaning of communications they receive. In Chapter 12 the results of Dyer's (1991) experiments were described. Bees left the hive when the returning scout indicated that the food was beside a lake. But they did not leave the hive when they were informed that food was near the middle of the lake. Thus honeybees appear to interpret the meaning of the dance – possibly by identifying the potential location of food on cognitive map – and then decide whether it is worth making the journey (Pearce, 2000).

(7) *Productivity.* Language can be used to produce an infinite variety of new messages from a limited vocabulary. Productivity is also called 'creativity'. In particular, productivity of language enables the use of analogies, and by the use of analogies children adopt rules of grammar, or syntax. Productivity is a powerful property of language that makes it an open-ended system. The use of intermediary languages enables animals to meet this criterion. One challenge is to explore whether any natural animal communication system meets this requirement.

(8) *Traditional (cultural) transmission.* Human language is not inborn, or at least, completely inborn. Although humans probably have a genetically based capacity for language, they must learn, or acquire, their native language from other speakers. This is different from many animal communication systems in which the animal is born knowing their entire system. At the same time, as was described in Parts VII and VIII, cultural transmission plays a great role in the development of communication in some species of birds, monkeys, whales and perhaps some others. We have also seen from Part VII that cultural transmission does have limits in some species. For example, with humans and finches, language acquisition does not happen easily or ever fully after a certain age. Finches have been tested by removing them from other finches until they are grown up; the test finches do make calls (so it is to some degree innate) but never to full capacity (so it is to some degree transmitted culturally). With humans, several cases of children raised in isolation due to unfortunate circumstances show that although some language can be learned, not nearly as much can once a certain age is passed. It is known that language acquisition occurs best at an early age, primarily before puberty, and the capacity for it decreases after that age. Some communication systems are fully innate, however, with no cultural transmission. For example, cowbirds (who lay their eggs in other birds' nests), when born, will give cowbird calls, not the calls of whatever bird they are raised by.

Chapter 30

Direct dialogue with animals: language-training experiments

The use of intermediary languages for studying 'linguistic' and intellectual potential of animals can be considered a revolutionary approach which has changed the general concept of animal intelligence. Only 30 years ago it would have been difficult to imagine that animals could learn to associate arbitrary signs with meaning, to generate new symbols with new meanings and to use these signs to communicate simple statements, requests and questions; to refer to objects and events displaced in time and space; to classify novel objects into appropriate semantic categories; and to transmit their knowledge to peers and offspring.

There are many excellent books and reviews written by researchers who have carried out projects on teaching sign languages to apes (Patterson and Linden, 1981; Savage-Rumbaugh, 1986; Gardner and Gardner, 1989; Savage-Rumbaugh and Lewin, 1994; Fouts, 1997), dolphins (Herman *et al.*, 1984) and an African grey parrot (Pepperberg, 1999). Here I briefly describe how this method has influenced development of studying of animal language behaviour and intelligence.

Attempts to enter into negotiations with animals began from dialogues between humans and apes. As was mentioned in Part I, Robert Yerkes made a suggestion that was not followed up for 40 years: perhaps apes could learn gesture sign language. Yerkes failed to teach chimpanzees to speak and he concluded that apes cannot speak because they lack the tendency to reinstate auditory stimuli – in other words to imitate sounds. Indeed, several attempts to teach apes to repeat human words were mostly unsuccessful. Furness (1916, reviewed in Ristau and Robbins, 1982) taught a female orang-utan to produce vocally 'papa', 'cup' and 'th' over 11 months of instruction. Hayes and Hayes (1952) achieved a little more with their chimpanzee Viki who was taught to pronounce a few words, 'mama', 'papa', 'cup' and 'up', by the use of positive reinforcements. At first, Viki was rewarded with food for making any vocalisation, and in this way she was taught how to produce sound on demand. It took 5 months before Viki was able to produce vocalisations that were recognisable in any way to humans, the first one being a hoarse 'ah' sound. This was later developed to 'mama'. However, the Hayes recorded that Viki had considerable difficulty pronouncing words, would often become confused and pronounce or use the words incorrectly. Also she held her lip while 'talking'. It has since been discovered that upper respiratory systems with a vocal apparatus enabling human speech differ in apes and humans and, what is not less important, the neural mechanism developed for encoding and decoding phonemic communication in humans is specific and much different from those in apes (Lieberman, 1984).

Once it was apparent that apes could not learn to speak, other variants of direct dialogue were elaborated. It is worth noting that such dialogues were based on apes' propensity for categorization, one of the most important properties of carriers of the developed language.

30.1 'Token language'

One of the variants of a mute dialogue with animals was the 'token language' elaborated by

Wolf (1936). In general, tokens have been described as secondary rewards that can be exchanged for any primary reward (such as food, drink, toys, etc.). They have mainly been used in studies of operant behaviour. Nevertheless, in his classic experiments Wolf (1936) demonstrated chimpanzees as possibly being able to understand symbolic meaning of tokens and use them to reach specific goals. In his experiments tokens were differently awarded and thus differed in their purchasing power. For instance, inserting a blue token into a slot an animal received two bananas, while a white token gave only one. A black token could be exchanged for any kind of food whereas a yellow one for drink only. In another series of experiments blue tokens being inserted into the hole gave apes the chance to return to their living cage from an enclosure, and the yellow one gave an opportunity to play with a human tutor. When a rat appeared in an enclosure, chimpanzees who hated rats abandoned all their entertainments and used blue tokens quickly in order to get to the home cage. Cowles (1937) revealed that chimpanzees can 'work' for tokens and they can even accumulate several tokens before exchanging them. Later 'token dialogue' was successfully used in several experiments (Kelleher, 1958; Schastnyi and Firsov, 1961; Firsov, 1972).

Recently Sousa and Matsuzawa (2001) have followed the same paradigm using a computer setting. Three chimpanzees were provided with a vending machine for inserting tokens (Japanese coins) and three touch-sensitive panels. The panels were connected to computers that controlled a discrimination task using tokens as a reward. Discrimination tasks were intellectually costly and included discrimination within several sets each of ten visual stimuli. The apes demonstrated a high level of accuracy, suggesting that the tokens were almost equivalent to direct food rewards. The apes also saved tokens before exchanging them for food. The chimpanzees performed tasks to collect tokens with the objective of exchanging them for food and thus planning their proximate behaviour. The authors suggest that chimpanzees use tokens as specific tools. In sum, these results support Wolf's suggestion that apes can understand symbolic meanings of tokens.

30.2 The use of intermediary gesture languages to speak with primates

Anthropoids can move their hands freely and fast, and sign language became the next obvious choice in the development of direct dialogue between humans and other species. Ulanova (1950) possibly was the first to teach a primate to use gesture signs to denote desired things (Fig. 30.1). She taught a rhesus macaque to make finger gestures in order to obtain rewards of different kinds. Different gestures corresponded to different food items (nuts, pieces of apples, bread, strawberry and radish) and drinks (coffee, tea and milk). A monkey was first taught to extend its hands to a human tutor and then the tutor moulded its hands and fingers to the right position and reinforced correct signs. It took from 152 to 576 trials to shape each gesture sign. The highest accuracy was achieved with the use of signs 'bread' and 'apple'.

Several projects trying to teach apes developed human language started in 1966, from the project of R. A. and B. T. Gardner who began teaching a chimpanzee, Washoe, the use of American Sign Language (ASL or AMESLAN), a gestural form of communication used by deaf people. The Gardners based their experiments on the cross-fostering paradigm which includes treating the chimpanzee infant like a human child in all its living arrangements, 24 hours a day, every day of a year. The same paradigm has been applied later in other projects connected with teaching apes different forms of language.

Using ASL, the Gardners followed the example of parents of deaf children by using an especially simple and repetitious register of ASL and by making signs on the youngster's body to capture her attention. The people who cared for and taught Washoe used sign language almost exclusively in her presence. The performance of Washoe can be compared with that of deaf children. The fascinating results of this project are described in the papers and books by the Gardners (Gardner and Gardner, 1969, 1980, 1998) and in Linden's (1974) popular book *Apes, Men and Language*.

Fig. 30.1 Symbols used for expressing the need for different types of food in Ulanova's experiments with a rhesus macaque: nut, strawberry, bread loaf, apple, radish. (Adapted from Ulanova, 1950.)

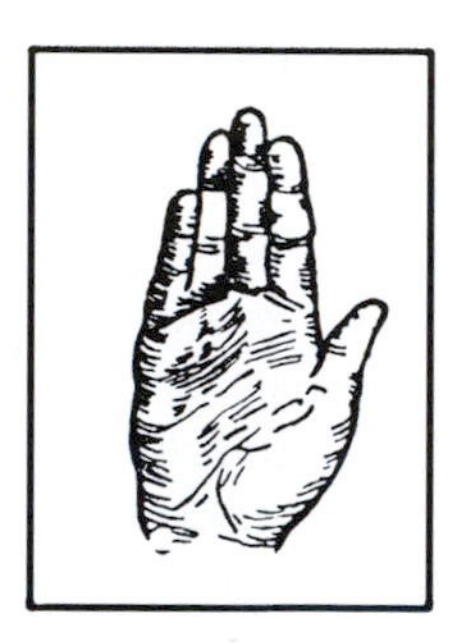
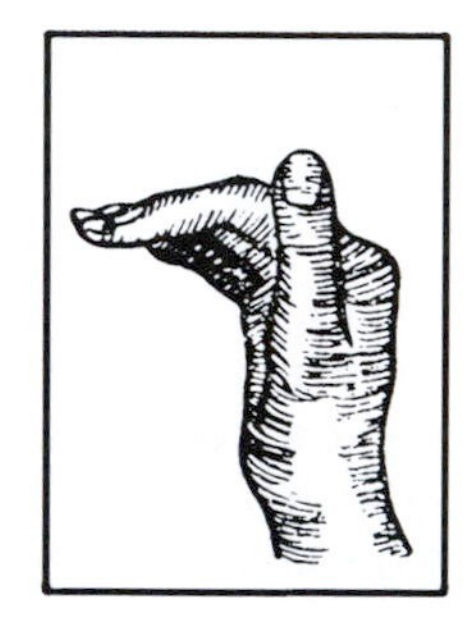

Washoe learned signs slowly at first but after 4 years of training she could use about 160 signs that included nouns, verbs and pronouns. Washoe was able to transfer her signs spontaneously to a new member of a class of referents; for example, she used the word 'more' in a wide variety of contexts (not just for more tickling, which was the first referent); she also used 'open' with doors, tins and nuts. She was able to use combination of signs, starting with two signs and then progressing to four- or five-sign 'sentences', for instance, 'give me tickle', 'open food drink' (meaning 'open the fridge'). One famous example is the novel sequence 'baby in my drink' which was produced in response to being shown her cup with a doll in it. Gardner and Gardner (1975) noted that Washoe used the 'dog' sign as a reaction to the sound of barking by an unseen dog. They also reported that Washoe began to use combinations of signs spontaneously after learning only about eight or ten of them. She thus has invented 'new' words: once she was signing 'Water' – 'Bird' when she saw a swan for the first time.

Some years after the start of the study (and after the death of her own baby) Washoe adopted an infant chimpanzee called Loulis in whose presence no human would use any signs. Despite this he had a repertoire of about 50 signs after 5 years. Some of these were learned by imitating Washoe and other signing apes in the research centre, but there was also evidence that Washoe moulded his hands into the correct position (Fouts *et al.*, 1982). She succeeded in teaching Loulis at least two words: 'food' and 'chair'. This seems to be an evidence of cultural transmission. The development of many phrase types with three signs was recorded for Loulis. An example is 'hurry you tickle'.

The Gardners soon extended their experiments to several other chimpanzees. All of them were born in American laboratories and each was taken into the cross-fostering experiment within a few days of birth. Young chimpanzees were raised in an environment of people who communicated by ASL between themselves and with chimpanzees. As a result, the chimpanzees purposely signed to friends and to strangers; they signed to each other and to themselves, to dogs and to cats, toys, tools, even to trees. Because the chimpanzees continued to use sign language without any input from humans, the Gardners concluded that once introduced, sign language is robust and self-supporting (Gardner and Gardner, 1989).

Another chimpanzee who was systematically taught sign language was Nim. This chimpanzee was educated by Terrace who was sceptical about the linguistic success of apes. Terrace believed that there were simpler explanations for many of the reported interpretations of these apes' language use. Although Terrace admitted that the apes had achieved something significant, he compared their behaviour to that of pigeons that are taught to peck different colours in a certain order (Terrace, 1979). He also believed that the apes used signs only to receive rewards from their human trainers. When Terrace set up his own experiment with Nim, he found some exciting and important details in chimpanzees' way of using language. Nim learned 125 signs, and a record of over 19 000 utterances he made was kept. Nim was observed practising his signs in the absence of their referents (Terrace, 1979). He often signed 'dirty' when he had to go to the toilet or 'sleep' when he was bored and wanted a change. He used the signs 'bite' and 'angry' to express his feelings, and he tended not to attack if he perceived that his warning was heeded; this is an important substitution of an arbitrary word for a physical action, displaying Hockett' s property of specialisation (the speaker does not act out messages).

However, Terrace found little evidence of Nim showing any ability to use language. When using more than two signs Nim showed no evidence of the use of grammar and he could not combine words to create new meanings on his own. Detailed analysis of video records showed that Nim interrupted his trainers more than human children interrupt their parents. Besides, a part of imitation among utterances produced by Nim increased with age whereas in children inverse negative relationship takes place. Terrace suggested that if we are going to say that apes can create a sentence, we must eliminate the other possible explanations for the utterances. Terrace is not as much of a sceptic as some others make him out to be, though; he believes that the conditions under which Project Nim operated were not ideal, and that future projects might have more success if they were able to instil a greater motivation to sign into their subjects.

These early projects stimulated interest in 'speaking apes'. Fouts spent several years educating young chimpanzees raised in laboratories or adopted by families. He found a great individual variability in apes' motivation and capability and discovered interesting abilities in some of them. For example the chimpanzee Elli was shown to be able to 'translate' nouns from English to ASL by herself, and Fouts proved this by special experiments (Fouts, 1997).

Similar studies using ASL were carried out using gorillas and an orang-utan. Koko is a female gorilla who has been trained by Patterson at Stanford University in California. Koko has a working vocabulary of over 500 signs and has emitted over 400 more (Patterson, 1978, 1981; Patterson *et al.*, 1988). Koko uses more words per utterance on average that Nim did, and a great deal of creativity, spontaneity and structure characterise her utterances. She also rhymes and jokes; on one occasion she used a metaphor of an elephant to refer to herself when she pretended a long tube was her 'trunk'. She also signed 'white tiger' to refer to a zebra, 'eye hat' for mask, 'cookie rock' for sweet, 'state roll', and 'elephant baby' when shown a Pinocchio doll. These characteristics of Koko's utterances show the property of productivity, in which a speaker says something never heard or said before and is understood by the audience. She signed 'me cry there' when she saw a picture of a gorilla in a bath, apparently a cry of sympathy, since she herself hates being bathed. In a fury, she once signed 'Penny toilet dirty devil' when she was angry with Penny Patterson, her trainer. Being hot-tempered but noble-minded, Koko apologised for a biting incident. When shown a bite mark on Penny's arm she signed 'Sorry bite scratch wrong bite'. 'Why bite?' queried Penny. 'Because mad', Koko replied. 'Why mad?' asked Penny. 'Don't know', responded the gorilla. Koko was later joined by another gorilla, Michael (also educated by ASL), and demonstrated the 'displacement' feature of her language by referring to him when he was not present. Chantek, an orang-utan, learned about 150 different signs, and used them spontaneously and without undue repetition. Chantek internalised the minimal value system, using signs for 'good' and 'bad' in appropriate contexts (Miles, 1993, 1994).

30.3 The use of alternative artificial languages to communicate with apes

Several alternative approaches to teaching apes a language have been elaborated based on special devices, and these have allowed a more objective comparison of the language abilities of animals.

Premack (1971, 1976) elaborated an artificial language in which the words were plastic figures, which varied in shape, size, texture and colour. Plastic words were arbitrary. For example, a red square meant 'banana', a black T-shaped figure denoted the colour 'yellow', and a pale blue star meant 'insert'. Sentences could be formed by placing the tokens, which had a metal backing, in a vertical line on a magnetic board. Sarah, Premack's brightest chimpanzee, was taught nouns, verbs, adjectives, pronouns and quantifiers; she was also taught same–different, negation and compound sentences. Subsequently, Sarah was taught to perform complex tasks by constructing sentences and following instructions composed of plastic words. If she created a correct sentence in response to an answer, then the trainer would place a word representing 'Correct' on the board, praise the chimpanzee verbally and perhaps give her a small reward. In trials when the sentence was wrong, a word representing 'Incorrect' was placed on the board, and the trainer said something like 'No, you dummy.' To test Sarah's view of words, Premack presented her with an apple and a set of features (for example, round versus square and red versus green). Then she was presented with her 'word' for apple, and the same set of features. She chose the correct features for both the real apple and her word for apple, a light-blue plastic triangle (Premack, 1976).

Sarah became proficient in the use of about 130 words, including complicated ideas such as 'colour of', 'same', 'different' and 'if ... then'. The questions she was presented with were formed by putting the token for 'query' in front of a sentence:

QUERY RED COLOUR OF APPLE (are apples red?)

'Reading' the sentences Sarah could answer questions or obey instructions such as 'Sarah insert apple red dish banana green dish.' In one task two coloured cards might be placed one on top of the other, on the table, and a sentence on the board would ask 'Query red on green' (is red on green?). Sarah correctly answered by placing 'yes' or 'no' token on the board depending on which card was placed on top. Premack (1983) concluded that, like human words, plastic words used by chimpanzees possessed both representational and communicative functions.

Premack's technique has been developed further in a project supervised by Rumbaugh (1977), in which the symbols serving as words – or 'lexigrams' as they were called – were displayed on a keyboard connected to a computer. Pressing a key resulted on its symbol being projected onto a screen above the consol. The language of lexigrams, each of which represented one word, was elaborated in the Yerkes Regional Primate Research Center in Atlanta, Georgia, and called Yerkish. The symbols were arbitrary and formed by combinations of geometric figures on different colour backgrounds, For example, a small solid circle inside a larger diamond on a purple background was a symbol for Lana, the first chimpanzee who was trained to use this variant of language. Lana used the symbols primarily to ask for things ('Please machine give apple Lana'). The purpose of the computer was to keep a record of Lana's statements and to dispense films, slides, music, food and drinks, when requested. However, Lana's requests not always could be met; for instance, she once asked 'Please machine tickle Lana.'

Lana started using 'No' as a protest (for example, when someone else was drinking a Coke and she did not have one) after having learned it as a negation ('it is not true that ...'). Lana acquired many linguistic-type skills for which she had received no specific training, which showed her ability to abstract and generalise. For example, she spontaneously used 'this' to refer to things for which she had no name, and she invented names for things by combining lexigrams in new ways. Lana's ability to generalise showed that her system has semanticity, that is, she understands that a symbol refers to a certain type of object, not just one particular thing. For example, she was taught the word 'more' in connection with an extra ration of fruit juice. Within a few days, she was readily attaching the symbol for 'more' to other types of food and drink ('more milk', 'more bread'). Lana also showed some evidence of creativity. For example, she was taught the words 'put' and 'in' in connection with putting a ball into a bowl or box. Soon after, Tim, one of her trainers, was late with her morning drink of milk. Lana spontaneously made the request 'Tim put milk in machine.' This shows not only creativity, but also displacement. In addition Lana used descriptive phrases such as 'banana which is green' for cucumber or 'apple which is orange' for orange (Rumbaugh and Gill, 1977).

Later two other chimpanzees, Sherman and Austin, educated in Yerkish, demonstrated the social use of language (Savage-Rambaugh, 1986; Rumbaugh and Savage-Rumbaugh, 1994). They had been trained to ask for tools to open food containers using lexigrams and to request and share food with one another. Similar results were demonstrated by chimpanzees in experiments of Schastnyi and Firsov (1961). Two chimpanzees were housed in adjacent cages. One of them was hungry but was lucky in having a toy in his cage. Another chimpanzee was fed and after that had been given bananas. When a hungry individual held a token that meant 'food' to his mate, the satisfied guy accepted a token in exchange for a banana. A short time later he 'asked' for a toy in exchange for a token and again the request was granted. These chimpanzees had never been taught to exchange by food and toys, so they acted spontaneously.

One of the most interesting studies of language learning in primates is that of a bonobo (*Pan paniscus*) called Kanzi (Rumbaugh and Savage-Rumbaugh, 1994; Savage-Rumbaugh and Brakke, 1996). Initially, researchers began by studying Kanzi's foster mother Matata with lexigrams but she failed to learn Yerkish well. However, even though no attempt was made to teach little Kanzi, he started to use lexigrams to request and name things. As soon as Kanzi started to use the lexigram keyboard, he received special training. Instead of taking part in formal training sessions for a restricted amount of time each day, he received constant attention and spent his time as a full member of the human group that worked in laboratory. Humans both spoke with him and communicated on a portable lexigram keyboard. Using this keyboard Kanzi has communicated with humans more and more successfully. For instance, he would press the lexigram for 'tickle' and then point to the person he wanted to tickle him. He followed strict order of action first and object second, just the opposite of the rule in English of the noun first then the verb second. This difference in communication indicates that not only does he seem to be imposing rules for how he 'speaks' but that he made up this rule which is contrary to how he had been spoken to or taught, indicating productivity of language use (see Fig. 30.2).

Kanzi went further by demonstrating his ability to discriminate on the basis of placement of an object. He was asked to bring the ball into the room from the outside. There was a ball inside already. The researchers wanted to observe whether or not he understood the difference between the balls and the words (inside and outside). He went directly to the ball outside and brought it in. This action (together with his reactions to many similar events) indicated that not only did he understand the difference between the two balls and the two words, but he also understood where the object was versus where he was.

Kanzi continued to receive training in the use of lexigrams, but these were now associated with the appropriate speech sounds. He was able to make associations between the sound of an English word and the picture of the object the word represented. For example, Kanzi listened to the voice coming from headphones he had on that asked him to give the picture of a specific object (e.g. a mushroom) to the person in the room with him. This person was not informed about what Kanzi was told to do, in order to avoid

Fig. 30.2 Chimpanzees using lexigrams on portable keyboards to communicate with their trainers (a) Kanzi with E. Sue Savage-Rumbaugh; (b) Nyota using the lexigram board. (Photographs Copyright © 2006 Great Ape Trust; courtesy of Duane Rumbaugh.)

giving him any cues to the correct response. Kanzi was very successful at the task of associating word to picture.

After several years of training, a series of different carefully controlled tests showed that Kanzi could both understand and produce sentences. For example, he was able to follow instructions such as 'Make the doggy bite the snake', operating with his dolls. In general, Kanzi displayed ability to request activities that involve other individuals like 'Person 1 tickle Person 2.' All the other language-trained apes only use themselves as an agent or a recipient of actions or objects. Kanzi was accurate in his understanding of location of items. For instance, if he was asked to get a pine cone from the fridge, he did so even if there was another pine cone in full view.

An important difference between Kanzi and the primates in previous studies is that much of his communication is spontaneous and not made in response to his trainers. He used lexigrams not to ask something but just to comment his actions like children do. For instance, he indicated the 'blanket' lexigram and began to play hide-and-seek, covering his head with the blanket. When eating an apple or a piece of melon, he would point to the corresponding lexigrams thus making his comments on the situation, not asking for food (Savage-Rumbaugh *et al.*, 1986, 1993, 1998).

Researchers claim that Kanzi's comprehension skills are equivalent to those of a child of 2.5 years of age and that Kanzi's communications, made with lexigrams and gestures, were structured according to rules, so Kanzi used a simple version of grammar (a 'proto-grammar'). For sentences containing an agent and an action, the action was consistently placed second, whereas for sentences containing an object and an action, the action came first (Rumbaugh and Savage-Rumbaugh, 1994; Savage-Rumbaugh and Lewin, 1994).

Savage-Rumbaugh and her colleagues have raised other chimpanzees and bonobos in a similar fashion to Kanzi with similar success and, given a failure with older individuals, have suggested that there is a sensitive period early in primate development when apes need to be exposed to language. Researchers believe that apes assimilating intermediary language are communicating very complex things. It is worth to note that many of the questions raised by critics about the linguistic abilities of apes are not relevant to Kanzi and other bonobos. It is still unclear whether their achievements can be attributed to species difference between *Pan troglodytes* and *Pan paniscus* or to peculiarities of the experimental paradigm.

30.4 English serves as an intermediary language: studies with parrots

The parrot Alex is undoubtedly a star performer among animals in the number of questions he is able to answer skilfully with the use of an intermediary language. When his human partner Irene Pepperberg shows him a paper triangle and asks: 'What shape?', he answers 'Three-corner'. Being shown five sticks dyed red and asked 'What colour?' he says 'Ros'. Then being asked 'How many?' he says 'Five'. Being shown a purple metal key and a larger green plastic key and asked 'Alex, how many?' the parrot says 'Two'. The same two keys are held up with a different question: 'Which is bigger?' The parrot stares, pauses, then says 'Green key.' Next is a wooden stick. 'What material?' Again the long pause, again a correct answer: 'Wood'. Alex is clearly responding to the question itself, as well as to the objects. He understands 'different' and 'same' and can answer questions about relationships: Show him a blue-dyed cork and blue key and ask 'What's the same?' he will answer 'Colour'. Show him two identical squares of rawhide and ask 'What's different?' and he will say 'None'. Substitute a pentagon for one square, and he will answer 'Shape'.

The African grey parrot *Psittacus erithacus* has been taught by Pepperberg to use English as the intermediary language. Human language is a non-species-specific code for a bird. Parrots are famous for their capacity to mimic human speech, and they often use human words in relevant context being conditioned (sometimes self-conditioned) to associate definite words with repeatable situations, for instance, to say 'Good bye' when somebody is leaving the room.

Mowrer (1950, 1954) was one of the first to study functional communication in birds under controlled behaviouristic procedures, using standard techniques. He studied mynahs *Gracula religiosa*, magpies *Pica pica*, crows (species not reported), and several psittacids (budgerigars *Melopsittacus undulatus*; a yellow-headed parrot *Amazona ochrocephala*; and a grey parrot). He used the operant methodology, but although his birds were trained according to principles of association and reward, they were neither socially isolated nor placed in operant chambers for their sessions. Mowrer introduced several different words and phrases, and the reward for all vocalisations was food. The idea was that, after a bird emitted vocalisations with some frequency, it could be trained to produce the utterance only in the original, appropriate context (on the appearance of the trainer), by providing the food only when the vocalisation was emitted in such a situation (such as saying 'Hello' when the trainer appeared). Mowrer's birds acquired few vocalisations. The use of food rewards that directly related neither to the task being taught nor to the skill being targeted probably delayed or possibly prevented learning. Most probably, the birds confounded the label of the object or action to be taught with that of the unrelated food reward (Pepperberg, 1981, 1983).

Pepperberg has treated Alex in a social environment from his early age, as primatologists have done with their bonobos. Alex has been her single subject for many years as he has been involved in social relations, play and training during nearly all periods of his activity. Alex was trained first to speak the names of objects (e.g. 'key', 'grain', 'paper' and so on). Whenever Alex named an object correctly, he was praised and then allowed to eat it or to play with it.

The language-training tests used by Pepperberg are based on the adaptation of a technique developed by Todt (1975), who found that the grey parrot learned phrases most quickly from two trainers: one formed a bond with the parrot, the other acted as both a 'rival' and a model. For the parrot to gain the attention of its 'mate' it had to learn to mimic simple phrases used repeatedly by the model/rival. In Pepperberg's study no one took the role of Alex's mate. Trainers took turns 'training' each other to name objects while Alex watched and listened; eventually he joined in. Alex would be in a position that allowed him to see two trainers. One trainer then asked the other, who adopted the role of a parrot, to name an object; if the trainer's reply was correct, then he or she was praised and allowed to play with the object.

Being trained in this manner during many years – more than 20 up to now – Alex has become the most educated bird in the world. He uses more than 100 English words correctly to refer to all the objects in his laboratory environment that play a role in his life including his 15 special foods, his gym, the shower, the experimenter's shoulder, and more than 100 other things. After Alex had learnt to use the numbers one to six and had learnt that a triangle is 'three-cornered' and a square is 'four-cornered', he spontaneously and creatively called a football 'two-corner' and a pentagon 'five-corner'. In tightly controlled experiments Alex was shown many objects in various combinations, and he answered correctly an astonishing number of questions regarding these objects, such as 'What object is blue?' 'What shape is wood?' 'How many are wool?' Since Alex never knows what questions he will be asked next, he must be able to carefully attend to and understand each question, and he must be ready at all times to answer questions regarding anything he has ever learned (Pepperberg, 1983, 1987, 1990, 1999).

Evidence that Alex is capable of understanding and creating sentences comes from his use of some phrases in the context of his living situations. At the outset of his training Alex was unhappy in novel places and consequently spent most of the time in either his cage or his gym, which contained a collection of rods and ropes. When in his cage he was often asked 'Wanna go gym?' and this frequently produced 'Yes' in reply. After a while he spontaneously uttered the phrase 'Wanna go gym' by himself and was immediately carried to it. He then modified this phrase to 'Wanna go gym–no' when he was in the gym and appeared to want to leave it. As he gained in confidence Alex would sit on chairs, shelves and a trainer's knee. During this time he often heard the names of these perches, but he never heard them in conjunction with the phrase 'Wanna go'. Despite this constraint Alex started to say phrases 'Wanna go chair', and if he was taken to a different place he responded either with a 'No' or with a repeat of the request (see Pearce, 2000).

Recently Pepperberg and her colleagues included several other grey parrots into their studies, focusing with them mainly on social and cognitive aspects of communication. Among new birds there are two that were raised in the laboratory, and they are very cooperative. Alex, by contrast, can be far from easy. He may refuse to answer questions, shouting 'No!' and turning his back. He may repeat the wrong answer stubbornly or demand some other item: 'Want corn!' When Alex was asked to name one from six objects on a tray the one that was green, he named the other five - everything *but* the desired answer - and then tipped the tray onto the floor. This may be the reason why he gets 'only' 80 per cent correct on his tests (see Kaufman, 1991).

30.5 Dialogue with marine mammals ... and with at least one dog

Many poetic and philosophical texts, from Plutarch to Shakespeare, bring to us the everlasting desire of humans to talk with dolphins, these friendly, social, intelligent but rather enigmatic creatures.

In the mid-1960s, Batteau began research into the possibilities of developing a 'man–dolphin communicator'. The pioneering studies began with a male bottlenosed dolphin, and were the preliminary phase of research into the feasibility of producing man-to-dolphin and dolphin-to-man translators. Batteau used electronic whistles produced by a generator. These whistles, ranging between 4 and 12 kHz, were emitted by an automatic set-up called Vocal-Trainer. Simultaneously the apparatus analysed and compared the animals' responses with the signals emitted. By conditioning the animals he showed after a few months of experimentation that dolphins could be trained to associate a body motion with an underwater acoustical signal. Later they could be trained to reply with whistles to acoustic signals of great complexity and to imitate them, copying pitch contours very accurately.

A major flaw in this approach, however, was that individual sounds were not associated with individual semantic elements, such as objects or actions, but instead functioned as holophrases (complexes of elements). For example, a particular whistle sound instructed the dolphin to 'hit the ball with your pectoral fin'. Another sound instructed the dolphins to 'swim though a hoop'. Unlike a natural language, there was no unique sound to refer to 'hit', 'ball', 'hoop', 'pectoral fin', or any other unique semantic element. Hence, there was no way to recombine sounds (semantic elements) to create different instructions, such as 'hit the hoop (rather than the ball) with your pectoral fin'.

Unfortunately, Batteau drowned during a morning swim in 1967. His death cut short what was the first established interspecies communication demonstration. The final report (Batteau and Markey, 1967) concluded that devices were constructed to translate articulated vowel sounds into sinusoidal whistles and to provide real-time visual displays of the frequency modulated whistles. Two dolphins were found able to respond distinctly to 35 response-demand messages embedded in five-word spoken sentences. However, because of the mentioned flaw in the approach to the construction of a language, the experiment was not a valid test of dolphins linguistic capabilities.

Herman and colleagues (Herman, 1980, 1986, 1990; Herman *et. al.*, 1984, 1994, 1999) studied dolphins' linguistic skills focusing on their language comprehension rather than on language production. They concentrated on doplhins' receptive competencies, mainly on their capabilities of processing both semantic and syntactic information. The primary syntactic device used in the studies was word order. Dolphins were shown to be capable of understanding that word order changes meaning.

Researchers taught two dolphins in parallel. One dolphin, Phoenix, was taught to respond to acoustic signals (short computer-generated noises), while the other, Akeakamai, was taught to respond to gestures (head and arm movements of a trainer by the pool). Both sets of signals were designed to act as a type of a language with lexical components (words) representing objects, actions and modifiers and a set of rules or syntax to combine the signals (grammar). The words that the dolphins learnt allowed a large range of sentences to be generated and the animals' comprehension was then tested by analysing their responses. For example, a sentence that means 'take the ball to the frisbee' should lead to a different response than 'take the frisbee to the ball', even though the same signals are used.

During the course of her training Akeakamai was tested with 193 novel sentences when all the

13 objects (such as a frisbee, a basket, a ball, a stream of water in the pool, a stationary speaker, and others, including the second dolphin) were in the pool so the objects provided no clue as to how Akeakamai should respond. The dolphin demonstrated understanding of the rules that structured the artificial language and the ability to use these rules to interpret novel sequences of signals. After experiencing many two- and three-word sentences, Akeakamai was suddenly given one containing four words, and she responded correctly. The implication of this finding is that she was able to use the rules relevant to the shorter sentences to understand a more complex syntactic structure. As Pearce (2000) states, this is precisely the sort of skill that should be demonstrated by an animal that is grammatically competent.

Akeakamai was demonstrated as being able to understand logical extensions of a syntactic rule spontaneously and to extract a semantically and syntactically correct sequence from a longer anomalous sequence of language gesture given by the human (Herman *et al.*, 1994; Herman, 2002). To perform this extraction, the dolphin in some cases had to conjoin non-adjacent terms in the sequence. For example, the anomalous string glossed as 'Water speaker frisbee fetch' violates syntactic rules in that there is no rule that accommodates three object names in a row. However, embedded in this sequence are two semantically and syntactically correct three-item sequences 'Water frisbee fetch' (bring a frisbee to the stream of water) and 'Speaker frisbee fetch' (bring the frisbee to the underwater speaker). In sequences of this type, the dolphin almost always extracted one or the other correct three-item sequences and operated on that implicit instruction. Herman (2002) suggests that the dolphin utilises its implicitly-learned mental representation or schema of the grammar of the language, to include not only word-order rules but also the semantic rules determining which items are transportable and which are not (neither the stream of water nor the underwater speaker affixed to the tank wall could be transported). There was no explicit training given for these rules.

The dolphins have also met the displacement criterion in numerous trials. In some trials the instruction related to an object that was hidden from view (spatial displacement), and the dolphin had to find it before responding. On others the command was issued as much as 30 seconds before the related objects were thrown into the pool (temporal displacement). One version of this type of tests consisted in placing all the objects but one in the tank and then giving a command that related to the missing object. Akeakamai would often search for up to nearly a minute for the missing item and then stop, without responding to the other objects. She also rapidly learned to press a paddle to indicate that the designated item was missing.

Dolphins' ability to understand whether things are present or missing gave the possibility of testing whether they are capable of symbolic reference. For this purpose, Herman and Forestell (1985) constructed a new syntactic frame consisting of an object name followed by a gestural sign called as 'Question'. For example, the two-item gestural sequence called as 'basket question' asks whether a basket is present in the dolphin's habitat. The dolphin could respond 'Yes' by pressing a paddle to her right or 'No' by pressing a paddle to her left. Over a series of such questions, with the particular objects present being changed over blocks of trials, the dolphin was as accurate at reporting that a named object was absent as she was at reporting that it was present. These results gave a clear indication that the gestures assigned to objects were understood referentially by the dolphin, i.e. that the gestures acted as symbolic references to those objects.

In terms of language receptive competencies, Herman and his colleagues have shown dolphins capable of (1) successfully processing the semantic and syntactic features of a command system; (2) learning syntactic rules; (3) understanding novel sentences; (4) object labelling; (5) reporting; (6) independence of sensory modalities in learning elaborate commands, demonstrating linguistic comprehension.

The result from similar studies on sea lions which revealed much the same findings as with dolphins can be found in Schusterman and co-authors' publications (Schusterman and Krieger, 1986; Schusterman and Gisiner 1988; Schusterman *et al.*, 2002). Similar to the studies on dolphins, the researchers focused their efforts on teaching three sea lions to relate particular gestural signals to objects (such as bats, balls and rings), modifiers (large, small, black and white), and actions (such as fetch, tail touch and flipper-touch). These signals could be combined into over 7000 different combinations, each instructing the animal to carry out a specific behavioral sequence.

A California sea lion named Rocky might have been able to learn to understand sentence forms similar to those understood by the dolphin Akeakamai. Rocky was able to carry out gesture instructions effectively for simpler types of sentences requiring an action to an object. The object was specified by its class membership (e.g. 'ball') and in some cases also by its colour (black or white) or size (large or small). Rocky was able to understand relational sentences requiring that one object be taken to another object. More complicated instructional sequences required the sea lion to press one of two paddles to indicate whether an object was present or absent. The most complicated instructions required the sea lion to select one object in the tank and bring it to another. These 'relational' sequences could include up to seven signs; for example, the gesture sequence 'large white cone, black small ball fetch' instructed the sea lion to bring the black small ball to the large white cone. The sea lions were eventually able to respond appropriately to familiar as well as novel combinations of signs with a great deal of accuracy.

These reports suggest that the sea lion was capable of semantic processing of symbols and, to some degree, of syntactic processing. A shortcoming of the sea lion work, however, was the absence of contrasting terms for relational sentences, such as the distinction between 'fetch' (take to) and 'in' (place inside of or on top) demonstrated for the dolphin Akeakamai. Additionally, unlike the dolphin, the sea lion's string of gestures were given discretely, each gesture followed by a pause during which the sea lion looked about to locate specified objects before being given the next gesture in the string. In contrast, gesture strings given to the dolphin Akeakamai were without pause, analogous to the spoken sentence in human language. Further, Rocky did not show significant generalisation across objects of the same class (e.g. different balls), but unlike the dolphin seemed to regard a gesture as referring to a particular exemplar of the class rather than to the entire class. Thus, although many of the responses of the sea lion resembled those of the dolphin, the processing strategies of the two seem to be different, and the concepts developed by the sea lion appear to be more limited than those developed by the dolphin (Herman, 2002).

It is interesting to note that at least one dog (a mongrel dog, Sofia) has displayed the ability to understand requests composed of two independent information items which had to be combined in the correct performance. Ades *et al.* (2005) conducted series of experiments on human–dog communication through arbitrary signals. The first experiment dealt with action–object sentences (such as 'point ball', 'fetch bottle'), the second one with action–place sentences, in which action terms were verbal ('point' and 'fetch') and place terms were gestural signals. Training consisted of a discrimination phase, in which verbal or gestural signals were associated to the relevant behaviours; a sequential request phase in which object and action (Experiment I) and place and action (Experiment II) requests were presented one at a time; and a simultaneous request phase, in which Sofia had to carry out action + object and object + place commands presented as sentences. Results indicate that Sofia and probably dogs in general have the ability to perceive and to keep in memory more than one request information, as chimpanzees and other linguistically trained animals do, but also that there are limits to such arbitrary signal cognitive processing. A third experiment was performed to test Sofia's ability to produce arbitrary signals through the use of an electronic keyboard with arbitrary geometric symbols for 'water', 'food', 'cage', 'petting', etc. Results indicate that Sofia pressed the keys in an intentional way, that is, according to motivational context and to presence of relevant stimuli. Use of the symbols was associated with glancing directed to the experimenter, an indication of communicative intent.

Chapter 31

A battle for the Rosetta Stone: attempts to decipher animals' signals

The Rosetta Stone is the famous key to the ancient Egyptian language. It was carved in 196 BC, and later was discovered by the French soldiers who came with Napoleon. The stone contained words in three types of writing: Egyptian hieroglyphs, demotic, which is a shorthand version of Egyptian hieroglyphic writing, and Greek. By translating the Greek section, Jean-François Champollion and Thomas Young in the 1820s were able to crack the code of the stone and to decipher what the hieroglyphs meant. Before that it was not possible to understand ancient Egyptian texts despite the abundance of texts in this language. Decoding an animal language is perhaps no simpler a task, and we can not hope to find a Rosetta Stone.

The problems underlying the construction of animal dictionaries have been discussed in details during the last few decades (Theberge and Falls, 1967; Reznikova and Ryabko, 1990, 1993; Ryabko, 1993; Hauser, 2000). Many workers have tried to decode animal languages by looking for 'letters', 'words' and 'phrases' and by compiling 'dictionaries'. With such an approach, it often remains unclear which sounds and gestures have something to do with the language and which do not, and there are also some technical difficulties connected with the high mobility of animals and often with their inaccessibility for recording signals (Ryabko and Reznikova, 1996).

Theberge and Pimlot (1969) noted that when studying wolves' ability to produce and distinguish subtle details of acoustic signals they were challenged by the problem of understanding a foreign culture's sounds, lacking a relevant dictionary and any ideas about this culture. Wolves form stable packs and, as many other social animals, they live in a world of dominance interactions. They use vocalisations extensively both when communicating with other pack members and with wolves in other packs. These researchers suggested that wolves were able to transfer the information by changing certain units of their acoustic communication, but the only 'word' they managed to decipher was a 'sound of loneliness' which wolves produce when placed in isolation, anxious to join others.

Since then, many distinct acoustic signals have been found in wolves, African wild dogs (Robbins, 2000), bottlenose dolphins (Janik, 2000), primates (Snowdon *et al.*, 1982) and others. Acoustic vocalisation in some species of birds and mammals often has a hierarchical structure, with notes grouped into syllables, syllables grouped into phrases, and phrases grouped into a song with a linear array of phrases. These data enable researchers to try to understand the meaning of animal signals and to test whether that species' communications exhibit a language format.

For example, dolphins produce various types of sounds, including clicks, burst-pulse emissions and whistles. Clicks are used for echolocation, burst-pulse sounds may indicate the dolphin's emotional state, ranging from pleasure to anger, whereas whistles may be used for communication. During the 1960s, researchers attempted to determine whether the whistle vocalisations could be a form of language. Investigators recorded whistles from many dolphins in many different situations, but failed to demonstrate sufficient complexity in the vocalisations to support the idea that dolphins are capable of referring symbolically to another individual, or to some other object or event in the environment (Herman, 1991). Although dolphins have demonstrated a great potential in the use of an artificial intermediary language, there is no evidence that dolphins do have a natural language. Is it hopeless to attempt to decipher at least fragments of natural animal languages?

Animals often behave similarly in repeatable situations, and if these repeatable behaviours are elicited by distinctive repeatable signals, these behaviours could serve as keys for cracking the animals' species-specific codes. Decoding the function and meaning of wild communications is a notoriously difficult problem. To date, there are two natural communication systems that have been partly deciphered – the acoustic communication of vervet monkeys and the dance language of honeybees. In both cases expressive and distinctive signals correspond to repeatable and frequently occurring situations in the context of animals' life.

31.1 The dance language of honeybees

The dance language of the honeybee is the most complex animal natural 'language' that has been decoded, at least partly. Successful forager honeybees (*Apis mellifera*) are able to recruit other bees to a distant food source (or water, or resin, or a new nest site) by specific 'dance' movements together with other components such as odours and sounds. As Manning (1979) notes, after Karl von Frisch described signals in the bees' hive as elements of complex language, the world must admit that information in a symbolic form is available not only to humans but also to such a modern creature as the honeybee.

The suggestion that honeybees can inform their nestmates about the location of a rich nectar source when they arrive back at the hive was made by Aristotle and has since been repeated more than once. Aristotle himself documented a honeybee's ability to recruit her nestmates to a good food source but did not speculate on how the communication took place. He and other naturalists did, however, observe that a bee who finds a new food source returns to the nest and dances for her sisters, rather than feasting alone. Pliny reportedly constructed a hive that had a window made from transparent horn, through which he could watch the dancing bees. The hypothesis about transferring information among honeybees was first given in the scientific literature by Spitzner in 1788. Famous explorers of social insects in the nineteenth century such as Dujardin, Emory and Lubbock also suggested that honeybees possess a kind of language that enables them to inform their nestmates about locations of the food. Nevertheless it was difficult to explain how insects could do that.

The dance language is an uncommon form of natural communication by which animals inform each other about remote objects. This form of communication is sometimes called 'distant homing' and only single cases have been described in the scientific literature. Among them are the chimpanzees in Menzel's (1973a, b) experiments which hypothetically passed each other information about hidden objects, and ants that inform their nestmates about the location of food at a distance, lacking the possibility of using smell trails and other modes of information transfer other than contact with a scouting individual (Reznikova, 1983, 2004). In terms of modern theory of social learning distant homing may be called 'social learning from an information centre'.

Honeybees must possess a powerful communication system in order to inform conspecifics about the exact location of a desired remote object. Von Frisch suggested that in the dance language, an abstract, or symbolic, code is used to transmit the information about the direction and distance of a desired object. In contrast to ants with their difficult-to-handle movements of antennae, forelegs and maxillae, honeybees demonstrate distinctive movements in the darkness of the hive. First von Frisch (1923) suggested that the bees have language but only as a stimulus response. Later he identified smell of the food source as being of primary importance for transferring information among bees (Frisch, 1937). After a series of special experiments with changing locations of artificial feeders accompanied by observations within a glass-walled hive, von Frisch came to the conclusion that the bees give others exact data about the source of food by means of dance language.

Von Frisch (1947, 1967) described three forms of dances. A simple 'round dance' serves as a recruiting signal for rich food sources close to the hive (50–100 m). Being activated by the round dance, bees leave the hive and search for nectar in the vicinity of the hive. At greater distances (100–150 m) a 'stickle dance' is performed. The 'waggle dance' informs the bees about distant sites, which could be located up to 12 km away but more often are located not farther than 3–5 km from the live. The waggle dance is named from the waggling run in which the dancer wags her body from side to side and emits sounds by vibrating her wings. In the waggle dance, the dance path takes on a figure-of-eight shape: the dancer runs in a straight line (the waggling run) and circles back, alternating between the left and the right return path. The direction of the waggling run relative to the vertical (gravity) on the comb indicates the direction to the food relative to the sun's azimuth in the field. If the scouting dancer waggled while facing straight upward, towards the 12 on a clock face, then the food could be found in the direction of the sun; if she waggled 60° to the left of 12, facing the 10 on a clock face, then the food lay 60° to the left of the sun. In addition, von Frisch noticed that how fast the dancer completed her circuits corresponded to the distance between the hive and the feeding site: the closer the food, the more frenzied her pace. A few follower bees keep close contact with the dancer, and these bees may be recruited to visit the food. Von Frisch and his colleagues made detailed accounts of the dance language. With a stopwatch close at hand, they could observe the dance, decipher its meaning and then locate the food supply of which it spoke.

Years after von Frisch interpreted the symbolism of the dances, other components of the bees' communication system have become understood and appreciated. Esch (1961) and Wenner (1962) found that dancing bees make sounds during their waggling run. Later it was found that the sounds the foragers produced could vibrate the combs under their feet as they danced (Eskov, 1979; Tautz, 1996; Tautz *et al.*, 2001). The comb vibrations might then advertise the dance to those bees who could not otherwise see the forager. Levchenko (1976) revealed the effective combinations of tactile, acoustic and odour components of the communication system in honeybees.

Suggesting, as von Frisch did, that bees have a symbolic language aroused the scepticism of some other scientists, who continue to this day to favour von Frisch's earlier idea that scent alone guides bees. Gould (1974, 1975) punctured the odour hypothesis. He showed that a forager can dispatch her nestmates to a site she has never visited. Such a feat would be impossible if the recruits relied on odours alone to track down a feeding site. The event could occur, however, if the searching bees gave priority to the information they received from the dance. In these experiments, Gould placed a bright light in the hive, which the dance followers mistook for the sun. In doing so, they interpreted the dances erroneously. These misdirected bees most often searched in the field using the misaligned dance information and seemed to ignore other cues such as odour. Gould concluded that they evidently preferred the message given in the dance to the other signals. Finally, the experiments using an artificial model of a bee confirmed von Frisch's hypothesis: the dances do indeed represent a sophisticated form of communication (see Kirchner and Towne, 1994).

It is a natural idea to communicate with the bees through the mediation of an artificial bee. Indeed, a crucial test for any communication system is to modify it experimentally. The idea of using a 'false bee' that dances and indicates the distance and direction of a target the other bees had never visited, was suggested a long time ago. In one of his pioneering papers devoted to the successful use of a robotic bee

Michelsen (1993) cites 'The future of biology' by J. B. S. Haldane (1927): 'We may be able to tell our bees to fertilize those apple trees five minutes fly to the south-east. To do this we should presumably need a model bee to make the right movements, and perhaps the right noise and smell.' Since 1957 several attempts have been made to recruit bees by means of mechanical models of dancing bees (Steche, 1957; Esch, 1964; Gould, 1976; Lopatina, 1971; Levchenko, 1976).

Michelsen and colleagues (Michelsen, 1993, 1999; Michelsen *et al.*, 1990) built a bee of brass and beeswax with a piece of a razor blade to represent the wings. Their robot bee could buzz its wings at the requisite 280 cycles per second (Hz), waggle its bottom and dance in a circle. It could also deliver drops of sugar-water to the dance watchers. Using a computer to control the robot, the researchers set their robotic bee to dancing, attempting to dissect the components of the dance. Dancing a normal waggle dance, this bee, like any robot, was a bit clumsy (Fig. 31.1). Nevertheless, recruits in the hive used the information to fly to a target. On another day, researchers sent the bees in the opposite direction. The bees found the new targets regardless of the wind direction, correctly following messages from the robot bee that never left the hive itself. These studies allow us to reject the hypothesis that bees use the smell of the new food source as the main source of information in favour of the hypothesis of the use of dance parameters to decide where to fly.

Fig. 31.1 A robot bee in a hive. (Adapted from a booklet of the Centre for Sound Communication, Institute of Biology, Odense University, Denmark; courtesy of A. Michelsen.)

The honeybees' dance language is considered symbolic in many respects. For instance, the exact relation between the velocity of the waggling run and the distance to the desired site is determined by local agreements, or dialects. For the bees studied by von Frisch, each waggle of the dance indicated that the desired site was located an extra 45 metres from the hive, while for bees in Italy this figure is 20 metres, and for Egyptian bees it is 12 metres (Boch, 1956). It is important that all bees in the hive follow the same rule. The dance language meets the arbitrary criterion. For instance, the bees can in some cases use the direction to the desired site relative to the north azimuth instead of the sun's azimuth (Gould and Gould, 1988). The dance language exhibits the displacement feature because the dancers can inform other bees about evidences distant both in space and time. The scouting bee keeps the sun's trajectory in mind over several hours and corrects her dance in accordance with the daily movement of the sun.

It is still an open question to what extent bees can display creativity and flexibility of communication. It is known that the bees can communicate within specific limits. At the same time they 'speak' about other things than flowers. For example, it has been noted that 'quartermasters' use figures of dances in order to indicate the location of a new nest site. They dance directly on the surface of a swarm of bees intending to relocate to a new place (Lindauer, 1961). It has been revealed recently that these signals for house-hunting are performed by a small proportion of the older (and possibly more experienced) bees. They perform waggle dances accompanied by the vibration signals (Lewis and Schneider, 2000).

Honeybees are not only capable of changing topics of their communication but also of producing and interpreting signals under circumstances not previously encountered. Lindauer, at that time a Ph.D. student and then an assistant of von Frisch, recalled their joint experiment in 1948 when they moved an observation hive from its normal vertical position to a horizontal position and were surprised to see that bees dancing on a horizontal comb could indicate a direction to a food source (see Seeley *et al.*, 2002).

The role of cultural transmission and social learning is weak in the functioning of honeybees' dance language. Lopatina (1971) demonstrated that young bees have to complete their innate stereotypes with the gained experience in order to receive messages from dancers. There have been, however, many experiments proving that details of the way in which the dance is performed and interpreted are determined genetically (Gould and Gould, 1988; Seeley, 1995). For instance, dialects in the dance language of bees from different races do not reflect any influence of social learning. If the pupae of one race are placed into the colony of another, then when the bees hatch, they persistently behave in a manner that is appropriate to their real sisters and they are not influenced at all by their foster sisters. Studies over the last decades (Rinderer and Beaman, 1995; Johnson *et al.*, 2002) have shown that each of the transition points in honeybees' dances (round–stickle, stickle–waggle) is controlled by a single gene showing simple Mendelian inheritance. Researchers are in agreement now that the honeybees' dance is a complex behaviour under simple genetic control. However, it is possible that honeybees are capable of flexible transposition of inherited elements of their communication in conformity with a changeable environment and new challengeable tasks.

31.2 Semantic vocalisations in animals: words without a language?

As we have already seen earlier in this Part, nonhuman primate infants are able to extract relevant structural elements from continuous human speech. This finding once more enables researchers to concentrate on the question as to what extent animals' calls are word-like.

The main methods for studying wild vocalisations are based on quantitative comparative analysis of signals and on playback experiments. Researchers record vocalisations that animals emit in distinct living situations, then digitise tapes by computer programs, and break them down into frequency and timing. Changes in frequency and time are measured, and this data is re-entered into the computer and analysed to see if there are differences between signals. During playback experiments tapes are played for animals either in their natural form, or artificially changed, for example, slowed or containing added elements.

Primates are good candidates to have dictionaries being compiled for them by human researchers. After Garner (1892), who tried to decipher acoustic signals of several primate species with the use of a phonograph, many primatologists worked in this field. During the second part of twentieth century lists of possibly meaningful signals have been compiled for several primate species such as signals of greetings, invitations for playing, sex, sharing food, predator alarm, intergroup treat, affiliation, infant distress, discontent, quarrel, nocturnal roll-call, cooing with infants and so on (Tikh, 1950; Goodall, 1968; Green, 1975; Marler and Tenaza, 1977; Firsov, 1983; Snowdon, 1986, 1999; Deryagina *et al.*, 1989).

A representative vervet–English dictionary has been compiled by Struhsaker (1967). He found that vervet monkeys in the wild emitted 25 discrete calls referring to different objects and situations. Nevertheless, as was already known for other primates, the majority of these signals were not sufficiently distinct and frequently used. But then three 'words' seemed to be understandable not only by monkeys but also by human observers. Three different sounds were emitted by vervets for three different predators (leopards, eagles and snakes), resulting in three different reactions or escape responses. For example, in response to the snake warning call, the troupe of vervets would stand up on their hind legs in the open and look around on the ground to find the snake. If the predator were an eagle or a leopard, standing up in the open would be very dangerous. In response to the leopard warning call, the members of the troupe would climb the nearest tree up to the top where the heavy leopard cannot follow them. This would be just the wrong thing to do to avoid an eagle, and not optimal in case of a snake either. In response to the eagle warning call, the members of the troupe would run into a nearby bush or under the lower branches of a nearby tree. This behaviour is again a good response only to the type of predator that the particular call signals. The calls appear to function as representational, or semantic, signals.

While this finding indicated to some researchers in the field that a specific warning was called out, depending on what kind of predator was seen, others suggested possible alternative explanations seeing an animal's call as an attempt to refer to an object, rather than as a simple emotional reaction was a radical departure from past thought.

Seyfarth *et al.* (1980) and later Seyfarth and Cheney (1990) in a similar study tried to eliminate alternative possibilities. They recorded many alarm calls and arranged to play them back to a vervet troupe. To do this, experimenters waited until the monkeys were quiet and had no real dangers about. The monkeys responded to playbacks as they had to the original calls. The first monkey to respond, at least, could not be imitating, because there was no one to imitate. And the troupe's response related to the call itself, not to a real danger, because there was no danger.

The researchers went on to analyse the more subtle vervet calls, called grunts. With practice, humans could detect differences in the grunts, but previous researchers had assumed these differences to depend on different situations, not on different meanings. Seyfarth and Cheney (1997) tested this assumption, again with playback experiments. Vervets responded differently and consistently to different grunts, no matter what context the researchers chose for playback. When they heard grunts recorded from a dominant male, the vervets moved away from the speaker. On the other hand, when they heard the grunts of a subordinate to a dominant male, they never moved. These experiments do not assign a meaning to the grunts. But they do show that the grunts convey some meaning; they are not all equivalent to each other.

Seyfarth and Cheney (1997) addressed the question about whether animals can refer to meanings of 'words' rather than to their acoustic properties. For instance, we consider the words 'mouth' and 'mouse' rather different although there is an acoustic similarity between them; at the same time the words 'mouth' and 'eater' sound different but they can mean the same in some contexts ('extra mouth to feed'). Vervets seem to use 'synonyms' in their vocalisations. For instance, they have at least three acoustically different calls that are given as response to neighbouring groups. Playback experiments have demonstrated that listeners treat these calls as roughly equivalent. Researchers have used a habituation–dishabituation technique as a way to determine whether vervets found two sounds they make (wrrs and chutters) were similar or dissimilar. They habituated vervets to one sound and observed their response to another sound. The results suggested that, in fact, these two acoustically different sounds were responded to in the same manner, leading to the belief that their referents were similar, that is, talking about the same thing.

An optimistic interpretation of these results is that not only are non-human primates making word–object representation associations but that they are discriminating on the basis of their referents and their meanings. Nevertheless, although these findings have dramatically changed common thoughts about acoustic communications in animals which were believed to be only an expression of their current emotional status, the authors accentuate limits of vervet communication and its great difference with the sophisticated language of humans (Seyfarth and Cheney, 2003, 2004). To be considered 'language-like', vervets' acoustic communication does not meet several important criteria. For instance, vervets' communication, in contrast to that of humans and even honeybees, does not possess displacement feature. The honeybees are able to inform each other about things that are distant in space and time, whereas monkeys can only 'talk' about a predator they currently see. Possessing a small number of calls in their vocal repertoire that serve as semantic labels for objects monkeys are able neither to create new labels for other objects and events in their environment nor to combine their present calls.

There is a little evidence at present for vocal learning in non-human primates. Infant vervet monkeys seem innately predisposed to give alarm calls with the acoustic features of adult eagle alarms in response to flying birds. Even when monkeys are reared in environments different from the ones they would normally experience, call production seems to be species-specific. This applies to the communications not only of vervets but also of other primate species.

For example, cross-fostering experiments with Japanese and rhesus macaques (Owren and Rendall, 2003) and squirrel monkeys (Newman and Simmes, 1982) showed that infants continue to give species-specific calls in contexts when their adoptive mothers and foster siblings give their specific calls. Firsov (1983) conducted cross-fostering experiments in which infant chimpanzees were adopted by human 'mothers' who gave them artificial feeding and round-the-clock care. Firsov recorded the age development of vocalisations in chimpanzees and revealed that all species-specific vocalisations appeared in adopted socially naive infant chimpanzees at the same age as in wild animals.

Nevertheless, this does not mean that all aspects of vocal communication in primates are innate and inflexible. Both vocal usage and call comprehension involve some learning. In the study with macaques cited above, infants learned to recognise and respond to calls of their adopted mothers, and vice versa. Infant vervets are initially quite indiscriminate about the sorts of aerial objects that elicit the eagle alarm call, and they often respond inappropriately to alarm calls of others. Over the period of growing up, however, they gradually learn to restrict their air alarms to the few raptor species, and they also learn the correct response to different alarm calls (Seyfarth and Cheney, 1997, 2002).

Earlier in this chapter we have already encountered the natural idea that language behaviour of animals should be based on complex forms of social relationships and in many cases on cooperation in groups. In a separate experiment, Cheney and Seyfarth (1990) played recorded alarm calls of young vervets. Adult females responded by looking at the infant's mother, not at the infant. This behaviour implies they can both identify an infant by its distinctive call and can understand the relationship between a mother and an offspring. Vervets also use intergroup calls to mark the status of immigrant males, giving intergroup calls to a male when he first enters a group and gradually shifting to other call types as he becomes more integrated into the social structure.

These findings show that vocal communication in non-human primates is based on the animals' estimation of the current social environment rather than on simple reflectively emitted calls. There are many data in this field concerning acoustic communication in other primate species. For example, white-faced capuchins emit food-associated calls when they find food such as fruits, eggs or caterpillars. Results of observations of social interactions after food discovery showed that individuals who had called when they discovered food were less likely to be approached by others who were in visual contact than those who remained silent. Furthermore, individuals who called when approached by higher-ranking animals were less likely to receive aggression than individuals who did not call. Therefore, food-associated calls may function to announce food ownership, thereby decreasing aggression from other individuals (Perry, 1997; Perry and Manson, 2003; Gros-Louis, 2004).

Despite a close relation between style of communication and style of social life in primates, a variety of results argue that, in marked contrast to humans, non-human primates do not produce vocalisations in response to their perception of another individual's ignorance or need for information. For example, adult vervet monkeys do not 'teach' their infants to react to different predators by distinct calls; instead, they inadvertently inform infants about the relation between alarm call type and predator simply by producing alarm calls selectively themselves (Seyfarth and Cheney, 2003). It is not that feedback is absent in communication within troupes. Young vervets seem to learn to produce signals in appropriate context much faster when their mothers 'confirm' them by responding with the same signal. This has been demonstrated by Hauser (1988) who spent 2500 hours recording vocal communications of 36 mother–infant dyads. Of special interest is that four infants were regularly punished by their mothers who spanked them when the infants made wrong calls. Being once punished, the infants improved their vocalisations the next time. Hauser (1988) noted that nearly all of those infants did not reach 1 year of age, so it was impossible to follow the development of their language behaviour.

Since the vervet studies, human comprehension of animal dictionaries has been expanded to several other species. For example, West African Diana monkeys *Cercopithecus diana* produce acoustically distinct alarm calls in response to leopards and crowned-hawk eagles, their main predators (Zuberbühler, 2000). Playback experiments have shown that the monkeys treat these vocalisations as semantic signals, in the sense that they compare signals according to their meanings. Similarly to vervet monkeys, Diana monkeys also can use synonyms in their vocal communication. They appear to judge a leopard's growl, a male Diana monkey's leopard alarm call and a female Diana monkey's leopard alarm call as designating the same class of danger, even though these calls are acoustically distinct (Zuberbühler *et al.*, 1999).

Diana monkeys are able to 'translate' alarm calls of sympatric chimpanzees and Campbell's monkeys *Cercopithecus campbelli* which also have distinct calls for different predators. Diana monkeys respond to these calls with their own corresponding alarm calls. It is interesting to note that monkeys can react adequately to combinations of elements of calls experimentally compiled by researchers. Zuberbühler (2002) suggests that these results indicate that these parts of signals are meaningful only in conjunction with Campbell's alarm calls and this supports a hypothesis about syntactic abilities of monkeys.

Comparative studies have demonstrated several non-primate species as being capable of conveying complex information through a vocal 'language' comparable to the communicative systems of vervet monkeys. Slobodchikoff *et al.* (1991) identified prairie dog calls for four predators – human, hawk, coyote and domestic dog. Prairie dogs even can speed up or slow down their signals depending on whether the predator is running or walking through their colony.

Meerkats *Suricata suricatta*, a South African mongoose, give one alarm call type for mammalian predators, primarily jackals and wild cats, a second alarm call type for avian predators, primarily the martial eagle and a third alarm call type for snakes such as the Cape cobra and to faecal, urine or hair samples of predators (Manser, 2001). Developmental studies suggest that the pups possess inherited vocalisations but they definitely need to learn how to respond appropriately to the different alarm call types (Hollén and Manser, 2006) (Fig. 31.2).

Playback experiments have revealed that not only alarm calls but also food calls may be functionally referential in the sense that they encode information about the kind, quantity and availability of food. The evidence that food calls provide sufficient information to evoke anticipatory feeding behaviour from conspecifics has been obtained from playback experiments with domestic chickens (Evans and Marler, 1991; Evans and Evans, 1999). Ravens emit structurally discrete yells when they approach rich but defended food sources, and thus attract conspecifics (Heinrich, 1999). Bugnyar *et al.* (2001) experimentally exposed a group of free-ranging ravens to six feeding situations. These researchers thus were able to model distinct repeatable situations and to examine what calls emitted by ravens corresponded to which kind of food, and of what quantity and availability. The different use of long and short yells relative to food availability suggests that short 'who' calls provide information about the caller, whereas long 'haa' calls may also provide information about the food itself. 'Haa' calling rates varied with the type of food. This makes ravens' 'haa' call a potential candidate for a functionally referential signal. As with vervets and meerkats, the appropriate usage of referential signals in ravens appears to be learned in juveniles. In contrast,

Fig. 31.2 Meerkats *Suricata suricatta* use a graded alarm call system where the acoustic structure of calls combines information on the type of predator approaching and the level of urgency (Manser, 2001; Hollén and Manser, 2006). For pups, recognising and responding appropriately to these calls at an early age may be crucial for survival. (Copyright © Adam Seward: Earth in Focus; courtesy of A. Seward.)

developmental studies in cotton-top tamarins have not revealed a gradual adjustment to context (Roush and Snowdon, 1994).

Of course, the data referred in this section do not pretend to completeness. Nevertheless, one can see that many interesting results have already been obtained by those researchers who have tried to decipher animals' acoustic signals. We know now that these signals have some parallels with language, and the dichotomous view that animal signals must be either referential or emotional is false because they can easily be both (Seyfarth and Cheney, 2003). However, our knowledge about the degrees of divergence between the potential and actual power of species' 'languages' is largely constrained by current methodology. It is a quite natural idea to face ravens and dolphins with artificially repeatable living situations and then to try to decipher their signals. However, only limited data have been obtained so far and it seems unlikely that prairie dogs use more words than ravens in their real life as we have seen from current experimental data.

Chapter 32

A dialogue with a black box: using ideas and methods of information theory for studying animal communication

It is natural to use an information theory approach in investigating animal language behaviour, because this theory presents general principles and methods for the study and development of effective and reliable communication systems. Shannon (1948) developed the basis of information theory at the end of the 1940s. The fundamental role of this theory was appreciated immediately, not only in the technology of information transmission, but also in the study of natural communication systems. In particular, in the 1950s and 1960s the entropies (degree of uncertainty and diversity) of most European languages were estimated. It was revealed that in all human languages the length of a message correlates with the quantity of information that is contained in this message. It means that one can lodge on two pages twice as much information as on one page.

What is a bit according to Shannon? If we make a 'heads-or-tails' trial with a 'fair' coin, both outcomes have equal probabilities. Uncertainty of this trial equals one bit, and if somebody informs us about the outcome, he or she transfers one bit of information. In general, if a trial has n equiprobable outcomes, and we will be informed about its result, we receive $\log_2 n$ bits of information. Imagine that you know that your friend lives in a house with two entrances, two storeys and two flats on each storey, and you do not know in which flat your friend lives. There are 8 flats, and the probability of finding your friend in anyone of them equals 1/8. If somebody informs you at which flat you should call, he or she transfers $\log_2 n = 3$ bits of information. In this case it is easy to follow each bit: information about one of the two entrances, then about one of the two storeys, and last about one of the two flats.

The considered notion of information forms the basis of Shannon's mathematical theory of communication. Later information theoretical ideas entered the fields of robotics, linguistics, psychology, physiology, etc. For example, human reaction time under experimental conditions turned out to be proportional to the uncertainty present in the experiment (Yaglom and Yaglom, 1967).

Surprisingly, biological applications of information theory have been incorporated only in a few studies. Thus, Wilson (1971) used the information theory approach to estimate the quantitative volume of the ability of the honeybees and the ants *Solenopsis saevissima* to memorise the location of a food source.

Reznikova and Ryabko (1990, 1994, 2003) suggested a new approach to studying animal communication based on the ideas of information theory. They investigated ants' communication system as a means of transfering concrete quantitatively measurable information. The main point of this new approach is not to decipher signals but to concentrate just on the process of transmission of a measured amount of information.

Ants are good candidates for studying the potential power of animals' communicative system. The signal activity of ants has attracted the attention of many researchers. Ants are known to be able to use a large variety of communication means for attracting their nestmates to a food source (Hölldobler and Wilson, 1990). It has remained unclear for a long time whether they have a distant homing system. In this connection, the so-called tactile (or antennal) 'code' has been discussed. A hypothesis regarding the existence of such an information transmission system in ants was put forward as early as the nineteenth century by Wasmann (1899). However, numerous attempts to decipher ants' 'language' have not given the desired results. Attempts were made to decipher ants' phrases such as 'I need food.' Many works were devoted to analysing the hypothetical antennal code. The sequence of antennal movements was studied in different aspects and situations: when ants share food, during the recruitment to food, and during social interactions. It was concluded that antennal movements have no structural unity of signals and replies (Lenoir and Jaisson, 1982; Bonavita-Cougourdan and Morel, 1984; Hölldobler, 1985). Indeed, in contrast to those of honeybees, ants' movements are so rapid fine and complex, that the question of the existence of a developed tactile 'language' in ants is rather difficult to answer.

Based on a dialogue with an ant family treated simply as a 'black box' and asking them to transfer a definite amount of information, we have revealed in a few highly social ant species the ability to transmit potentially unlimited numbers of messages. This approach has allowed us to estimate the rate of information transmission in ants, which turned out to be not high (approximately one bit per minute). We also succeeded in studying some properties of ant intelligence, namely, their ability to memorise and use simple regularities, thus compressing the information available. One can suggest that the experimental schemes described can be used to study the communication systems of other animals.

32.1 Ants on the binary tree

The main point of the suggested approach is that the experiments provide a situation in which ants have to transmit information quantitatively known to the researcher in order to obtain food. This information concerns the sequence of turns toward a trough of syrup. We used the laboratory set-up called a 'binary tree' where each 'leaf' of the tree ends with an empty trough with the exception of one (randomly chosen) filled with syrup. The simplest design was the tree with two leaves. It represents one binary choice that corresponds to one unit of information introduced by Shannon, the bit. In this situation a scouting animal must transmit one bit of information to other individuals: to go to the right (R) or to the left (L). In other experiments the number of forks in one branch increased to six. Hence, the number of bits necessary to choose the correct way is equal to the number of bifurcations, in this case six, and three in Fig. 32.1.

The experiments with binary tree were conducted with nine laboratory colonies of five ant species. Ants were housed in plastic transparent nests that made it possible to observe their contacts. All the ants were labelled with an individual colour mark. The composition of ants' groups was revealed during preliminary stages of the experiments. In some species the small cliques within the colonies were discovered which were composed of a 'scout' and between five and eight 'recruits' (foragers). A total of 335 scouts was used in the main trials. A scout was placed on a trough containing food, and then it returned to the nest by itself. The scout had to make up to four trips before it was able to mobilise the group of foragers. In all the cases of mobilisation we measured (in seconds) the duration of the contact between the scout and the

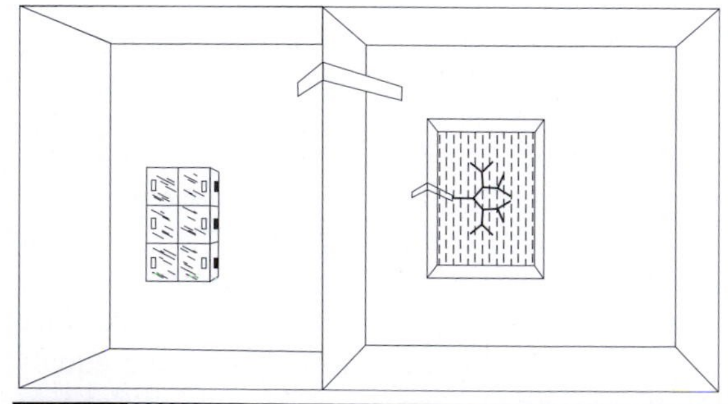

Fig. 32.1 The laboratory arena and the maze 'binary tree'.

foragers. When the group began moving to the maze the scout was removed with tweezers and the foragers had to search for the food by themselves. To prevent access to the food by a straight path, the set-up was placed in a water bath, and the ants reached the initial point of the binary tree by going over a bridge.

The experiments were devised so as to eliminate all possible ways helpful for finding food, except information contact with the scout. To avoid the use of an odour track, the experimental set-up was replaced by an identical one when the scout was in the nest contacting to its group. The fresh maze contained all troughs full with water in order to avoid the possible impact of the smell of syrup. If the group reached the correct leaf of the binary tree they were immediately presented with the food (Fig. 32.2).

The long-term experiments revealed information transmission based on distant homing within small constant cliques consisting of a scout and foragers in three species. Not all of the scouts managed to memorise the way to the correct leaf on the maze. The number of such scouts decreases with the complication of the task. In the case of two forks, all active scouts and their groups (up to 15 per colony) were working, while in the case of six forks, only one or two coped with the task. It turned out that able scouts could transmit information on several different sequences of turns during one daily experiment.

Evidence of information transmission from the scouts to the foragers came from two sets of data: firstly, from the statistical analysis of the number of findings of the goal without error by a group, and secondly, from special series of control experiments with 'uninformed' (naive) and 'informed' foragers.

The statistical analysis of number of errorless findings of the goal was carried out by comparing hypothesis H_0 (ants find the 'right' leaf randomly) with hypothesis H_1 (they find the goal thanks to information obtained), proceeding from that the probability of a chance finding of the correct way with i number of forks is $(1/2)^i$. Different series of experiments (338 trials in sum) were analysed, separately for two, three, four, five and six forks. In all cases H_0 was rejected in favour of H_1, $P < 0.001$ (Ryabko and Reznikova, 1996).

The special experiments were performed in which naive ants were tested in the maze. Searching results were compared in the ants that had and had not previously had contact with the scout (the informed and naive scouts, respectively). The naive and informed ants were allowed to search for the food for 30 minutes. In more agile *F. pratensis* the time spent on

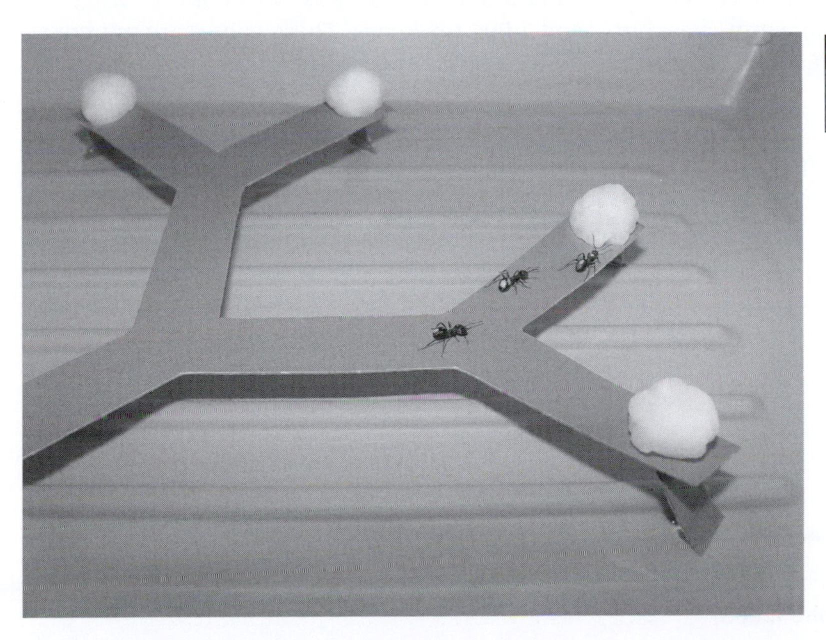

Fig. 32.2 Ants on the binary tree maze. Now all troughs are full with water. (Photograph by I. Yakovlev.)

searching the trough by informed and uninformed individuals was compared (Reznikova and Ryabko, 2003; Novgorodova, 2005, 2006). For example, in *F. pratensis* almost all naive foragers were able to find food on their own but they spent 10–15 times more than those ants which entered the maze after the contact with the successful scout. Average values as well as the numbers of samples are given in Table 32.1. For every trial Wilcoxson's nonparametric test was used (see Hollander and Wolf, 1973) to test hypothesis H_0 (data from both samples follow the same distribution) against H_1 (they follow different distributions) at significance level of 0.01. It turned out that the duration of searching time is essentially more in those ants that previously had not had contact with the successful scout.

In sum, the obtained data confirm information transmission in three ant species and exclude any orientation mechanism, except the use of information transmission by the scouts.

No information transmission based on distant homing was observed in the other two species. In singly foraging *F. cunicularia* the foragers learned the way towards the maze, while making up to 30 trips per day, but they did not try to recruit other members of their colony. Not more than five ants were active in the maze per day. *Myrmica rubra* workers used olfactory cues, but when the maze was changed, they had to do without odour trails. In these cases they resorted to only solitary foraging, just as *F. cunicularia*.

Evaluation of information transmission rate in ants is based on the fact that the quantity of information (in bits), necessary for choosing the correct way toward the maze, equals i, the number of forks. One can assume that this duration (t) was $ai + b$, where i is the number of forks, a is a coefficient of proportionality equal to the rate of information transmission (min/bit) and b is an introduced constant, since ants can transmit information not related directly to the task, for example, the simple signal 'food'. Besides, it is not ruled out that a discovering ant might transmit, in some way, the information on its way back to the nest, using acoustic or some other means of communication. In this connection, it is important that the way back from the maze to the nest was in all experiments approximately the same and, therefore, the time before the antennal contact with the foragers in the nest, which the scout could hypothetically use for the message transmission, was approximately the same and did not depend on the number of forks.

From the data obtained, the parameters of linear regression and the sample correlation coefficient (r) can be evaluated. The rate of the information transmission (a) derived from the equation $t = ai + b$, was 0.738 for *F. sanguinea* and 1.094 for *F. rufa*. These values are not considered species-specific constants, they probably

Table 32.1. *Comparison of duration of search for the trough by 'uninformed' (U)* F. pratensis *ants and individuals that previously had contact with the successful scout ('Informed', I); July–August, 2003*

Sequence of turns	Ants (U/I)	Mean duration (s)	Number in sample	*P*
RRRR	U	345.7	9	<0.01
	I	36.3	9	
LLLL	U	508.0	9	<0.01
	I	37.3	9	
LRRL	U	118.7	7	<0.01
	I	16.6	7	
RLLR	U	565.9	7	<0.01
	I	16.3	7	

Source: Reznikova and Ryabko (2003).

vary. Note that the rate of information transmission is relatively small in ants.

In order to estimate the potential productivity of ants 'language', let us count the total number of different possible ways to the trough. In a simplest binary tree with one fork there are two leaves and therefore two different ways. In a tree with two forks there are 2^2 ways, with three forks, 2^3 ways, and with six forks, 2^6 ways; hence, the total number of different ways is equal to $2+2^2+2^3+\ldots+2^6=126$. This is the minimal number of messages the ants must possess in order to pass the information about the food placed on any leaf of the binary tree with six forks.

Another series of experiments was based on a basic concept of Kolmogorov complexity and was intended to check whether in highly social ant species individuals possess such an important property of intelligent communicators as the ability to quickly grasp the regularities and to use them for coding and 'compression' of information. Thus the length of the text should be proportional to the complexity of the information. This idea is the basic concept of Kolmogorov complexity. This concept is applied to words (or text) composed of the letters of an alphabet, for example, of an alphabet, consisting of two letters, L and R. Informally, the complexity of a word (and its uncertainty) equals the length of its most concise description, according to Kolmogorov. For example, the word 'LLLLLLLL' can be represented as '8L', the word 'LRLRLRLR' as '4LR', while the 'random' word of shorter length 'LRRLRL' probably cannot be expressed more concisely, and this is the most complex of the three.

Reseachers analysed the question of whether ants can apply simple 'text' regularities for compression (here 'text' means the sequence of the turns toward the maze). As proved by Kolmogorov (1965), there is no algorithmically computable quantitative measure of text complication. So, strictly speaking, we can only verify whether ants have the 'notion' of simple and complex sequences.

In the special series of experiments, ants *F. sanguinea* were presented with different sequences of turns. Comparison of the main hypothesis H_0 (the time for information transmission does not depend on the text complexity) with hypothesis H_1 (this time actually does depend on it) showed that H_0 should be rejected, in other words, the more time ants spent on information transmission, the more information – according to Kolmogorov – was contained in the message. It is interesting that the ants began to use regularities to compress only quite large 'texts'. Thus, they spent from 120 to 220 seconds to transmit information about random turn patterns on the maze with five and six forks and from 78 to 135 seconds when turn patterns were regular. On the other hand, there was no essential significance when the length of sequences was less than 4 (Table 32.2).

32.2 Evaluation of ants' 'language'

The majority of models that describe processes of self-organisation in colonies of social insects consider cognitive skills redundant for effective information transmission. However, another set of arguments comes from the data concerning the excellent learning abilities of social insects. The bulk of the results obtained concerns orientation and memory, mainly in bees and wasps. Probably, studying communicative means is one of the most effective tools to comprehend the limits of intelligence in social insects at the individual level. For this, appropriate species have to be chosen as well as an adequate set of tasks for the colony to solve.

From the early experiments of Schneirla (1946) and Thorpe (1950) it was known that some ants and solitary wasps perform almost as well as rats and dogs in maze-learning and detour tasks. Mazokhin-Porshnyakov (1969; Mazokhin-Porshnyakov and Kartsev, 1984) experimentally demonstrated that the honeybees and social wasps are capable of abstraction, extrapolation and solving rather sophisticated discrimination tasks at the level of dogs and monkeys. In particular, several individually trained bees were able to distinguish between chains consisting of paired and unpaired small figures and thus capable for concept formation. Giurfa *et al.* (1999) found bees to be able to form concepts about symmetry versus asymmetry.

Table 32.2. *Duration of information transmission on the way to the trough by* F. sanguinea *scouts to foragers*

Number[a]	Sequences	Mean duration (s)	SD	Number of experiments
1	LL	72	8	18
2	RRR	75	5	15
3	LLLL	84	6	9
4	RRRRR	78	8	10
5	LLLLLL	90	9	8
6	RRRRRR	88	9	5
7	LRLRLR	130	11	4
8	RLRLRL	135	9	8
9	LLR	69	4	12
10	LRLL	100	11	10
11	RLLR	120	9	6
12	RRLRL	150	16	8
13	RLRRRL	180	22	6
14	RRLRRR	220	15	7
15	LRLLRL	200	18	5

Note:
[a] Numbers 1–8 regular turn pattern; numbers 9–15 random turn pattern.
Source: Ryabko and Reznikova (1996).

As we have already seen in Parts IV, V and VI, bees, wasps and ants possess several specific mechanisms which help them in the efficient perception and learning (Collett *et al.*, 2003; Menzel and Müller, 1996). For example, memory of the landmarks passed on flights between the hive and the feeding place is organised sequentially in honeybees, so they have to 'count landmarks' (Chittka and Geiger, 1995). Behavioural flexibility allows social insects to switch their learning patterns in accordance with changeable environment. For example, when the bees are prevented from learning landmark cues on arrival at a hive, they match for learning them during specific 'turn-back-and-look' flight manoeuvres (Lehrer and Bianko, 2000).

Ants are known to do many clever things such as using cognitive maps (Wehner and Menzel, 1990), learning by observation (Reznikova, 1982, 2001) and even counting (Reznikova and Ryabko, 2001). Similarly to bees, from which only one species, the honeybee, possesses symbolic language, from about 10 000 ant species, only few highly social species use a sophisticated system of communication that has been revealed by applying methods of information theory. Among these highly social species red wood ants (*Formica* s. str.) belong to the most intelligent ones. Rosengren and Fortelius (1987) characterised them as 'replete ants' storing not lipids in their fat-bodies but habitat information in their brains. In comparison to many sympatric species, red wood ants have hundreds of times more individuals in their colonies and huge feeding territories. They face complex tasks every day; for example, in order to obtain honeydew, their basic food, red wood ants have to memorise and possibly pass each other information about locations of thousands of aphid colonies within such a huge three-dimension space as a tree is for an ant (Reznikova and Novgorodova, 1998). Tasks of this kind require communicative means for distant homing. It is natural to expect that ant species with a different colony design develop communication systems of different levels of complexity. Thus, Robson and Traniello

(1998) found a major difference between mass-recruitment ant species that are characterised by collective actions of simple individuals, and the group-retrieving ones. In the latter case, a colony design is based on complexity rather than simplicity. The process of recruitment is so organised that the removal of the discovering ant leads to the dissolution of the retrieval group.

Elaborating a possible approach for the comparative investigation of communication, we found out that the most promising situation for observing task distribution and behavioural flexibility at the individual level and for evaluating the potential properties of communication in group-retrieving ants is to force them to solve a complex search problem. In this situation hidden processes of information transmission would become observable.

Using ideas from information theory to design experiments enables researchers not only to reveal information transmission from scouts to foragers within cliques but also to estimate the rate of this process and to suggest that group-retrieving ants are able to memorise and use simple regularities, thus compressing the information available. This, in turn, gives the possibility of assuming flexibility of the communication system in highly social ant species based on individual cognitive capabilities that help colonies to solve many daily, but non-standard, problems.

The hypothesis of the flexibility of ants' 'language' is supported by the results of the 'counting' experiments by Reznikova and Ryabko (2000, 2001) which were described in Part VI. It was suggested that ants are able to invent several new 'words' in order to denote frequently encountered landmarks. Applying the ideas of Kolmogorov complexity to study ants' 'language' enabled us to consider ants to be able to 'compress' both perceived information and transferred messages on the basis of grasped regularities of serial turns to the desired goal. All these data suggest that group-retrieving ant species possess a flexible communication system which makes the process of transferring information less costly for them.

CONCLUDING COMMENTS

During the last few decades, the combined efforts of scientists applying different experimental approaches have revealed some features of communication systems of non-human species that were earlier attributed exclusively to humans. Among them one can list animals' ability to use referential signals organised by 'proto-grammar' rules, to transfer messages in abstract 'symbolic' form, to create messages about things and events distant in time and space, to 'translate' messages of other species and to extract meaningful parts from strangers' signals.

However, one can find many points of discontinuity with the communicative practice of animals. Although members of several species demonstrate understanding of grammatical rules when using artificial intermediary languages, there is no evidence of syntax in the natural communication of animals. There is also little evidence for learning and modification in the natural signals of animals. There is much to be done to reveal the evolutionary roots of such sophisticated system of communication as human language.

Each of three methodological approaches that have been discussed in this part has its specific power. Direct deciphering of animals' communication is the oldest of the methods under consideration. By now some of the researchers who have tried to crack wild codes by means of recording signals and playback experiments have become masters of at least segments of King Solomon's Ring. Dictionaries, although very fragmentary, have been compiled for several species of mammals, birds and insects. The decoded 'words' concern alarm calls, calls for cohesion, and signals about food. The honeybees' dance language remains the most complex among animals' communication system that has been decoded up to date. Apart from this unique system, other communications are difficult to decipher because of many methodological barriers, among which low repeatability of standard living situations in the wild is important to note.

A number of studies with captive animals that have been trained with human-designed artificial communication systems has revolutionarily changed our conception of animals' linguistic abilities and language-related cognitive skills. Language-training experiments are of great help for discovering the potentials of animal language behaviour and for studying the roots of human predisposition for the development of sophisticated language. Animals from very widely separated taxa that were taught very different artificial languages have met the same criteria of language rules and demonstrated their competence in syntax and semantics. But the evolutionary significance of these animals' ability would be difficult to evaluate if species' naturally existing syntactic abilities that have evolved in the context of their natural communicative behaviour had not been revealed. At the same time, the fact that only narrow and limited syntactic abilities have been discovered in natural communicative systems restricts our opportunity to judge about the potential power of species' language behaviour and related cognitive abilities.

For this, the use of ideas from information theory opens new horizons. This approach is designed to study quantitative characteristics of natural communicative systems and important properties of animal intelligence. Applying this method, there is no need to crack animals' codes. All we need is to ask animals to pass messages of definite lengths and complexities. By measuring the time duration that the animals spend on transmitting messages with desired conditions, we can judge the potentials of their communicative system. By means of this approach it has been demonstrated that communicative system of highly social ants meets several fundamental criteria of language such as productivity, specialisation and displacement.

From the view of information theory at least two important standards should be added to the list of criteria that characterise language. Firstly, the length of a message should correlate with the quantity of information contained in this message. Secondly, the ability to grasp the regularities and to use them for coding and 'compression' of information should be considered one of the most important properties of language and its carriers' intellect. These properties were revealed in the language behaviour of *Formica* ants. The use of ideas and methods of information theory led to the discovery of flexibility of communication systems in ants which may be considered one of the most complex properties of animal 'languages'. It is a challenge to apply the information theory approach to the study of communication in a wide variety of social animals.

Part X

Social life and social intelligence in the wild

Alpha children wear grey. They work much harder than we do, because they're so frightfully clever. I'm really awfully glad I'm a Beta, because I don't work so hard. And then we are much better than the Gammas and Deltas.

Aldous Huxley, *Brave New World*

After discussing animal language behaviour in the previous part, in this, final Part of the book, we will consider the question *who* speaks, if 'speaks' at all. It is impossible to understand specific communication systems ignoring specific codes of social interrelations. Some of these codes are controlled by flexible and intelligent behaviour whereas others are innate. Together with mechanisms of maintaining social structure such as hierarchy and nepotism, which can be considered universal for many social animals, there are many specific variants of sociality, and in some species several variants of sociality of different levels of complexity fit within species-specific 'toolbars'. Display of social behaviour in animals is diverse, and one can find thousands of interesting papers and hundreds of books devoted to evolutionary, ecological and ethological aspects of social life in the wild. In this Part I only briefly analyse the role of intelligence in functioning of animal societies.

Chapter 33

Diversity of social systems in animals

Social systems in animals can be described in terms of social organisation, i.e. characteristic grouping patterns, mating systems and variants of the social relationships among individuals. It is not easy to navigate the diversity of social systems in nature. Analogous variants of social structures develop independently in distant classes of the animal kingdom. At the general level of consideration, some forms of territorial, aggressive, breeding and parental behaviour can be displayed similarly in some species of birds, mammals and insects. For instance, dragonflies display characteristics of territorial defence and hunting behaviour similar to those of vertebrates. At the same time, closely related species often possess essentially different social organisation. Thus, four species of great apes have lived in tropical forests since the Miocene and share many lines of life histories; nevertheless they show dramatic variations in their social systems (for a review see van Schaik *et al.*, 2004). In contrast, the large-cat family, which has a cosmopolitan distribution and a great diversity of habitats, retains relative uniformity of socio-demographic systems in many (but surely not all) species (Guggisberg, 1975; Green, 1991).

The social flexibility of a species is often associated with its ability to occupy different habitats (Lott, 1991). For example, rodents such as marmots, prairie voles, striped mice and some others can be monogamous or solitary promiscuous, or polygynous or cooperatively breeding depending on their habitat (Rogovin, 1992; Roberts *et al.*, 1998; Schradin and Pillay, 2004; Randall *et al.*, 2005). Many ant species possess multiple social and foraging organisations, depending on their ecological variables, from solitary foraging in small families to 'professional' division of labour in million-strong colonies (Reznikova, 1979, 1980, 1999). Using quantitative comparisons, Clutton-Brock (1974) and Clutton-Brock and Harvey (1977) showed that species differences in group size, ranging behaviour and sexual dimorphism among primates are consistently related to ecological variables, such as diet type, timing of activity and breeding system.

There are many ways to rate types of animal communities based on the characteristic features of their social systems. For example, species like elephants, dolphins and some primates live in so-called *fission–fusion societies* in which social groups meet up and then disperse again on a regular basis. Many fossorial and semi-fossorial rodents live in stable groups often sharing the work of making burrows and using common trails and holes. When considering social styles of life one can find more in common between pumas and tortoises than between pumas and lions. Analogous social structures in very distant species could have been developed under the influence of rather different evolutionary events and ecological circumstances. At the same time there is not a broad spectrum of variants of social roles of individuals within groups. This lightens the work of those who are willing to classify animal societies.

The sociobiological approach elaborated by Edward Wilson (1975) allows one to describe animal societies based on their style of social life, which may be the same in taxonomically distant species. Sociobiology is not primarily related to gathering new kinds of data; rather, it is a way of looking at biological phenomena related to social behaviour from a comprehensive and explicitly evolutionary perspective.

Here we consider briefly only two of many existing approaches for such a classification. According to the first of them, all communities can be subdivided into two large types, namely, anonymous and individualised communities. Although very simple, this approach is useful when a prediction of the division of relationships within communities is needed. The second approach for classification is based on the 'levels of sociality' that can be distinguished in groups of animals.

33.1 Anonymity versus individual recognition in animal communities

There are many reasons for individual recognition in animal societies including territoriality, reciprocal altruism, monogamous pairing, and the maintenance of hierarchy. Lorenz's (1966) at first sight paradoxical view that 'aggression is impossible without love' is really based on the notion of address space. Individuals are able to sort out their relationships if they know each other 'personally'. The concept of Machiavellian intelligence, presented by de Waal (1982) and Byrne and Whiten (1988), refers to the intellectual ability to deal with social situations and to change tactics creatively as the game progresses. Such a system of social navigation is characteristic for primate societies, and is based on members' awareness of the characters and habits of their mates and their ability to predict the possible reactions of individuals to events and the actions of others. Broom (2003) considers individual recognition between members of societies of group-living species a necessary condition for the development of morality and religion.

From the experimental scientist's point of view, individual recognition is a cognitive process whereby an animal becomes familiar with a conspecific and later discriminates it from other familiar individuals.

Individually distinct cues have been demonstrated in a variety of mammals as well as in several bird species (for a review see Mateo, 2004, 2006).

There are many studies of social species which indicate that they discriminate between individuals of their own species and in some cases remember them even if they have been absent for months or even years. Among primates, humans are good at remembering other members of their own species. An average person can recognise (in the sense of knowing whether they have seen it before) about 2000 faces. Judging by differential behavioural (usually vocalisations or threat) responses in face-matching tasks chimpanzees could do this at least for 50 different examples of faces (Parr and de Waal, 1999; Parr *et al.*, 2000).

At the same time, complex social behaviour can be based on mechanisms that do not require individual recognition. Even to act congruently within a team there is no need for individuals to know each other personally. It is possible to interact as a team member being informed only about the functions of agents. Anonymous communities such as fish shoals and bird flocks are capable of complex coherent behaviour. In particular, they can act as coordinated groups against predators. For instance, flocks of starlings when meeting with a sparrowhawk group together, push on to meet the predator, flow around it and then gather again after its tail. For their turn, predators usually do not attack an individual within a group; instead, they wait for a chance to catch a member of the group that has got lost.

Were nature arranged more simply one could suggest that anonymity is typical for primitive societies, and that increasing complexity of social organisation in the animal kingdom corresponds to the increasing of complexity of both nervous systems and behaviour in general. In reality, the picture is variegated. For example, in the class of birds, swans, geese and ravens enjoy intimate personal interrelations whereas at least some species of storks and herons are known not to be able to recognise each other.

A mouse can remember for 5–7 days the odours of another mouse with whom it has only been in contact for 2 minutes and mice are able to distinguish between 12 individual conspecifics presented simultaneously. A rat can only remember each rat it encounters for about an hour (Eichenbaum, 2004). Like many social insects, rats strongly distinguish members of their colony from strangers and react to other rats in a very aggressive manner whereas individual recognition seems to play a little part in their social life (Lorenz, 1966). It is worth noting that discrimination between nestmates and strangers can be based on rather complex processes. For example, nestmate recognition in the social Hymenoptera is by colony scent, which turns out to be a complex Gestalt of hydrocarbons absorbed into the outer cuticle of the exoskeleton, shared by food exchange and grooming, learned by imprinting, and largely independent of kinship in composition (Wilson and Hölldobler, 2005).

Invertebrates are usually thought to be incapable of individually identifying conspecifics. Nevertheless, there is at least some evidence of complex behavioural interactions between individuals which are probably based on individual recognition. This is not astonishing when one considers those feats of intelligence of the social hymenopterans that were described earlier in this book.

Field experiments with *Formica pratensis* ants showed that 'frontier guards' in neighbouring ants' families identify each other at least as members of small 'guard groups' (Reznikova, 1974, 1982). This was investigated in the context of permanent units of ants patrolling the bounds of their feeding territories. To test the hypothesis that members of these groups recognise each other, ants were attached by thin 'leashes' to inscribed object-plates. Inscriptions informed the researchers which ants had been taken from the same or from an alien family; some of the alien ants were frontier guards (taken from the corresponding plots of ants' territories) hypothetically acquainted with the guards of the family being tested, whereas others were strangers taken from remote areas of the feeding territory of the alien family (Fig. 33.1). It turned

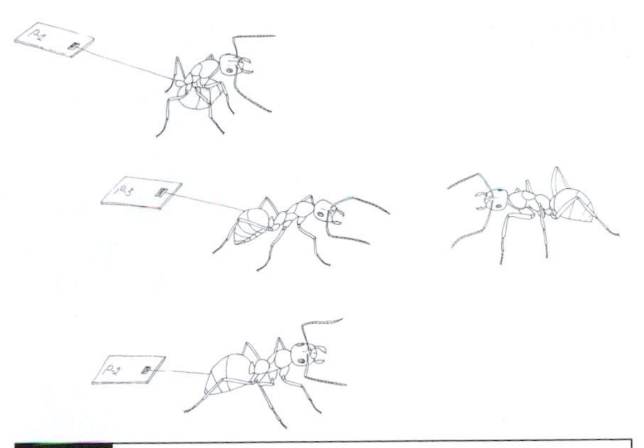

Fig. 33.1 Schematic representation of the experiment in which ants distinguished between acquainted guards and strangers. (Drawings by S. Panteleeva.)

out that the ants do not touch members of their own family, excitedly feel all over alien but possibly acquainted guards from the neighbouring family, and kill all ants that had come from the remote areas of the neighbouring territory. Since ants definitely can not read inscriptions, it is very likely that they can easily distinguish between acquainted guards and strangers. Besides *F. pratensis* several other ant species could be expected to possess the same system of recognition; the great majority of ants use a much more simple 'strangers–no strangers' system of identification.

Similar experiments have revealed the same effect in vertebrates species such as oscines (Falls and Brooks, 1975), salamanders (Temeles, 1994), beavers (Bjǿrkøyli and Rosell, 2002) and others. This system of identification underlies the so-called *dear enemy hypothesis* (DEH): personally acquainted neighbours establish boundaries, split resources, and form a temporal society in which balance is based on sharing informative signals (Godard, 1991; Fox and Baird, 1992). Many social species such as monkeys, hyenas, wild dogs and others behave in a way that indicates recognition of members of neighbouring social groups.

There is some evidence that individual 'face-control' based on a higher level of recognition exists in invertebrate species that interact in a stable prolonged cohesion. The first example came from desert woodlice. These crustaceans burrow in pairs and recognise each other by

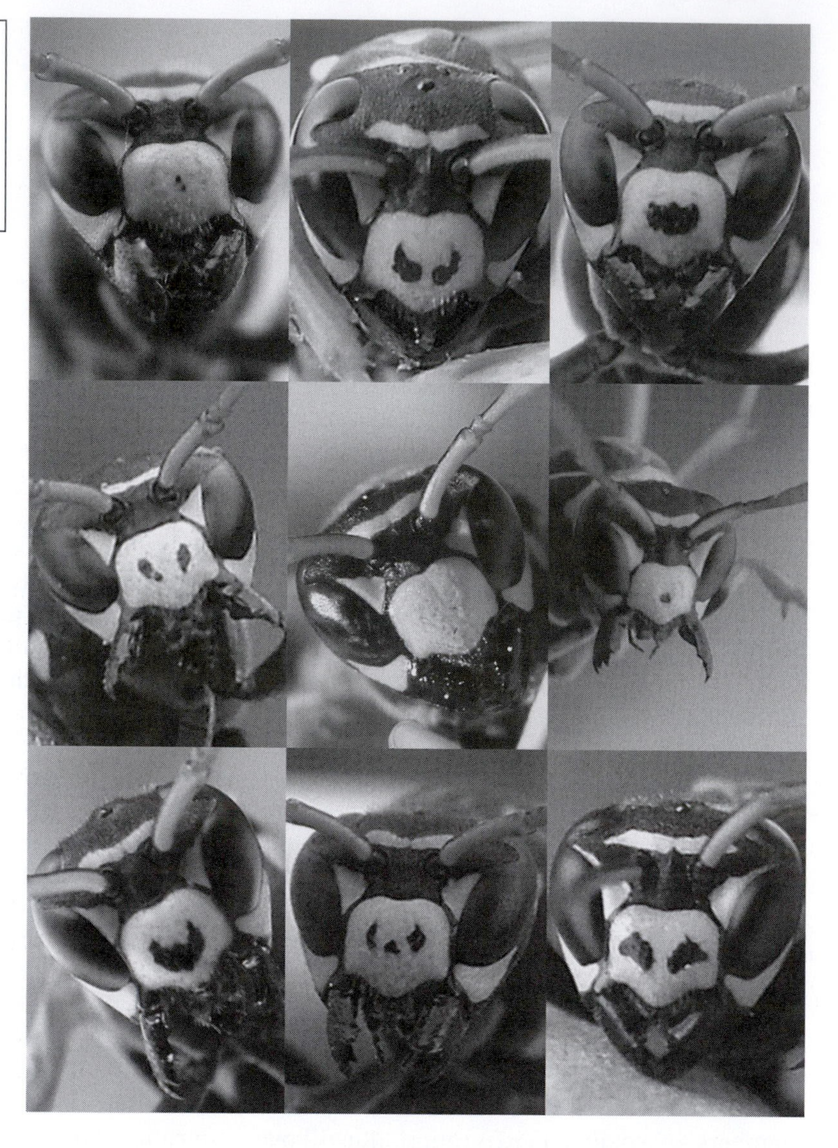

Fig. 33.2 In the paper wasp *Polistes fuscatus* queens and workers use yellow facial and abdominal patterns to visually identify individual nest mates (Tibbetts, 2002). (Photograph by E. A. Tibbetts; Courtesy of E. A. Tibbetts.)

tactile examination of the location of the aciculae on their partners' body (Marikovsky, 1969; Linsenmaier, 1987). In the paper wasp *Polistes fuscatus* queens and workers use yellow facial and abdominal patterns to visually identify individual nest mates (Fig. 33.2). Individuals whose yellow markings were experimentally altered with paint received more aggression than control wasps painted in a way that did not alter their markings. In addition, aggression to wasps with experimentally altered markings declines significantly over time, suggesting that it is the unfamiliarity of the new markings rather than something else about the new markings that caused the aggression (Tibbetts, 2002; Tibbetts and Dale, 2004). Ants of highly social species working in 'teams' or 'cliques' possibly use individual recognition to work effectively in complex situations (Reznikova and Ryabko, 2003). We will consider this case in Chapter 35.

In long-living social species tight cohesion in societies can be based on highly developed capacities for social recognition. For instance, African elephants can recognise and remember 600 other individuals (Poole and Moss, 1983). Playback experiments carried out in Kenya by McComb *et al.* (2000, 2001) have established that female elephants may be able to recognise about 100 different individuals by their acoustic signals and in some cases even remember them for 12 years. Discriminatory abilities in this case seem to improve with age. Female elephants travel around in small social groups led by a matriarch. These groups may meet up with a dozen or more other groups at different times while traversing the ranges of their habitat. To prevent hostilities and reduce possible stress or panic it is important for the elephants to know whether the group they are meeting is a potential threat or not. Researchers found that the older matriarchs (of about 55+ years old) were able to respond in these playback experiments with a greater degree of association between the calls they recognised and the exact degree of familiarity they had with the caller.

There are several limitations of individual recognition in animals' societies. For example, as was noted in Part IX, in playback experiments with vervet monkeys, when the call of one mother's infant is played to the troupe, many of monkeys will look in the direction of the mother whose baby has called. Recognising and responding to relationships between other individuals is a very advanced level of social evolution (Cheney and Seyfarth, 1990). The laughing hyena represents one non-primate species with a high level of group organisation that would appear to demand individual recognition skills. Indeed, playback experiments have confirmed that mothers in this species respond differentially to the 'whoop' calls produced by their own cubs. There is also some indication that they can recognise the voices of other individuals within the clan.

The basic capacity for individual social recognition seems to be widespread in animals. We do not know yet whether an ant can distinguish a face of another familiar ant if its portrait is rotated in front of it from front to profile but sheep seem to be capable of this. Kendrick and colleagues (Kendrick and Baldwin, 1987; Kendrick *et al.*, 1995, 2001) applied several behavioural approaches to examine whether sheep can distinguish between categories of individuals and the extent to which they can actually identify specific individuals by their faces. Researchers constructed a choice maze apparatus that allowed sheep to choose between face images in order to gain access to the real individual whose face-picture had been seen. The sheep were shown pairs of faces that had different attractions to them (i.e. sheep versus human; familiar versus unfamiliar animal or breed; male versus female). The sheep chose sheep faces over human ones and familiar sheep faces over unfamiliar ones. Another way of testing the potential for using mental imagery is to train the animals to discriminate between pairs of familiar or unfamiliar faces by the frontal image and then to switch suddenly the images used to profile views (or vice versa). The sheep continue to get the task right when the face views are changed. This shows both that they know it is the act of recognising a particular individual rather than a particular view of their face that gets them a reward (i.e. true individual recognition) and that they may be able to mentally rotate the image of the face from front to profile in order to match a front view with a profile one. Another set of experiments showed that sheep can divide pictures of conspecifics' faces into emotionally distinct categories. They can distinguish whether faces had horns and how big they were – an important index of dominance and gender; whether faces were of members of the same breed and how familiar they were – sheep prefer the company of their own breed and are known to strike up long-term individual friendships; whether faces were from species that could pose a threat – humans and dogs. This face recognition system in sheep is mainly designed for identifying categories of individuals that have a specific emotional significance. It implies close interactions between the brain systems dealing with detection of faces and those associated with making emotional responses. Interestingly it is often just this link that appears to break down in human individuals with schizophrenia and autism.

Cognitively demanding individual recognition in animals' societies can be implemented on different kinds of individual traits including olfactory cues. It has been experimentally demonstrated on Belding's ground squirrels, group-living burrowing rodents found in the

Fig. 33.3 Belding's ground squirrel inspecting the odour profile of its relative. (Copyright © Cornell University.)

mountainous regions of the western United States. They live up to 12 years, so there is potential for them to interact repeatedly and recognise individuals. These animals produce a number of cues that are individually distinct, including odours from oral, dorsal, pedal and anal glands and from the ears (Fig. 33.3). Mateo (2006) has used habituation–dishabituation tasks (see Chapter 6) to determine which odours are individually distinct. In this task, an animal is repeatedly presented with a particular stimulus (here, an individual's odour) until it habituates to it, and then an animal is presented with a novel stimulus (here, another individual's odour) to determine whether the animal dishabituates to it, indicating discrimination of the two stimuli. A variant of the task was used to determine whether ground squirrels form multiple representations of familiar individuals. That is, does the knowledge of an odour from a familiar individual generalise to other odours of that individual (produced from other glands). Subjects were habituated to one odour collected from Individual 1, and were then tested with the odour from a second source collected from Individual 1 and the same odour source from Individual 2. If ground squirrels recognise individuals as a whole, rather than just remember their separate odours, they should generalise among several odours of an individual and dishabituate (show an increased response) only to odours of another individual. The results obtained suggest that ground squirrels incorporate multiple odours into their memories of conspecifics and thus form a mental representation of familiar individuals.

In sum, there are many studies not only of such acknowledged highly social and intellectual species as primates, elephants and dolphins, but also of pigs, sheep, dogs, pigeons, ground squirrels and some others which indicate that they can discriminate between individuals of their own species and other species, can behave in a consistently different way towards various members of their social group, and can use the information in a way which suggests that they have concepts of individuals and of some of their qualities (Broom, 2003). Knowing whether species members live in anonymous or in individualised social space we can judge their behaviour more exactly and even use this for practical purposes. For instance, Albright (1978) recommended that herds of 100 or more cattle should be split into smaller groups on the basis that it is difficult for the cattle to recognise and remember more than 70 to 100 individuals. Anthropologist Dunbar (1996) suggests that human beings should reach a 'natural' cognitive limit when group size reaches about 150. There is extensive empirical evidence of social groupings of about this size in the anthropological literature. Many companies organise working space based on these projections of social limits.

33.2 Levels of sociality in animal communities

The highest level of sociality in animals is *eusociality*. The word 'eusocial' is derived from the Greek 'eu' meaning 'good'. Eusociality is defined by Wilson (1971, 1975) as the state where 'individuals of the same species cooperate in caring for the young; there is reproductive division of labour, with more or less sterile individuals working on behalf of fecund individuals; and there is an overlap of at least two generations in life stage capable of contributing of colony labour, so that offspring assist parents during some period of their life'. Three important characteristics are essential for a species to be called eusocial. There are: (1) cooperative brood care, (2) overlap between generations and (3) reproductive caste differentiation. Reproductive caste differentiation means division of labour into reproductive- and non-reproductive castes.

The last of the eusocial traits, the existence of a subordinate or even completely sterile worker caste, is the rarest of the three. It is also by far the most significant with reference to the further evolutionary potential of social life, for when individuals can be turned into specialised working machines, an intricate division of labour can be achieved and a complicated social organisation becomes attainable even with a relatively simple repertoire of individual behaviour. Colonies of social insects impress us with their strength and organisation, for example, the gigantic colonies of African driver ants, with more than 20 million workers, that live in large excavated soil cavities and hunt in armies for a wide range of arthropod and small vertebrate prey (Hölldobler and Wilson, 1990).

Eusociality had been earlier considered a social structure unique to insects. Michener and Lin (Michener, 1969; Lin and Michener, 1972) provided a classification of various levels of sociality in insects that was later adopted by Wilson (1975). Indeed, all levels of sociality have analogues in the social systems of other animals. Semi-social and eusocial systems that were earlier considered specific for social insects have been recently discovered in other, although few, species. Here is a classification of social systems extended to arthropods (see Gadagkar, 1987, 1992).

(1) *Solitary*. This term characterises the absence of any extent of sociality so that members of a species may not interact with each other at all except during courtship and mating.

(2) *Subsociality*. The adults care for their own young for some period of time. Subsociality is widespread. Examples may be found among crustaceans, spiders, mites, scorpions, millipedes, centipedes, cockroaches, crickets, bugs and beetles.

(3) *Communality*. Members of the same generation use the same composite nest without cooperation in brood care. Many spiders and digger wasps can serve as examples of such social system.

(4) *Quasi-sociality*. This is one step higher that communality, because it involves cooperative brood care. Several species of spiders and euglossine bees are good examples of this. Thus, quasi-sociality includes one of the three features of eusociality, namely, cooperative brood care.

(5) *Semi-sociality*. A situation that incorporates two features of eusociality, namely, cooperative brood care and reproductive caste differentiation (workers and queens) but lacks overlappimg of generations. Many wasps and bees are semi-social. For instance, most polistine wasp colonies are merely semi-social at the beginning of the colony cycle. Only after the first daughter emerges and begins to work for the colony is there overlapping of generations to qualify for the title eusocial.

(6) *Parasociality*. This is a relatively new term which includes communality, quasi-sociality and semi-sociality but excludes subsociality. This term is useful because there appear to have been two routes in the evolution of eusociality, namely, the subsocial route and the parasocial route.

(7) *Eusociality*. It is a common practice to distinguish two levels: 'Primitively eusocial' refers to the cases such as many wasps and bees where there is little or no morphological caste differentiation, and as a consequence, there is often considerable flexibility in social roles that a given animal may adopt. Reversal of roles from queen to worker and vice versa is also sometimes seen. 'Highly eusocial' refers to the most advanced societies where there is clear-cut morphological caste differentiation and little, if any, flexibility in the social roles that adult animals may adopt. Caste determination involves rather complicated nutritional and hormonal mechanisms. All termites, most ants and many bees and wasps such as the honeybees and the vespine wasps are highly eusocial.

It has become clear in the last decades that complex social systems are more widespread in the animal kingdom than researchers had expected. Sociality has been discovered in such unexpected creatures as lizards and spiders. Lizards possess variety and sophistication in social organisation. Territorial males beat up intruders and have harems of females in the defended area (see Fox *et al.*, 2003). Some spider species live in colonies from a few to several hundred individuals and cooperate in web building, prey capture and nest maintenance in contrast to solitary species that are often aggressive and cannibalistic towards other spiders. Social spiders have evolved independently in Africa, the Middle East, the Americas and Australia (Avilés and Tufico, 1998; Evans, 1999; Jones and Parker, 2000).

There are many particular classifications of social systems created for different taxonomical groups of species. For example, all birds exhibit social behaviour, from pairing territoriality to common breeding. Brown (1978) suggested a general classification of avian communal systems. Detailed classification of communal breeding systems in birds is considered in the book edited by Koenig and Dickinson (2004). For ungulates Baskin (1976) suggested the following classification of social systems: companies based on personal acquaintance grouping around a dominating individual (typical for argali), parcels that syndicate several companies (typical for bison), harems in which animals defend boundaries of communal territories (saiga, horses), and troops, in which males possessing local harems are tied by relations of dominance (some species of antelopes).

Eusociality recently has been expanded to a few more groups of organisms. In insects there are eusocial gall aphids (Aoki, 1977, 2003; Foster, 1990; Benton and Foster, 1992; Abbot *et al.*, 2001; Strassmann and Queller, 2001), gall-forming thrips (Thysanoptera) and a social weevil *Austroplatypus incompertus* (Kent and Simpson, 1992) (for reviews see Gadagkar, 1993; Choe and Crespi, 1997). In non-insects, eusociality has been described only for very specific rodents and shrimps. Naked mole rats live in complex underground tunnel systems in Africa and animals in the same nest are closely related, while only one female (the queen) reproduces, although workers, normally sterile, can ovulate when removed from the nest, presumably due to a lack of inhibition from the queen (see Sherman *et al.*, 1991). Snapping shrimp *Synalpheus regalis* lives inside sponges and each 'colony' has 200–300 individuals, but only one queen reproduces; again the caste is probably not fixed – the workers remain totipotent and can potentially become queens when the queen shrimp is removed (Adler, 1996; Duffy, 1996). From the anthropomorphic point, it seems that 'eu' does not sound good for sterile workers which are 'enslaved' by members of the reproductive caste. This is, however, a question of special discussion among students of eusocial organisms: who enslaves whom. We will analyse some details of social structures of eusocial species further in this Part.

Besides truly eusocial species, many species of mammals and birds called *cooperative breeders* share the first and third characteristics of eusociality and partly the second in a sense that some members of a society serve as 'helpers' in raising offspring and may not breed at all during their life. Nevertheless, vertebrate cooperative breeders are not eusocial because none but the naked mole rat has evolved a true sterile caste that works for the more fertile members of the community.

Further we will consider in detail behavioural aspects of the division of labour into reproductive and non-reproductive groups within communities.

Chapter 34

If one must be sacrificed, why me? Evolutionary and behavioural aspects of altruism in animals

The aim of this chapter is to analyse, although very briefly, the roots of altruistic behaviour in animals in order to imagine the reasons for social and cognitive specialisation in animal communities.

An individual animal can play different roles in communities depending on its sex, age, relatedness, rank, and last but not least, intelligence. An individual's path to the top of a hierarchically organised community may be paved by highly developed individual cognitive skills. A classic example came from Goodall's (1971) book *In the Shadow of Man*: Mike, the young chimpanzee, gained top rank at once by making a terrible noise with empty metal jerrycans stolen from the researchers' camp.

At the same time, the upper limit of the individual's self-expression may be imposed by the specific structure of communities. There are several variants of division of social roles, from division of labour in kin groups to the fine balance between altruism and 'parasitism' within groups of genetically unrelated individuals. Task allocation in animal communities can impose restrictions on the display of members' intelligence. For instance, rodents, termites and ants condemned to digging or baby-sitting or suicide defending can not forage, scout or transfer pieces of information even if they are intelligent enough to do this. Furthermore, subordinate members of cooperatively breeding societies sacrifice their energy to dominant individuals serving as helpers or even as sterile workers.

The famous Russian writer Lev Tolstoy in his novel *Anna Karenina* focused attention on several dramatic dilemmas in women's lives and among them the dilemma: to give birth to children or to stay in a family as a perpetual helper. In many novels of the nineteenth century the lives of members of a facultative 'sterile caste', governesses, were described: intelligent but poor members of society who often devoted their whole lives to caring for the offspring of rich ones (recall, for example, Charlotte Brontë's Jane Eyre). As we will see later in this Part, baby-sitting is one of the most costly and essential tasks in animal communities, including human ones. Some members of communities serve as helpers that are physically able to breed, but most never will.

To be serious, analysing the problem of the division of roles in animal societies, we face the paradox of *altruism* – that is, the situation in which some individuals subordinate their own interests and those of their immediate offspring in order to serve the interests of a larger group beyond their offspring. That altruistic behaviour is possible did not always seem natural for biologists. Darwin's followers, and among them Thomas Henry Huxley, the most enthusiastic populariser of natural selection as a factor of evolution, concentrated mainly on inter- and intraspecies competition, arguing that 'the animal world is on about the same level as the gladiator's show' (Huxley,1893), and thus nature is an arena for pitiless struggle between self-interested creatures. This concerns also human beings, although Darwin himself discussed the idea of how altruism can evolve in human societies in *The Descent of Man* (1871). Kropotkin (1902) was one of the first thinkers who countered these arguments and considered mutual aid as a factor of evolution, in particular of human evolution. He viewed cooperation as an ancient animal and human legacy.

In contemporary evolutionary biology, an organism is said to behave altruistically when its behaviour benefits other organisms, at a cost to itself. The costs and benefits are measured in terms of reproductive fitness, or expected number of offspring. So by behaving altruistically, an organism reduces the number of offspring it is likely to produce itself, but increases the number that other organisms are likely to produce.

Eusociality can be considered an extreme form of altruism in animal communities because sterile members of a group sacrifice the opportunity to produce their own offspring in order to help the alpha individuals to raise their young. Evolution favours individuals whose inherited predisposition enables them to behave in ways that maximise their reproductive success. What induces individuals to engage in behaviour that decreases their individual fitness?

Analysis of these problems became possible on the basis of ideas of gene dominance and fitness outlined by Ronald Fisher (1925, 1930). Haldane (1932, 1955) suggested that an individual's genes can be multiplied in a population even if that individual never reproduces, providing its actions favour the differential survival and reproduction of collateral relatives, such as siblings, nieces and cousins, to a sufficient degree. This hypothesis later came to be known as *kin selection*, the phrase coined by Maynard Smith (1964). These ideas can be illustrated by the following simple construction. Suppose an organism produces offspring some of which are reproducing, while others are non-reproducing but help greatly in caring for the reproducing ones. Compare this strategy with producing only offspring that reproduce. For an individual offspring it is advantageous to reproduce itself, but since it has the genes of its parent, it will follow the same strategy, that is, produce only reproducing offspring. Since we supposed that a non-reproducing child helps greatly in caring for the others, we can see that the average number of grandchildren will be greater if some of the offspring are non-reproducing. Note that here we assume that all offspring (reproducing and non-reproducing) have the same genes, and have shown that it can be advantageous for the population that an individual with some probability (or better to say, under some circumstances) becomes non-reproducing. In this construction altruism is directed at a certain groups of nearest relatives (parents, siblings, etc.). However, sometimes models of less direct altruism are also considered. A popular (although somewhat speculative) example concerns behaviour in populations of wild rabbits. It is assumed that some rabbits drum with their hind legs when they see a predator instead of running immediately to the nearest hole. Being warned by this alarm signal, other rabbits have time to flee. Of course, this does not mean that the drumming rabbit makes a decision to sacrifice its own life to the rest of community (Fig. 34.1). It simply acts in accordance with its inherited behavioural programme. Some members of a group of rabbits give alarm drums when they see predators (because they have a hypothetical "drumming gene"), but others (that lack such a gene) do not. By selfishly refusing to give an alarm signal, a rabbit can reduce the chance that it will itself be attacked, while at the same time benefiting from the alarm signals of others. However, it is possible to show that, under certain conditions, if there are sufficiently many relatives among the recipients of the altruistic behaviour, then altruistic behaviour is promoted within the population. For details and discussion of this model and accompanying ideas see: Hamilton (1964, 2001), Maynard Smith (1964, 1982), Eibl-Eibesfeldt (1970), Dawkins (1976), Grafen (1984, 2006), McFarland (1985), Axelrod *et al.* (2004) and Rice (2004).

In the 1960s and 1970s two theories emerged which tried to explain evolution of altruistic behaviour: *kin selection* (or *inclusive fitness*) theory, due to Hamilton (1964), and the theory of *reciprocal altruism*, due primarily to Trivers (1971, 1974) and Maynard Smith (1974).

Fig. 34.1 A bunny preparing to sacrifice himself. (Cartoon by P. Ryabko.)

The main mechanism of kin selection is *nepotism*, that is, preferential treatment for kin. Many social species including humans form nepotistic alliances to keep the flag of family interests flying. There is much evidence that animals behave nepotistically when facing vital problems in their life. For example, pig-tailed macaques, when helping group members who are attacked, do so most readily for close relatives, less readily for more distant relatives, and least readily for non-relatives (Massey, 1977). To do so, animals must recognise their relatives, but there is no strong correlation between nepotism and recognition ability. For example, Mateo's (2002, 2004) data on closely related species of ground squirrels support a hypothesis that kin favouritism and recognition capacities can evolve independently, depending on variation in the costs and benefits of nepotism for a given species. A highly nepotistic species, *Spermophilus beldingi*, produces odours from two different glands that correlate with relatedness ('kin labels'). Using these odours ground squirrels make accurate discriminations among never-before-encountered unfamiliar kin. A closely related species *S. lateralis* similarly produces kin labels and discriminates among kin, although it shows no evidence of nepotistic behaviour.

The ability to discriminate between kin and non-kin is found in many species, and is due either to the innate recognition of character traits associated with relatedness, or to the recognition of specific individuals with whom they have grown up (Wilson, 1987; Hepper, 1991; see also Fletcher and Michener, 1987 and Hepper, 2005). Nepotism is not always clearly altruistic and does not necessarily require genuine cognitive skills. For instance, most young plains spadefoot toads are detritivorous and congregate with kin. Some of the tadpoles become carnivorous, and such individuals live more solitarily and at least when satiated prefer to eat non-kin than kin, reducing the damage they might otherwise do to the survivorship of their relatives (Pfennig, 1992). Cannibalistic tiger salamander larvae *Ambystoma tigrinum* also discriminate kin and preferentially consume less-related individuals (Pfennig *et al.*, 1999). Genetic analyses of numerous fish species have shown that shoals formed by larvae often consist of closely related kin (Krause *et al.*, 2000). Recent experiments have shown that juvenile zebrafish can recognise and prefer their siblings to unrelated conspecifics based on olfactory cues (Mann *et al.*, 2003).

Chimpanzees possibly solve much more complex problem of kin recognition. The mechanisms underlying male cooperation in chimpanzee communities are still enigmatic (van Hooff and van Schaik, 1994). Chimpanzees live in unit groups, whose members form temporary parties that vary in size and composition. Females usually leave their natal groups after reaching sexual maturity whereas males do not disperse (Goodall, 1971, 1986; Ghiglieri, 1984; Nishida, 1990). Male chimpanzees develop strong bonds with others in their communities being engaged in a variety of social behaviour. Field observations together with DNA analysis have shown that such affiliations join together males of close rank and age rather than males belonging to the same matrilines (Mitani *et al.*, 2000); Lukas *et al.*, 2005; Mitani and Watts, 2005). It is worth noting that females give birth to a single offspring only once every 5–6 years, so brothers obviously should have an essential disparity in years. Do chimpanzees bias their behaviour to non-kin? Although current evidence indicates that Old World monkeys are unable to discriminate paternal relatives (Frederikson and Suckett, 1984; Erhart *et al.*, 1997), a recent study suggests that chimpanzees may be able to identify kin relationships between others on the basis of facial features alone, overmatching humans in sorting photographs by features of family relatedness (Parr and de Waal, 1999). As Mitani *et al.* (2002) note, this raises the intriguing possibility that male chimpanzees might be able to recognise their paternal relatives.

The importance of kinship for the evolution of altruism is widely accepted today, on both theoretical and empirical grounds. However, altruism is not always kin-directed, and there are many examples of animals behaving altruistically towards non-relatives.

The theory of reciprocal altruism is an attempt to explain the evolution of altruism among non-kin. *Reciprocity* involves the non-simultaneous exchange of resources between unrelated individuals. The basic idea is straightforward: it may benefit an animal to behave altruistically towards another, if there is an expectation of the favour being returned in the future: 'If you scratch my back, I'll scratch yours.' In his now classic paper 'The evolution of reciprocal altruism', Trivers (1971) argued that genes for cooperative and altruistic acts might be selected if individuals differentially distribute such behaviours to others that have already been cooperative and altruistic towards the donor. The cost to the animal of behaving altruistically is offset by the likelihood of this return benefit, permitting the behaviour to evolve by natural selection. This evolutionary mechanism is termed *reciprocal altruism*.

A study of blood-sharing among vampire bats suggests that reciprocation does indeed play a role in the evolution of this behaviour in addition to kinship (Wilkinson, 1984, 1990). Vampire bats *Desmodus rotundus* typically live in groups composed largely of females, with a low coefficient of relatedness. It is quite common for a vampire bat to fail to feed on a given night. This is potentially fatal, for bats die if they remain without food for more than a couple of days. On any given night, bats donate blood (by regurgitation) to other members of their group who have failed to feed, thus saving them from starvation. Since vampire bats live in small groups within large colonies and associate with each other over long periods of time, the preconditions for reciprocal altruism – multiple encounters and individual recognition – are likely to be met. Wilkinson's study showed that bats tend to share food with their close associates, and are more likely to share with those who have recently shared with them. These findings provide a confirmation of reciprocal altruism theory.

Maynard Smith (1974, 1989) suggested that cooperative behaviour can be an evolutionarily stable strategy, that is, a strategy for which no mutant strategy has higher fitness (see details in Part VII). His concept is based on game theory which, in turn, attempts to model how organisms make optimal decisions when these are contingent on what others do.

Cognitive aspects of reciprocal altruism are the source of much debate. Indeed, cooperation based on reciprocal altruism requires certain basic cognitive prerequisites, among which are repeated interactions, memory and the ability to recognise individuals. Experimental evidence that reciprocal altruism relies on cognitive abilities, making current behaviour contingent upon a history of interaction, comes from primate studies. For example, de Waal and co-authors (de Waal and Berger, 2000; de Waal and Brosnan, 2006) made a pair of brown capuchins work for food by pulling bars to obtain rewards. They found that monkeys share rewards obtained by joint effort more readily than rewards obtained individually. De Waal (1997, 2005) also demonstrated a strong tendency to 'pay' for grooming by sharing food in captive chimpanzees who based their 'service economy' on remembering reciprocal exchanges.

In many examples of cooperation among non-related animals such as grooming and food sharing behaviour in primates, or cooperative hunting in lions, wolves, hyenas and chimpanzees, it is still under discussion whether this can be interpreted in terms of reciprocal altruism. Several alternative concepts exist which explain evolution of altruistic and cooperative behaviour (Dawkins, 1982; Clutton-Brock and Parker, 1995; Sober and Wilson, 1998; Grafen, 1984, 2006). One can find detailed analysis of evolutionary aspects of cooperative behaviour in many interesting books and papers, in particular, in a chapter in the book *Behavioural Ecology* written by Emlen (1991), as well as in *Cooperative Breeding in Mammals* edited by Solomon and French (1996), and in Dugatkin (1997, 2005).

In this book we will consider several examples of social specialisation in animal communities in order to imagine how animal intelligence could fall into patterns of social specificity.

Chapter 35

Intelligence in the context of the functional structure of animal communities

In recent experimental works devoted to animal cognition authors routinely indicate not only the weight, age and sex of members of the species investigated but also their rank in a local conspecific community or at least in a laboratory group. This is necessary because results obtained in experimental trials often depend on individual's social role. Modern experimental cognitive ethology does not consider members of populations equal; instead, their membership in different functional structures is considered an important factor that impacts their intellectual potentials.

The main theme of this chapter is social specialisation in animal communities. In Part VII behavioural and cognitive specialisation of individuals within populations was described mainly from the viewpoint of individuals' inherited predisposition for certain types of behaviour as well as certain forms of learning. Here we are interested in an individual's specialisation that can be based on its social role within the local community. In some situations behavioural, social and cognitive specialisation can be congruent. Perhaps in such situations individuals are lucky to be in harmony with their mentality and environment. Maybe this is the formula for happiness. It is an intriguing problem for cognitive ethologists: is there room for intelligence within the framework of social specialisation in animal communities?

There are many gradations of social specialisation, from rigid caste division to constitutional and (or) behavioural bias towards certain roles in groups accomplishing certain tasks.

35.1 Caste division and polyethism in eusocial communities

The system of caste division was first described for social insects. Wheeler (1928) was the first to propose a detailed description of a caste system in social insects based on anatomy with no fewer than 30 categories. Hölldobler and Wilson (1990) define a *caste* as a group that specialises to some extent in one or more roles. A *role* means a set of closely linked behavioural acts (for example, queen care). Broadly characterised, a caste is any set of a particular morphological type, age group or physiological state (such as inseminated versus barren) that performs specialised labour in the colony. A *physical caste* is distinguished not only by behaviour but also by distinctive anatomical traits. A *temporal caste*, in contrast, is distinguished by age. The term *task* is used to denote a particular sequence of acts which serves to accomplish a specific purpose, such as foraging or nest repair. Finally, the division of labour by the allocation of tasks among various castes is often referred to as *polyethism*, a term apparently first employed by Weir (1958).

A good example of division of labour in eusocial communities based on caste differentiation is the existence of soldiers, that is, a specialised caste of workers that defend the colony against intruders (for a review see Judd, 2000). Termites, social aphids, social thrips and some ants produce special castes of soldiers. Some species of ants as well as eusocial shrimps and naked mole rats show a distinct polymorphism among workers with larger individuals specialised as guards. In some species of bees and wasps guards differ from other colony members only in their aggressive behaviour but not morphologically (Fig. 35.1).

Let us consider several examples of animal social systems based on caste determination and polyethism.

Eusocial insects

Eusociality is displayed in three main insect orders: Hymenoptera (ants, bees and wasps), Isoptera (termites) and Homoptera (aphids). We consider here only a rough schema of the processes of caste regulation in them. There is a great diversity of species: ants include about 11 000 species and termites about 2300 species. Different taxa have different numbers of castes, and different degrees of caste specification. There is a huge body of literature devoted to analysis of displays of eusociality in social insects (for reviews see: Hamilton, 1964, 2001; Ratnieks, 1988; Page *et al.*, 1989; Boomsma and Grafen, 1991; Gadagkar, 1994, 2001; Heinze *et al.*, 1994; Crozier and Pamilo, 1996; Queller *et al.*, 1997; Kaib, 1999; Reznikova, 2003; Wilson and Hölldobler, 2005).

As has already been mentioned, ants, bees and wasps belong to the haplodiploid group Hymenoptera (it should be noted that Hymenoptera is a large group and the majority of hymenopterans are not social). The termites (which are most closely related to cockroaches and mantids), in contrast to the Hymenoptera, exhibit diploidy. The strategy of eusociality arose once in an ancestral termite, whilst it arose several times in the Hymenoptera. Recently, some species of aphids have been found to be eusocial, with many separate origins of the state. This is explicable due to their partially asexual mode of reproduction. Most aphids that are related within a colony are members of the same clone. When social aphids form a gall (a special structure on a plant) and concentrate there, some soldiers will not reproduce. This form of eusociality tends to be restricted to a few soldiers, because the sterile forms only defend and do not care for the young. Therefore, there is less potential for the development of advanced societies.

In general, in social insects most members of a community sacrifice their own reproductive potential to provide food and protection for the few reproductive members and their offspring. The so-called *primer* pheromone causes long-term physiological changes in nestmates within

Fig. 35.1 Soldiers in different eusocial species. (a) A major worker ('soldier') and a minor worker in *Camponotus saxatilis* (photograph by S. Panteleeva and I. Yakovlev); (b) a soldier (with large mandibles) and two minor workers in termites (photograph by T. Judd; courtesy of T. Judd); (c) a bumblebee defending a nest entrance (photograph by T. Oganesov; courtesy of T. Oganesov); (d) aphid soldiers attacking a moth larva (photograph by H. Shibao; courtesy of H. Shibao); (e) a defending snapping shrimp (photograph by A. Bray; courtesy of A. Bray); (f) a major worker (soldier) naked mole-rat (photograph by P. W. Sherman; courtesy of P. W. Sherman).

Fig. 35.1 (cont.)

a colony by controlling their endocrine and/or reproductive systems. The primer pheromone is usually dispersed by only one or a few individuals ('queens') and may regulate sexuality and caste expression. In contrast, chemical signals that cause immediate behavioural changes in conspecifics are defined as *releaser* pheromones and are produced by numerous nestmates (Wilson, 1971). The social organisation in colonies depends on the control of the proportion of different castes, and on efficient recognition and communication system.

Apis mellifera, the honeybee, has the best-studied system of caste differentiation. Differences in caste-specific behaviour have been understood for many years (Michener, 1974), but recent molecular studies have shed new light on the mechanisms by which it occurs. In honeybees,

Fig. 35.1 (cont.)

the primary determination is between worker bees and gyne (future queens). Gynes are given a special diet that activates queen-specific development. Workers assume different roles in the nest as they age, a pattern known as *temporal polyethism*. Young workers stay in the nest, and as they age they replace foragers, and are replaced by younger workers within the nest. The timing of the progression through the tasks is not fixed. The progression can be delayed, or even reversed, if young workers die. Over the winter, the progression is also delayed, so that there are workers to staff the hive early in the spring.

Some ants also have an age-correlated division of labour. In ants with multiple worker castes, different morphological types assume different tasks (usually soldiers versus workers),

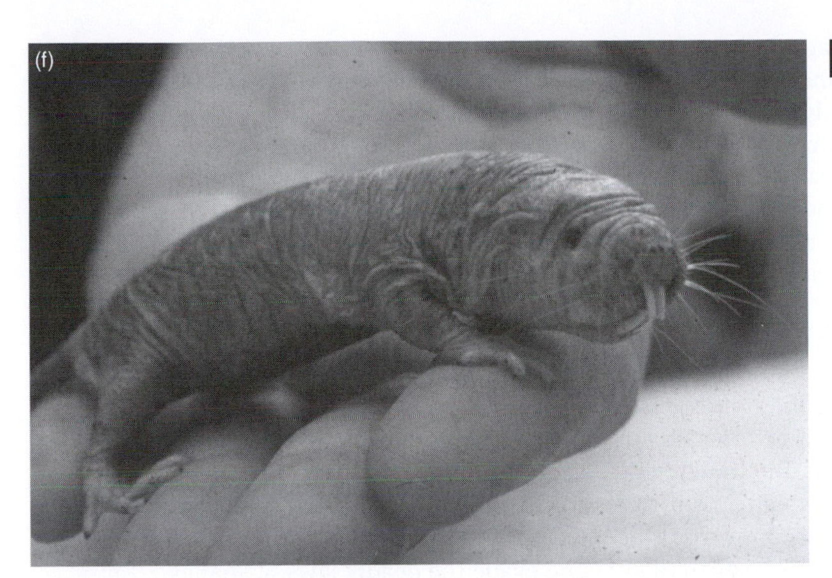

Fig. 35.1 (cont.)

but within each morphological type, work is divided in a temporal fashion.

In termites, in contrast to hymenopterans, the only adults present in colonies are the king and the queen. This one adult caste is initially winged (alate). Termite queens typically become physogastric, due to an enormous growth of the fat-bodies and ovaries, while the males remain relatively small. Indeed, the termite queen looks awfully fat and large in comparison to workers. For example, in the African termite *Macrotermes subhyalinus*, the queen's body becomes so swollen with eggs that she is incapable of movement. When fully engorged, she may be 14 cm long (more than ten times as long as a worker termite), and capable of producing up to 30 000 eggs per day.

The second true caste in termites comprises the soldiers. They are always non-reproductive and are more sclerotised and more heavily pigmented than workers. They also have highly sclerotised and powerful mandibles, which make them suitable for colony defence. Soldiers cannot feed themselves and have to be fed by workers. In some species members of the subcaste 'minor soldiers' serve as scouts and leaders for workers, being more sensitive than workers to trial pheromones. Soldier termites can regulate their own numbers by inhibiting the larval development of other soldiers. Worker termites may be more or less differentiated, depending on the evolutionary status of the species. In primitive species, social tasks are accomplished by unspecialised larvae or nymphs. In the more highly developed Termitidae, and some other termites groups, workers constitute a true caste, specialised in morphology and behaviour and permanently excluded from the nymphal development pathway. In theory, each nymph can be developed to an alate and leave the natal colony. In some species the workers are dimorphic, having large and small forms; in the Macrotermitinae the larger workers are the males and the smaller workers the females. Workers accomplish different tasks and subtasks in the colony. For example, in the termite *Hodotermes mossambicus*, one set of workers climbs up grass stems, cuts off pieces of grass, and drops them to the ground below (subtask 1) while the second set of workers transports the material back to the nest (subtask 2). The lifetime of termites is amazingly long for insects. Sterile workers live for 2–4 years while the primary sexuals live for at least 20 and perhaps 50 years (for details see Emerson, 1952; Grassé, 1982–1986; Kaib *et al.*, 1996; Eggleton, 2000).

Social aphids introduce a whole new direction in the evolution of eusociality and behavioral ecology. As was mentioned before, aphids exhibit diploidy and they reproduce both sexually and parthenogenetically, so they have the ability to produce genetically identical individuals. In these clonal stages large colonies are formed consisting of genetically identical individuals. Aphids are the only colonial species that exhibit eusocial behaviour (Alexander *et al.*, 1991). Aoki (1977) was the first to discover that the aphid *Colophina clematis* produces instars that defend the colony from intruders. Since then many species of social aphids have been described in the two families Pemphigidae and Hormaphididae (Stern and Foster, 1996). Social aphids produce galls, which are tough pockets artificially induced in a plant by the aphids. All of the alates and reproductive destined instars are normally found inside the galls The individuals on the outside are the soldiers which defend the gall from any predator that would destroy this nest and its contents. Stern and Foster (1996) have described several types of soldiers based on physical characteristic and behaviour.

Eusocial rodents

In the same way as calling termites 'white ants' one can call naked mole-rats *Heterocephalus glaber* 'mammalian termites'. These unique eusocial mammals share many features with termites. They spend virtually their entire lives in the total darkness of underground burrows, they are very small (7–8 cm long, and weighing 25–40 g), and, what is the most important, they are eusocial. Besides, as with termites, the mole-rats' high-cellulose diet is rather hard to digest, and their stomachs and intestines are inhabited by bacteria, fungi and protozoa that help break down the vegetable matter. The similarity to insects is intensified by the fact that the naked mole-rat is virtually cold-blooded; it cannot regulate its body temperature at all and requires an environment with a specific constant temperature in order to survive. These eusocial rodents cooperate to thermoregulate. By huddling together in large masses, they slow their rate of heat loss. They also behaviorally thermoregulate by basking as needed in their shallow surface tunnels, which are warmed by the sun.

These amazing creatures are neither moles nor rats. Like rats, they are rodents, but they are more closely related to porcupines and chinchillas. *Heterocephalus glaber* has been known since 1842, but it was only in 1981 that Jarvis discovered their eusocial organisation which is believed to be unique among mammals. Since then, this species has been intensively studied (see Jarvis, 1981; Sherman *et al.*, 1991; Reeve, 1992; Jarvis *et al.*, 1994; Bennett and Faulkes, 2005). The reader can see a fragment of the experimental colony housed in Sherman's laboratory in Fig. 35.2.

There are essential ecological reasons for which naked mole-rats have broken many mammalian rules and evolved an oddly insect-like social system. These animals are ensconced in the arid soils of central and eastern Ethiopia, central Somalia and Kenya, where they must continually dig tunnels with their enlarged front teeth in search for sporadic food supplies, and evade the deadly jaws of snakes.

Naked mole-rats live in well-organised colonies, with up to 300 members in a group (20 to 30 is usual). A dominant female (the queen), who outweighs the others by up to 20 g, leads the colony. The queen is the only female that breeds,

Fig. 35.2 Laboratory colony of naked mole-rats. (Photograph by P. W. Sherman; courtesy of P. W. Sherman.)

and she breeds with one to three males. When a female becomes a queen she actually grows longer, even though she is already an adult, by increasing the distance between the vertebrae in her spine (Fig. 35.3). These animals are extremely long-lived; in captivity some mole-rats have lived to 25 years old. One naked mole-rat queen produced more than 900 pups in her 12-year lifetime in a laboratory colony. The young are born blind and weigh only about 2 g. The queen nurses them for the first month, and then the other members of the colony feed them with faeces (again like termites) until they are old enough to eat solid food.

The breeding female (the queen) suppresses the breeding of all the other females in the colony. She sometimes leaves her nest chamber to check on her workers and to keep them unfertilised by pheromone control as well as by swoops and bites thus demonstrating that they should

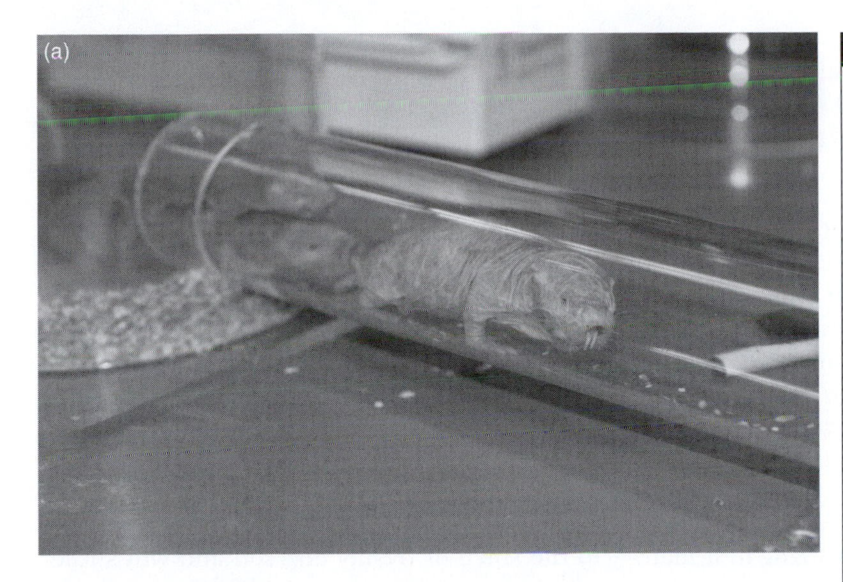

Fig. 35.3 Naked mole-rat (*Heterocephalus glaber*). The Catania laboratory (see Catania and Remple, 2002) of the Vanderbilt University Department of Biological Sciences performs research on the organisation of mammalian sensory systems, including that of *H. glaber*. (a) In their natural environment in Africa, naked mole-rats build an extensive network of underground tunnels; in captivity, they move frequently through artificial tunnels; (b) the powerful jaw muscles of *H. glaber* make up a significant fraction of the body mass; (c) the queen (shown here lactating) is the only reproductive female in the colony. (Images from Bioimages home. © 2003 Steve Baskauf http://www.cas.vanderbilt.edu/bioimages/animals/mammalia/naked-mole-rat.htm.)

Fig. 35.3 (cont.)

not 'think' about anything but digging tunnels and defending the colony from snakes and newcomers. The worker males are also suppressed, although they do produce some sperm. When the queen dies, several of the larger females fight, sometimes to death, to become queen. They can regain their fertility quickly.

The majority of workers (both males and females) spend their entire lives working for the colony (Fig. 35.3). Workers cooperate in burrowing, gathering food and bringing nest material to the queen and non-workers. They use their teeth to chisel earth and to create piles of soil. There is a great deal of branching and interconnection of tunnels, with the result that a colony's total tunnel length can add up to 4 km. Tunnels connect nest chambers, toilet areas and food sources. Burrowing is the only way these animals find food, since they do not travel above ground. Some colony members 'farm' succulent tubers that are formed by many of the plant species that grow in arid areas. They generally bore through the tuber, eating mainly the interior flesh while leaving the thin epidermis intact. This behaviour may allow the plant to remain healthy for some time, indeed even to continue growing, thereby providing a long-term food resource for the colony. Judd and Sherman (1996) studied captive colonies in order to determine whether successful foragers recruit colony mates, like many eusocial insects do. It has been revealed that individuals that find a new food source typically give a special vocalisation on their way back to the nest, wave the food around once they get there, and lay an odour track for nestmates to follow.

Whilst most offspring become workers, some continue to grow and become colony defenders. Their main duty is to defend the colony against predators. In particular, rufous-beaked snakes *Rhamphiophis oxyrhynchus rostratus* are attracted to the smell of freshly dug soil and will slither into burrows through molehills in search of a rodent meal. Soldier mole-rats fight back with their teeth and attempt to block the entrance with dirt. If everything fails, a soldier will directly attack the snake, sometimes sacrificing its own life while others escape.

Should a breeder die, just one of defenders will become reproductive to replace it. They can occasionally disperse to found a new colony with an unrelated member of the opposite sex.

In general, caste differentiation in mole-rats bears a strong resemblance (of course, merely superficial) to the termites' system. The sterility in the working females is only temporary, and not genetic. As in termites, there are castes of fertilised queens and kings and unfertilised workers and soldiers, and workers descend from 'nymphs', that is, under-grown members of the colony. The lifespan of mole rats is unparalleled among small rodents just as the lifespan of termites is unparalleled among insects. It is possible that these long-lived animals will surprise experimenters with their cognitive abilities.

Eusocial shrimps

Tiny marine coral-reef Crustacea offer a new data about the ecology and evolution of eusociality. Colonies of the social snapping shrimp *Synalpheus regalis* share several features with those of eusocial insects and cooperatively breeding vertebrates (Duffy, 1996). This shrimp inhabits the internal canals of tropical sponges, living in colonies of up to a several hundreds of individuals. Colonies consist of close kin groups containing adults of at least two generations which cooperatively defend the host sponge using their large and distinctive snapping claws, and in which invariably only a single female breeds. Irreversible caste differentiation is governed by the queen that typically sheds her large snapping claw and regrows a second minor-form chela, rendering her morphologically unique among the members of the colony. It is still not completely known how the queen achieves social control over the sexual maturation of other colony members. Both genetic data and colony structure confirm that many offspring remain in the natal sponge through adulthood. Colonies consist largely of full-sib offspring of a single breeding pair which 'reigns' for most or all of the colony's life. In captive colonies researchers have regularly observed a large male in association with the queen behaving aggressively towards other large males who approach her. The inference of monogamy from genetic data suggests that the queen associates with a single male for a prolonged period. There is a strong competition for suitable nest sites and a shrimp attempting to disperse and breed on their own would have low success. Colony members discriminate between nestmates and others in their aggressive behaviour. Laboratory experiments have revealed behavioural division of labour within colonies. Large males shoulder the burden of defence, leaving small juveniles free to feed and grow, and the queen free to feed and reproduce (see Fig. 35.1e). Such size- and age-related polyethism in shrimps has many similarities with polyethism in social insects (Duffy *et al.*, 2002).

Considering the intellectual potential of social shrimps, Duffy (2003) refers to Darwin's (1871) note that 'the mental powers of the Crustacea are probably higher than might be expected'. Social shrimps demonstrate coordinated behaviour. For example, they pick up dead colony members and push them out of their sponge dwelling. Recent experiments suggested coordinated snapping, during which a sentinel shrimp reacts to danger by recruiting other colony members to snap at intruders. The phenomenon of 'mass snapping' begins by rhythmic snapping of one individual, following by rapid recruitment of many others. The initial one-to-one confrontation elicits a snap response from the defender. Colony members join in with a cacophony of snapping thus providing an unequivocal signal that the sponge is already colonised. This distinctive behaviour is the first evidence for coordinated communication in the social shrimp and represents yet another remarkable convergence between social shrimps, insects and vertebrates (Tóth and Duffy, 2005).

Summarising the data on caste division of labour within communities of eusocial organisms we have to admit that the correlation between cognitive and morphological specialisation in these animals has not yet been completely described. Even in ants and bees which have been intensively studied for more than 100 years, it remains unclear what effect caste determination has on their intelligence. Further we will consider a more gentle system of division of labour in animal societies that perhaps leaves more room for cognitive activities.

35.2 Cooperative (communal) breeding: helpers sacrifice their intelligence for breeders

Eusociality can be considered an extreme of cooperative breeding systems being based on irreversible caste determination. However, many vertebrate species possess more flexible social systems which are based on facultative division of labour and temporal limits on breeding for some members of communities. In cooperative breeding vertebrates, a dominant pair usually produces the majority of the offspring, whereas the cost of caring for offspring is shared by non-breeding subordinates. In certain cooperative breeding animals one or a few dominant females are the only ones capable of breeding; the subordinates do not have the proper hormone levels to be fertile although they are physiologically equipped for the task.

There is still a great controversy in the literature as to what extent cooperative breeding can be explained in terms of kin selection theory. Results so far are mixed: while some studies have produced evidence supporting the association between kinship and contributions to cooperative activities, others have found no consistent association between contributions to helping behaviour and variation in relatedness (for reviews see Clutton-Brock *et al.*, 2000, 2002).

Cognitive aspects of cooperative breeding are intriguing and have not been studied enough. Serving as helpers for the 'royal family' young animals gain experience that can be useful for them in the future when they establish their own families. Nevertheless, in many cases helpers have no chance to have their own offspring. Somehow or other, a cooperative breeding system enables helpers to sacrifice their intelligence for other members of the community. It is possible that these helping individuals accomplish a wider variety of tasks and under more risky circumstances than those who have the opportunity to raise their young being given every support by helpers. Several examples will give us an impression of how division of labour occurs within communities that are based on communal breeding.

In birds about 3 per cent (approximately 300 species) of species are known as cooperative breeders. Helpers (also called auxiliaries) at the nest were first described by Skutch in 1935. It was not until the mid 1960s, however, with the advent of modern behavioural ecology, that widespread attention began to focus on cooperatively breeding species (see Emlen, 1991, 1995).

Cooperative systems often appear to arise when environmental constraints force birds into breeding groups because the opportunities for younger birds to breed independently are severely limited. Limitations may include a shortage of territory openings, a shortage of sexual partners or unpredictable availability of resources. That cooperative breeding is a common strategy in arid and semi-arid regions of Africa and Australia lends strong support to this line of reasoning. For some species the role of ecology is not completely clear (Arnold and Owens, 1999).

Cooperative breeding may be viewed primarily as a means by which young adults put off the start of their own breeding in order to maximise their lifetime reproductive output, and in the process occasionally promote genes identical to their own via kin selection. There are two types of cooperative arrangements: those in which mature non-breeders help to protect and rear the young, but are not parents of any of them, and those in which there is some degree of shared parentage of offspring. Cooperative breeders may exhibit shared maternity, shared paternity or both.

The best-studied North American cooperative breeders, the scrub jay, gray-breasted (Mexican) jay, groove-billed ani and acorn woodpecker provide good examples of communal breeding (see Ehrlich *et al.*, 1988).

Scrub jays in Florida reside in permanent, group-defended territories. Woolfenden and Fitzpatrick (1984) have found that groups consist of a permanently bonded monogamous pair and one to six helpers, generally the pair's offspring of previous seasons. About half the territories are occupied by pairs without helpers, and most other pairs have only one or two helpers. Although pairing and breeding can occur after one year spent as a helper, birds often spend several years as non-breeding auxiliaries. Males may remain in this subsidiary role for up to six years; females generally disperse and pair after one or two years of helping. Helpers participate in all non-sexual activities except nest construction, egg laying and incubation. Pairs with helpers are more successful – they fledge one and a half times more young than pairs without helpers.

Like the Florida scrub jays, the closely related gray-breasted jays live in permanent group-defended territories, and breeding adults are monogamous. Brown (1970, 1974) has shown that the cooperative system of this species is more complex than that of its south-eastern relative in several ways. Gray-breasted jay groups are much larger, ranging from eight to 18 individuals; thus, they usually include offspring from more than just the preceding year. Within each group, two and sometimes three breeding pairs nest separately but simultaneously each season, and some interference among them often occurs. Interference usually involves theft of nest-lining materials, but can include tossing eggs out of the nest by a female from a rival nest. Although the laying female does all the incubating, she is fed on the nest both by her mate and by auxiliaries. Nestlings receive more than half of their feeding from auxiliaries.

An acorn woodpecker group of communal breeders is composed of up to 15 members whose territories are based on the defence and maintenance of granaries in which they store acorns (Stacey and Koenig, 1984, 1990; see also Koenig and Dickinson, 2004). Groups consist largely of siblings, their cousins, and their parents. Some of the sexually mature birds are non-breeding helpers. Within each group, up to four males may mate with one (or occasionally two) females, and all the eggs are laid in a single nest. Thus paternity and sometimes maternity of the communal clutch is shared.

In mammals more than 100 species have been described as cooperative breeders, and among them are cooperative canids (such as wolves and wild dogs), lions, mongooses (meerkats, dwarf mongooses), primates (marmosets and tamarins) and several species of rodents and shrews. As was noted earlier, some rodent species possess facultative communality in dependence of their habitat and many ecological factors.

The painted hunting dog (African wild dog) *Lycaon pictus* provides a good example of obligate cooperative breeding. These dogs live in packs of up to 20 adults, in which most of the time only the dominating ('alpha') pair breeds. The remaining adults are reproductively suppressed and help to raise the pups; they must wait to breed until their circumstances improve, either through the death of a higher-ranking female or by finding a mate with an unoccupied te[illegible] (Burrows, 1992; Fuller *et al.*, 1992). Babysitting [illegible] a costly task, and this includes: watching pups to prevent loss, alerting them to danger (lions, hyenas), protecting them from smaller predators or alien dogs and moving them under cover in heavy rain. Other members of the pack are also involved in caring for common babies: they feed pups with regurgitated meat when they return from successful hunting. Babysitting is not an obligatory duty for pack members, as they can choose between hunting and guarding the young. Researchers have observed situations in which a dog returned to a den to babysit after encountering a predator close by (Malcolm and Marten, 1982). At the same time, *Lycaon* hunt cooperatively and babysitting draws a member of a pack away from hunting where both efficiency and the risk of losing prey to kleptoparasites depend on the size of the party (Gorman *et al.*, 1998). It is worth noting that in contrast to queens in eusocial communities that are specialised baby-machines, the breeding female in wild dogs, as in other cooperative carnivores, is often an experienced hunter, and her presence in the hunting pack may increase the efficiency of the enterprise. Besides, there is a threshold for the group size to survive. Smaller packs need to hunt more often to feed their pups, especially when using a pup guard (Courchamp *et al.*, 2002).

Another impressive example of obligatory communal breeding in mammals comes from small arboreal monkeys, marmosets and tamarins of the family Callitrichidae endemic to the northern half of South America. Within the family, cooperative breeding strategies are widespread and virtually all species are characterised by small territorial groups of approximately four to 15 individuals, in which reproduction is monopolised by one or a small number of dominant individuals. Typically one dominant female breeds, normally producing dizygotic (not genetically identical) twins. An important role of helpers in the group is to assist in the care of the dominant female's offspring. This is principally by sharing the burden of carrying the relatively bulky twin infants around their arboreal habitat. Each group member helps rear the young, which involves food sharing, caring and defence against predators (Snowdon and Soini, 1988; McGrew and Feistner, 1992).

The life history of Callitrichidae can serve as an example of cooperative breeding in groups consisting both of related and unrelated individuals. Helping behaviour in these primates is thus possibly governed by mechanisms of reciprocal rather than kin altruism. This raises a question as to what extent cognitive abilities allow these small primates to calculate reciprocity in their groups. Hauser *et al.* (2003) have conducted experiments on food sharing within groups of cotton-top tamarins concentrating on psychological mechanisms of reciprocity. The design of experiments was based on the animals' tool-using abilities (see Part VI). The apparatus consisted of a tray with an inverted L-shaped tool. When food was on the actor's side, pulling the tool's stem brought the food within reach. As with the experimental paradigm used in many experiments on social learning (see Part VIII) where researchers trained several animals to be the demonstrators of new skills, here again stooge 'altruists' and 'defectors' were specially trained to pull pieces of food to their partners or to themselves. Results clearly showed that tamarins discriminate between altruistic and selfish actions, identify and recall conspecifics by their cooperativeness and give more food to those who give food back. A special series of experiments also demonstrated that tamarins give food to genetically unrelated conspecifics even though they obtain no immediate benefit from doing so. Tamarins therefore have the psychological capacity for reciprocally mediated altruism.

The ability to estimate a partner's cooperativeness and remember the history of inter-individual relationships is particularly important for those communal breeders that incorporate both kin and non-kin in their communities, and whose altruistic acts are costly. This is well illustrated by the experiments of Clutton-Brock *et al.* (2000) on individual contributions to babysitting in a cooperative mongoose, the meerkat.

Meerkats are desert-adapted animals living in groups of three to 25 animals that typically include a dominant female who is responsible for more than 75 per cent of all breeding attempts, a dominant male who fathers most of the offspring born in the group and a number of helpers of both sexes. A dominant female controls the presence of subordinate adult females in the group. During the first month of the pups' life babysitters usually remain at the burrow with young all day while the rest of the group is foraging and feed little or not at all during their period of babysitting (Clutton-Brock *et al.*,

Fig. 35.4 Meerkats: where many are vigilant (a) one can have a rest (b). (a, photograph by M. Manser; courtesy of L. Hollén; b, photograph by L. Hollén.)

Fig. 35.4 (cont.)

Fig. 35.5 A daily procedure for feeding and weighing meerkats. (Photograph by L. Hollén.)

1999; Doolan and McDonald, 1999) (Fig. 35.4). Clutton-Brock *et al.* (2000) have shown how costly babysitting is: the helpers suffer substantial weight loss. It is important that large differences in contributions exist between helpers. These differences are correlated with such characteristics of group members as age, sex and weight, but, surprisingly, not with their kinship to the young being raised. In field experiments researchers regularly provided some group members with food (boiled eggs) matching them with controls of the same sex and age (Fig. 35.5). It turned out that feeding essentially increased contribution of helpers to babysitting. So a regular salary may increase and equalise the individual contributions of cooperators. However, in natural situations meerkats cannot rely on donations from above; rather they depend on their ability to distinguish between more and less conscientious cooperators.

35.3 Working in teams: wild professional profiles

In group-living animals division of labour is sometimes based on coordinated activities of group members. In relatively rare cases individuals form groups in which the members stay together for extended periods to accomplish a certain task. Such groups are called teams or cliques (Hölldobler and Wilson, 1990; Anderson and Franks, 2001).

For example, when working together to dig tunnels, naked mole-rats line up nose-to-tail and operate like a conveyor belt. A digger mole-rat at the front uses its teeth to break through new soil. Behind the digger, sweepers use their feet and the fine hairs between their toes to whisk the dirt backwards. At the back of the line a trailing member of the group kicks the dirt up onto the surface of the ground, creating a distinctive volcano-shaped molehill. One of the folk names of naked mole-rats is 'sand puppies'. There are several other creatures that join efforts of group members to survive in the running sand. Desert ants *Cataglyphis pallida* demonstrate the same manner of coordinated working, digging tunnels like a conveyor belt.

There are several examples of hunting teams in vertebrates (for a review see Anderson and Franks, 2001). Usually individuals coordinate efforts so that one or more individuals chase the prey, or flush it from hiding, while others head off its escape. For instance, in chimpanzees, some group members chase and surround the prey (usually juvenile baboons) forcing it to climb a tree while at the same time other chimpanzees climb adjacent trees ready to capture the prey when it attempts to leap across to escape. In African wild dogs some individuals chase the prey, and can change leaders during the chase (Goodall, 1968; van Lawick and van Lawick-Goodall, 1970).

The most organised teams in animal societies are based on a discrete division of labour that may be called 'professional specialisation'. Stander (1992) has shown that lion teams can be organised in such a way that an individual will tend to stick to a particular position (subtask) during the hunting on successive hunts. That is, some lions can be classed as 'wingers', individuals who always tend to go around the prey and approach it from the front or from the side, while others are better classified as 'centres', individuals who remain chasing directly behind the prey. Perhaps the most organised hunting teams in vertebrates occur in Galapagos and Harris' hawks (Faaborg *et al.*, 1995). Hawks hunt cooperatively with several birds simultaneously swooping on their prey, which would be such animals as wood rats, jackrabbits and other birds. However, if the prey item finds cover, some birds land and surround it, while one or two hawks will walk or fly into the vegetation to kill the prey. Once the prey is killed, all the birds feed together on the carcass.

Until recently, the existence of teams within insect colonies, possibly based on individual identification, has not been found. According to Hölldobler and Wilson (1990), ants do not appear to recognise each other as individuals. Indeed, their classificatory ability is limited to recognition of nestmates, different castes such as majors and minors, the various growth stages among immature nestmates, and possibly also kin groups within the colony. There are, however, several examples showing elements of team task distribution. In swarm-raiding army ants, large prey items are transported by the structured teams which include members of different castes (Franks, 1986). In the desert ant *Pheidole pallidula*, minor workers pin down intruding ants and later major workers arrive to decapitate the intruders (Detrain and Deneubourg, 1997). Robson and Traniello (1998, 2002) found complex relations between discovering and foraging individuals in group-retrieving ant species; removal of the discovering ant during the process of recruitment led to dissolution of the retrieval group.

The question of constant membership and individual recognition within groups of workers in ant colonies has so far remained obscure. Reznikova and Ryabko's (1990, 1994, 2003) findings on teams in ants are connected with the discovery of the existence of complex communicative system in group-retrieving ant species (see Fig. 19.3). As was described in detail in Chapter 32, such a communication system is based on scout–foragers informative contacts in which each scout transfers messages to a small (five animals on average) constant group of foragers and does not pass the information to other groups. The ants thus work as coordinated groups which may be called teams. Does this necessarily mean that they recognise each other as individuals? Indeed, it is possible that the animals rely on recognition of specialists' roles rather than their personal traits.

Donald Michie (personal communication) has referred to his experience as a rugby player (he was a famous cryptographer who used to play games not only with machines: see Maynard Smith and Michie, 1964). Being a scrum-half, he was always confident of his ability to spot his opposite number (that is, another scrum-half) when meeting an opposing team socially before the game. To be adapted to the scrum-half's specialist role, one must typically be small, resilient and agile, not necessarily a fast runner. The only other typically agile team member is the fly-half, but he has also to be a fast accelerator and need not be resilient. A year later he might still recognise one of that same team's forwards, for example, but not remember the face of the scrum-half.

One can find it hard to say that ants are able to recognise each other personally. That a scout can distinguish members of its own team from members of another team is not the same thing as individual recognition. Continuing the use of the metaphor from football, one can imagine a team manager who might be able to distinguish players of his own team from those of a different team (for example by the patterns of their shirts), but was not yet able to distinguish same-team players one from another.

We have not yet identified reliable behavioural signs in ants indicating personal recognition like the well-known 'eyebrow flash' in humans (see Eibl-Eibesfeldt, 1989); neither have we been able to train ants to distinguish between pictures of different individuals as in Kendrick *et al.*'s (2001) experiments with sheep (see Chapter 33). Nevertheless, we can be confident on at least circumstantial evidence that group-retrieving ant species possess personalised teams as functional structures within their colonies.

The first evidence comes from ontogenetic studies. Reznikova and Novgorodova (1998) observed the ontogenetic trajectories of 80 newly hatched *Formica sanguinea* ants in one of the laboratory colonies and watched the process of the shaping of teams. There were 16 working teams in that colony which mastered mazes. From 80 individually marked naive ants, 17 entered seven different working teams, with one to four individuals in each. Only three became scouts, two of them starting as foragers joining two different teams and one starting as a scout at once. The three new groups were composed of workers of different ages, mainly from reserve ones. The age at which the ants were capable of taking part in the working groups as foragers ranged from 18 to 30 days, and the ants could become scouts at the age of 28 to 36 days. Constancy of membership was examined in two colonies of *F. sanguinea* and *F. polyctena*. In a separate experiment researchers isolated all team members from nine scouts. Three scouts appeared to mobilise their previous acquaintances and attract new foragers, four scouts were working alone, and two ceased to appear in the arenas. In another experiment scouts from five *F. polyctena* teams were removed. It was possible to see foragers from those groups on the arenas without their scouts. Different foragers were placed on the trough with the food 15 times, but after their return to home they contacted other ants only rarely and occasionally. These results suggest that the formation of teams in group-retrieving ants is a complex process which is based on extensible relations and possibly includes individual identification.

Another evidence of existence of teams in ants is based on division of labour within groups of aphid tenders discovered in red wood ants. It is well known that ants look after symbiotic aphids, protect them from adverse conditions, and in return, ants 'milk' the aphids, whose sweet excretions are one of the main sources of carbohydrate for adult ants (Mordvilko, 1936; Bradley and Hinks, 1968). In an ant family, there is a group of ants dealing with aphids

Fig. 35.6 'Professional specialisation' in an ants' working team: (a) a 'shepherd' milking aphids and a 'guard' (with open mandibles) protecting an aphid colony; (b) a 'transit' ant receiving the food from a shepherd in order to transport it to the nest; a 'guard' is also present here. (Photographs by T. Novgorodova.)

(trophobionts), which has a constant composition. Reznikova and Novgorodova (1998) were the first to describe a system of intricate division of labour (professional specialisation) in trophobionts: 'shepherds' only look after aphids and milk them, 'guards' only guard the aphid colony and protect them from external factors, 'transit' ants transfer the food to the nest, and 'scouts' search for the new colonies (see Fig. 35.6). This professional specialisation increases the efficiency of trophobiosis. When ants were experimentally forced to change their roles, much food was lost. The ants belonging to the same aphid tending group distinguish at least two to three shepherds from two to three guards within this group. Such professional specialisation has only been found in the same species that exhibited the complex communication system in the experiments of Reznikova and Ryabko (1994, 2003).

Chapter 36

What sort of intelligence is required to navigate social landscapes?

The title of this chapter is borrowed from the chapter 'The structure of social knowledge in monkeys' by Seyfarth and Cheney (2002) in *The Cognitive Animal*. Animals that live in groups and are connected by close social relationships can navigate the social landscape relying upon their abilities to recognise associated relatives of their own as well as of their neighbours, to continuously track the position, social behaviour, foraging and sexual success of other individuals, to memorise and, at least partly, to predict social events. Highly social vertebrates, and pre-eminently primates, are endowed with these abilities. There is a growing body of evidence of an association between social complexities and cognitive abilities in animals. Volumes have been written about social life in the wild. Here we will briefly consider several themes concatenated by experimental approaches for studying animal social problems.

36.1 Hierarchies and roles

In the wild, a great deal of diversity of hierarchically organised animal communities exists, and specific structures are often initialised by evolutionary histories and species predispositions as well as environmental conditions.

Individual dominance, or the 'pecking order', serves as a universal mechanism of maintaining an order within animal communities. The history of the discovery of the pecking order by the Norwegian zoologist Thorleif Schjelderup-Ebbe (1922, 1935) is fascinatingly described by de Waal (2001). From when he was a boy, Schjelderup-Ebbe had kept detailed notebooks recording who pecked whom in the flock of domestic chickens. He was particularly interested in exceptions to the hierarchy, so-called 'triangles', in which hen A is master over B, and B over C, but C over A. So, from the start, like a real scientist, he was interested in not only the regularities but also the irregularities of the rank order. The social organisation that he discovered is now so obvious to us that we cannot imagine how anyone could have missed it, but no one had described it before. After Schjelderup-Ebbe received his degree in zoology, he published the chicken observations of his youth, introducing the term *Hackordnung*, German for pecking order. His classic paper, which appeared in 1922, describes dominants as 'despots' and demonstrates the elegance of hierarchical arrangements in which every individual has its place.

The basis of the linear dominance is individual recognition. In a linear hierarchy, a single individual generally dominates all the others. His immediate subordinate dominates all others except him, and so on. The last animal in the order is subordinate to all the other animals and may lead a frenetic life indeed. Certain features associated with threat may determine one's position in hierarchy. In chickens, if the comb and wattle (structures used in aggressive signalling) are removed from a hen, she will begin to lose more fights and may end up at the end of the pecking order (Marler and Hamilton, 1966). Although not smart in solving many cognitive problems (as we have seen earlier in this book), chickens display remarkable abilities to remember their partners in a flock. Moreover, they can recognise members of several flocks and estimate social circumstances. In the experiments, hens were placed into different groups that were formed in such a way that a certain chicken occupied different levels in them. Linear hierarchy turned out to be important for public tranquillity within flocks. The experimenters compared two sets of groups: in one set of groups dominants were removed each week and replaced by other individuals in order to disturb the social order, whereas in another set of groups social structure was not disturbed. In control groups clashes were observed more rarely, chickens ate more and laid many more eggs (Guhl and Allee, 1944).

In the 1930s, very impressive results were obtained concerning variants of social ranking in animal communities, in particular, by Lorenz (1931, 1937) on ranking in jackdaws and other birds, and Zuckerman (1932) and Maslow (1936) on dominance in monkeys. A concept of domination had been elaborated, in which it was possible for a dominant individual to do all things against the convenience of other members of a group. Correspondingly, subordinating individuals have to adjust all their behaviours (such as sexual, feeding and social activities) to suit the dominant's 'opinion'.

The discovery of the pecking order influenced ecological and evolutionary sciences because the idea that a dominating individual has immediate access to limited resources was concordant with the view tracing back to Thomas Henry Huxley (1893) of the animal world as the 'gladiator's show'. Wynn-Edwards (1962) and Lack (1966) suggested that hierarchies may act as 'social guillotines' when food becomes scarce by causing the quick elimination of certain, less vigorous, segments of the population.

More recently it has been shown that although the linear dominance model is a useful construct for analysing social organisation within groups of animals, there are limitations in this model concerning linearity of rank itself as well as correlation between ranking and living standards in animal communities.

First of all, even in controlled laboratory conditions when researchers use the method of pair-wise interaction (very popular in physiological experiments), it is difficult to find a correlation between parameters of ranking when animals struggle for different limited resources such as food, water, refuges and sexual partners. So it is impossible to reveal an absolute leader. In the wild, qualities of leadership do not always depend on the animal's size or even on its fighting prowess. In some species not only are individuals ranked according their own attributes, but also according to whose mates they are and whose offspring. Besides, factors other than pugilistic skill may determine one's place in the pecking order, and, as we will see further, intelligence is not the least among these factors. Finally, even if one reaches the top rank this does not automatically reflect one's fitness. For instance, it is known that the highest-ranking males of polygamous species do not necessarily fertilise more females than other males.

Dominance concepts have been largely replaced by the concepts of *roles*, that is, a behavioural set including the sum of reactions on environmental and social stimuli. For example, in macaque communities there is a control male, central subgroup males (competitive with others in that group, but dominant to peripheral males), peripheral males, isolate males, and central and peripheral females. Such roles are not fixed and immutable, and changes of position in the role hierarchy are usually correlated with changes in general behaviour. Males may shift their roles many times in a lifetime, while females are socially far more stable. Their stability thus gives them a certain 'baseline' importance in determining the roles of males of the group (Crook, 1970; Clutton-Brock, 1974, 2002). Poyarkov (1986) described the role hierarchy in packs of stray dogs. There are certain members of the pack who are responsible for making an itinerary for the whole group to search for food, whereas other dogs are responsible for the efficient running of contacts with other packs, and with humans.

Certain roles, as would be expected, are characteristic of some species and not of others. For instance, the harem 'overlord' exists in hamadryas baboons *Papio hamadryas hamadryas*, but not in gelada baboons *Theropithecus gelada* or in macaques (Rowell, 1966). At the same time, as has already been noted, many species that belong to remote orders show similarities in their social structures. For example, many species of Old World monkeys typically live in groups that include several matrilineal families arranged in a stable, linear, dominant order (Silk, 1999). Hyenas also live in social groups comprising matrilines in which offspring inherit their mother's dominance rank (Smale *et al.*, 1997).

However, within the species-specific profile of social structure variants can exist depending on animals' social experience as well as their personal attributes. Animals of the same species can use different social strategies in similar situations. In order to compare various dominance styles, patterns of conflicts and reconciliation have been used as criteria over last decades (de Waal and Yoshihara, 1983; Thierry, 1985; Butovskaya *et al.*, 1996; de Waal, 2005).

Flexibility of social behaviour and learning may help animals to change social strategies in concrete situations. This can be illustrated by an experiment of Cords and Thurnheer (1993). Pairs of longtail macaques were trained to obtain rewards by acting in a coordinated fashion: the only way to obtain popcorn would be for two monkeys to sit side by side at a dispenser, a procedure that attached significant benefits to their relationship. After this training, subjects showed a three times greater tendency to reconcile after an induced fight than subjects that had not been trained to cooperate.

One can assume that strong and despotic linear hierarchy is the inherited 'blank' that is characteristic for many social animals. An acquired ability to estimate and use social affiliations makes social structures much more flexible and often more effective in changeable environment.

This hypothesis can be supported by experimental comparison of dominance styles within groups of socially naive and experienced animals. For example, Anderson and Mason (1978) experimentally formed two groups of rhesus macaques: one was composed of young individuals raised in social isolation carried by 'artificial mothers', whereas the other, control, group was composed of young animals raised in a normal social environment. Experimenters then forced the animals to compete for drinking bowls. In the control group composed of socially experienced macaques animals displayed flexible and crafty behaviour. They could, for instance, form alliances and 'ask' the dominant male for protection in a conflict that was framed just to draw the dominant's attention away from the desired drinking bowl. Socially naive monkeys formed a stable linear pecking order ranking individuals strongly in accordance to their fighting prowess.

A similar experiment was conducted in our laboratory by Harkiv (1997). He formed two laboratory families of *Formica sanguinea* ants. The first group was composed of 'wild' ants from a basic natural ant-hill. The second group included socially naive ants hatched from cocoons (taken from the same ant-hill) under laboratory controlled conditions. Ants in both groups were individually labelled by colour marks and housed in laboratory transparent nests so it was easy to observe their contacts within the nests and in large arenas where both families had the opportunity to walk and search for food. It turned out that although individual relationships in natural ant families are not so complex as in primates, behavioural patterns of members of this group are characterised by flexibility, lack of self-confidence and propensity to cooperate. The absence of clashes proves that such tactics were efficient. A strong linear hierarchy was established in the group composed of socially naive ants, after numbers of aggressive encounters during which the ants sorted their relationships out, and the primary problem was which ant has rights to enter the arena first.

In general, we can suggest that linear ranking reflects one-dimensionality and unambiguity of individual interrelations within communities whereas complexity and flexibility of dominance styles in animal communities are connected with a variety of living problems. The more diverse are the problems to be solved by members of communities, the fuller the colour of the social palette in animal societies.

36.2 Social intelligence in animals

Since the second part of the twentieth century a growing body of field data about social life in the wild has led researchers to the idea that social animals should display advanced cognitive abilities within specific domains related to social living and that intelligence is not a monolithic functional entity but includes a number of specialised mental abilities to cope with life in complex and changeable social environments. Thus, to the primary components of intelligence, such as the ability for flexible problem-solving and the ability to cope with novel situations, we can add the ability for solving social problems.

According to the *social intelligence hypothesis*, which was first articulated by Jolly (1966) and Humphrey (1976), complex social interactions (including cooperation, competition, manipulation and deception) can occur when animals live in large and stable social groups. After spending 3 months with Dian Fossey and her gorillas in Rwanda (see Fossey, 1983), Humphrey wrote a review essay in 1976 entitled 'The social function of intellect', on the evolution of cognitive skills. He argued that primate and human intelligence is an adaptation to social problem-solving, well suited to forward planning in social interactions but less suited to non-social domains. These subforms of intelligence assumed the name 'Machiavellian intelligence' after the sixteenth-century Italian politician and author, Niccolò Machiavelli. It provides individuals or groups with a means of social manipulation in order to attain particular goals. In 1532 Machiavelli published his book *The Prince*. Giving somewhat cynical recommendations to an aspiring prince, he was prescient in his realisation that an individual's success is often most effectively promoted by seemingly altruistic, honest and prosocial behaviour. According to Machiavelli's realpolitik, a popular leader had to give the impression of being sincere, trustworthy and merciful. To retain his power, however, a prince can set himself above all moral rules and use cunning, lies and force. Skill in deception and maintaining alliances are two of a prince's most important properties.

'Machiavellian intelligence' seemed an appropriate metaphor that inspired primatologists to make explicit comparisons between the animal social strategies and some of the advice offered five centuries earlier.

De Waal, in his book *Chimpanzee Politics* (1982), describes how clever high-ranking chimpanzees are at manipulating others. Byrne and Whiten (1988) propose that the ability to use other individuals as tools, manipulating the social environment in order to meet preconceived goals, is an important factor in the evolution of primate intelligence. In order to compete successfully within groups, apes and monkeys have to recognise who outranks whom, who is closely bonded to whom, and who is likely to be allied to whom.

Skilfulness in navigating social landscape is based on an advanced ability that seems to be unique to primates, that is, the ability to keep track of how other animals relate to each other and thus to recognise the close relationships that exist among individuals (Cheney and Seyfarth, 2003; Kitchen *et al.*, 2005).

Experimental evidence for animals' ability for tracking social and kin relations came from the laboratory study performed by Dasser (1988) on captive long-tailed macaques *Macaca fascicularis*. The monkeys were shown a pair of slides of members of the group, and their task was to identify another pair of photographs which 'matched' the first one. The first pair could be, for example, a mother and a daughter, two sisters, or two unrelated individuals. The macaques quickly learned to identify the right kinship patterns. The experiment indicates that they do not just recognise their own offspring and siblings, but that they also keep track of other individuals' kinship relations. For example, in one test, Dasser trained a female to choose between slides of one mother–offspring pair and slides of two unrelated individuals.

Having been trained to respond to one mother–offspring pair, the monkey was then tested with 14 novel slides of different mothers and offspring paired with an equal number of novel pairs of unrelated animals. In all tests, she correctly selected the mother–offspring pairs. Dasser suggests that the monkeys can use the abstract category to classify pairs of individuals that was analogous to our concept of 'mother–child affiliation'.

The experiments of Parr and de Waal (1999) demonstrated that chimpanzees are able to judge about mother–offspring relationships by comparing pairs of photographs of mothers and sons and mothers and daughters. Surprisingly, within the mother–offspring category, the chimpanzees could find similarities between mothers and sons much better than between mothers and daughters. The authors suggest that facial similarities are more noticeable to chimpanzees in males in view of their male philopatric society and the tendency towards 'political' alliances in which males incur great risk on behalf of other males (de Waal, 1982). Phenotypic matching might assist the recognition of subsets of related males who tend to support each other.

A number of naturalistic studies have suggested that monkeys recognise the close associates of other group members. For example, playback experiments using the contact calls of rhesus macaques have demonstrated that females not only distinguish the identities of different signallers but also categorise signallers according to matrilineal kinship (Rendall *et al.*, 1996). In playback experiments with vervet monkeys Cheney and Seyfarth (1980) found that when females were played the scream of an unrelated juvenile, they were more likely to look towards that juvenile's mother than towards other females. Also Cheney and Seyfarth (1990, 2003) argue that vervets can perform vendettas: they prefer to attack relatives of the individuals who have attacked their own relatives.

Knowledge of the relationship between other group members, the so-called *third-party relationships*, plays a particularly important role in formation of coalitions, helping individuals to predict who will support or intervene against them when they are fighting with particular opponents, and to assess which potential allies will be effective in coalitions against their opponents (Tomasello and Call, 1997). There is much evidence that monkeys and apes cultivate relationships with powerful supporters. Silk (1999) has demonstrated that male bonnet macaques put their knowledge of their own relationships with other males and their knowledge of relationships among other males to good use when they initiate coalitions. By selectively soliciting males that had most frequently supported them and animals that outranked them and their opponents, males focused their recruitment efforts on the candidates that were most likely to intervene on their behalf and those whose support was most likely to be effective in defeating their opponents. They avoided soliciting top-ranking males that were more loyal to their opponents than to themselves. To do this they have to have some knowledge of the pattern of support amongst other individuals, another kind of third-party knowledge.

Although not so well studied as monkeys and apes, several non-primate species also show the ability to acquire information about many different individual social relationships. Male dolphins form dyadic and triadic alliances when competing over access to females, and allies with the greatest degrees of partner fidelity are most successful (Connor *et al.*, 1992, 2001; Mann *et al.*, 2000). Analysis of patterns of alliance formation in hyenas suggests that they do monitor other individual's interactions and extrapolate information about other animals' relative ranks from their observations. During competitive interactions over meat, hyenas often solicit support from other, uninvolved individuals. When choosing to join ongoing skirmishes, hyenas that are dominant to both of the contestants almost always support the more dominant of the two individuals. When the ally is intermediate in rank between the two opponents, it inevitably supports the dominant individual. These data enable researchers to suggest that hyenas are able to infer transitive rank relations among other group members. However, unlike monkeys, they showed no evidence for recognising third-party relationships

(Engh *et al.*, 2005). Observational evidence suggests that colonial white-fronted bee eaters *Merops bullockoides* may recognise other individuals and kin groups and associate these groups with specific feeding territories (Emlen *et al.*, 1995).

Summarising data about animals that are able to track social events and manipulate their social partners as tools, we come to the question of to what extent non-humans understand the properties of these 'living tools'. Let us consider this problem in the next chapter.

Chapter 37

Theory of Mind

A 'Theory of Mind' (often abbreviated to TOM) is a specific cognitive ability to understand others as intentional agents, that is, to interpret their minds in terms of theoretical concepts of intentional states such as beliefs and desires. It is called a 'theory' because we can never actually directly know about another's mind, and there is no objective way either to verify the contents of another's consciousness or to access their motivations and desires. It is commonplace in philosophy (see Davidson, 1984; Dennett, 1987) to see this ability as intrinsically dependent upon our linguistic abilities. Language provides us with a representational medium for meaning and intentionality: thanks to language we are able to describe other people's and our own actions in an intentional way. It may be worth mentioning that theory of mind is the essential part of the Buddha's teaching.

In their 1978 paper 'Does the chimpanzee have a theory of mind?' Premack and Woodruff argued that experimental evidence of chimpanzees' understanding of human behaviour could be interpreted as detection of intentions. They suggested that their chimpanzee, Sarah, predicted the actions of a man by deducing his 'intentions' and 'motivations' and that she reacted according to her predictions. Thus the researchers raised the question of whether great apes, like humans, have the ability to make inferences about the intentions, desires, knowledge and states of minds of other animals. This question has become one of dominant problems in cognitive ethology. Theory of mind has recently been considered one of subcomponents of social intelligence both in humans and non-humans (Whiten, 1993, 1999). The criterion for this is having an understanding that others possess mental states that accommodate ideas and accounts of the world that are different to their own, enabling the animal to make sophisticated predictions about others' actions and motivations (Gómez, 1994; Byrne, 1995; Heyes, 1998). The presence of such a capacity in non-human species leads to the conclusion that it is possible to investigate theory of mind as a biological endowment independently of language.

In the discussion of children's cognitive development, the question of when they develop a theory of mind has come into focus (Gopnik and Astington, 1988; Gopnik and Meltzoff, 1997). Children show a precocious ability to understand the intentions and external manifestations of another's mind such as gaze direction, attention and pretence. Nevertheless, it is still open to discussion at what age children develop a full-fledged theory of mind.

There are several types of behaviour that are said to be representative of an animal that has a theory of mind, and among them are self-awareness, knowledge attribution, perspective-taking and deception. These types of behaviour are closely related to Machiavellian intelligence, and some researchers consider social intelligence (and especially Machiavellian intelligence) an integral part of a theory of mind. Let us consider several sorts of evidences attributed to theory of mind in animals.

37.1 Perspective-taking

Perspective-taking is said to be one of the essential manifestations of a theory of mind because it is supposed to require knowledge that one individual has a different perspective to another and thus requires mental concepts of the self and others.

A major step towards having an idea of others' minds consists in being able to put oneself in their position to see *how they see the world*. This can be described as detaching oneself from one's own perspective. Small children are incapable of this, as many experiments have shown. A classical example is Piaget's 'three mountains test' (Piaget and Inhelder, 1956). In this experiment one puts three 'mountains', one of them much bigger than the other two, in a triangle on a table. The child who is tested sits in front of the small mountains and can see the big one behind them. A doll is placed on the other side of the table with its face towards the big mountain. The child is asked to draw what the doll can see from its side of the table. A child at the 'pre-operational level' (Piaget's term) draws the scene from his own perspective, regardless of where the doll is sitting. However, a child at the 'concrete operational level' (roughly from the age of 7 years) can imagine how the doll sees the mountains and draws the scene from the right perspective. When the child has reached this stage, it can imagine the world from different angles, regardless of the perceptions it has at a particular moment. Later experiments have shown that small children are better at imagining what others see in a certain situation if they themselves have previously been in that situation.

Long before animals' social behaviour began to be considered in terms of theory of mind, ethologists argued that social cues, such as the orientation of the head and eyes of conspecifics, play an important role in the daily interactions of community members. Gaze-following, looking where somebody else is looking, is one mechanism for extracting such information from the behaviour of others in primates including humans (Chance, 1962; Goodall, 1968; Scaife and Bruner, 1975).

Povinelli and his colleagues capitalised on captive chimpanzees' tendency to beg from their keepers to ask the question: do chimpanzees understand the seemingly elementary fact that people see? Seven young chimpanzees took part in a complex series of longitudinal studies. In one series of experiments Povinelli and Eddy (1996) tested whether chimpanzees are able to understand where others are looking by putting a human opposite a chimpanzee with a transparent screen between them. The apes had no trouble in looking at the spot to which the experimenter was directing his gaze, even if it was behind the chimpanzee. In another experiment using buckets, the experimenters stood or sat in the same place, but one held a large bucket on her shoulder and could see the animal, while the other experimenter covered her head with a bucket and could not see the animal. In the similar screen condition, one experimenter held a cardboard screen on her shoulder while the other held it so that it completely obscured her face from the chimpanzee. In one recent variant of the experiment, a screen with an opaque lower half was set up between the human and the chimpanzee. When the experimenter looked at a point on the opaque part of the screen, the ape would lean forward to try to see what was on the other side. This shows that it is not just the direction of the gaze that the apes follow, but also that they understand that the gaze is directed towards a certain point in the surroundings (Barth *et al.*, 2005).

Recently data on non-primate species have been obtained in this field. In particular, it has been revealed that dogs can use a variety of experimenter-given cues such as pointing, head direction and eye direction to locate food hidden in one of several containers. Experimental studies by Miklósi *et al.* (2000) have demonstrated that dogs would look at the human if they were faced with an insoluble problem situation, for example, if they could not attain access to some food, which was hidden out of their reach. In such situations dogs preferentially look at their owner, and also alternate their gaze between where the food is hidden and the owner. Dogs seem to be not less sensitive to visual cues of attention in humans than chimpanzees.

Fig. 37.1 An elephant has to choose whether to beg food from a person who can see them or a person who can not. (Photograph by M. Nissani; courtesy of M. Nissani and D. Nissani.)

Soproni *et al.* (2001) provided evidence that dogs discriminate between persons looking into a container (containing a reward) or looking above the container. While the former behaviour of the human could be interpreted as displaying attention, the latter cue could signal inattentive behaviour on the part of the human. Moreover, dogs can discriminate between the attentional focus of the human even if the human does not pay attention to them. In a test series Soproni *et al.* (2002) studied the ability of dogs to recognise human attention in different experimental situations (ball-fetching game, fetching objects on command, begging from humans). The attentional state of the humans was varied along two variables: (1) facing versus not facing the dog; (2) visible versus non-visible eyes. Researchers suggest that dogs have an understanding of the role of the human's face orientation in social interactions but that they pay less attention to whether such facing behaviour is accompanied by the visibility of eyes. Dogs show an overall performance in the test situations that is even somewhat better than previously reported for apes, but this difference could be due to differences in levels of socialisation with humans and also to experimental conditions.

To rule out the possibility that dogs performed better than apes because they had much more experience with humans, Hare *et al.* (2002) tested two groups of domestic dog puppies with different experience with humans. One group lived with human families and the other group was litter-reared. All puppies could use human pointing and head direction cues and there were no differences in response to experimenter-given cues between the groups. Therefore, it seems that extensive interaction with humans is not necessary for dogs to be able to use such cues. Some authors hypothesise that the process of domestication is important for effective communication between humans and dogs (Frank, 1980; Miklósi *et al.*, 2000). However, the results obtained from similar experimental procedures in dolphins (Herman *et al.*, 1999) and fur seals (Scheumann and Call, 2004) reinforce the idea that effective interspecies communication can occur without a history of domestication and without formal training if animals have experienced extensive daily interactions with humans.

Nissani (2004) has adopted Povinelli and Eddy's (1996) experimental paradigm 'seeing is knowing' in a study of elephants; in particular, he used such variants as 'the buckets condition' and 'screen variations' (Fig. 37.1). Two zoo elephants obtained even higher scores than chimpanzees in situations in which they had to choose whether to beg food from a person who could see them or a person who could not. The author himself, however, gives a very careful estimate of these results. It may be worthwhile to note here that if elephants possess theory of mind, then these data complete a character of an elephant as a rather strange intellectual animal. Drawing to the close of the book, we can attribute the same to members of many species. As far as elephants are concerned, the results of simple discrimination tests cannot be readily reconciled with the widespread view that elephants possess exceptional intelligence. They mastered these tasks no faster than rats and bees (Nissani *et al.*, 2005). But then, elephants display excellent memory and can remember trained discriminations for several years. One female Asian elephant retained trained discriminations in her memory and readily solved the problem after 8 years (Markowitz *et al.*, 1975). This is in conformity with elephants' ability to navigate extensive social networks based on individual recognition of a number of members of their community and applying flexible communicative means (McComb *et al.*, 2000; Poole *et al.*, 2005). Elephants surprisingly do not enjoy excellent memory beyond the age of 20 to 30 years. This is the more puzzling since elephants in their forties are at their prime and can do many clever things in the wild. To examine this effect, further experiments are needed (Nissani *et al.*, 2005). Returning to the problem of experimental investigation of theory of mind in animals we should admit on the basis of the data described above that if it possesses theory of mind an organism can claim membership in one but not in all clubs of intellectuals.

37.2 Deceptive tactics

At the behavioural level, deception constitutes misinterpretation of a situation by one individual as a consequence of the acts of another one (Krebs and Dawkins, 1984; Byrne and Whiten, 1988). Deceptive tactics in animals and humans are tightly related with knowledge about what 'game partners' see and do not see, that is, with the capacities for perspective-taking. Besides, subjects have to understand that 'seeing is knowing', in other words, that the visual information that their partner had acquired gave him knowledge of what he had seen, for instance, where the reward was hidden. Such understanding is related to another type of behaviour, knowledge attribution, which will be considered later in this section.

Byrne and Whiten (1991) have compiled a catalogue of naturalistic evidence of intentional deception in primates which by 1990 involved over 250 individual observations of such behaviour. Examples include cases in which female primates hid from dominant males and suppressed their usual copulation calls whilst mating with a subordinate male; diverting the attention of dominant individuals by giving false alarm calls; distracting the dominant male with threats of aggression while the other mates with one of his females – and then reversing their roles.

The first experimental attempt to test chimpanzees' ability to deceive was performed by Woodruff and Premack (1979). They let four chimpanzees become acquainted with two trainers, one of whom was kind and helpful to them ('cooperative' trainer), while the other, 'competitive' trainer treated the chimpanzees aggressively. During the experiment the chimpanzees one by one were placed in such a position that they were able to observe a third person, a laboratory assistant, who hid food under one of two containers that the animals could not reach. The assistants then left the room, and one of the two trainers came in. Since the chimpanzee was in a cage, it could not reach the cups, but it was able to show the trainer where the food was. If it was the cooperative trainer and the chimpanzee pointed to the right cup, the trainer picked up the reward and gave it to the chimpanzee. If it pointed to the wrong cup, the trainer looked under it and then went out. If it was the competitive trainer and the chimpanzee pointed to the right cup, the trainer took the food and ate it himself. But if the chimpanzee pointed to the wrong cup, the nasty trainer looked under it and then went to the corner to sit and sulk. When this happened, the chimpanzee was let out of the cage and could retrieve the food itself from under the right cup. In this situation the chimpanzee was rewarded if it pointed to the wrong cup. All the chimpanzees needed at least 50 attempts before they could systematically distinguish which behaviour was most appropriate for which trainer. Since it takes such a long time, the question is whether they understand anything about how the trainers think when the chimpanzees point. During their training subjects had ample time to learn that certain responses in the presence of the cooperative trainer always resulted in reward, whereas different responses resulted in reward in the presence of competitive trainer. Discrimination of this sort can more readily be explained as a form of conditioning. However, sets of results obtained in other experiments concerning knowledge attribution in non-humans (see details below) indirectly support the hypothesis that the chimpanzees can assimilate intentional deceptive tactics.

Recently experimental evidence of deception has come from studies on corvids, birds that have a complex social life and wide traditions of pilfering.

Emery and colleagues (Emery and Clayton, 2001; Emery *et al.*, 2004) looked at how scrub jays reacted to the possibility that some neighbour would steal their food cache. Some of the jays in the laboratory had criminal histories of snitching food that another bird had buried. The researchers let the 'criminal' jays as well as those with clean records hide treats. When the birds stashed their seeds in private, they didn't take an opportunity to move their treasure to a new hiding place. However, if the researchers let another bird get near enough to watch the caching, the criminal jays took the next opportunity to recover the treat and hide it in a different place. The birds with no experience of thievery didn't re-cache.

Bugnyar and Kotrschal (2002, 2004) reported deceptive behaviour of a common raven leading a competitor away from food in a social foraging task. In the experiment, four individuals had to search and compete for hidden food at colour-marked clusters of artificial food caches. A subordinate male, Hugin, found and exploited most of the food. As a result, the dominant male, Munin, displaced him from the already opened boxes. The subordinate male then developed a pattern, when the loss of reward to the dominant got high, in which he would move to unrewarded clusters and open boxes there. This diversion often led the dominant to approach the unrewarded clusters and the subordinate then had a head start for exploiting the rewarded boxes. Subsequently, however, the dominant male learned not to follow the subordinate to unrewarded clusters and eventually started searching for the reward himself. These interactions between the two males illustrate the ravens' potential for deceptively manipulating conspecifics.

Hugin's behaviour enabled the experimenters to compare ravens' misleading behaviour with similar behaviour in other species. The first of them was described by Goodall (1986) for a wild chimpanzee that was socially restrained from feeding at an artificial food dispenser. The subject repeatedly induced leaving of the whole group by walking off the site but then circled back and got more food. In addition, Hugin's behaviour resembled the results of the experiments with chimpanzees (Menzel, 1973a, b; Hirata and Matsuzawa, 2001) and mangabeys (Coussi-Korbel, 1994) in which a particular subject was informed of the location of hidden food. The informed subjects repeatedly lost the food to a dominant conspecific and, as a consequence, began misleading the competitor. Ring-tailed lemurs tested with a similar procedure failed to use deceptive tactics consistently although the informed subject apparently did benefit from accidental manoeuvres that distracted the competitor (Deaner *et al.*, 2000). Subordinate domestic pigs, in contrast, showed misleading-like behaviours in response to exploitation by dominants even though these moves hardly affected their own foraging success (Held *et al.*, 2000). The authors suggest a parsimonious explanation that misleading conspecifics is based on the manipulation of the others' behaviour rather than on a consideration of others' mental state.

37.3 Knowledge attribution

Knowledge attribution includes several forms of behaviour that are related to subject's ability to 'put itself in somebody's shoes', or to be able mentally to put itself in the place of others. This ability is based on perspective-taking described above as well as discrimination of external manifestations of other's intentions and desires such as gaze directions and body movements. Here we will consider several experimental paradigms aimed to test these abilities in animals.

Do animals know what others do and do not see?

The criterion 'seeing is knowing', as a part of knowledge attribution, is considered a minimum requirement for having a theory of others' minds.

To test whether or not chimpanzees meet this criterion, Povinelli and co-authors conducted series of experiments (Povinelli *et al.*, 1992a; Povinelli and Prince, 1998). One of them was done in two steps. In the first step a trainer came in and hid a reward under one of two cups, but since these were behind a screen the chimpanzee could not see under which one the food was hidden. Yet the chimpanzee could understand that it was hidden, as the trainer's hands were empty afterwards. Then the screen was removed so that the chimpanzee could see the cups. Another trainer came in and the two trainers each pointed to a different cup. The chimpanzee was only allowed to look under one of them. To begin with, the chimpanzees chose at random, but after a while they more often looked under the cup that the trainer who had hidden the food was pointing to. The long learning time agrees entirely with the results of Woodruff and Premack's experiment. The second phase of Povinelli's experiment began when the chimpanzees had learned to choose correctly in the first phase. Now both trainers were present when the reward was hidden behind the screen, but it was hidden by a third person. One trainer looked on passively, while the other had a bucket over his head so that he could see nothing. Just as in the first phase, the screen was removed, the bucket was lifted from the trainer's head, and the trainers each pointed to a different cup. Two of the four chimpanzees who were tested were able immediately to choose the cup pointed out by the trainer who had not had a bucket on his head. A third chimpanzee could manage it after a little training.

Povinelli and DeBlois (1992) repeated the second part of the experiment with children. From the age of about 3 years, the children were able to choose the cup pointed out by the trainer who had seen where the reward was hidden. At earlier ages they chose cups more or less at random. A later experiment, however, cast doubt on whether the chimpanzees' success was due to the ability to link 'seeing' with 'knowing'. This experiment utilised the natural tendency of chimpanzees to beg for food by extending their hands. A chimpanzee sat behind a transparent screen, but it could stick its hand out through a hole towards the trainer. First the experimenters checked that the begging gesture worked by placing two keepers in front of the hungry chimpanzee, one of them with food, the other with a piece of wood. The chimpanzee understood the situation correctly and begged only from the keeper with the food reward. The experiment then went on to study whether the apes could distinguish a trainer who knew that there was food from one who did not know. One trainer was blindfolded while the other had his mouth gagged. It turned out that the chimpanzees could not tell which of the trainers it was worth begging from. A number of variants of the experiment were tried. In one variant, one of the trainers held his hands over his eyes while the other had his hands over his ears; in another variant one trainer had a bucket over his head while the other had a bucket on his shoulder; in a third form the two trainers sat with their backs to the chimpanzee, but one of them was looking over his shoulder and could see the food. In none of these cases could the chimpanzees determine which trainer it was best to beg from. After continued training the chimpanzees were able to distinguish the trainers correctly. Yet it seemed as if the only significant factor was whether the trainer's face was visible or not; it did not matter whether the trainer's eyes were open or shut. Despite the successes in the first experiment, the probable conclusion is that the chimpanzees cannot clearly understand that 'seeing is knowing'.

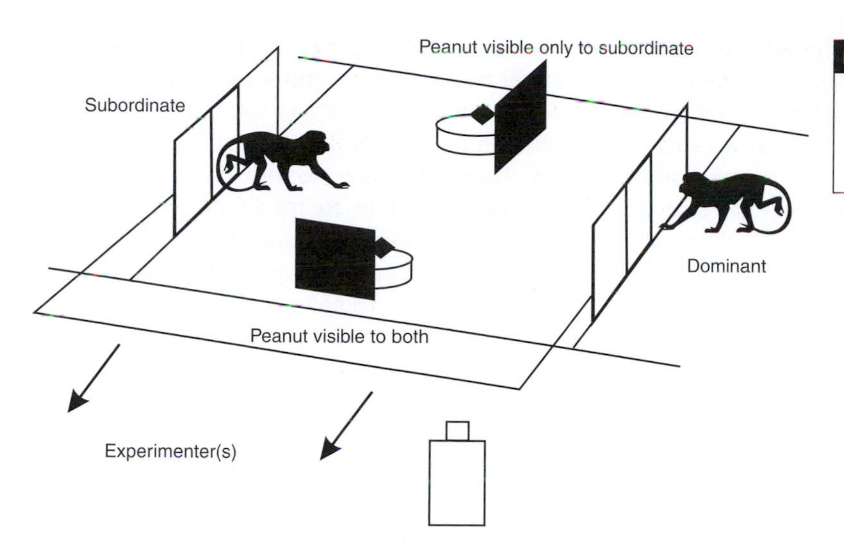

Fig. 37.2 Experimental set-up for studying monkeys' sensitivity to perception of what another individual does or does not see. (Adapted from Hare *et al.*, 2003.)

Hare *et al.* (2001) devised a new test to examine whether chimpanzees know what other chimpanzees do and do not see. This test is based on the concept of social dominance that possibly makes a situation more natural. Two individuals, one dominant to the other, were placed in rooms on the opposite sides of a test room were food was positioned in different ways. They were then released into the test room to retrieve the food. For example, two pieces of food were placed between the subjects so that one subject could see only one of the food pieces (the food was hidden behind a small occluder). The results showed that subordinates preferentially retrieved food hidden from the dominant, while dominants preferentially retrieved food visible to the subordinate. In follow-up studies one subject was given a slight head start over the other. The delay forced subjects to decide which of the two pieces to approach before they saw where the competitor would go. Summarised results enabled researchers to conclude that chimpanzees know what conspecifics do and do not see and that they can use this information to formulate effective social strategies. Using the same experimental paradigm with capuchin monkeys, Hare *et al.* (2003) showed that capuchins perform differently from chimpanzees when competing for food. While subdominant capuchin monkeys are extremely sensitive to the behaviour of dominants, there was little evidence that they assess what conspecifics do and do not see when approaching and retrieving food (Fig. 37.2).

Role-taking and role-reversal

Role-taking in animals was first experimentally tested by Premack and Woodruff (1978). A language-trained chimpanzee, Sarah, was shown videotapes of human actors attempting to solve different problems – for example, attempting to escape from a locked cage, trying to obtain out-of-reach food, or to switch on a tape recorder by a keystroke when the machine was unplugged. The final image of each videotape sequence was put on hold, and Sarah was offered a choice of two photographs to place beside the video monitor. Both of these represented the actor in the problem situation, but only one of them showed the actor taking a course of action that would solve the problem. Sarah consistently chose the photographs representing problem solutions, and this was interpreted as evidence that she attributed mental states to the actor. It was argued that if Sarah did not ascribe beliefs and desires to the actor then she would see the video as an undifferentiated sequence of events rather than a problem (see Premack, 1983, 2004).

Close examination of the published reports of the videotape experiments suggests that for any given problem Sarah could have responded on the basis of familiarity, physical matching and/or formerly learned associations. For example, when the actor was trying to reach food that was horizontally out of reach, matching could have been responsible for Sarah's success because a horizontal stick was prominent in both the final frame of the videotape and the photograph depicting a solution. Similarly, when the actor was shivering and looking wryly at a broken heater, Sarah may have selected the photograph of a burning roll of paper rather than an unlit or spent wick because she associated the heater with the red–orange colour of fire. Taken together, however, the results of Premack and Woodruff's videotape experiments are not subject to a single, straightforward non-mentalistic interpretation, and in this respect they are apparently unique in the literature on theory of mind in primates. Thus, according to this standard, no advance has been made on the original studies of theory of mind in primates (Heyes, 1993, 1998).

Povinelli *et al.* (1992a) have shown that pairs of chimpanzees can reverse roles after being trained to perform complementary parts of a single task. This experiment involved ensuring that one chimpanzee was able to see where food was located but could not reach it, whereas the other chimpanzee was unable to see where the food was located but was in a position to reach it. Four chimpanzees were initially trained either to choose from an array of containers the one to which an experimenter was pointing (cue detection task), or to observe food being placed in one of the containers and then to point at the baited receptacle (cue provision task). Once criterion performance had been achieved on the initial problem, each chimpanzee was confronted with the other problem, and for three of the four animals this switch did not result in a significant decline in choice accuracy. This result was tentatively interpreted as evidence of 'cognitive empathy' or 'role-taking', that is, the ability to adopt the viewpoint of another individual (Povinelli *et al.*, 1992a). Using the same experimental procedure with rhesus monkeys, Povinelli *et al.* (1992b) have shown that although monkeys coped well with the task each on its initial place, they did not match a criterion of 'cognitive empathy'.

Further experiments with monkeys have brought mixed results, and some of them are positive regarding the ability of some species to infer several kinds of mental states, such as knowledge, goals and intentions in others. In particular, monkeys were able to modify their behaviour according to another individual's knowledge and intention. For example, capuchin monkeys were shown to be able to solve cooperative problems requiring different roles to be played by two players. Capuchins learned to discriminate between a 'knower', who inspected the box with food, and a 'guesser', who did not (Hirata and Matsuzawa, 2001; Kuroshima *et al.*, 2002, 2003).

False-belief tests

Wimmer and Perner (1983) devised a false-belief test, which was used to evaluate the ability of a subject to ascribe definite but false beliefs to another. Baron-Cohen *et al.* (1985) later simplified Wimmer and Perner's test so they could compare autistic, Down's syndrome and normal children at different ages. As one of the variants of this test, the Sally–Ann test (Baron-Cohen *et al.*, 1985) is aimed at examining whether subjects express perspective knowledge and understand what others can and can not see. This test is presented as a simple story. There are two puppets, Sally and Ann. Sally has a marble, which she keeps in a basket. Then Sally leaves the room, and while she is away Ann takes the marble out of the basket and hides it in the box. Sally comes back into the room. The child subject is then asked the question: 'Where will Sally look for her marble?' Older children say that she will look in the basket, because although they know the marble is in the box, they know that Sally doesn't know it has been moved from the basket, and they can distinguish Sally's (false) belief from their own (true) belief. Younger children, on the other hand, and autistic children, do not distinguish between the two. They simply say that Sally will look in the box. The false-belief test, therefore, explores the change that happens as common sense develops.

Research using the Sally–Ann test on language-taught chimpanzees aged between 5 and 6 years has shown that none passed the test (Call *et al.*, 1994). However, Hauser (1998) has developed an 'ape version' of the test which he has used to test for a theory of mind in cotton-top tamarin monkeys and pre-verbal human children (about 2 years of age). The 'ape version' involves a subject observing an actor watching where an object is hidden. When the object has been hidden a screen is raised to prevent the actor seeing another person moving the object to a different location but does not prevent the observer from seeing where the object is relocated. The screen is then removed and the actor either looks in the new location or the old location. Hauser reasoned that if an individual was aware that the actor could not know about the new location they should respond differently when they observe the actor looking in the new location and the old location. The observer with a theory of mind should spend more time staring at the actor who looks in the new location rather than the actor who looks in the old location. This is because the former violates the observer's knowledge that the actor cannot know the new location due to the observer has not seen the object being moved. It turned out that both monkeys and pre-verbal children stare significantly longer when the actor looks in the new location than when the actor looks in the old location. This supports the author's hypothesis that the subjects passed the test.

37.4 Self-awareness

It seems natural that subjects possessing a capacity for discriminating others' mental states should be aware of corresponding native data, that is, they should possess self-awareness.

Self-awareness refers to consciousness of the self as a separate individual and occupies the next level above conscious awareness of events and objects in the world (Cartwright, 2002). It is a debated question whether non-human animals are conscious and how such functioning has evolved (Griffin, 1984, 1992, 2001; Bekoff and Allen, 2002; Broom, 2003).

The objective study of self-recognition, with a mirror and a mark applied to the face, was conducted independently by a comparative psychologist Gallup (1970) for use with chimpanzees and monkeys, and by a clinical child psychologist Amsterdam (1972) for use with infant humans (for reviews see Anderson and Gallup, 1999; Bard *et al.*, 2006).

Gallup (1970) developed a mirror test as an experimental paradigm for examining manifestations of self-awareness in animals based on ideas that the use of the mirror for self-directed behaviour depends on self-awareness. Chimpanzees, after being given the chance to play with a mirror for 10 days, had a mark placed on their foreheads while they were anaesthetised. Later they seemed to have no awareness of the mark until given access to a mirror, after which they began to touch their foreheads, using the mirror to guide their hands as they touched the mark on their own faces (rather than the mark on the mirror image). Gallup interpreted this behaviour as indicating self-recognition by the animals. Since then, only a few species have passed this test and among them are chimpanzees (but not macaques and tamarins: see Gallup, 1977; Hauser *et al.*, 2001), orang-utans (Swartz *et al.*, 1999), cetaceans (Delfour and Marten, 2001; Reiss and Marino, 2001) and grey parrots (Pepperberg *et al.*, 1995). Human children tend to fail this test until they are 3–4 years old. Members of several other species tested, for instance dogs and 2-year-old children, usually react to a mirror in fear or with curiosity, or simply ignore it, while birds often attack their own reflections (for a detailed review see Keenan *et al.*, 2003).

In some experiments chimpanzees (Gallup *et al.*, 1995) grey parrots (Pepperberg *et al.*, 1995) and children of 3 years of age (Amsterdam, 1972; Zazzo, 1979) were shown as being able to find hidden objects using a mirror to guide their behaviour. There is some evidence that animals that have not passed the mirror test for self-recognition are nevertheless able to use information provided by mirrors. Monkeys can use a mirror to locate a plastic flower suspended above their heads by means of a specially adapted collar (Anderson, 1984; Itakura, 1987). Povinelli (1989) describes occasions when an elephant carefully guided its trunk with the help of a mirror in order to retrieve a carrot that was not otherwise visible.

Marten and Psarakos (1995) developed an experimental paradigm combining the use of a mirror and television tracks in order to adopt this experimental technique for studying self-awareness in dolphins. When a dolphin looks in a mirror it often opens its mouth and moves its head around in some rhythmic fashion. In a mark test, as it does not have a hand to touch its mark, the dolphin manoeuvres its body in various postures to see it. These behaviours are likely to be contingency-checking and self-examination, respectively. However, another possible interpretation is that the dolphin is using postures to interact 'socially' with what he perceives to be another dolphin in the mirror. To distinguish self-examination from social behaviour, researchers put a video camera aimed into the tank next to the television monitor. A real-time image is displayed on the dolphin's television, allowing the dolphin to look at itself as in a mirror; this material is videotaped. The dolphins were exposed to alternating sessions of mirror mode and of playback mode. If the dolphin perceives its mirror mode television image as another dolphin, then whether it is watching mirror mode or playback mode, the dolphin should act the same way – possibly interacting socially with the 'television dolphin', or perhaps just observing it, but behaving the same in both modes. The results for the adult male were clear-cut: in mirror mode he spent quite a bit of time opening his mouth wide and moving his head in various rhythmic ways, whereas he never did this during playback mode. The results for juveniles and babies, however, did not show such clear-cut differences between the viewing modes, although the babies seemed to spend more time in front of the television during mirror mode.

Some of the dolphins who participated in the mirror mark tests were also exposed to the mirror mode/playback test. The video camera was located next to the television, to the dolphins' left, and thus favoured showing the left side when they looked at the television. The dolphins were marked on their right sides to see if they would counter this bias by manoeuvring the marked side into view. Three of the adults used the mirror to visually inspect their bodies where the mark was located. One of the most important results from this research comes from a combination of the mirror mark tests and the television tests. Keola, the only adult to have both tests, behaved significantly differently when the television was in mirror mode as compared to playback mode. His behaviour strongly suggested that he did not perceive the television image as another dolphin in either mode and was not interacting with it 'socially'. When he was marked with a dot, he immediately positioned himself in front of the mirror so that the mark would be visible to him and then engaged in rapid, extreme, postures. It is very likely that he was examining his mark. The only reasonable alternative is that he might have been posturing socially to what he believed to be another dolphin in the mirror. Since this 'social' explanation is rendered not very likely by the television tests, the data suggest that this dolphin examined his mark in the mirror during his mark test, and took his mirror image to be himself.

No single test proves self-recognition in bottlenose dolphins. The tests were developed mainly from primate research paradigms, and their limitations for the interpretation of dolphin behaviour are apparent. Nevertheless, the data taken together make a compelling case for self-recognition in this species.

There are some specific limitations to interferences about mirror use in animals. For example, although at least one specific gorilla, Koko, passed this test (Patterson and Cohn, 1994), several others did not (Shillito *et al.*, 1999). This is probably because gorillas consider eye contact an aggressive gesture and normally try to avoid looking each other in the face. At the same time it is worth noting that not all chimpanzees show evidence of mirror self-recognition (Swartz *et al.*, 1999). Again, negative results with mirror self-recognition in apes and elephants may be ascribable to small samples. There is also debate as to the value of the test as applied to animals that rely primarily on senses other than vision, such as dogs.

It is important to note that the relative weakness of the mirror self-recognition experimental paradigm can be attributed to the fact that

Fig. 37.3 A capuchin monkey confronted with its own reflection in a mirror. (Photograph by M. Dindo; courtesy of F. B. M. de Waal and M. Dindo.)

researchers often take an all-or-nothing view of self-awareness without considering intermediate stages. De Waal and colleagues (2005) have elaborated experimental tests that allow them to demonstrate capuchin monkeys as possibly being able to reach a level of self–other distinction intermediate between seeing their mirror image as another being and recognising it as self. Members of relatively large groups of capuchins were tested, and the researchers used the monkeys' social interrelations in their experiment. They formalised scenarios of monkeys' contacts with friendly, hostile and unfamiliar conspecifics as well as with mirror reflections. During a series of tests each capuchin entered a test chamber, where it was presented with three different situations. In the first one, the monkey saw an unfamiliar monkey of the same sex on the other side of a glass barrier and behind a mesh screen. In the second scenario, the capuchin saw a monkey of the same sex that it was familiar with. Finally, it confronted its own reflection in a mirror behind the screen (Fig. 37.3). When capuchins saw another monkey they behaved quite naturally, for example, when meeting with an unfamiliar partner, males made threatening gestures, and females looked nervous and avoided eye contact. Instead, the reflection got treated as a special phenomenon, generally eliciting curiosity and friendly overtures from females and a mix of distress and fear from males. It is likely that capuchin monkeys possess a greater understanding of the mirror's illusory qualities than is generally assumed (de Waal *et al.*, 2005).

It is possible that some animals catch differences between behavioural responses from 'normal' intruders and illusory ones that are their own mirror reflections. It is likely that treatment if the reflection as a special phenomenon can be attributed to species outside primates. For instance, Oliveira *et al.* (2005) have shown in their experiments with cichlid fish *Oreochromis mossambicus* that fighting males react to a mirror-image challenger differently than to a 'normal' intruder. In natural situations androgen concentrations are changed under salient social situations. Decades of various researchers' studies have shown that among fish, birds and mammals, winners' androgen concentrations surge and losers' dip. These hormonal changes improve or diminish, respectively, an animal's chance of winning its next fight. Moreover, just watching a fight boosts androgens in spectator fish. This finding echoed a study of androgens in sports fans watching soccer or basketball (Oliveira, 2004). However, androgen concentrations are not increased in fishes that are fighting their own image in a mirror, despite their explicitly aggressive behaviour towards the virtual intruder. These results indicate that the hormonal response normally triggered in male contests is not induced by signals delivered by a mirror-image challenger. This enable us to hypothesise that the male fish perceives its mirror reflection as something different from 'another fish'.

Recently Bard *et al.* (2006) integrated the approaches of developmental and comparative psychologists concerning the mirror test. They refer to Piagetian developmental stages (see Chapter 13). Stage 6 in object permanence would suggest that the self is conceived as something permanent in time and space. The mirror test shows, additionally, that self can exist and can be represented: the mirror image of the self is a representation of the self as an iconic symbol. Bard *et al.* (2006) suggest that passing of the mark test is based on being able to concurrently represent the self in multiple forms: the acting self is understood to exist at the same time as the visually reflected acting self in the mirror, an iconic symbolic capacity. So, these authors suggest, mirror-self recognition may be based on specific aspects of mental representation, the cognitive ability to symbolise. It has been revealed in the cited study that when comparable testing procedures and assessment criteria are used, chimpanzees and human infants perform comparably.

The exact nature of animal awareness about the self has not yet been resolved by experiments, and results of mirror tests, although sophisticated, do not fully reflect animals' mental states. In humans, mirror recognition is only a precursor to a continually developing capacity for self-awareness and self-evaluation. Paraphrasing the title of Nagel's (1974) paper 'What is it like to be a bat?' we can admit that we do not know what it is like to be a conscious non-human animal.

The Russian psychologist Lev Vygotsky (1962) showed that a significant moment in the development of the human individual occurs when language and practical intelligence converge. It is when thought and speech come together that children's thinking is raised to new heights and they start acquiring truly human characteristics. Language becomes a tool of thought allowing children increasingly to master their own behaviour.

Results obtained on language-trained apes can expand our knowledge of awareness in animals. For example, when a bonobo Kanzi was taught a sign for himself and a sign for 'bad' and then was shown the two signs together, the bonobo was visibly distressed. This is likely to indicate self-awareness because neither sign alone had such an effect (Savage-Rumbaugh and Lewin, 1994).

CONCLUDING COMMENTS

The development of ethology, a science that is rooted in the observation of animal behaviour, has helped researchers to understand that animals use cognition in the natural environment. In nature animals confront problems that differ markedly from one niche to another, and they exhibit intelligence best when it serves their own purposes rather than when the experimenters expect them to perform some task. To navigate social landscape, animals need a surplus of intelligence that overcomes their immediate survival needs, such as eating, avoiding predators, feeding offspring, etc., and this surplus intelligence might have been advantageous for social manipulation, deception and cooperation.

There is much work to be done to evaluate the role of intelligence in maintaining cooperative behaviour. We can assume that cooperation that is based on reciprocal altruism requires more advanced cognitive skills than altruism towards kin, because reciprocity requires remembering and discounting levels of cooperativeness among individuals. Specific cognitive adaptations can be expected in some species such as specific concentration of attention and calculation of mutual aids.

Only recently have cognitive ethologists selected species for their experiments taking into account specific sociality. There is a growing body of evidence that highly social animals are more adept at classifying items according to their relative relations, making abstract discriminations about relations between relations, forming concepts about relations between objects independently of their physical features as well as concepts of numerosity and transivity. With reference to data on social learning and information transferring in animals that were considered in Parts VIII and IX, we should expect that members of group-living species have more opportunities to learn during their lifetime than more solitary animals.

However, we should not expect to find a linear correlation between social complexity and levels of intelligence in non-human species. Although experiments based on pair comparison of intellectual abilities in group-living and solitary living species have brought some positive results, we should take into consideration that animals that live solitarily in complex and risky environments rely on their own memory and learning skills and may enjoy freedom from restrictions and obligations imposed upon them by their possibly narrow roles within a community.

In general, our understanding of animal intelligence will improve with great strides if experimenters arm themselves with knowledge about evolutionary histories and characteristics of niches of the species in hand as well of their sensory equipment, inherited predisposition for particular types of learning, and sociality.

References

Abbot, P., Withgott, J.H. and Moran, N.A. (2001) Genetic conflict and conditional altruism in social aphid colonies. *Proceedings of the National Academy of Sciences of the USA*, **98**, 12068–12071.

Able, K.P. (1980) Mechanisms of orientation, navigation, and homing. In *Animal Migration, Orientation, and Navigation*, ed. S.A. Gauthreaux. New York: Academic Press, pp. 284–373.

Abramson, C.I. (2003) Charles Henry Turner: contributions of a forgotten African American to honey bee research. *American Bee Journal*, **143**, 643–644.

Adams, D.K. (1929) Experimental studies of adaptive behavior in cats. *Comparative Psychological Monographs*, **6** (1), 1–128.

Ades, C., Ramos, D., Rossi, A.P. and Suenaga, L.V. (2005) Human–dog /dog–human communication through arbitrary signals. *Abstracts 29th International Ethological Conference*, Budapest, Hungary, p. 12.

Adler, T. (1996) A shrimpy find: communal crustaceans. *Science News*, **149**, 359.

Adrien, J.L., Martineau, J., Barthelemy, C., *et al.* (1995) Disorders of regulation of cognitive activity in autistic children. *Journal of Autism and Developmental Disorders*, **25**, 249–263.

Agin, V., Dickel, L., Chichery, R. and Chichery, M.P. (1998) Evidence for a specific short-term memory in the cuttlefish, *Sepia*. *Behavioural Processes*, **43**, 329–334.

Agras, S., Sylvester, D. and Oliveau, D. (1969) The epidemiology of common fears and phobia. *Comprehensive Psychiatry*, **10**, 151–156.

Ainslie, G. (1974) Impulse control in pigeons. *Journal of the Experimental Analysis of Behavior*, **21**, 485–489.

Ainsworth, M. (1982) Attachment: retrospect and prospect. In *The Place of Attachment in Human Behaviour*, ed. C.M. Parkes and J. Stevenson-Hinde. New York: Basic Books, pp. 185–201.

Aitchison, J. (1983) *The Articulate Mammal: An Introduction to Psycholinguistics*. London: Hutchinson.

Akins, C.K. and Zentall, T.R. (1998) Imitation in Japanese quail: the role of reinforcement of the demonstrator's response. *Psychonomic Bulletin and Review*, **5**, 694–697.

Albright, J.L. (1978) Optimal group size for high producing cows appear to be near 100. *Hoard's Dairyman*, 25 April 1978, 534–535.

Alcock, J. (2005) *Animal Behavior: An Evolutionary Approach*, 11th edn. Sunderland, MA: Sinauer Associates.

Alexander, R.D., Noonan, K.M. and Crespi, B.J. (1991) The evolution of eusociality. In *The Biology of the Naked Mole-Rat*, ed. P.W Sherman, J.U.M. Jarvis and R.D. Alexander. Princeton, NJ: Princeton University Press, pp. 3–44.

Allen, C. (1998) Assessing animal cognition: ethological and philosophical perspectives. *Journal of Animal Science*, **76**, 42–47.

Allen, C. (2004) Is anyone a cognitive ethologist? *Biology and Philosophy*, **19**, 589–607.

Allen, C. and Bekoff, M. (1997) *Species of Mind: The Philosophy and Biology of Cognitive Ethology*. Cambridge, MA: MIT Press.

Allison, J. and Timberlake, W. (1974) Instrumental and contingent saccharin-licking in rats: response deprivation and reinforcement. *Learning and Motivation*, **5**, 231–247.

Altbäcker, V. (2004) Experimenter as a predator. *Abstracts 2nd European Conference on Behavioural Biology*, Groningen University, the Netherlands, p. 100.

Altbäcker, V., Hudson, R. and Bilko, A. (1995) Rabbit-mothers diet influences pups later food choice. *Ethology*, **99**, 107–116.

Amsterdam, B. (1972) Mirror self-image reactions before age two. *Developmental Psychobiology*, **5**, 297–305.

Anderson, C. and Franks, N.R. (2001) Teams in animal societies. *Behavioral Ecology*, **12**, 534–540.

Anderson, C.O. and Mason, W.A. (1978) Competitive social strategies in groups of deprived and experienced rhesus monkeys. *Developmental Psychobiology*, **11**, 289–299.

Anderson, J.R. (1984) Monkeys with mirrors: some questions for primate psychology. *International Journal of Primatology*, **5**, 81–98.

Anderson, J.R. (1985) *Cognitive Psychology and Its Implications*, 2nd edn. New York: W.H. Freeman.

Anderson, J.R. and Gallup, G.G., Jr (1999) Self-recognition in non-human primates. In *Animal Models of Human Emotions and Cognition*, ed. M. Haug and R.E. Whalen. Washington, DC: American Psychological Association, pp. 175–194.

Anglin, J.M. (1977) *Word, Object, and Conceptual Development*. New York: W.W. Norton.

Anokhin, K.V., Mileusnic, R., Shamakhina, I.Y. and Rose, S.P.R. (1991) Effects of early experience on *c-fos* gene expression in the chick forebrain. *Brain Research*, **544**, 101–107.

Anokhin, P. K. (1961) A new conception of the physiological architecture of conditioned reflex. In *Brain Mechanisms and Learning*, ed. J. F. Delafresnay. Oxford, UK: Blackwell, pp. 189–229.

Anokhin, P. K. (1968) Nodular mechanisms of functional system as self-regulatory apparatus. *Progress in Brain Research*, **22**, 230–251.

Anokhin, P. K. (1974) *Biology and Neurophysiology of Conditioned Reflex and Its Role in Adaptive Behavior*. Oxford, UK: Pergamon Press.

Aoki, S. (1977) *Colophina clematis* (Homoptera, Pemphigidae), an aphid species with soldiers. *Kontyu*, **45**, 276–282.

Aoki, S. (2003) Soldiers, altruistic dispersal and its consequences for aphid societies. In *Genes, Behavior, and Evolution in Social Insects*, ed. T. Kilkuchi, N. Azumi and S. Higashi. Sapporo, Japan: Hokkaido Univerity Press, pp. 201–215.

Arnold, K. E. and Owens, I. P. F. (1999) Cooperative breeding in birds: the role of ecology. *Behavioral Ecology*, **10**, 465–471.

Aumann, T. (1990) Use of stones by the black-breasted buzzard, *Hamirostra melanosternon*, to gain access to egg contents for food. *Emu*, **90**, 141–144.

Avilés, L. and Tufico, P. (1998) Colony size and individual fitness in the social spider *Anelosimus eximius*. *American Naturalist*, **152**, 403–418.

Axelrod, R., Hammond, R. A. and Grafen, A. (2004) Altruism via kin-selection strategies that rely on arbitrary tags with which they coevolve. *Evolution*, **58**, 1833–1838.

Axelsen, B. E., Anker-Nilssen, T., Fossum, P., Kvamme, C. and Nøttenstad, L. (2001) Pretty patterns but a simple strategy: predator–prey interactions between juvenile herring and Atlantic puffins observed with multibeam sonar. *Canadian Journal of Zoology*, **79**, 1586–1596.

Bachevalier, J. and Mishkin, M. (1989) Mnemonic and neuropathological effects of occluding the posterior cerebral artery in *Macaca mulatta*. *Neuropsychologia*, **27**, 83–105.

Baddeley, A. (1986) *Working Memory*. Oxford, UK: Clarendon Press.

Baker, M. C. and Thompson, D. B. (1985) Song dialects of white-crowned sparrows: historical processes inferred from patterns of geographic variation. *Condor*, **87**, 127–141.

Baker, R. R. (1984) *Bird Navigation: The Solution of a Mystery?* London: Hodder & Stoughton.

Balda, R. P. (1980) Recovery of cached seeds by a captive *Nucifraga caryocatactes*. *Zoological Tierpsychology*, **52**, 331–446.

Balda, R. P. and Kamil, A. C. (1992) Long-term spatial memory in Clark's nutcracker, *Nucifraga columbiana*. *Animal Behaviour*, **44**, 761–769.

Balda, R. P., Pepperberg, I. M. and Kamil, A. C. (eds.) (1998) *Animal Cognition in Nature*. San Diego, CA: Academic Press.

Baldwin, J. M., (1894) The origin of right-handedness. *Popular Science Monthly*, **44**, 606–615.

Baldwin, J. M., Cattell, J. M. and Jastrow, J. (1898) Physical and mental tests. *Psychological Review*, **5**, 172–179.

Banfield, E. J. (1908) *The Confessions of a Beachcomber*. London: T. Fisher Unwin. Reprinted (2004) Sydney, NSW: Sydney University Press.

Baptista, L. F. (1975) Song dialects and demes in sedentary populations of the white-crowned sparrow (*Zonotrichia leucophrys nuttalli*). *University of California Publications in Zoology*, **105**, 1–52.

Bard, K., Todd, B. K., Bernier, C., Love, J. and Leavens, D. A. (2006) Self-awareness in human and chimpanzee infants: what is measured and what is meant by the mark and mirror test? *Infancy*, **9**, 191–219.

Barnett, S. A. (1958) An analysis of social behaviour in wild rats. *Proceedings of the Zoological Society of London*, **130**, 107–152.

Barnett, S. A. (1970) *Instinct and Intelligence*. Harmondsworth, UK: Penguin.

Baron-Cohen, S., Leslie, A. M. and Frith, U. (1985) Does the autistic child have a 'theory of mind'? *Cognition*, **21**, 37–46.

Barth, J., Reaux, J. E. and Povinelli, D. J. (2005) Chimpanzees (*Pan troglodytes*) use of gaze cues in object-choice tasks: different methods yield different results. *Animal Cognition*, **8**, 84–92.

Baskin, L. M. (1976) *Behaviour of Ungulates*. Moscow: Nauka. (In Russian.)

Bastian, J. (1967) The transmission of arbitrary environmental information between bottlenosed dolphins. In *Animal Sonar Systems: Biology and Bionics*, vol. 2, ed. R. G. Busnel. Jouy-en-Josas, France: NATO Advanced Study Institute/Laboratorie de Physiologie Acoustique, pp. 807–873.

Bateson, M. and Kacelnik, A. (1997) Starling's preferences for predictable and unpredictable delays to food: risk-sensitivities or time-discounting? *Animal Behaviour*, **53**, 1129–1142.

Bateson, P. P. G. (1966) The characteristics and context of imprinting. *Biological Reviews*, **41**, 177–220.

Bateson, P. P. G. (1978) Early experience and sexual preferences. In *Biological Determinants of Sexual Behaviour*, ed. J. B. Hutchison. Chichester, UK: John Wiley, pp. 29–53.

Bateson, P. P. G. (1979) How do sensitive periods arise and what are they for? *Animal Behaviour*, **27**, 470–486.

Bateson, P. P. G. (1983) Rules for changing the rules. In *Evolution from Molecules to Men*, ed. D. S. Bendall. Cambridge, UK: Cambridge University Press, pp. 483–507.

Bateson, P. P. G. (ed.) (1991) *The Development and Integration of Behaviour*. Cambridge, UK: Cambridge University Press.

Bateson, P. P. G. (2000) What must be known in order to understand imprinting? In *The Evolution of Cognition*, ed. C. Heyes and L. Huber. Cambridge, MA: MIT Press, pp. 85–102.

Bateson, P. P. G. (2001) Where does our behaviour come from? *Journal of Biosciences*, **26**, 561–570.

Bateson, P. P. G. (2003) The promise of behavioural biology. *Animal Behaviour*, **65**, 11–17.

Bateson, P. P. G. (2005). The return of the whole organism. *Journal of Biosciences*, **30**, 31–39.

Bateson, P. and Martin, P. (2000) *Design for a Life: How Behavior and Personality Develop*. New York: Simon & Schuster.

Batteau, D. W. and Markey, P. R. (1967) *Man/Dolphin Communication*, Final Report: 15 December 1966–13 December 1967. Available online: www.helsinki.fi/~lauhakan/whale/literature/biblion.html/

Beck, B. B. (1980) *Animal Tool Behavior: The Use and Manufacture of Tools by Animals*. New York: Garland STPM Press.

Beck, M. and Galef, B. G. (1989) Social influences on the selection of a protein-sufficient diet by Norway rats (*Rattus norvegicus*). *Journal of Comparative Psychology*, **103**, 132–139.

Bekhterev, V. M. (1926) *Reflexologie des Menschen*. Leipzig: Felix Meiner.

Bekhterev, V. M. (1928) *General Principles of Human Reflexology*. Trans. E. and W. Murphy (1973) New York: Arno Press.

Bekhterev, V. M. (1921) *Collective Reflexology*, Part 1. Trans. E. Lockwood, ed. L. H. Strickland (1994) New York: Nova Science Publishers.

Bekoff, M. C. (1995) *Readings in Animal Cognition*. Cambridge, MA: MIT Press.

Bekoff, M. C. (2002) *Minding Animals: Science, Nature, Spirituality, and Heart*. New York: Oxford University Press.

Bekoff, M. C. and Allen, C. (2002) The evolution of social play: interdisciplinary analyses of cognitive processs. In *The Cognitive Animal: Empirical and Theoretical Perspectives on Animal Cognition*, ed. M. Bekoff, C. Allen and G. M. Burghardt. Cambridge, MA: MIT Press, pp. 426–436.

Bekoff, M., C. Allen and G. M. Burghardt (eds.) (2002) *The Cognitive Animal: Empirical and Theoretical Perspectives on Animal Cognition*. Cambridge, MA: MIT Press.

Bennett, A. T. (1996) Do animals have cognitive maps? *Journal of Experimental Biology*, **199**, 219–224.

Bennett, N. C. and Faulkes, C. G. (2005) *African Mole-Rats: Ecology and Eusociality*. Cambridge, UK: Cambridge University Press.

Benton, T. G. and Foster W. A. (1992) Altruistic housekeeping in a social aphid. *Proceedings of the Royal Society of London B*, **247**, 199–202.

Beran, M. J. (2001) Summation and numerousness judgements of sequentially presented sets of items by chimpanzees (*Pan troglodytes*). *Journal of Comparative Psychology*, **115**, 181–191.

Beran, M. J. (2004) Long-term retention of the differential values of arabic numerals by chimpanzees (*Pan troglodytes*). *Animal Cognition*, **7**, 86–92.

Beran, M. J. and Beran, M. M. (2004) Chimpanzees remember the results of one-by-one addition of food items to sets. *Psychological Science*, **15**, 94–99.

Beran, M. J. and Rumbaugh, D. M. (2001) 'Constructive' enumeration by chimpanzees (*Pan troglodytes*) on a computerised task. *Animal Cognition*, **4**, 81–89.

Beran, M. J., Beran, M. M., Harris, E. H. and Washburn, D. A. (2005) Ordinal judgements and summation of nonvisible sets of food items by two chimpanzees (*Pan troglodytes*) and a rhesus macaque (*Macaca mulatta*). *Journal of Experimental Psychology: Animal Behavior Processes*, **31**, 351–362.

Bernstein, S. and Bernstein, R. (1969) Relationship between foraging efficiency and the size of the head and component brain and sensory structures in the red wood ant. *Brian Research*, **16**, 85–104.

Berthold, P., Wiltschko, W., Miltenberger, H. and Querner, U. (1990) Genetic transmission of migratory behavior into a nonmigratory bird population. *Experientia*, **46**, 107–108.

Beusekom, G. van. (1948a) Some experiments on the optical orientation of British hunting wasps. *Proceedings of Linnean Society of London*, **160**, 12–37.

Beusekom, G. van (1948b) Some experiments on the optical orientation in *Philanthus triangulum* Fabr. *Behaviour*, **1**, 195–225.

Bhatt, R. S., Wasserman, E. A., Reynolds, W. F., Jr and Knauss, K. S. (1988) Conceptual behavior in pigeons: categorisation of both familiar and novel examples from four classes of natural and artificial stimuli. *Journal of Experimental Psychology: Animal Behavior Processes*, **14**, 219–234.

Biederman, I. (1972) Perceiving real-world scenes. *Science*, **177**, 77–80.

Binet, A. (1905) New methods for the diagnosis of the intellectual level of subnormals. *L'Analyse Psychologique*, **12**, 191–244. Trans. E.S. Kite (1916) in *The Development of Intelligence in Children*. Vineland, NJ: Publications of the Training School at Vineland.

Binet, A. and Simon, T. (1905) *The Development of Intelligence in Children*. Baltimore, MD: Williams & Wilkins. Reprinted (1983) Salem, NH: Ayer Co.

Bingman, V.P. and Able, K.P. (2002) Maps in birds: representational mechanisms and neural bases. *Current Opinion in Neurobiology*, **12**, 745–750.

Bingman, V.P., Bagnoli, P., Ioalè, P. and Cassini, G. (1984) Homing behaviour of pigeons after telencephalic ablations. *Brain, Behaviour and Evolution*, **24**, 94–108.

Bischof, N. (1978) On the phylogeny of human morality. In *Morality as a Biological Phenomenon*, ed. G. Stent. Berlin: Abakon, pp. 53–74.

Bischof-Köehler, D. (1985) Zür Phylogenese menschlicher Motivation (On the phylogeny of human motivation). In *Emotion und Reflexivität*, ed. L.H. Eckensberger and E.D. Lantermann. Vienna: Urban & Schwarzenberg, pp. 3–47.

Bitterman, M.E., Menzel, R., Fietz, A. and Schafer, S. (1983) Classical conditioning of proboscis extension in honeybees (*Apis mellifera*). *Journal of Comparative Psychology*, **97**, 107–119.

Bjørkøyli, T. and Rosell, F. (2002) A test of the dear enemy phenomenon in the eurasian beaver. *Animal Behaviour*, **63**, 1073–78.

Blagosklonov, K.N. (1974) Individual differences in behaviour of flycatchers. In *Ecological and Evolutionary Aspects of Animal Behaviour*, ed. P. Manteifel. Moscow: Nauka, pp. 119–126. (In Russian.)

Bloom, F.E., Lazerson, A. and Hofstadter, L. (1985) *Brain, Mind, and Behavior*. New York: W.H. Freeman.

Blough, D.S. (2002) Measuring the search image: expectation, detection and recognition in pigeon visual search. *Journal of Experimental Psychology: Animal Behavior Processes*, **28**, 397–405.

Boal, J.G., Wittenberg, K.M. and Hanlon, R.T. (2000) Observational learning does not explain improvement in predation tactics by cuttlefish (Mollusca: Cephalopoda). *Behavioural Processes*, **52**, 141–153.

Boch, R. (1956) Die Tänze der Bienen bei nahen und fernen Trachtquellen. *Zeitschrift für vergleichende Physiologie*, **40**, 289–320.

Boesch, C. (1991) Teaching among wild chimpanzees. *Animal Behaviour*, **41**, 530–532.

Boesch, C. (1995) Innovation in wild chimpanzees. *International Journal of Primatology*, **16**, 1–16.

Boesch, C. (1996) The question of culture. *Nature*, **379**, 207–208.

Boesch, C. (2003) Is culture a golden barrier between human and chimpanzee? *Evolutionary Anthropology*, **12**, 26–32.

Boesch, C. and Boesch, H. (1983) Optimisation of nut-cracking with natural hammers by wild chimpanzees. *Behaviour*, **83**, 256–286.

Bolnick, D.I., Svanback, R., Fordyce, J.A., *et al.* (2003) The ecology of individuals: incidence and implications of individual specialization. *American Naturalist*, **161**, 1–28.

Bonavita-Cougourdan, A. and Morel, L. (1984) Les activités antennaires au cours des contacts trophallactiques de la fourmi *Camponitus vagus* Scop.: ont-elles valeur de signal? *Insectes Sociaux*, **31**, 113–131.

Boomsma, J.J. and Grafen, A. (1991) Colony-level sex ratio selection in eusocial Hymenoptera. *Journal of Evolutionary Biology*, **3**, 383–407.

Borgia, G. (1995) Complex male display and female choice in the spotted bowerbird: specialised functions for different bower decorations. *Animal Behaviour*, **49**, 1291–1301.

Borgia, G. and Coleman, S. (2000) Co-option of male courtship signals from aggressive display in bowerbirds. *Proceedings of the Royal Society of London B*, **267**, 1735–1740.

Borgia, G. and Presgraves, D. (1998) Coevolution of elaborated male display traits in the spotted bowerbird: an experimental test of the threat reduction hypothesis. *Animal Behaviour*, **56**, 1121–1128.

Bovet, D. and Vauclair, J. (1998) Functional categorization of objects and of their pictures in baboons (*Papio anubis*). *Learning and Motivation*, **29**, 309–322.

Bovet, D. and Vauclair, J. (2000) Picture recognition in animals and in humans: a review. *Behavioral and Brain Research*, **109**, 143–165.

Bovet, D. and Vauclair, J. (2001) Judgement of conceptual identity in monkeys. *Psychonomic Bulletin and Review*, **8**, 470–475.

Bovet, D., Vauclair, J. and Blaye, A. (2005) Categorization and abstraction abilities in 3-year-old children: a comparison with monkey data. *Animal Cognition*, **8**, 53–59.

Bowlby, J. (1982) *Attachment and Loss*, vol. 1, *Attachment*, 2nd edn. New York: Basic Books. Reprinted (1999) with foreword by Allan N. Schore.

Boycott, B.B. and Young, J.Z. (1955) A memory system in *Octopus vulgaris*. *Proceedings of the Royal Society of London B*, **143**, 449–479.

Boyd, R. and Richerson, P.J. (2005) *The Origin and Evolution of Cultures*. Oxford, UK: Oxford University Press.

Boysen, S.T. (1993) Counting in chimpanzees: nonhuman principles and emergent properties of number. In *The Development of Numerical Competence: Animal and Human Models*, ed. S.T. Boysen and E.J. Capaldi. Hillsdale, NJ: Lawrence Erlbaum, pp. 39–59.

Boysen, S.T. and Hallberg, K.I. (2000) Primate numerical competence: contributions toward understanding nonhuman cognition. *Cognitive Science*, **24**, 423–443.

Boysen, S.T., Berntson, G.G., Hannan, M.B. and Cacioppo, J.T. (1996) Quantity-based interference and symbolic representation in chimpanzees (*Pan troglodytes*). *Journal of Experimental Psychology: Animal Behavior Processes*, **22**, 76–86.

Bradley, G.A. and Hinks, J.D. (1968) Ants, aphids, and jack pine in Manitoba. *Canadian Entomologist*, **100**, 40–50.

Bradshaw, G.A., Schore, A.N., Brown, J.L., Poole, J.H. and Moss, C.J. (2005) Elephant breakdown: social trauma – early disruption of attachment can affect the physiology, behaviour, and culture of animals and humans over generations. *Nature*, **433**, 801.

Braithwaite, V.A. and Girvan, J.R. (2003) Use of waterflow direction to provide spatial information in a small-scale orientation task. *Journal of Fish Biology A*, **63**, 74–83.

Brannon, E.M. and Terrace, H.S. (1998) Ordering of the numerosities 1 to 9 by monkeys. *Science*, **282**, 746–749.

Breland, K. and Breland, M. (1961) The misbehavior of organisms. *American Psychologist*, **16**, 681–684.

Breuer, T., Ndoundou-Hockemba, M. and Fishlock, V. (2005) First observation of tool use in wild gorillas. *PLoS Biology*, **3**, 2041–2043.

Brockmann, J. (2001) 50 years of International Ethological Conferences. *Contributions 27th International Ethological Conference*, ed. R. Apfelbach, M. Fendt, S. Krämer and B.M. Siemers. Tübingen: Blackwell Science, p. 1.

Brodbeck, D.R. (1994) Memory for spatial and local cues: a comparison of a food-storing and non-storing species. *Animal Learning and Behaviour*, **22**, 119–133.

Brodbeck, D.R. and Shettleworth, S.J. (1995) Memory for the location and color of a compound stimulus: comparison of a food-storing and a non-storing bird species. *Journal of Experimental Psychology: Animal Behavior Processes*, **21**, 64–77.

Brogden, W.J. (1939) Sensory pre-conditioning. *Journal of Experimental Psychology*, **25**, 323–332.

Broom, D.M. (1981) *Biology of Behaviour*. Cambridge, UK: Cambridge University Press.

Broom, D.M. (1996) Animal welfare defined in terms of attempts to cope with the environment. *Acta Agriculturae Scandinavica Section A: Animal Science Supplement*, **27**, 22–28.

Broom, D.M. (ed) (2001) *Coping with Challenge: Welfare in Animals including Humans*. Berlin: Dahlem University Press.

Broom, D.M. (2003) *The Evolution of Morality and Religion*. Cambridge, UK: Cambridge University Press.

Brower, L.P. (1985) New perspectives on the migration biology of the monarch butterfly, *Danaus plexippus* L. In *Migration: Mechanisms and Adaptive Significance, Contributions in Marine Science*, vol. 27, *Marine Science Institute*, ed. M.A. Rankin. Austin, TX: University of Texas, pp. 748–785.

Brown, C. and Laland, K. (2002) Social learning of a novel avoidance task in the guppy, *Poecilia reticulata*: conformity and social release. *Animal Behaviour*, **64**, 41–47.

Brown, C. and Laland, K. (2003) Social learning in fishes: a review. *Fish and Fisheries*, **4**, 280–288.

Brown, G.S. (1994) Spatial association learning by rufous hummingbirds (*Selasphorus rufus*): effects of relative spacing among stimuli. *Journal of Comparative Psychology*, **108**, 29–35.

Brown, J.L. (1970) Cooperative breeding and altruistic behaviour in the Mexican jay, *Aphelocoma ultramarina*. *Animal Behaviour*, **18**, 366–378.

Brown, J.L. (1974) Alternate routes to sociality in jays: with a theory for the evolution of altruism and communal breeding. *American Zoologist*, **14**, 63–80.

Brown, J.L. (1978) Avian communal breeding systems. *Annual Review of Ecology and Systematics*, **9**, 123–155.

Bshary, R., Wickler, W. and Fricke, H. (2002) Fish cognition: a primate's eye view. *Animal Cognition*, **5**, 1–13.

Buckley, K.W. (1982) The selling of a psychologist: John Broadus Watson and the application of behavioral techniques to advertising. *Journal of the History of the Behavioral Sciences*, **18**, 207–221.

Buckmaster, C.A., Eichenbaum, H., Amaral, D.G., Suzuki, W.A. and Rapp, P.R. (2004) Entorhinal cortex lesions disrupt the relational organisation of memory in monkeys. *Journal of Neuroscience*, **24**, 9811–9825.

Budaev, S.V., Zworykin, D.D. and Mochek, A.D. (1999) Individual differences in parental care and behavioural profile in the convict cichlid: a correlational study. *Animal Behaviour*, **58**, 195–202.

Bugnyar, T. and Kotrschal, K. (2002) Scrounging tactics in free-ranging ravens. *Ethology*, **108**, 993–1009.

Bugnyar, T. and Kotrschal, K. (2004) Leading a conspecific away from food in ravens, *Corvus corax*. *Animal Cognition*, **7**, 69–76.

Bugnyar, T., Kijne, M. and Kotrschal, K. (2001) Food calling in ravens: are yells referential signals? *Animal Behaviour*, **61**, 949–958.

Burns, J. G. and Thomson, J. D. (2006) A test of spatial memory and movement patterns of bumblebees at multiple spatial and temporal scales. *Behavioral Ecology*, **17**, 48–55.

Burrows, R. (1992) Rabies in wild dogs. *Nature*, **359**, 27.

Butovskaya, M., Kozintsev, A. and Welker, C. (1996) Conflict and reconciliation in two groups of crab-eating monkeys differing in social status by birth. *Primates*, **37**, 259–268.

Butterworth, B. (1999) *The Mathematical Brain*. London: Macmillan.

Byrne, R. W. (1995) *The Thinking Ape*. Oxford, UK: Oxford University Press.

Byrne, R. W. (2002) Imitation of novel complex actions: what does the evidence from animals mean? *Advances in the Study of Behavior*, **31**, 77–105.

Byrne, R. W. and Byrne, J. M. E. (1993) Variability and standardisation in the complex leaf-gathering tasks of mountain gorillas (*Gorilla g. beringei*). *American Journal of Primatology*, **31**, 241–261.

Byrne, R. W. and Russon, A. E. (1998) Learning by imitation: a hierarchical approach. *Behavioral and Brain Sciences*, **21**, 667–721.

Byrne, R. W. and Tomasello, M. (1995) Do rats ape? *Animal Behaviour*, **50**, 1417–1420.

Byrne, R. W. and Whiten, A. (1988) *Machiavellian Intelligence: Social Expertise and the Evolution of Intellect in Monkeys, Apes and Humans*. Oxford, UK: Claredon Press.

Byrne, R. W. and Whiten, A. (1991) Computation and mindreading in primate tactical deception. In *Natural Theories of Mind*, ed. A. Whiten. Oxford, UK: Blackwell, pp. 127–141.

Cain, P. and Malwal, S. (2002) Landmark use and development of navigation behaviour in the weakly electric fish *Gnathonemus petersii* (Mormyridae; Teleostei). *Journal of Experimental Biology*, **205**, 3915–3923.

Caldwell, C. A. and Whiten, A. (2002) Evolutionary perspectives on imitation: is a comparative psychology of social learning possible? *Animal Cognition*, **5**, 193–208.

Caldwell, C. A., Whiten, A. and Morris, K. D. (1999) Observational learning in the marmoset monkey, *Callithrix jacchus*. In *Proceedings of the AISB'99 Symposium on Imitation in Animals and Artifacts*, Society for the Study of Artificial Intelligence and Simulation of Behaviour, Edinburgh, UK, pp. 27–31.

Call, J. (2000) Representing space and objects in monkeys and apes. *Cognitive Science*, **24**, 397–422.

Call, J. (2002) A fish-eye lens for comparative studies: broadening the scope of animal cognition. *Animal Cognition*, **5**, 15–16.

Call, J. and Tomasello, M. (1994) The social learning of tool use by orangutans (*Pongo pygmaeus*). *Human Evolution*, **9**, 297–313.

Call, J. and Tomasello, M. (1995) The use of social information in the problem-solving of orangutans (*Pongo pygmaeus*) and human children (*Homo sapiens*). *Journal of Comparative Psychology*, **109**, 308–320.

Call, J., Hare, B. A. and Tomasello, M. (1994) Chimpanzee gaze following in an object-choice task. *Animal Cognition*, **1**, 89–99.

Calvin, W. H. (1989) *The Cerebral Symphony: Seashore Reflections on the Structure of Consciousness*. New York: Bantam Books.

Candolin, U. and Voigt, H.-R. (2003) Do changes in risk-taking affect habitat shifts of sticklebacks? *Behavioral Ecology and Sociobiology*, **55**, 42–49.

Cannon, W. B. (1932) *The Wisdom of the Body*. New York: W. W. Norton.

Caro, T. M. (1981) Predatory behaviour and social play in kittens. *Behaviour*, **76**, 1–24.

Caro, T. M. (1987) Cheetah mothers' vigilance: looking out for prey or for predators? *Behavioral Ecology and Sociobiology*, **20**, 351–361.

Caro, T. M. (1994) *Cheetahs of the Serengeti Plains: Group Living in an Asocial Species*. Chicago, IL: University of Chicago Press.

Caro, T. M. (2005) *Antipredator Defenses in Birds and Mammals*. Chicago, IL: University of Chicago Press.

Caro, T. M. and Hauser, M. D. (1992) Is there teaching in non-human animals? *Quarterly Review of Biology*, **67**, 151–174.

Cartwright, J.-A. (2002) *Determinants of Animal Behavior*. New York: Routledge.

Catania, A. C. (1998) *Learning*, 4th edn. Upper Saddle River, NJ: Prentice-Hall.

Catania, A. C. and Laties, V. G. (1999) Pavlov and Skinner: two lives in science. *Journal of the Experimental Analysis of Behavior*, **72**, 455–461.

Catania, K. C. and Remple, M. S. (2002) Somatosensory cortex dominated by the representation of teeth in the naked mole-rat brain. *Proceedings of the National Academy of Sciences of the USA*, **99**, 5692–5697.

Cate, C. ten and Bateson, P. (1989) Sexual imprinting and a preference for 'supernormal' partners in Japanese quail. *Animal Behaviour*, **38**, 356–358.

Cate, C. ten and Vos, D. (1999) Sexual imprinting and evolutionary processes in birds. *Advances in the Study of Behavior*, **28**, 1–31.

Cavalli-Sforza, L. L. (1997) Genes, peoples and languages. *Proceedings of the National Academy of Sciences of the USA*, **94**, 7719–7724.

Chance, M. R. A. (1962) An interpretation of some agonistic postures: the role of 'cut-off' acts and postures. *Symposium of the Zoological Society of London*, **8**, 71–89.

Chappell, J. and Kacelnik, A. (2004) Selection of tool diameter by New Caledonian crows *Corvus moneduloides*. *Animal Cognition*, **7**, 121–127.

Chapuis, N. (1987) Detour and shortcut abilities in several species of mammals. In *Cognitive Processes and Spatial Orientation in Animal and Man*, vol. 1, *Experimental Animal Psychology and Ethology*, ed. P. Ellen and C. Thinus-Blanc. Dordrecht the Netherlands: Martinus Nijhoff, pp. 97–106.

Chapuis, N. and Scardigli, P. (1993) Shortcut ability in hamsters. *Animal Learning and Behaviour*, **21**, 255–265.

Charnov, E. L. (1976) Optimal foraging: the marginal value theorem. *Theoretical Population Biology*, **9**, 129–136.

Charrier, I., Mathevon, N. and Jouventin, P. (2003) Vocal signature recognition of mothers by fur seal pups. *Animal Behaviour*, **65**, 543–550.

Chen, S., Swartz, K. B. and Terrace, H. S. (1997) Knowledge of the ordinal position of list items in rhesus monkeys. *Psychological Science*, **8**, 80–86.

Cheney, D. L. and Seyfarth, R. M. (1980) Vocal recognition in free-ranging vervet monkeys. *Animal Behaviour*, **28**, 362–367.

Cheney, D. L. and Seyfarth, R. M. (1990) *How Monkeys See the World: Inside the Mind of Another Species*. Chicago, IL: University of Chicago Press.

Cheney, D. L. and Seyfarth, R. M. (1998) Why animals don't have language. In *The Tanner Lectures on Human Values*, ed. G. B. Pearson. Salt Lake City, UT: University of Utah Press, pp. 175–209.

Cheney, D. L. and Seyfarth, R. M. (2003) The recognition of other individuals' social relationships. In *Kinship and Behavior in Primates*, ed. B. Chapais and C. M. Berman. Oxford, UK: Oxford University Press, pp. 347–364.

Cheng, K. (1986) A purely geometric module in the rat's spatial representation. *Cognition*, **23**, 149–178.

Cheng, K. (2000) How honeybees find a place: lessons from a simple mind. *Animal Learning and Behaviour*, **28**, 1–15.

Cheng, K. (2005) Reflections on geometry and navigation. *Connection Science*, **17**, 5–21.

Cheng, K. and Spetch, M. L. (1998) Mechanisms of landmark use in mammals and birds. In *Spatial Representation in Animals*, ed. S. Healy. Oxford, UK: Oxford University Press, pp. 1–17.

Cheng, K., Peña, J., Porter, M. A. and Irwin, J. D. (2002) Self control in honeybees. *Psychonomic Bulletin and Review*, **9**, 259–263.

Chevalier-Skolnikoff, S. (1976) The ontogeny of primate intelligence and its implications for communicative potential: a preliminary report. *Annals of the New York Academy of Sciences*, **280**, 173–211.

Chevalier-Skolnikoff, S. and Liska, J. (1993) Tool use by wild and captive elephants. *Animal Behaviour*, **46**, 209–219.

Chittka, L. and Geiger, K. (1995) Can honeybees count landmarks? *Animal Behaviour*, **49**, 159–164.

Chittka, L., Williams, M. N., Rasmussen, H. and Thomson, J. D. (1999) Navigation without vision: bumblebee orientation in complete darkness. *Proceedings of the Royal Society of London B*, **266**, 45–50.

Choe, J. C. and Crespi, B. J. (eds.) (1997) *The Evolution of Social Behavior in Insects and Arachnids*. Cambridge, UK: Cambridge University Press.

Chomsky, N. (1968) *Language and Mind*. New York: Harcourt Brace.

Chomsky, N. (1986) *Knowledge of Language: Its Nature, Origins, and Use*. New York: Praeger.

Clark, R. B. (1960) Habituation of the polychaete *Nereis* to sudden stimuli. II. Biological significance of habituation. *Animal Behaviour*, **8**, 92–103.

Clarke, M. F. and Kramer, D. L. (1994) Scatter-hoarding by a larder-hoarding rodent: intraspecific variation in the hoarding behavior of the eastern chipmunk, *Tamias striatus*. *Animal Behaviour*, **48**, 299–308.

Clayton, D. A. (1978) Socially facilitated behavior. *Quarterly Review of Biology*, **53**, 373–391.

Clayton, N. S. (1995) Development of memory and the hippocampus: comparison of food-storing and non-storing birds on a one-trial associative memory task. *Journal of Neuroscience*, **15**, 2796–2807.

Clayton, N. S. and Dickinson, A. (1998) Episodic-like memory during cache recovery by scrub jays. *Nature*, **395**, 272–274.

Clayton, N. S. and Krebs, J. R. (1994) Memory for spatial and object-specific cues in food-storing and non-storing species of birds. *Journal of Comparative Physiology A*, **174**, 371–379.

Clayton, N. S., Bussey, T. J. and Dickinson, A. (2003a) Can animals recall the past and plan for the future? *Nature Reviews: Neuroscience*, **4**, 685–691.

Clayton, N. S., Yu, K. S. and Dickinson, A. (2003b) Interacting cache memories: evidence for flexible memory use by western scrub jays (*Aphelocoma californica*). *Journal of Experimental Psychology: Animal Behavior Processes*, **29**, 14–22.

Clutton-Brock, T. H. (1974) Primate social organization and ecology. *Nature*, **250**, 539–542.

Clutton-Brock, T. H. (2002) Breeding together: kin selection and mutualism in cooperative vertebrates. *Science*, **296**, 69–72.

Clutton-Brock, T. H. and Harvey, P. (1977) Primate ecology and social organisation. *Journal of Zoology, London*, **183**, 1–39.

Clutton-Brock, T. H. and Parker, G. A. (1995) Punishment in animal societies. *Nature*, **373**, 209–216.

Clutton-Brock, T. H, O'Riain, M. J., Brotherton, P. N. M., *et al.* (1999) Selfish sentinels in cooperative mammals. *Science*, **284**, 1640–1644.

Clutton-Brock, T. H., Brotherton, P. N. M., O'Rian, M. J., *et al.* (2000) Individual contributions to baby-sitting in a cooperative mongoose *Suricatta suricatta*. *Proceedings of the Royal Society of London B*, **267**, 201–205.

Clutton-Brock, T. H., Russell, A. F., Sharpe, L. L., *et al.* (2002) Evolution and development of sex differences in cooperative behavior in meerkats. *Science*, **297**, 253–256.

Collen, P. and Gibson, R. J. (2001) The general ecology of beavers (*Castor* spp.) as related to their influence on stream ecosystems and riparian habitats, and the subsequent effects on fish: a review. *Reviews in Fish Biology and Fisheries*, **10**, 439–461.

Collett, M., Collett, T. S., Chameron, S. and Wehner, R. (2003) Do familiar landmarks reset the global path integration system of desert ants? *Journal of Experimental Biology*, **206**, 877–882.

Connor, R. C., Smolker, R. A. and Richards, A. F. (1992) Two levels of alliance formation among bottlenose dolphins (*Tursiops* sp.). *Proceedings of the National Academy of Sciences of the USA*, **89**, 987–990.

Connor, R. C., Heithaus, M. R. and Barre, L. M. (2001) Complex social structure, alliance stability and mating access in a bottlenose dolphin 'super-alliance'. *Proceedings of the Royal Society of London B*, **268**, 263–267.

Cook, M. and Mineka, S. (1990) Selective associations in the observational conditioning of fear in rhesus monkeys. *Journal of Experimental Psychology: Animal Behavior Processes*, **16**, 372–389.

Cook, R. G. (ed.) (2001) *Avian Visual Cognition*. Medford, MA: Comparative Cognition Press. Available online: www.pigeon.psy.tufts.edu/avc/.

Cook, R. G. and Katz, J. S. (1999) Dynamic object perception in pigeons. *Journal of Experimental Psychology: Animal Behavior Processes*, **25**, 194–210.

Cook, R. G. and Tauro, T. L. (1999) Object-goal positioning influences spatial representation in rats. *Animal Cognition*, **2**, 55–62.

Cook, R. G., Brown, M. F. and Riley, D. A. (1985) Flexible memory processing by rats: use of prospective and retrospective information in the radial maze. *Journal of Experimental Psychology: Animal Behavior Processes*, **11**, 453–469.

Cook R. G., Kelly, D. M. and Katz, J. S. (2003) Successive two-item same–different discrimination and concept learning by pigeons. *Behavioural Processes*, **62**, 125–144.

Coolen, I., Day, R. L. and Laland, K. N. (2003) Species difference in adaptive use of public information in sticklebacks. *Proceedings of the Royal Society of London B*, **270**, 2413–2419.

Coon, D. J. (1994) 'Not a creature of reason': the alleged impact of Watsonian behaviorism on advertising in the 1920s. In *Modern Perspectives on John B. Watson and Classical Behaviorism*, ed. J. T. Todd and E. K. Morris. Westport, CT: Greenwood Press, pp. 37–63.

Cords, M. and Thurnheer, S. (1993) Reconciliation with valuable partners by long-tailed macaques. *Ethology*, **93**, 315–325.

Cornell, E. H. and Heth, C. D. (2004) Memories of travel: dead reckoning within the cognitive map. In *Human Spatial Memory: Remembering Where*, ed. G. Allen. Mahwah, NJ: Lawrence Erlbaum, pp. 191–215.

Courchamp, F., Rasmussen, G. S. A. and Macdonald D. W. (2002) Small pack size imposes a trade-off between hunting and pup-guarding in the painted hunting dog *Lycaon pictus*. *Behavioural Ecology*, **13**, 20–27.

Coussi-Korbel, S. (1994) Learning to outwit a competitor in mangabeys. *Journal of Comparative Psychology*, **108**, 164–171.

Couzin, I. D., Krause, J., Franks, N. R. and Levin, S. A. (2005) Effective leadership and decision making in animal groups on the move. *Nature*, **433**, 513–516.

Cowles, J. T. (1937) Food-tokens for learning by chimpanzees. *Comparative Psychology Monographs*, **14**(5), 1–96.

Crook, J. H. (1970) The socio-ecology of primates. In *Social Behaviour in Birds and Mammals*, ed. J. H. Crook. New York: Academic Press, pp. 103–166.

Crowcroft, P. (1966) *Mice All Over*. London: G. T. Foulis.

Crowcroft, P. and Rowe, F. P. (1963) Social organization and territorial behavior in the wild house mouse

(*Mus musculus* L.). *Proceedings of Zoological Society of London B*, **140**, 517–531.

Crozier, R. H. and Pamilo, P. (1996) Sex allocation in social insects: problems in prediction and estimation. In *Evolution and Diversity of Sex Ratio in Insects and Mites*, ed. D. L. Wrensch and M. A. Ebbert. New York: Chapman & Hall, pp. 369–383.

Cruze, W. W. (1935) Maturation and learning in chicks. *Journal of Comparative and Physiological Psychology*, **19**, 371–409.

Curio, E. (1988) Cultural transmission of enemy recognition by birds. In *Social Learning: Psychological and Biological Perspectives*, ed. T. R. Zentall and B. G. Galef, Jr. Hillsdale, NJ: Lawrence Erlbaum, pp. 75–97.

Curtis, H. S. (1900) Automatic movements of the larynx. *American Journal of Psychology*, **11**, 237–39.

Custance, D. M., Whiten, A. and Bard, K. A. (1995) Can young chimpanzees imitate arbitrary actions? Hayes and Hayes (1952) revisited. *Behaviour*, **132**, 839–58.

Cuvier, F. (1825) Essai sur la domesticité des mammifères, précédé de considérations sur les divers états des animaux, dans lesquels il nous est possible d'étudier leurs actions. *Mémoires du Muséum d'histoire naturelle*, **13**, 415–416.

Damasio, A. R. (1994) *Descartes' Error: Emotion, Reason and the Human Brain*. New York: Gosset & Putnam Press.

D'Amato, M. R. and Colombo, M. (1988) Representation of serial order in monkeys (*Cebus apella*). *Journal of Experimental Psychology*, **14**, 131–139.

D'Amato, M. R. and Worshman, R. W. (1972) Delayed matching in the capuchin monkey with brief sample durations. *Learning and Motivation*, **3**, 304–312.

Danchin, E., Giraldeau, L. A., Valone, T. J. and Wagner, R. H. (2004) Public information: from nosy neighbors to cultural evolution. *Science*, **305**, 487–491.

Darmaillacq, A.-S., Dickel, L., Chichery, M.-P., Agin, V. and Chichery, R. (2004) Rapid taste-aversion learning in adult cuttlefish, *Sepia officinalis*. *Animal Behaviour*, **68**, 1291–1298.

Darwin, C. (1840) *Zoology of the Voyage of the* Beagle. London: John Murray. Reprinted (1989) ed. J. Browne and M. Neve. New York: Penguin.

Darwin, C. (1859) *On the Origin of Species*. London: John Murray. Facsimile reprint (1964) with an introduction by Ernst Mayr. Cambridge, MA: Harvard University Press.

Darwin, C. (1871) *The Descent of Man, and Selection in Relation to Sex*, 2 vols. London: John Murray.

Darwin, C. (1872) *The Expression of the Emotions in Man and Animal*. London: John Murray. Reprinted (1998) with an introduction, afterword, and commentaries by Paul Ekman. New York: Oxford University Press.

Darwin, C. (1873) On the origin of certain instincts. *Nature*, **7**, 417–418.

Darwin, C. and Wallace, A. R. (1858) 'On the tendency of species to form varieties; and on the perpetuation of varieties and species by natural means of selection' by Charles Darwin and Alfred Wallace; communicated by Sir Charles Lyell and Joseph D. Hooker to the LSL meeting of 1 July 1858. *Journal of the Proceedings of the Linnean Society, Zoology*, **3**, 45–62.

Dasser, V. (1988) A social concept in Java monkeys. *Animal Behaviour*, **36**, 225–230.

Davenport, R. K. and Rogers, C. M. (1968) Intellectual performance of differentially reared chimpanzees. I. Delayed response. *American Journal of Mental Deficiency*, **72**, 674–680.

Davey, G. C. L. (1989) *Ecological Learning Theory*. London: Routledge.

Davey, G. C. L. (1995) Preparedness and phobias: specific evolved associations or a generalized expectancy bias? *Behavioral and Brain Sciences*, **18**, 289–325.

Davidson, D. (1984) *Inquiries into Truth and Interpretation*. Oxford, UK: Clarendon Press.

Davis, H. P. and Pérusse, R. (1988) Numerical competence in animals: definitional issues, current evidence, and a new research agenda. *Behavioral and Brain Sciences*, **11**, 561–615.

Davies, H. P. and Squire, L. (1984) Protein synthesis and memory: a review. *Psychology Bulletin*, **96**, 518–559.

Dawkins, R. (1976) *The Selfish Gene*. Oxford, UK: Oxford University Press.

Dawkins, R. (1982) *The Extended Phenotype*. San Francisco, CA: W. H. Freeman.

Dawson, B. V. and Foss, B. M. (1965) Observational learning in budgerigars. *Animal Behaviour*, **13**, 470–474.

Deaner, R. O., Nunn, C. L. and van Schaik, C. P. (2000) Comparative tests of primate cognition: different scaling methods produce different results. *Brain, Behaviour and Evolution*, **55**, 44–52.

De Blois, S. T., Novak, M. A. and Bond, M. (1998) Object permanence in orangutans (*Pongo pygmaeus*) and squirrel monkeys (*Saimiri sciureus*), *Journal of Comparative Psychology*, **112**, 137–152.

Deecke, V. B., Ford, J. K. B. and Spong, P. (2000) Dialect changes in resident killer whales: implications for vocal learning and cultural transmission. *Animal Behaviour*, **60**, 629–638.

Dehaene-Lambertz, G. and Dehaene, S. (1994) Speed and cerebral correlates of syllable discrimination in infants. *Nature*, **370**, 292–295.

Delfour, F. and Marten, K. (2001) Mirror image processing in three marine mammal species: killer whales (*Orcinus orca*), false killer whales (*Pseudorca crassidens*) and California sea lions (*Zalophus californianus*). *Behavioural Processes*, **53**, 181–190.

DeLillo, C. and Visalberghi, E. (1994) Transfer index and mediational learning in tufted capuchins (*Cebus apella*). *International Journal of Primatology*, **15**, 275–287.

Delius, J. D. (1994) Comparative cognition of identity. In *International Perspectives on Psychological Science*, vol. 1, ed. P. Bertelson, P. Eelen, and G. D'Ydewalle. Hillsdale, NJ: Lawrence Erlbaum, pp. 25–40.

Delius, J. D. and Habers, G. (1978) Symmetry: can pigeons conceptualize it? *Behavioral Biology*, **22**, 336–342.

Dennett, D. (1987) *The Intentional Stance*. Cambridge, MA: MIT Press.

Denny, M. R. (1958) The 'Kamin effect' in avoidance conditioning. *American Psychologist*, **13**, 419–425.

Dépy, D., Fagot, J. and Vauclair, J. (1997) Categorization of three-dimension stimuli by humans and baboons: search for prototype effects. *Behavioural Processes*, **39**, 299–306.

Dépy, D., Fagot, J. and Vauclair, J. (1999) Processing of above/below categorical spatial relations by baboons (*Papio papio*). *Behavioural Processes*, **48**, 1–9.

Deryagina, M. A., Butovskaya, M. L. and Semenov, A. G. (1989) Evolutionary changes in communication of primates and hominids (in connection with origin of speech). *Biological Prerequisites for Anthroposociogenesis*, **1**, 98–129. (In Russian.)

Detrain, C. and Deneubourg, J. L. (1997) Scavenging by *Pheidole pallidula*: a key for understanding decision-making systems in ants. *Animal Behaviour*, **53**, 537–547.

Dewsbury, D. A. (1978) *Comparative Animal Behavior*. New York: McGraw-Hill.

Dewsbury, D. A. (1985) *Leaders in the Study of Animal Behavior*. Lewisburg, PA: Bucknell University Press.

Diamond, A. (1990) Developmental time course in human infants and infant monkeys, and the neural bases of inhibitory control in reaching. *Annals of the New York Academy of Sciences*, **608**, 637–676.

Diamond, I. T., Chow, K. L. and Neff, W. D. (1958) Degeneration of caudal medial geniculate body following cortical lesions ventral to auditory area II in cat. *Journal of Comparative Neurology*, **109**, 349–362.

Dias, R., Robbins, T. W. and Roberts, A. C. (1997) Dissociable forms of inhibitory control within prefrontal cortex with an analog of the Wisconsin Card Sort Test: restriction to novel situations and independence from 'on-line' processing. *Journal of Neuroscience*, **17**, 9285–9297.

Dickel, L., Boal, J. G. and Budelmann, B. U. (2000) The effect of early experience on learning and memory in cuttlefish. *Developmental Psychobiology*, **36**, 101–110.

Dickinson, A., Campos, J., Varga, Z. L. and Balleine, B. (1996) Bidirectional instrumental conditioning. *Quarterly Journal of Experimental Psychology B*, **48**, 289–306.

Diez-Chamzio, V., Sterio, D. and Mackintosh, N. J. (1985) Blocking and overshadowing between intra-maze and extra-maze cues: a test of the independence of locale and guidance learning. *Quarterly Journal of Experimental Psychology B*, **37**, 235–253.

Dilger, W. C. (1962) The behavior of lovebirds. *Scientific American*, **206**, 88–98.

Dlussky, G. M., Volcit, O. V. and Sulhanov, A. V. (1978) Organisation of group foraging in ants of the genus *Myrmica*. *Zoological Journal*, **57**, 65–77. (In Russian with English summary.)

Domarus, E. von (1944) The specific laws of logic in schizophrenia. In *Language and Thought in Schizophrenia*, ed. J. S. Kasanin. Berkeley, CA: University of California Press, pp. 135–137.

Doolan, S. P. and MacDonald, D. W. (1999) Co-operative rearing by slender tailed meerkats (*Suricata suricatta*) in the southern Kalahari. *Ethology*, **105**, 851–866.

Doré, F. Y., Fiset, S., Goulet, S., Dumas, M.-C. and Cagnon, S. (1996) Search behaviour in cats and dogs: interspecific differences in working memory and spatial cognition. *Animal Learning and Behaviour*, **24**, 142–149.

Douglas-Hamilton, I. and Douglas-Hamilton, O. (1975) *Among the Elephants*. New York: Viking.

Duffy, J. E. (1996) Eusociality in a coral-reef shrimp. *Nature*, **381**, 512–514.

Duffy, J. E. (2003) Eusociality in sponge-dwelling shrimps. In *Genes, Behavior, and Evolution in Social Insects*, ed. T. Kilkuchi, N. Azumi and S. Higashi. Sapporo, Japan: Hokkaido University Press, pp. 217–252.

Duffy, J. E., Morrison, C. L. and Macdonald, K. S. (2002) Colony defense and behavioral differentiation in the eusocial shrimp *Synalpheus regalis*. *Behavioral Ecology and Sociobiology*, **51**, 488–495.

Dugatkin, L. A. (1992) Tendency to inspect predators predicts mortality risk in the guppy (*Poecilia reticulata*). *Behavioral Ecology*, **3**, 124–127.

Dugatkin, L. A. (1997) *Cooperation among Animals: An Evolutionary Perspective*. New York: Oxford University Press.

Dugatkin, L.A. (2005) *Principles of Animal Behavior*. New York: W.W. Norton.

Dugatkin, L.A. and Godin, J.-G.J. (1993) Female mate choice copying in the guppy (*Poecilia reticulata*): age-dependent effect. *Behavioral Ecology*, **4**, 289–292.

Dujardin, F. (1850) Mémoire sur le système nerveux des insectes. *Annales des sciences naturelles – zoologie et biologie animale*, **14**, 195–206.

Dukas, R. (2004) Evolutionary biology of animal cognition. *Annual Review of Ecology, Evolution and Systematics*, **35**, 347–374.

Dukas, R. and Real, L.A. (1993) Effects of recent experience on foraging decisions by bumblebees. *Oecologia*, **94**, 244–246.

Dunbar, R. (1996) *Grooming, Gossip and the Evolution of Language*. London: Faber & Faber.

Durrant, J.R. (1986) The making of ethology: the Association for the Study of Animal Behaviour. *Animal Behaviour*, **34**, 1601–1616.

Dyer, F.C. (1987) Memory and sun compensation in honey bees. *Journal of Comparative Physiology A*, **160**, 621–633.

Dyer, F.C. (1991) Bees acquire route-based memories but not cognitive maps in a familiar landscape. *Animal Behaviour*, **41**, 239–246.

Dyer, F.C. (1996) Spatial memory and navigation by honeybees on the scale of the foraging range. *Journal of Experimental Biology*, **199**, 147–154.

Ebbinghaus, H. (1885) *Über das Gedächtnis*. Leipzig, Germany: Duncker & Humbolt. Translated (1964) *Memory: A Contribution to Experimental Psychology*. New York: Dover.

Ebenholtz, S.M. (1963) Serial learning: position learning and sequential associations. *Journal of Experimental Psychology*, **66**, 353–362.

Edwards, C.A., Jagielo, J.A., Zentall, T.R. and Hogan, D.E. (1982) Acquired equivalence and distinctiveness in matching to sample by pigeons: mediation by reinforcer-specific expectancies. *Journal of Experimental Psychology: Animal Behavior Processes*, **8**, 244–259.

Eggleton, P. (2000) Global patterns of termite diversity. In *Termites: Evolution, Sociality, Symbioses, Ecology*, ed. T. Abe, D.E. Bignell and M. Higashi. Dordrecht, the Netherlands: Kluwer Academic Publishing, pp. 25–51.

Ehrlich, P.R., Dobkin, D.S. and Wheye, D. (1988) *The Birder's Handbook*. New York: Simon & Schuster.

Ehrenfels, C. von (1890) Über Gestaltqualitäten. *Vierteljahrsschrift für wissenschaftliche Philosophie*, **14**, 249–292.

Eibl-Eibesfeldt, I. (1961a) The interactions of unlearned behavioural patterns and learning in mammals. In *Brain Mechanisms and Learning*, ed. J.F. Delafresnage. Oxford, UK: Blackwell Scientific Publications, pp. 53–73.

Eibl-Eibesfeldt, I. (1961b) Über den Werkzeuggebrauch des Spechtfinken *Camarhynchus pallidus* (Sclater and Salvin). *Zoological Tierpsychology*, **18**, 343–346.

Eibl-Eibesfeldt, I. (1967) Concepts of ethology and their significance for the study of human behaviour. In *Early Behaviour: Comparative and Developmental Approaches*, ed. H.W. Swenson. New York: John Wiley, pp. 127–146.

Eibl-Eibesfeldt, I. (1970) *Ethology: The Biology of Behavior*. New York: Holt, Rinehart and Winston.

Eibl-Eibesfeldt, I. (1971) *Love and Hate: The Natural History of Behavior Patterns*. New York: Holt, Rinehart and Winston.

Eibl-Eibesfeldt, I. (1979) *The Biology of Peace and War*. New York: Viking Press.

Eibl-Eibesfeldt, I. (1989) *Human Ethology*. New York: Aldine de Gruyter.

Eibl-Eibesfeldt, I. and Salter, F.K. (eds.) (1998) *Indoctrinability, Ideology and Warfare: Evolutionary Perspectives*. Oxford, UK: Berghahn Books.

Eichenbaum, H. (2004) Hippocampus: cognitive processes and neural representations that underlie declarative memory. *Neuron*, **44**, 109–120.

Eimas, P.D. (1985) The perception of speech in early infancy. *Scientific American*, **252**, 45–52.

Emerson, A.E. (1952) The biogeography of termites. *Bulletin of the American Museum of Natural History*, **99**, 217–225.

Emery, N.J. and Clayton, N.S. (2001) Effects of experience and social context on prospective caching strategies in scrub jays. *Nature*, **414**, 443–446.

Emery, N.J., Dally, J. and Clayton, N.S. (2004) Western scrub-jays (*Aphelocoma californica*) use cognitive strategies to protect their caches from thieving conspecifics. *Animal Cognition*, **7**, 37–43.

Emlen, S.T. (1991) The evolution of cooperative breeding in birds and mammals. In *Behavioural Ecology: An Evolutionary Approach*, 3rd edn, ed. J. Krebs and N.B. Davies. Oxford, UK: Blackwell, pp. 301–337.

Emlen, S.T. (1995) An evolutionary theory of the family (review article). *Proceedings of the National Academy of Sciences of the USA*, **92**, 8092–8099.

Emlen, S.T., Wrege, P.H. and Demong, N.J. (1995) Making decisions in the family: an evolutionary perspective. *American Scientist*, **83**, 148–157.

Emmerton, J. (2001) Birds' judgements of number and quantity. In *Avian Visual Cognition*, ed. R.G. Cook. Available online: www.pigeon.psy.tufts.edu/avc/emmerton/.

Emmerton, J., Lohmann, A. and Niemann, J. (1997) Pigeons' serial ordering of numerosity with visual arrays. *Animal Learning and Behaviour*, **25**, 234–244.

Engh, A.L, Siebert, E.R., Greenberg, D.A. and Holekamp, K.E. (2005) Patterns of alliance formation and post-conflict aggression indicate spotted hyaenas recognize third party relationships. *Animal Behaviour*, **69**, 209–217.

Ens, B.J., Piersma, T., Wolff, W.J. and Swarts, L. (eds.) (1990) Homeward bound: problems waders face when migrating from the Banc d'Arguin, Mauritania, to their northern breeding grounds in spring. *Ardea* (special issue) **78**, 1–36.

Epstein, R. (1987) Reflections on thinking in animals. In *Language, Cognition, and Consciousness: Integrative Levels*, ed. G. Greenberg and E. Tobach. Hillsdale, NJ: Lawrence Erlbaum, pp. 19–29.

Epstein, R., Kirshnit, C.E., Lanza, R.P. and Rubin, L.C. (1984) 'Insight' in the pigeon: antecedents and determinants of an intelligent performance. *Nature*, **308**, 61–62.

Erhart, C., Coelho A. and Bramblett, C. (1997) Kin recognition by paternal half-siblings in captive *Papio cynocephalus*. *American Journal of Primatology*, **43**, 147–157.

Esch, H. (1961) Über die Schallerzeugung beim Werbetanz der Honingbiene. *Zeitschrift für Vergleichende Physiologie*, **45**, 1–11.

Esch, H. (1964) Beiträge zum Problem der Entfernungwesung in den Schwanzeltänzen der Honigbiene. *Zeitschrift für Vergleichende Physiologie*, **48**, 534–546.

Eskov, E.K. (1979) *Acoustic Signalling in Social Insects*. Moscow: Nauka. (In Russian.)

Etienne, A.S. (1973) Searching behaviour towards a disappearing prey in the domestic chick as affected by preliminary experience. *Animal Behaviour*, **21**, 749–761.

Etienne, A.S. and Jeffery, K.J. (2004) Path integration in mammals. *Hippocampus*, **14**, 180–192.

Evans, C.S. and Evans, L. (1999) Chicken food calls are functionally referential. *Animal Behaviour*, **58**, 307–319.

Evans, C.S. and Marler, P. (1991) On the use of video images as social stimuli in birds: audience effects on alarm calling. *Animal Behaviour*, **41**, 17–26.

Evans, R.I. (1976) *The Making of Psychology*. New York: Knopf.

Evans, T.A. (1999) Kin recognition in a social spider. *Proceedings of the Royal Society of London B*, **266**, 287–292.

Evans, W.E. and Bastian, J. (1969) Marine mammal communication: social and ecological factors. In *The Biology of Marine Mammals*, ed. H.T. Andersen. New York: Academic Press, pp. 425–475.

Faaborg, J., Parker, P.G., Delay, L., *et al.* (1995) Confirmation of cooperative polyandry in the Galapagos hawk (*Buteo galapagoensis*). *Behavioral Ecology and Sociobiology*, **36**, 83–90.

Fabre, J.H. (1879) *Souvenirs Entomologiques*, 1 Sér. Paris: Delagrave.

Fagot, J., Wasserman, E.A. and Young, M.E. (2001) Discriminating the relation between relations: the role of entropy in abstract conceptualization by baboons (*Papio papio*) and humans (*Homo sapiens*). *Journal of Experimental Psychology: Animal Behavior Processes*, **17**, 316–328.

Falls, J.B. and Brooks, R.J. (1975) Individual recognition by song in white-throated sparrows. II. Effects of location. *Canadian Journal of Zoology*, **53**, 1412–1420.

Farrah, D. (1976) Picture memory in the chimpanzee. *Perceptual and Motor Skills*, **25**, 305–315.

Fawcett, T.W., Skinner, A.M.J. and Goldsmith, A.R. (2002) A test of imitative learning in starlings using a two-action method with an enhanced ghost control. *Animal Behaviour*, **64**, 547–556.

Fersen, L.V. and Lea, S.E.G. (1990) Category discrimination by pigeons using five polymorphous features. *Journal of the Experimental Analysis of Behavior*, **54**, 69–84.

Fiaschi, V., Farina M. and Ioalé, P. (1974) Homing experiments on swifts *Apus apus* L. deprived of olfactory perception. *Monitore Zoologico Italiano (Nuova Serie)*, **8**, 235–244.

Finger S. (1994) *Origins of Neuroscience*. New York: Oxford University Press.

Firsov, L.A. (1972) *Anthropoid Memory*. Leningrad, USSR: Nauka. (In Russian.)

Firsov, L.A. (1977) *Behaviour of Anthropoids under Natural Conditions*. Leningrad, USSR: Nauka. (In Russian.)

Firsov, L.A. (1983) Pre-verbal language in apes. *Journal of Evolutionary Biochemistry and Physiology*, **19**, 381–389. (In Russian.)

Firsov, L.A. and Plotnikov, V. Yu. (1981) *Vocal Behaviour in Anthropoids*. Leningrad, USSR: Nauka. (In Russian.)

Fisher, E. (1939) Habits of the southern sea otter. *Journal of Mammalogy*, **20**, 21–36.

Fisher, J. and Hinde, R.A. (1949) The opening of milk bottles by birds. *British Birds*, **42**, 347–357.

Fisher, R.A. (1925) *Statistical Methods for Research Workers*. New York: Hafner.

Fisher, R.A. (1930) *The Genetical Theory of Natural Selection*. Oxford, UK: Clarendon Press.

Fleay, D.H. (1937) Nesting habits of the brush turkey. *Emu*, **36**, 153–163.

Fletcher, D.J.C, and Michener, C.D. (eds.) (1987) *Kin Recognition in Animals*. New York: John Wiley.

Flourens, M.-J.-P. (1924) *Recherches expérimentales sur les propriétés et les fonctions du système nerveux dans les animaux vertébrés*. Paris: Crévot.

Ford, J.K.B. and Fisher, H.D. (1983) Group-specific dialects of killer whales (*Orcinus orca*) in British Columbia. In *Communication and Behavior of Whales*, ed. R. Payne. Boulder, CO: Westview Press, pp. 129–161.

Forel, A. (1874) Les Fourmis de la Suisse. *Nouvelles Mémoires de la Société Helvétique des Sciences Naturelles Zürich*, **26**, 1-200.

Fossey, D. (1983) *Gorillas in the Mist*. Boston, MA: Houghton Mifflin.

Foster, W.A. (1990) Experimental evidence for the effective and altruistic colony defense against natural predators by soldiers of the gall-forming aphid *Pemphigus spyrothecae* (Hemiptera: Pemphigidae), having sterile soldiers. *Behavioral Ecology and Sociobiology*, **27**, 421–430.

Fouts, R.S. (1997) *Next of Kin: My Conversations with Chimpanzees*. New York: William Morrow.

Fouts, R.S., Hirsch, A.D. and Fouts, D.H. (1982) Cultural transmission of a human language in a chimpanzee mother–infant relationship. In *Psychobiological Perspectives: Child Nurturance*, vol. 3, ed. H.E. Fitzgerald, J.A. Mullins and P. Page. New York: Plenum Press, pp. 159–196.

Fox, E.A., Sitompul, A.F. and van Schaik, C.P. (1999) Intelligent tool use in wild Sumatran orangutans. In *The Mentalities of Gorillas and Orangutans: Comparative Perspectives*, ed. S.T. Parker, R.W. Mitchell and H.L. Miles. Cambridge, UK: Cambridge University Press, pp. 99–116.

Fox, S.F. and Baird, T.A, (1992) The dear enemy phenomenon in the collared lizard, *Crotaphytus collaris*, with a cautionary note on experimental methodology. *Animal Behaviour*, **44**, 780–782.

Fox, S.F., McCoy J.K. and Baird T.A. (eds.) (2003) *Lizard Social Behavior*. Baltimore, MD: Johns Hopkins University Press.

Fragaszy, D. and Perry, S. (eds.) (2003) *The Biology of Traditions: Models and Evidence*. Cambridge, UK: Cambridge University Press.

Fragaszy, D., Fedigan, L. and Visalberghi, E. (2004) *The Complete Capuchin*. Cambridge, UK: Cambridge University Press.

France, R.L. (1997) The importance of beaver lodges in structuring littoral communities in boreal headwater lakes. *Canadian Journal of Zoology*, **75**, 1009–1013.

Frank, H. (1980) Evolution of canine information processing under conditions of natural and artificial selection. *Zoological Tierpsychology*, **53**, 389–399.

Franks, N.R. (1986) Teams in social insects: group retrieval of prey by army ants. *Behavioral Ecology and Sociobiology*, **18**, 425–429.

Frantsevich, L.I. (1993) Adaptation of the visual system for processing of communicative signals. In *Sensory Systems of Arthropods*, ed. K. Wiese, F.G. Gribakin, A.V. Popov and G. Renninger. Basel, Switzerland: Birkhäuser Verlag, pp. 233–241.

Frederikson, W. and Suckett, G. (1984) Kin preferences in primates, *Macaca nemestrina*: relatedness or familiarity? *Journal of Comparative Psychology*, **98**, 29–34.

Freeman, W.J. (1991) The physiology of perception. *Scientific American*, **264**, 78–85.

Freeman, W.J. (1995) *Societies of Brains: A Study in the Neuroscience of Love and Hate*. Hillsdale, NJ: Lawrence Erlbaum.

Freeman, W.J. (1999) *How Brains Make Up Their Minds*. London: Weidenfeld & Nicolson.

Freyd, J.J. (1983) The mental representation of movement when viewing static stimuli. *Perception and Psychophysics*, **33**, 575–581.

Frisch, K. von (1923) Über die Spräche der Bienen. *Zoologische Jahrbucher-Abteilung für Allgemeine Zoologie und Physiologie der Tiere*, **40**, 1–119.

Frisch, K. von (1937) The language of bees. *Science Progress*, **32**, 29–37. Reprinted in Wenner, A.M. (ed.) (1993) The language of bees. *Bee World*, **74**, 90–98.

Frisch, K. von (1947) The dances of the honey bee. *Bulletin of Animal Behaviour*, **5**, 1–32.

Frisch, K. von (1967) *The Dance Language and Orientation of Bees*. Cambridge, MA: Harvard University Press.

Frith, H.J. (1956a) Temperature regulation in the nesting mounds of the Mallefowl, *Leipoa ocellata*. *Australian Wildlife Research*, **1**, 79–95.

Frith, H.J. (1956b) Breeding habits of the family Megapodiidae. *Ibis*, **98**, 620–640.

Fritz, J. and Kortschal, K. (1999) Social learning in common ravens, *Corvus corax*. *Animal Behaviour*, **57**, 785–793.

Fritz, J., Bisenberger, A. and Kortschal, K. (2000) Stimulus enhancement in greylag geese: socially mediated learning of an operant task. *Animal Behaviour*, **59**, 1119–1125.

Fryday, S.L. and Greig-Smith, P.W. (1994) The effects of social learning on the food choice of the house sparrow (*Passer domesticus*). *Behaviour*, **128**, 281–300.

Fujita, K. (1983) Acquisition and transfer of a higher-order conditional discrimination performance in

Japanese monkey. *Japanese Psychological Research*, **25**, 1–8.

Funahashi, S., Bruce, C.J. and Goldman-Rakic, P.S. (1993) Dorsolateral prefrontal lesions and oculomotor delayed-response performance: evidence for mnemonic scotomas. *Journal of Neuroscience*, **13**, 1479–1497.

Fuller, T.K., Kat, P.W., Bulger, J.B., *et al.* (1992) Population dynamics of African wild dogs. In *Wildlife Populations*, ed. D.R. McCullough and R.H. Barrett. Amsterdam, the Netherlands: Elsevier, pp. 1125–1139.

Funk, M.S. (1996) Development of object permanence in the New Zeland parakeet (*Cyanoramphus auriceps*). *Animal Learning and Behaviour*, **24**, 375–383.

Funk, M.S. (2002) Problem-solving skills in young yellow-crowned parakeets (*Cyanoramphus auriceps*). *Animal Cognition*, **5**, 167–176.

Gadagkar, R. (1987) What are social insects? *International Union for Studying Social Insects Indian Chapter Newsletter*, 1–2, 3–4.

Gadagkar, R. (1992) The origin and evolution of social life in insects. *Bulletin of Sciences*, **6**, 31–35.

Gadagkar, R. (1993) And now ... eusocial thrips! *Current Science*, **64**, 215–216.

Gadagkar, R. (1994) Why the definition of eusociality is not helpful to understand its evolution and what should we do about it. *Oikos*, **70**, 485–488.

Gadagkar, R. (2001) *Social Biology of* Ropalidia marginata: *Toward Understanding the Evolution of Eusociality*. Cambridge, MA: Harvard University Press.

Gagnon, S. and Doré, F.Y. (1994) Cross-sectional study of object permanence in domestic puppies (*Canis familiaris*). *Journal of Comparative Psychology*, **108**, 220–232.

Galdikas, B.M.F. and Briggs, N. (1999) *Orangutan Odyssey*. New York: Abrams.

Galef, B.G., Jr (1980) Diving for food: analysis of a possible case of social learning in rats (*Rattus norvegicus*). *Journal of Comparative and Physiological Psychology*, **94**, 416–425.

Galef, B.G., Jr (1988) Imitation in animals: history, definition, and interpretation of data from the psychological laboratory. In *Social Learning: Psychological and Biological Perspectives*, ed. T.R. Zentall and B.G. Galef, Jr. Hillsdale, NJ: Lawrence Erlbaum, pp. 3–28.

Galef, B.G., Jr (1992) The question of animal culture. *Human Nature*, **3**, 157–178.

Galef, B.G., Jr (1993) Functions of social learning about foods by Norway rats: a causal analysis of effects of diet novelty on preference transmission. *Animal Behaviour*, **46**, 257–265.

Galef, B.G., Jr and White, D.J. (1998) Mate-choice copying in Japanese quail, *Coturnix coturnix japonica*. *Animal Behaviour*, **55**, 545–552.

Galef, B.G., Jr and White, D.J. (2000) Evidence of social effects on mate choice in vertebrates. *Behavioural Processes*, **51**, 167–175.

Galef, B.G., Jr, Kennett, D.J. and Stein, M. (1985) Demonstrator influence on observer diet preferences: effects of simple exposure and the presence of a demonstrator. *Animal Learning and Behaviour*, **13**, 25–30.

Gallistel, C.R. (1990) *The Organization of Learning*. Cambridge, MA: MIT Press.

Gallistel, C.R. (1998) Insect navigation: brains as symbol-processing organs. In *Conceptual and Methodological Foundations: An Invitation to Cognitive Science*, 2nd edn, vol. 4, ed. S. Sternberg and D. Scarborough. Cambridge, MA: MIT Press, pp. 1–51.

Gallistel, C.R. and Cramer, A.E. (1996) Computation on metric maps in mammals: getting oriented and choosing a multi-dimension route. *Journal of Experimental Biology*, **199**, 211–217.

Gallistel, C.R. and Gelman, R. (1992) Preverbal and verbal counting and computation. *Cognition*, **44**, 43–74.

Gallup, G.G., Jr (1970) Chimpanzees: self-recognition. *Science*, **167**, 86–87.

Gallup, G.G., Jr (1977) Absence of self-recognition in a monkey (*Macaca fascicularis*) following prolonged exposure to a mirror. *Developmental Psychobiology*, **10**, 281–284.

Gallup, G.G., Jr, Povinelli, D.J., Suarez, S.D., *et al.* (1995) Further reflections on self-recognition in primates. *Animal Behaviour*, **50**, 1525–1532.

Garcia, J. and Koelling, R.A. (1966) Relation of cue to consequence in avoidance learning. *Psychonomic Science*, **4**, 123–124.

Garcia, J., McGowan, B.K. and Green, K.E. (1972) Biological constrains on conditioning. In *Classical Conditioning II: Current Research and Theory*, ed. A.N. Black and W.F. Prokasy. New York: Appleton-Century-Crofts, pp. 3–27.

Garcia, J., Rusiniak, K.W. and Brett, L.P. (1977) Conditioning food-illness in wild animals: *Caveant canonici*. In *Operant–Pavlovian Interactions*, ed. H. Davis and H.M.B. Hurwitz. Hillsdale, NJ: Lawrence Erlbaum, pp. 273–316.

Gardner, B.T. and Gardner, R.A. (1975) Evidence for sentence constituents in the early utterances of child and chimpanzee. *Journal of Experimental Psychology: General*, **104**, 244–267.

Gardner, R. A. and Gardner, B. T. (1969) Teaching sign language to a chimpanzee. *Science*, **165**, 664–672.

Gardner, R. A. and Gardner, B. T. (1980) Comparative psychology and language acquisition. In *Speaking of Apes: A Critical Anthology of Two-Way Communication with Man*, ed. T. A. Sebok and J. T. A. Umiker-Sebok. New York: Plenum Press, pp. 287–329.

Gardner, R. A. and Gardner, B. T. (1989) *Teaching Sign Language to Chimpanzees*, Albany, NY: State University of New York Press.

Gardner, R. A. and Gardner, B. T. (1998) *The Structure of Learning*. Mahwah, NJ: Lawrence Erlbaum.

Garner, R. L. (1892) *The Speech of Monkeys*. New York: C. L. Webster.

Garner, W. R. (1974) *The Processing of Information and Structure*. Hillsdale, NJ: Lawrence Erlbaum.

Gellermann, L. W. (1933) Form discrimination in chimpanzees and two-year-old children. *Journal of Genetic Psychology*, **42**, 3–27.

Gelman, R. and Gallistel, C. R. (1978) *The Child's Understanding of Number*. Cambridge, MA: Harvard University Press.

Gelman, S. A. and Coley, J. D. (1990) The importance of knowing a dodo is a bird: categories and inferences in 2-year-old children. *Developmental Psychology*, **26**, 796–804.

Gemberling, G. A. and Domjan, M. (1982) Selective association in one-day-old rats: taste-toxicosis aversion learning. *Journal of Comparative and Physiological Psychology*, **96**, 105–113.

Gerber, B. and Menzel, R. (2000) Contextual modulation of memory consolidation. *Learning and Memory*, **7**, 151–158.

Ghiglieri, M. P. (1984) *The Chimpanzees of Kibale Forest*. New York: Columbia University Press.

Gibson, R. M. and Hoglund, J. (1992) Copying and sexual selection. *Trends in Ecology and Evolution*, **7**, 229–232.

Giebel, H. D. (1958) Visuelles lernvermogen bei einhufern. *Zoologisches Jahrbuch*, **67**, 487–489. Reprinted (1983) in *Horse Behavior*, ed. G. H. Waring. Park Ridge, NJ: Noyes, pp. 229–231.

Gilbert, D. B., Patterson, T. A. and Rose, S. P. R. (1989) Midazolam induces amnesia in a simple one-trial maze-learning task. *Pharmacology Biochemistry and Behavior*, **34**, 439–442.

Gillan, D. J. (1981) Reasoning in the chimpanzee. II. Transitive inference. *Journal of Experimental Psychology: Animal Behavior Processes*, **7**, 150–164.

Giraldeau, L. A., Valone, T. J. and Templeton, J. J. (2002) Potential disadvantage of using socially acquired information. *Philosophical Transactions of the Royal Society of London B*, **375**, 1559–1566.

Giurfa, M. (2003) Cognitive neuroethology: dissecting non-elemental learning in a honeybee brain. *Current Opinion in Neurobiology*, **13**, 726–735.

Giurfa, M. and Lehrer, M. (2001) Honeybee vision and floral displays: from detection to close-up recognition. In *Cognitive Ecology of Pollination*, ed. L. Chittka and J. D. Thomson. Cambridge, UK: Cambridge University Press, pp. 61–82.

Giurfa, M., Hammer, M., Stach, S., *et al.* (1999) Pattern learning by honeybees: conditioning procedure and recognition strategy. *Animal Behaviour*, **57**, 315–324.

Godard, R. (1991) Long-term memory of individual neighbours in a migratory songbird. *Nature*, **350**, 228–229.

Goelet, P., Castelluci, S., Schachner, S. and Kandel, E. R. (1986) The long and the short of long-term memory: a molecular framework. *Nature*, **322**, 419–423.

Goldman-Rakic, P. S. (1992) Working memory and the mind. *Scientific American*, **267**, 111–117.

Goldman-Rakic, P. S. (1995) Cellular basis of working memory. *Neuron*, **14**, 477–485.

Goldsmith, E. (1998) *The Way: An Ecological Worldview*. Athens, GA: University of Georgia Press.

Goldstein, K. (1939) *The Organism: A Holistic Approach to Biology Derived from Psychological Data in Man*. New edition (1998) with foreword by Oliver Sacks. New York: Zone Books.

Goller, F. and Esch, H. (1990) Comparative study of chill coma temperatures and muscle potentials in insect flight muscles. *Journal of Experimental Biology*, **150**, 221–231.

Gómez, J. C. (1994) Mutual awareness in primate communication: a Gricean approach. In *Self-Awareness in Animals and Humans*, ed. S. T. Parker, R. W. Mitchell and M. L. Boccia. Cambridge, UK: Cambridge University Press, pp. 61–80.

Goodall, J. (1964) Tool-using and aimed throwing in a community of free living chimpanzees. *Nature*, **201**, 1264–1266.

Goodall, J. (1968) The behaviour of free-living chimpanzees in the Gombe Stream area. *Animal Behaviour Monographs*, **1**, 161–311.

Goodall, J. (1970) Tool using in primates and other vertebrates. In *Advances in the Study of Behavior*, vol. 3, ed. D. S. Lehrman, R. A. Hinde and E. Shaw. New York: Academic Press, pp. 195–249.

Goodall, J. (1971) *In the Shadow of Man*. Boston, MA: Houghton Mifflin.

Goodall, J. (1986) *The Chimpanzees of Gombe: Patterns of Behavior*. Cambridge, MA: The Belknap Press of Harvard University Press.

Goodall, J. van Lawick- and van Lawick, H. (1966) Use of tools by the Egyptian vulture *Neophron pernopterus*. *Nature*, **212**, 1468–1469.

Gopnik, A. and Astington, J.W. (1988) Children's understanding of representational change, and its relation to the understanding of false belief and the appearance–reality distinction. *Child Development*, **59**, 26–37.

Gopnik, A. and Meltzoff, A.N. (1986) Relations between semantic and cognitive development in the one-word stage: the specificity hypothesis. *Child Development*, **57**, 1040–1053.

Gopnik, A. and Meltzoff, A.N. (1997) *Words, Thoughts, and Theories*. Cambridge, MA: MIT Press.

Gorman, M.L., Mills, M.G.L., Raath, J.P. and Speakman, J.R. (1998) High hunting costs make African wild dogs vulnerable to kleptoparasitism by hyenas. *Nature*, **391**, 479–481.

Goss-Custard, J.D., Clarke, R.T. and Durell, S.E.A. Le V. dit (1984) Rates of food intake and aggression of oystercatchers *Haematopus ostralegus* on the most and least preferred mussel *Mytilus edulis* beds of the Exe estuary. *Journal of Animal Ecology*, **53**, 233–235.

Göth, A. (2001) Innate predator recognition in Australian brush-turkey (*Alectura lathami*, Megapodidae) hatchlings. *Behaviour*, **138**, 117–136.

Göth, A. and Evans, C.S. (2004) Social responses without early experience: Australian brush-turkey chicks use specific visual cues to aggregate with conspecifics. *Journal of Experimental Biology*, **207**, 2199–2208.

Göth, A. and Jones, D. (2003) Ontogeny of social behavior in the megapode Australian brush-turkey (*Alectura lathami*). *Journal of Comparative Psychology*, **117**, 36–43.

Gottfredson, L.S. (1998). The general intelligence factor. *Scientific American Presents*, **9**(4), 24–29.

Gottlieb, G. (1961) The following response and imprinting in wild and domestic ducklings of the same species (*Anas platyrhynchos*). *Behaviour*, **18**, 205–228.

Gottlieb, G. (1971) *Development of Species Identification in Birds*. Chicago, IL: University of Chicago Press.

Gould, J.L. (1974) Honey bee communication. *Nature*, **252**, 300–301.

Gould, J.L. (1975) Honey bee recruitment: the dance language controversy, *Science*, **189**, 685–693.

Gould, J.L. (1976) The dance language controversy. *Quarterly Review of Biology*, **57**, 211–244.

Gould, J.L. (1982a) *Ethology: The Mechanisms and Evolution of Behavior*. New York: W.W. Norton.

Gould, J.L. (1982b) The map sense of pigeons. *Nature*, **296**, 205–211.

Gould, J.L. (1984) Natural history of honeybee learning. In *The Biology of Learning*, ed. P. Marler and H.S. Terrace. Berlin: Springer Verlag, pp. 149–180.

Gould, J.L. (1986) The locale map of honey bees: do insects have cognitive maps? *Science*, **232**, 861–863.

Gould, J.L. and Gould, C.G. (1988) *The Honey Bee*. New York: Scientific American Library. Revised edn (1995).

Gould, J.L. and Gould, C.G. (1994) *The Animal Mind*. New York: Scientific American Library.

Gould, J.L. and Marler, P. (1984) Ethology and the natural history of learning. In *The Biology of Learning*, ed. P. Marler and H.S. Terrace. Berlin: Springer Verlag, pp. 47–74.

Gould, J.L. and Marler, P. (1987) Learning by instinct. *Scientific American*, **256**, 74–85.

Gouteux, S., Vauclair, J. and Thinus-Blanc, C. (1999) Reaction to spatial novelty and exploratory strategies in baboons. *Animal Learning and Behaviour*, **27**, 323–332.

Grafen, A. (1984) Natural selection, Kin selection and group selection. In *Behavioural Ecology*, 2nd edn, ed. J.R. Krebs and N.B. Davis. Oxford, UK: Blackwell Scientific Publications, pp. 62–84.

Grafen, A. (2006) Optimisation of inclusive fitness. *Journal of Theoretical Biology*, **238**, 541–563.

Grant, D.S. (1976) Effect of sample presentation time on long-delay matching in the pigeon. *Learning and Motivation*, **7**, 580–590.

Grant, D.S. (1981) Short-term memory in the pigeon. In *Information Processing in Animals: Memory Mechanisms*, ed. N.E. Spear and R.R. Miller. Hillsdale, NJ: Lawrence Erlbaum, pp. 227–256.

Grant, D.S. and Kelly, R. (2001) Anticipation and short-term retention in pigeons. In *Avian Visual Cognition*, ed. R.G. Cook. Available online: www.pigeon.psy.tufts.edu/avc/grant/.

Grant, P.R. and Grant, B.R. (2002) Adaptive radiation of Darwin's finches. *American Scientist*, **90**, 130–139.

Grassé, P.P. (1982–1986) *Termitologia*, vol. 1, *Anatomie, physiologie, reproduction*, vol. 2, *Fondation des sociétés et construction*, vol. 3, *Comportement, socialité, écologie, évolution, systématique*. Paris: Masson.

Green, R. (1991) *Wild Cat Species of the World*. Plymouth, UK: Basset Publications.

Green, S. (1975) Dialects in Japanese monkeys: vocal learning and cultural transmission of locale-specific behavior? *Zoological Tierpsychology*, **38**, 304–314.

Griffin, A.S., Evans, C.S. and Blumstein, D.T. (2002) Selective learning in a marsupial. *Ethology*, **108**, 1103–1114.

Griffin, D. R. (1976) *The Question of Animal Awareness: Evolutionary Continuity of Mental Experience*. New York: Rockefeller University Press.

Griffin, D. R. (1978) Prospects for a cognitive ethology. *Behavioral and Brain Sciences*, **4**, 527–538.

Griffin, D. R. (1984) *Animal Thinking*. Cambridge, MA: Harvard University Press.

Griffin, D. R. (1992) *Animal Minds*. Chicago, IL: University of Chicago Press.

Griffin, D. R. (2001) *Animal Minds: Beyond Cognition to Consciousness*. Chicago, IL: University of Chicago Press.

Griffiths. D., Dickinson, A. and Clayton, N. S. (1999) Episodic and declarative memory: what can animals remember about their past? *Trends in Cognitive Science*, **3**, 74–80.

Gros-Louis, J. (2004) The function of food-associated calls in white-faced capuchin monkeys, *Cebus capucinus*, from the perspective of the signaller. *Animal Behaviour*, **67**, 431–440.

Gross, M. R. (1979) Cuckoldry in sunfishes (*Lepomis*: Centrarchidae). *Canadian Journal of Zoology*, **57**, 1507–1509.

Gross, M. R. (1984) Review of evolution and the theory of games. *Quarterly Review of Biology*, **59**, 172–173.

Grutter, A. S. (2004) Cleaner fish use tactile dancing as pre-conflict management strategy. *Current Biology*, **14**, 1080–1083.

Guggisberg, C. (1975) *Wild Cats of the World*. New York: Taplinger.

Guhl, A. M. and Allee, W. C. (1944) Some measurable effects of social organization in flocks of hens. *Physiological Zoology*, **17**, 320–347.

Guiton, P. (1959) Socialization and imprinting in brown leghorn chicks. *Animal Behaviour*, **7**, 26–34.

Gulz, A. (1991) The planning of action as a cognitive and biological phenomenon. Ph.D. thesis, Lund University, Sweden.

Gwinner, E. (1972) Endogenous timing factors in bird migration. In *Animal Orientation and Navigation*, ed. S. R. Galler, K. Schmidt-Koenig, G. J. Jacobs and R. E. Belleville. Washington, DC: NASA, pp. 321–338.

Hailman, J. P. (1967) The ontogeny of an instinct: the pecking response in chicks of the laughing gull (*Larus atricilla* L.) and related species. *Behavioural Supplements*, **15**, 1–196.

Haldane, J. B. S. (1927) *Possible Worlds and Other Essays*. London: Chatto & Windus.

Haldane, J. B. S. (1932) *The Causes of Evolution*. London: Longman and Green.

Haldane, J. B. S. (1955) *Population Genetics: New Biology*. New York: Penguin Books.

Hall, K. R. L. (1963) Observational learning in monkeys and apes. *British Journal of Psychology*, **54**, 201–226.

Hall, K. R. L. and Schaller, G. (1964) Tool-using behavior of the California sea otter. *Journal of Mammalogy*, **45**, 287–298.

Hamilton, C. R. (1988) Hemispheric specialisation in monkeys. In *Brain Circuits and Functions of the Mind*, ed. C. Trevarthen. Cambridge, UK: Cambridge University Press, pp. 181–195.

Hamilton, W. D. (1964) The genetical evolution of social behaviour. *Journal of Theoretical Biology*, **7**, 1–52.

Hamilton, W. D. (2001) *The Collected Papers of W. D. Hamilton*, vol. 2, *Narrow Roads of Gene Land*. Oxford, UK: Oxford University Press.

Hammer, M. (1993) An identified neuron mediates the unconditioned stimulus in associative olfactory learning in honeybees. *Nature*, **366**, 59–63.

Hammerschmidt, K., Freudenstein, T. and Juergens, U. (2001) Vocal development in squirrel monkeys. *Behaviour*, **138**, 1179–1204.

Hampton, R. R. (2001) Rhesus monkeys know when they remember. *Proceedings of the National Academy of Sciences of the USA*, **98**, 5359–5362.

Hampton, R. R. (2003) Metacognition and explicit representation in nonhumans. *Behavioral and Brain Science*, **26**, 346–347.

Hampton, R. R. and Sherry, D. F. B. (1994) The effects of cache loss on choice of cache sites in black-capped chickadees. *Behavioral Ecology*, **5**, 44–50.

Hanlon, R. T. and Messenger, J. B. (1996) *Cephalopod Behaviour*. Cambridge, UK: Cambridge University Press.

Hansson, L. (1986) Geographic differences in the sociability of voles in relation to cyclicity. *Animal Behaviour*, **34**, 1215–1221.

Hare, B., Call, J. and Tomasello, M. (2001) Do chimpanzees know what conspecifics know? *Animal Behaviour*, **61**, 139–151.

Hare, B., Brown, M., Williamson, C. and Tomasello, M. (2002) The domestication of social cognition in dogs. *Science*, **298**, 1634–1636.

Hare, B., Addessi, E., Call, J., Tomasello, M. and Visalberghi, E. (2003) Do capuchin monkeys (*Cebus apella*) know what conspecifics do and do not see? *Animal Behaviour*, **65**, 131–142.

Harkiv, W. A. (1997) Competition as a mechanism of division of labour in colonies of slave-making ants *Formica sanguinea* (Hymenoptera, Formicidae). *Zoological Journal*, **76**, 444–447. (In Russian with English summary.)

Harlow, H. F. (1949) The formation of learning sets. *Psychological Review*, **56**, 51–65.

Harlow, H. F. (1959) The development of learning in the rhesus monkey. *American Scientist*, **45**, 459–479.

Harlow, H. F. (1971) *Learning to Love*. San Francisco, CA: Albion.

Harlow, H. F. and Bromer, J. A. (1938). A test apparatus for monkeys. *Psychological Review*, **19**, 434–438.

Harlow, H. F. and Harlow, M. K. (1962) Social deprivation in monkeys. *Scientific American*, **207**, 136–146.

Harlow, H. F. and Settlage, P. H. (1934) Comparative behavior of primates. VII. Capacity of monkeys to solve patterned strings tests, *Journal of Comparative Psychology*, **18**, 423–435.

Harlow, H. F., Uehling, H. and Maslow, A. H. (1932) Comparative behaviour of primates. I. Delayed reaction tests on primates from lemur to the orang-utan. *Journal of Comparative Psychology*, **13**, 313–343.

Harris, B. (1979) Whatever happened to Little Albert? *American Psychologist*, **34**, 151–160.

Harrison, J. F., Fewell, J. H., Stiller, T. M. and Breed, M. D. (1989) Effects of experience on use of orientation cues in the giant tropical ant. *Animal Behaviour*, **37**, 869–871.

Hart, B. L., Hart, L. A., McCoy, M. and Sarath, C. R. (2001) Cognitive behaviour in Asian elephants: use and modification of branches for fly switching. *Animal Behaviour*, **62**, 839–847.

Hashiya, K. and Kojima, S. (2001) Acquisition of auditory-visual intermodal matching-to-sample by a chimpanzee: comparison with visual–visual intramodal matching, *Animal Cognition*, **4**, 231–240.

Hauser, M. D. (1988) How infant vervet monkeys learn to recognise starling alarm calls: the role of experience. *Behaviour*, **105**, 187–201.

Hauser, M. D. (1998) A non-human primate's expectations about object motion and destination: the importance of self-propelled movement and animacy. *Developmental Science*, **1**, 31–38.

Hauser, M. D. (2000) A primate dictionary? Decoding the function and meaning of another species' vocalizations. *Cognitive Sciences*, **24**, 445–475.

Hauser, M. D. (2001) What's so special about speech? In *Language, Brain and Cognitive Development: Essays in Honor of Jacques Mehler*, ed. E. Dupoux. Cambridge, MA: MIT Press, pp. 121–134.

Hauser, M. D. and Fitch, T. W. (2003) What are the uniquely human components of the language faculty? In *Language Evolution*, ed. M. Christiansen and S. Kirty. Oxford, UK: Oxford University Press, pp. 158–181.

Hauser, M. D. and Marler, P. (1992) How do and should studies of animal communication affect interpretations of child phonological development? In *Phonological Development*, ed. C. Ferguson, L. Menn and C. Stoel-Gammon. Baltimore, MD: York Press, pp. 663–680.

Hauser, M. D., Kralik, J. and Botto-Mahan, C. (1999) Problem solving and functional design features: experiments with cotton-top tamarins (*Saguinus oedipus*). *Animal Behaviour*, **57**, 565–582.

Hauser, M. D., Carey, S. and Hauser, L. B. (2000) Spontaneous number representation in semi-free-ranging rhesus monkeys. *Proceedings of the Royal Society of London B*, **267**, 829–833.

Hauser, M. D., Miller, C. T., Liu, K. and Gupta, R. (2001) Cotton-top tamarins fail to show mirror-guided self-exploration. *American Journal of Primatology*, **53**, 123–130.

Hauser, M. D., Chomsky, N. and Fitch, W. T. (2002a) The faculty of language: What is it, who has it, and how did it evolve? *Science*, **298**, 1569–1579.

Hauser, M. D., Santos, L. R., Spaepen, G. M. and Pearson, H. E. (2002b) Problem solving, inhibition and domain-specific experience: experiments on cottontop tamarins, *Saguinus oedipus*. *Animal Behaviour*, **64**, 387–396.

Hauser, M. D., Chen, K., Chen, F. and Chuang, E. (2003) Give unto others: genetically unrelated cotton-top tamarin monkeys preferentially give food to those who give food back. *Proceedings of the Royal Society of London B*, **270**, 2363–2370.

Hayes, K. J. and Hayes, C. (1952) Imitation in a home-raised chimpanzee. *Journal of Comparative Psychology*, **45**, 450–459.

Healy, S. (1998) *Spatial Representation in Animals*. Oxford, UK: Oxford University Press.

Hebb, D. O. (1949) *The Organization of Behavior*. New York: John Wiley.

Heinrich, B. (1995) An experimental investigation of insight in common ravens (*Corvus corax*). *Auk*, **112**, 994–1003.

Heinrich, B. (1999) *The Mind of the Raven*. New York: HarperCollins.

Heinrich, B. (2000) Testing insight in ravens. In *The Evolution of Cognition*, ed. C. Heyes and L. Huber. Cambridge, MA: MIT Press, pp. 289–305.

Heinroth, O. (1911) Beträge zur Biologie, näm entlich Ethologie und Physiologie der Anatiden. *Verhandlungen der V Internazionale Ornithologische Kongress*, Berlin, pp. 589–702.

Heinze, J. B., Hölldobler, B. and Peeters, C. (1994) Conflict and cooperation in ant societies. *Natürwissenschaften*, **81**, 489–497.

Helbig, A. J. (1994) Genetic basis of evolutionary change of migratory directions in a European passerine migrant, *Sylvia atricapilla*. *Ostrich*, **65**, 151–159.

Held, S., Mendl, M., Devereux, C. and Byrne, R.W. (2000) Social tactics of pigs in a competitive foraging task: the 'informed forager' paradigm. *Animal Behaviour*, **59**, 569–576.

Henry, S., Hemery, D., Richard, M.-A. and Hausberger, M. (2005) Human–mare relationships and behaviour of foals toward humans. *Applied Animal Behaviour Science*, **93**, 341–362.

Hepper, P.G. (1991) *Kin Recognition*. Cambridge, UK: Cambridge University Press.

Hepper, P.G. (ed.) (2005) *Kin Recognition*. Cambridge, UK: Cambridge University Press.

Heran, H. and Wanke, L. (1952) Beobachtungen über die Entfernungsmeldungen der Sammelbiennen. *Zeitschrift für vergleichende Physiologie*, **34**, 383–393.

Herenhahn, B.R. and Olson, M.H. (2001) *Theories of Learning*, 6 edn. Upper Saddle River, NJ: Prentice-Hall.

Herman, L.M. (1980) Cognitive characteristics of dolphins. In *Cetacean Behavior: Mechanisms and Functions*, ed. L.M. Herman. New York: John Wiley, pp. 363–429.

Herman, L.M. (1986) Cognition and language competencies of bottlenosed dolphins. In *Dolphin Cognition and Behavior: A Comparative Approach*, ed. R.J. Schusterman, J. Thomas and F.G. Wood. Hillsdale, NJ: Lawrence Erlbaum, pp. 221–251.

Herman, L.M. (1990) Cognitive performance of dolphins in visually guided tasks. In *Sensory Abilities of Cetaceans: Laboratory and Field Evidence*, ed. A. Thomas and R.A. Kastelein. New York: Plenum Press, pp. 455–462.

Herman, L.M. (1991) What the dolphin knows, or might know, in its natural world. In *Dolphin Societies: Discoveries and Puzzles*, ed. K. Pryor and K.S. Norris. Los Angeles, CA: University of California Press, pp. 349–364.

Herman, L.M. (2002) Vocal, social, and self-imitation by bottlenose dolphins. In *Imitation in Animals and Artifacts*, ed. C. Nehaniv and K. Dautenhaun. Cambridge, MA: MIT Press, pp. 63–108.

Herman, L.M. and Forestell, P.H. (1985) Reporting presence or absence of named objects by a language-trained dolphin. *Neuroscience and Biobehavioral Reviews*, **9**, 667–691.

Herman, L.M. and Thompson, R.K.R. (1982) Symbolic identity and probed delay matching of sounds in the bottlenosed dolphin. *Animal Learning and Behaviour*, **10**, 22–34.

Herman, L.M., Richards, D.G. and Woltz, J.P. (1984) Comprehension of sentences by bottlenosed dolphins. *Cognition*, **16**, 129–219.

Herman, L.M., Pack, A.A. and Wood, A.M. (1994) Bottlenose dolphins can generalize rules and develop abstract concepts. *Marine Mammal Science*, **10**, 70–80.

Herman, L.M., Abichandani, S.L., Elhajj, A.N., *et al.* (1999) Dolphins (*Tursiops truncatus*) comprehend the referential character of the human pointing gesture. *Journal of Comparative Psychology*, **113**, 1–18.

Hermer, L. and Spelke, E. (1994) A geometric process for spatial reorientation in young children. *Nature*, **370**, 57–59.

Hermer, L. and Spelke, E. (1996) Modularity and development: the case of spatial reorientation. *Cognition*, **61**, 195–232.

Herrnstein, R.J. (1990) Levels of stimulus control: a functional approach. *Cognition*, **37**, 133–166.

Herrnstein, R.J. and Loveland, D.H. (1964) Complex visual concept in the pigeon. *Science*, **146**, 549–551.

Hess, E.H. (1956) Natural preferences of chicks and ducks for objects of different colours. *Psychological Reports*, **2**, 477–483.

Hess, E.H. (1959) Imprinting. *Science*, **130**, 133–141.

Hess, E.H. (1964) Imprinting in birds. *Science*, **146**, 1128–1139.

Hessing, M.J.C., Hagelso, A.M., van Beek, J.A., *et al.* (1993) Individual behavioural characteristics in pigs. *Applied Animal Behaviour Science*, **37**, 285–295.

Heyes, C.M. (1993) Imitation, culture, and cognition. *Animal Behaviour*, **46**, 999–1010.

Heyes, C.M. (1998) Theory of mind in nonhuman primates. *Behavioral and Brain Sciences*, **21**, 101–115.

Heyes, C.M. (2001) Causes and consequences of imitation. *Trends in Cognitive Sciences*, **5**, 253–261.

Heyes, C.M. and Huber, L. (eds.) (2000) *The Evolution of Cognition*. Cambridge, MA: MIT Press.

Heyes, C.M., Jaldow, E. and Dawson, G.R. (1994) Imitation in rats: conditions of occurrence in a bidirectional control procedure. *Learning and Motivation*, **25**, 276–287.

Hilchenko, A.E. (1950) Forming of reactions to traits of relations (size ratio) in monkeys. In *Investigations of High Nervous Processing in Natural Experiments*, ed. V.P. Protopopov. Kiev, USSR: Gosmedizdat, pp. 80–92. (In Russian.)

Hilgard, E.R. and Marquis, D.G. (1940) *Conditioning and Learning*. New York: Appleton Century.

Hinde, R.A. (1970) *Animal Behavior: A Synthesis of Ethology and Comparative Psychology*. New York: McGraw-Hill.

Hinde, R.A. (1974) *Biological Bases of Human Social Behaviour*. New York: McGraw-Hill.

Hinde, R.A. and Tinbergen, N. (1958) The comparative study of species-specific behavior. In *Behavior and*

Evolution, ed. A. Roe and G. G. Simpson. New Haven, CT: Yale University Press, pp. 251–268.

Hirata, S. and Matsuzawa, T. (2001) Tactics to obtain a hidden food item in chimpanzee pairs (*Pan troglodytes*). *Animal Cognition*, **4**, 285–295.

Hobhouse, L. T. (1901) *Mind in Evolution*. Reprinted (1998) ed. R. H. Wozniak. New York: Thoemmes Continuum Press.

Hockett, C. D. (1960) The origin of speech. *Scientific American*, **203**, 96–99.

Hockett, C. D. (1963) The problem of universals in language. In *Universals of Language*, ed. J. H. Greenberg. Cambridge, MA: MIT Press, pp. 1–22.

Hoffman, H. S. (1996) *Amorous Turkeys and Addicted Ducklings: A Search for the Causes of Social Attachment*. Boston, MA: Authors Cooperative.

Holland, P. C. and Straub, J. J. (1979) Differential effect of two ways of devaluing the unconditioned stimulus after Pavlovian appetitive conditioning. *Journal of Experimental Psychology: Animal Behavior Processes*, **5**, 65–78.

Hollander, M. and Wolf, D. A. (1973) *Nonparametric Statistical Methods*. New York: John Wiley.

Hollard, V. D. and Delius, J. D. (1982) Rotational invariance in visual pattern recognition by pigeons and humans. *Science*, **218**, 802–804.

Hölldobler, B. (1985) Liquid food transmission and antennation signals in ponerine ants. *Israel Journal of Entomology*, **19**, 89–99.

Hölldobler, B. and Wilson, E. O. (1990) *The Ants*. Cambridge, MA: The Belknap Press of Harvard University Press.

Hölldobler, B. and Wilson, E. O. (1995) *Journey to the Ants*. Cambridge, MA: The Belknap Press of Harvard University Press.

Hollén, L. I. and Manser, M. B. (2006) Ontogeny of alarm call responses in meerkats (*Suricata suricatta*): the role of age, sex and nearby conspecifics. *Animal Behaviour*, **72**, 1345–1353.

Holst, E. von (1937) Vom Wesen der Ordnung im Zentralnervensystem. *Natürwissenschaften*, **25**, 625–531. Translated (1937) On the nature of order in the central nervous system. In *The Behavioral Physiology of Animals and Man: Selected Papers of E. von Holst*, vol. 1. Coral Gables, FL: Univerity of Miami Press.

Honig, W. K. and Thompson, R. K. R. (1982) Retrospective and prospective processing in animal working memory. In *The Psychology of Learning and Motivation: Advances in Research and Theory*, vol. 16, ed. G. H. Bower. New York: Academic Press, pp. 239–283.

Hooff, J. van and Schaik, C. van (1994) Male bonds: affiliative relationships among nonhuman primate males. *Behaviour*, **130**, 309–337.

Hornby, P. (1972) The psychological subject and predicate. *Cognitive Psychology*, **3**, 632–642.

Hörster, A., Curio, E. and Witte, K. (2000) No sexual imprinting on a red bill as a novel trait. *Behaviour*, **137**, 1223–1239.

Hothersall, D. (1995) *History of Psychology*. Boston, MA: McGraw-Hill.

Houk, J. and Geibel, J. (1974) Observation of underwater tool use by the sea otter, *Enhydra lutris* Linnaeus. *California Fish and Game*, **60**, 207–208.

Huber, L. (1995) On the biology of perceptual categorization. *Evolution and Cognition*, **2**, 121–138.

Huber, L. (1998) Perceptual categorization as the groundwork of animal cognition. In *Downward Processing in the Perception Representation Mechanism*, ed. C. Taddei-Ferretti and C. Musio. Hong Kong: World Scientific Press, pp. 287–293.

Huber, L. (2000) Psychophylogenesis: innovations and limitations in the evolution of cognition. In *The Evolution of Cognition*, ed. C. Heyes and L. Huber. Cambridge, MA: MIT Press, pp. 23–41.

Huber, L. (2002) Clever birds: keas learn through observation. *Interpretive Birding Bulletin*, **3**(4), 57–59.

Huber, L., Troje, N., Loidolt, M., Aust, U. and Grass, D. (2000) Natural categorization through multiple feature learning in pigeons. *Quarterly Journal of Experimental Psychology B*, **53**, 341–357.

Huber, L., Rechberger, S. and Taborsky, M. (2001) Social learning affects object exploration and manipulation on keas, *Nestor notabilis*. *Animal Behaviour*, **62**, 945–954.

Huffman, M. A. and Nishie, H. (2001) Stone handling, a two decade old behavioural tradition in a Japanese monkey troop. *Advances in Ethology, Contributions to 27th International. Ethological Conference*, p. 35.

Hull, C. L. (1932) The goal gradient hypothesis and maze learning. *Psychological Review*, **39**, 25–43.

Hull, C. L. (1943) *Principles of Behavior*. New York: Appleton-Century-Crofts.

Hull, C. L. (1952) *A Behavior System: An Introduction to Behavior Theory concerning the Individual Organism*. New Haven, CT: Yale University Press.

Humle, T. (1999) New record of fishing for termites (*Macrotermes*) by the chimpanzees of Guinea-Bissau (*Pan troglodytes verus*). *Pan Africa News*, **6**, 3–4.

Humphrey, N. K. (1976) The social function of intellect. In *Growing Points in Ethology*, ed. P. P. G. Bateson and R. A. Hinde. Cambridge, UK: Cambridge University Press, pp. 303–317.

Hunt, G. R. (1996) Manufacture and use of hook-tools by New Caledonian crows. *Nature*, **379**, 249–251.

Hunt, G. R. (2000) Human-like, population-level specialization in the manufacture of pandanus tools by New Caledonian crows *Corvus moneduloides*. *Proceedings of the Royal Society of London B*, **267**, 403–413.

Hunt, G. R. and Gray, R. D. (2003) Diversification and cumulative evolution in tool manufacture by New Caledonian crows. *Proceedings of the Royal Society of London B*, **270**, 867–874.

Hunt, G. R. and Gray, R. D. (2004) The crafting of hook tools by wild New Caledonian crows. *Proceedings of the Royal Society of London B*, **271**, 88–90.

Hunter, W. S. (1912) The delayed reaction in animals. *Behavioral Monographs*, **2**, 1–85.

Huxley, A. (1932) *Brave New World*. Reissued (2003) Piscataway, NJ: Maxnotes, Research & Education Association.

Huxley, J. (1942) *Evolution: The Modern Synthesis*. New York: Harper & Row.

Huxley, T. H. (1893) *Evolution and Ethics*. London: Macmillan.

Imanishi, K. (1952) The evolution of human nature. In *Man*, ed. K. Imanishi. Tokyo: Mainichi-shinbunsha, pp. 36–94.

Immelman, K. (1972) Sexual imprinting in birds. *Advances in the Study of Behavior*, **4**, 147–174.

Insley, S. J. (2000) Long-term vocal recognition in the northern fur seal. *Nature*, **406**, 404–405.

Ioalé, P., Gagliardo, A. and Bingman, V. P. (2000) Further experiments on the relationship between hippocampus and orientation following phase-shift in homing pigeons. *Behavioral and Brain Research*, **108**, 157–167.

Irwin, D. E. and Price, T. (1999) Sexual imprinting, learning and speciation. *Heredity*, **82**, 347–354.

Itakura, S. (1987) Use of a mirror to direct their responses in Japanese monkeys (*Macaca fuscata*). *Primates*, **28**, 343–352.

Itani, J. and Nishimura, A. (1973) The study of infrahuman culture in Japan: a review. In *Precultural Primate Behavior*, ed. E. W. Menzel. Basel, Switzerland: Karger, pp. 26–50.

Jacobs, L. F. (2003) The evolution of the cognitive map. *Brain, Behaviour and Evolution*, **62**, 128–139.

Jacobs, L. F. and Schenk, F. (2003) Unpacking the cognitive map: the parallel map theory of hippocampal function. *Psychological Review*, **110**, 285–315.

Jacobsen, C. F. (1936) Studies of cerebral function in primates. I. The functions of the frontal association areas in monkeys. *Comparative Psychology Monographs*, **13**, 30–60.

James, D. G. (1993) Migration biology of the Monarch butterfly in Australia. In *Biology and Conservation of the Monarch Butterfly*, ed. S. B. Malcolm and M. P. Zalucki. Los Angeles, CA: Natural History Museum, pp. 189–200.

James, W. (1890) *The Principles of Psychology*, 2 vols. New York: Henry Holt. Reprinted (1999) Bristol, UK: Thoemmes Press.

Janik, V. M. (2000) Food-related bray calls in wild bottlenose dolphins (*Tursiops truncatus*). *Proceedings of the Royal Society of London B*, **267**, 923–927.

Jarvis, J. U. M. (1981) Eusociality in a mammal: cooperative breeding in naked mole rat colonies. *Science*, **212**, 571–573.

Jarvis, J. U. M., O'Rian, M. J., Bennett, N. C. and Sherman, P. W. (1994) Mammalian eusociality: a family affair. *Trends in Ecology and Evolution*, **9**, 47–51.

Jaynes, J. (1969) The historical origins of 'ethology' and 'comparative psychology'. *Animal Behaviour*, **17**, 601–606.

Jenkins, S. H. and Breck, S. W. (1998) Differences in food-hoarding among six species of heteromyid rodents. *Journal of Mammalogy*, **79**, 1221–1233.

Jennings, H. S. (1906) *Behavior of the Lower Organisms*. New York: Academic Press.

Jitsumori, M. (2004) Categorisation and formation of equivalence classes in animals: studies in Japan on the background of contemporary developments. *Japanese Psychological Research*, **46**, 182–194.

Jitsumori, M. and Yoshihara, M. (1997) Categorical discrimination of human facial expression by pigeons: a test of the linear feature model. *Quarterly Journal of Experimental Psychology B*, **50**, 253–268.

Johnson, R. N., Oldroyd, B. P., Barron, A. B. and Crozier, R. H. (2002) Genetic control of the honey bee (*Apis mellifera*) dance language: segregating dance forms in a backcrossed colony. *Journal of Heredity*, **93**, 170–173.

Jolly, A. (1966) Lemur social behavior and primate intelligence. *Science*, **153**, 501–506.

Jones, T. and Kamil, A. C. (1973) Tool-making and tool-using in the northen blue jay. *Science*, **180**, 1076–1078.

Jones, T. C. and Parker, P. G. (2000) Cost and benefit of foraging associated with delayed dispersal in the spider *Anelosimus studiosus* (Araneae, Theridiidae). *Journal of Arachnology*, **28**, 61–69.

Jouventin, P. and Weimerskirch, H. (1990) Satellite tracking of wandering albatrosses. *Nature*, **343**, 746–748.

Judd, S. P. D. and Collett, T. S. (1998) Multiple stored views and landmark guidance in ants. *Nature*, **392**, 710–713.

Judd, T. M. (2000) Division of labour in colony defence against vertebrate predators by the social wasp *Polistes fuscatus*. *Animal Behaviour*, **60**, 55–61.

Judd, T. M. and Sherman, P. W. (1996) Naked mole-rats recruit colony mates to food sources. *Animal Behaviour*, **52**, 957–969.

Kacelnik, A., Chappell, J., Weir, A. A. S. and Kenward, B. (2004) Tool use and manufacture in birds. In *Encyclopedia of Animal Behavior*, vol. 3, ed. M. Bekoff. Westport, CT: Greenwood Press, pp. 1067–1069.

Kaib, M. (1999) Termites. In *Pheromones of Non-Lepidopteran Insects Associated with Agricultural Plants*, ed. R. J. Hardie and A. K. Minks. Wallingford, UK: CAB International, pp. 329–353.

Kaib, M., Husseneder, C., Epplen, C., Epplen, J. T. and Brandl, R. (1996) Kin-biased foraging in a termite. *Proceedings of the Royal Society of London B*, **263**, 1527–1532.

Kamil, A. C. and Balda, R. P. (1990) Spatial memory in seed-caching corvids. *Psychology of Learning and Motivation*, **26**, 1–25.

Kamil, A. C. and Cheng, K. (2001) Way-finding and landmarks: the multiple bearing hypothesis. *Journal of Experimental Biology*, **204**, 103–113.

Kamil, A. C. and Jones, J. E. (2000) Geometric rule learning by Clark's nutcrackers (*Nucifraga columbiana*). *Animal Behaviour*, **34**, 1289–1298.

Kamin, L. J. (1957) The retention of an incompletely learned avoidance response. *Journal of Comparative Physiology and Psychology*, **50**, 457–460.

Kamin, L. J. (1963) The retention of an incompletely learned avoidance response: some further analyses. *Journal of Comparative Physiology and Psychology*, **56**, 719–722.

Kandel, E. R. (2001) The molecular biology of memory storage: a dialogue between genes and synapses. *Science*, **294**, 1030–1038.

Karlson, P. and Lüscher, M. (1959) Pheromones: a new term for a class of biologically active substances. *Nature*, **183**, 55–56.

Kartsev, V. M. (1990) Visual searching of local food sources in social hymenoptera. In *Sensory Systems and Communication in Arthropods*, ed. F. G. Gribakin, K. Wiese and A. V. Popov. Basel, Switzerland: Birkhäuser Verlag, pp. 154–157.

Kartsev, V. M., Oganesov, T. and Kalinin, D. (2005) Learning at consequential stages of orientation in bumblebees and paper wasps. *Proceedings 3rd European Congress on Social Insects*, St Petersburg, Russia, p. 68.

Kaufman, E. L., Lord, M. W., Reese, T. W. and Volkmann, J. (1949) The discrimination of visual number. *American Journal of Psychology*, **62**, 498–525.

Kaufman, K. (1991) The subject is Alex. *Audubon Magazine*, Sept.–Oct., 52–58.

Kaufman, L. (1974) *Sight and Mind: An Introduction to Visual Perception*. New York: Oxford University Press.

Kawai, M. (1965) Newly acquired pre-cultural behavior of the natural troop of Japanese monkeys on Koshima Islet. *Primates*, **6**, 1–30.

Kaye, H. and Pearce, J. M. (1984) The strength of the orienting response during Pavlovian conditioning. *Journal of Experimental Psychology: Animal Behavior Processes*, **10**, 90–109.

Keenan, J., Gallup, G. G., Jr and Falk, D. (2003) *The Face in the Mirror: The Search for the Origins of Consciousness*. London: HarperCollins.

Keeton, W. T. (1974) The orientational and navigational basis of homing in birds. *Advances in the Study of Behavior*, **5**, 47–132.

Kelleher, R. T. (1958) Fixed-ratio schedules of conditioned reinforcement with chimpanzees. *Journal of the Experimental Analysis of Behavior*, **1**, 281–289.

Kellman, P. J. and Spelke, E. S. (1983) Perception of partly occluded objects in infancy. *Cognitive Psychology*, **15**, 483–524.

Kellog, W. N. and Kellog, L. A. (1933) *The Ape and the Child*. New York: McGraw-Hill.

Kelly, D. M., Spetch, M. L. and Heath, C. D. (1998) Pigeons' (*Columba livia*) encoding of geometric and featural properties of a spatial environment. *Journal of Comparative Psychology*, **112**, 259–269.

Kendrick, K. M. and Baldwin, B. A. (1987) Cells in the temporal cortex of conscious sheep can respond preferentially to the sight of faces. *Science*, **236**, 448–450.

Kendrick, K. M., Atkins, K., Hinton, M. R., *et al.* (1995) Facial and vocal discrimination in sheep. *Animal Behaviour*, **49**, 1655–1676.

Kendrick, K. M., Leigh, A. and Peirce, J. W. (2001) Behavioural and neural correlates of mental imagery in sheep using face recognition paradigms. *Animal Welfare*, **10**, 89–101.

Kent, D. S. and Simpson, J. A. (1992) Eusociality in the beetle *Austroplatypus incompertus* (Coleoptera: Scolytidae). *Naturwissenschaften*, **79**, 86–87.

Kenward, B., Weir, A. A. S., Rutz, C. and Kacelnik, A. (2005) Tool manufacture by naïve juvenile crows. *Nature*, **433**, 121–122.

Kilian, A., Yaman, S., von Fersen, L. and Gunturkun, O. (2003) A bottlenose dolphin discriminates visual stimuli differing in numerosity. *Learning and Behavior*, **31**, 133–142.

Kimble, G. A. and Wertheimer, M. (1998) *Portraits of Pioneers in Psychology*, vol. 3. Washington, DC: American Psychological Association.

Kimble, G.A., and Wertheimer, M. (2000) *Portraits of Pioneers in Psychology*, vol. 4. Washington DC: American Psychological Association.

Kimble, G.A., Wertheimer, M. and White, C.L. (1991) *Portraits of Pioneers in Psychology*, vol. 1. Washington, DC: American Psychological Association.

Kimble, G.A., Boneau, C.A. and Wertheimer, M. (1996) *Portraits of Pioneers in Psychology*, vol. 2. Washington, DC: American Psychological Association.

Kimura, D. (1979) Neuromotor mechanisms in the evolution of human communication. In *Neurobiology of Social Communication in Primates*, ed. H.D. Steklis and M.J. Raleigh. New York: Academic Press, pp. 197–219.

King, G.E. (1997) The attentional basis for primate response to snakes. *20th Annual Meeting of the American Society of Primatologists*, San Diego, CA.

Kinnaman, A.J. (1902) Mental life of two *Macacus rhesus* monkeys in captivity. *American Journal of Psychology*, **13**, 98–148.

Kipling, R. (1902) *Just So Stories*. Reissned (1996) London: HarperCollins.

Kirchner, W.H. and Towne, W.F. (1994) The sensory basis of the honeybee's dance language. *Scientific American*, **270**, 74–80.

Kitchen, D.M., Cheney, D.L. and Seyfarth, R.M. (2005) Male chacma baboons (*Papio hamadryas ursinus*) discriminate loud call contests between rivals of different relative ranks. *Animal Cognition*, **8**, 1–6.

Kline, L.W. (1899) Methods in animal psychology. *American Journal of Psychology*, **10**, 256–279.

Kloft, W. (1960) Die Trophobiose zwischen Waldamesien und Pflanzenläusen mit Untersuchungen über Wechselwirkungen zwischen Pflanzenläusen und Pflanzengeweben. *Entomophaga*, **5**, 43–54.

Klopfer, P.H. (1959) An analysis of learning in young Anatidae. *Ecology*, **40**, 90–102.

Klopfer, P.H. and Hailman, J.P. (1967) *An Introduction to Animal Behavior: Ethology's First Century*. Englewood Cliffs, NJ: Prentice-Hall.

Koenig, W.D. and Dickinson, J.L. (eds.) (2004) *Ecology and Evolution of Cooperative Breeding in Birds*. Cambridge, UK: Cambridge University Press.

Köhler, O. (1941) Vom Erlernen unbenannter Anzahlen bei Vögeln. *Naturwissenschaften*, **29**, 201–218.

Köhler, O. (1956) Thinking without words. *Proceedings of 14th International Congress of Zoology*, Copenhagen, pp. 75–88.

Köhler, W. (1918) Nachweis einfacher Structurfunktionen beim Schimpansen und beim Haushuhn. Abhandlungen der preussischen Akademie der wissenschaften 2, 1–101. Translated and condensed (1969) Simple structural functions in chimpanzees and chicken. In *A Source Book for Gestalt Psychology*, ed. W.D. Ellis. London: Routledge & Kegan Paul.

Köhler, W. (1925) *The Mentality of Apes*. London: Routledge & Kegan Paul. Reprinted (1976) New York: Liversidge.

Köhler, W. (1959) Gestalt psychology today. *American Psychologist*, **14**, 727–734.

Kolmogorov, A.N. (1965) Three approaches to the quantitative definition of information. *Problems in Information Transmission*, **1**, 1–7.

Konorski, J. (1967) *Integrated Activity of the Brain*. Chicago, IL: University of Chicago Press.

Koolhaas, J.M., Korte, S.M., De Boer, S.F., *et al.* (1999) Coping styles in animals: current status in behavior and stress-physiology. *Neuroscience and Biobehavioral Reviews*, **23**, 925–935.

Kotenkova, E. and Bulatova, N. (eds.) (1994) *The House Mouse* (Mus musculus). Moscow: Nauka. (In Russian with English summaries.)

Krachun, C. (2002) Numerical competence in nonhuman primates: a review of indicators. Available online: www.carleton.ca/iis/TechReports/.

Kralik, J.D. and Hauser, M.D. (2002) A nonhuman primate's perception of object relations: experiments on cottontop tamarins, *Saguinus oedipus*. *Animal Behaviour*, **63**, 419–435.

Kramer, G. (1953) Wird die Sonnenhohe bei der Heimfindeorientierung verwertet. *Journal für Ornithologie*, **94**, 201–219.

Krause, J., Hoare, D.J., Croft, D., *et al.* (2000) Fish shoal composition: mechanisms and constraints. *Proceedings of the Royal Society of London B*, **267**, 2011–2017.

Krebs, J.R. and Davies N.B. (eds.) (1997) *Behavioural Ecology: An Evolutionary Approach*, 3rd edn. Oxford, UK: Blackwell.

Krebs, J.R. and Dawkins, R. (1984) Animal signals: mind-reading and manipulation. In *Behavioral Ecology: An Evolutionary Approach*, ed. J.R. Krebs and N.B. Davies, 2nd edn. Oxford, UK: Blackwell Scientific Publications, pp. 380–402.

Krebs, J.R., Sherry, D.F., Healy, S.D., Perry, V.H., and Vaccarino, A.L. (1989) Hippocampal specialization of food-storing birds. *Proceedings of the National Academy of Sciences of the USA*, **86**, 1388–1392.

Krekling, S., Tellevik, J.M. and Nordvik, H. (1989) Tactual learning and cross-modal transfer of an oddity problem in young children. *Journal of Experimental Child Research*, **47**, 88–96.

Kroodsma, D.E. and Baylis, J.R. (1982) A world survey of evidence for vocal learning in birds. In *Ecology and Evolution of Acoustic Communication in Birds*, ed. D.E. Kroodsma, E.M. Miller and H. Quellet. Ithaca, NY: Cornell University Press, pp. 311–338.

Kropotkin, P. (1902) *Mutual Aid: A Factor in Evolution*. Reprinted (1998) London: Freedom Press.

Krushinskaya, N. (1966) Some complex forms of feeding behavior of nutcracker *Nucifraga caryocatactes*, after removal of old cortex. *Journal of Evolutionary Biochemistry and Physiology*, **2**, 563–568. (In Russian with English summary.)

Krushinsky, L.V. (1958) Extrapolation reflexes as the elementary basis of reasoning activity in animals. *Doklady Academii Nauk SSSR*, **121**, 762–765. (In Russian with English summary.)

Krushinsky, L.V. (1965) Solution of elementary logical problems by animals on the basis of extrapolation. In *Cybernetics of the Nervous System*, vol. 17, ed. N. Wiener and J.P. Shade. New York: Elsevier, pp. 280–308.

Krushinsky, L.V. (1977) *Biological Foundations of Reasoning: Evolutionary, Physiological and Genetical Aspects of Behaviour*. Moscow: Moscow University Press. (In Russian.)

Krushinsky, L.V. (1990) *Experimental Studies of Elementary Reasoning: Evolutionary, Phylogenetical and Genetic Aspects of Behavior*. New Delhi, India: American Publishing Co. Pvt. Ltd.

Krützen, M., Mann, J., Heithaus, M., *et al.* (2005) Cultural transmission of tool use in bottlenose dolphins. *Proceedings of the National Academy of Sciences of the USA*, **105**, 8939–8943.

Kuba, M., Meisel, D.V., Byrne, R.A., Griebel, U. and Mather, J.A. (2003) Looking at play in *Octopus vulgaris*. In *Coleoid Cephalopods through Time*, ed. K. Warnke, H. Keupp and S. Boletzki. *Berliner Paläobiologie (special issue)*, **3**, 63–169.

Kullberg, C. and Lind, J. (2002) An experimental study of predator recognition in great tit fledglings. *Ethology*, **108**, 429–441.

Kummer, H. and Goodall, J. (1985) Conditions of innovative behaviour in primates. *Philosophical Transactions of the Royal Society of London B*, **308**, 203–214.

Kuroshima, H., Fujita, K., Fuyuki, A. and Masuda, T. (2002) Understanding of the relationship between seeing and knowing by tufted capuchin monkeys (*Cebus apella*). *Animal Cognition*, **5**, 41–48.

Kuroshima, H., Fujita, K., Adachi, I., Iwata, K. and Fuyuki, A. (2003) A capuchin monkey (*Cebus apella*) recognises when people do and do not know the location of food. *Animal Cognition*, **6**, 283–291.

Kuwabara, M. (1957) Bildung des bedingten Reflexes von Pavlovs Typus bei der Honigbiene, *Apis mellifera*. *Journal of the Faculty of Science, Hokkaido University, Series VI, Zoology*, **13**, 458–464.

Lack, D.L. (1966) *Population Studies of Birds*. Oxford, UK: Clarendon Press.

Ladygina Koths, N.N. (1923) *Study of Cognitive Abilities in the Chimpanzee*. Moscow: Gosizdat. (In Russian.)

Ladygina Koths, N.N. (1935) *Infant Chimpanzee and Human Child*. Reprinted (2002) in *A Classic 1935 Comparative Study of Ape Emotions and Intelligence*, ed. F.B.M. de Waal, trans. B. Vekker. New York: Oxford University Press.

Ladygina Koths, N.N. (1959) *Construction and Tool Use in Great Apes*. Moscow: Academy of Science, USSR. (In Russian.)

Laland, K.N. (1994) Sexual selection with a culturally transmitted mating preference. *Theoretical Population Biology*, **45**, 1–15.

Laland, K.N. and Brown, G. (2002) *Sense and Nonsense*. Oxford, UK: Oxford University Press.

Laland, K.N. and Reader, S.M. (1999) Foraging innovation in the guppy. *Animal Behaviour*, **57**, 331–340.

Lancia, R.A. and Hodgdon, H.E. (1983) Observations on the ontogeny of behavior of hand-reared beavers (*Castor canadensis*). *Acta Zoologica Fennica*, **174**, 117–119.

Landsberg, J.W. and Weiss, J. (1976) Stress and increase in corticosterone level prevent imprinting in ducklings. *Behaviour*, **57**, 173–189.

Lansade, L., Bertrand, M. and Bouissou, M.-F. (2005) Effects of neonatal handling on subsequent manageability, reactivity and learning ability of foals. *Applied Animal Behaviour Science*, **92**, 143–158.

Lashley, K.S. (1929) *Brain Mechanisms and Intelligence: A Quantitative Study of Injuries to the Brain*. Chicago, IL: University of Chicago Press.

Lashley, K.S. (1938). Experimental analysis of instinctive behavior. *Psychological Review*, **45**, 445–471.

Lashley, K.S. (1950) In search of the engram. In *Symposia of the Society for Experimental Biology*, **4**, 454–482.

Lashley, K.S. (1951) The problem of serial order in behavior. In *Cerebral Mechanisms in Behavior*, ed. L.A. Jeffries. New York: John Wiley, pp. 112–136.

Lazareva, O.F., Smirnova, A.A., Bagozkaja, M.S., *et al.* (2004) Transitive responding in hooded crows requires linearly ordered stimuli. *Journal of Experimental Analysis of Behavior*, **82**, 1–19.

Lawick H. van and van Lawick-Goodall, J. (1970) *Innocent Killers*. London: Collins.

LeBas, N. R., Hockham, L. P. and Ritchie, M. G. (2004) Sexual selection in the gift-giving dance fly, *Rhamphomyia sulcata*, favours small males carrying small gifts. *Evolution*, **58**, 1763–1772.

LeCompte, G. K. and Gratch, G. (1972) Violation of a rule as a method of diagnosing infant's levels of object concepts. *Child Development*, **43**, 385–396.

Lefebvre, A. L. and Giraldeau, L. (1994) Cultural transmission in pigeons is affected by the numbers of tutors and bystanders present. *Animal Behaviour*, **47**, 331–337.

Lefebvre, L. and Palameta, B. (1988) Mechanisms, ecology and population diffusion of socially learned, food-finding behaviour in feral pigeons. In *Social Learning: Psychological and Biological Perspectives*, ed. T. Zentall and B. G. Galef, Jr. Hillsdale, NJ: Lawrence Erlbaum, pp. 141–163.

Lefebvre, L., Whittle, P., Lascaris, E. and Finkelstein, A. (1997) Feeding innovations and forebrain size in birds. *Animal Behaviour*, **53**, 549–560.

Lehrer, M. (1987) To be or not to be a colour-seeing bee. *Israel Journal of Entomology*, **21**, 51–76.

Lehrer, M. (1993) Why do bees turn back and look? *Journal of Comparative Physiology A*, **172**, 544–563.

Lehrer, M. (1996) Small-scale navigation in the honey bee: active acquisition of visual information about the goal. *Journal of Experimental Biology*, **199**, 253–261.

Lehrer, M. (ed.) (1997) *Orientation and Communication in Arthropods*. Basel, Switzerland: Birkhäuser Verlag.

Lehrer, M. (1998) Looking all around: honeybees use different cues in different eye regions. *Journal of Experimental Biology*, **201**, 3275–3292.

Lehrer, M. and Bianko, G. (2000) The turn-back-and-look behaviour: bee versus robot. *Biological Cybernetics*, **83**, 211–229.

Lehrer, M. and Collett, T. C. (1994) Approaching and departing bees learn different cues to the distance of a landmark. *Journal of Comparative Physiology A*, **175**, 171–177.

Lehrer, M., Wehner, R. and Srinivasan, M. V. (1985) Visual scanning behaviour in honey bees. *Journal of Comparative Physiology A*, **157**, 405–415.

Lehrer, M., Srinivasan, M. V., Zhang, S. W. and Horridge, G. A. (1988) Motion cues provide the bees visual world with a third dimension. *Nature*, **332**, 356–357.

Lenoir, A. and Jaisson, P. (1982) Evolution et rôle des communications antennaries ches les insectes sociaux. In *Social Insects in the Tropics*, ed. P. Jasson. Paris: Université Paris-Nord, pp. 157–180.

Levey, D. J., Duncan, R. S. and Levins, C. F. (2004) Use of dung as a tool by burrowing owls. *Nature*, **431**, 39.

Lewis, L. A. and Schneider, S. S. (2000) The modulation of worker behavior by the vibration signal during house hunting in swarms of the honeybee, *Apis mellifera*. *Behavioral Ecology and Sociobiology*, **48**, 154–164.

Levchenko, I. A. (1959) The return of bees to the hive. *Pchelovodstvo*, **36**, 38–40. (In Russian.)

Levchenko, I. A. (1976) *Transferring of Information about Coordinates of a Food Source in the Honey Bee*. Kiev, USSR: Naukova Dumka. (In Russian.)

Leyhausen, P. and Tonkin, B. (1979) *Cat Behavior: The Predatory and Social Behavior of Domestic and Wild Cats*. New York: Garland STPM Press.

Lieberman, P. (1984) *The Biology and Evolution of Language*. Cambridge, MA: Harvard University Press.

Limongelli, L., Boysen, S. and Visalberghi, E. (1995) Comprehension of cause and effect relationships in a tool-using task by common chimpanzees (*Pan troglodytes*). *Journal of Comparative Psychology*, **109**, 18–26.

Lin, N. and Michener, C. D. (1972) Evolution of sociality in insects. *Quarterly Review of Biology*, **47**, 131–159.

Lindauer, M. (1960) Time compensated sun orientation in bees. *Cold Spring Harbor Symposium on Quantitative Biology*, **25**, 372–377.

Lindauer, M. (1961) *Communication among Social Bees*. Cambridge, MA: Harvard University Press.

Lindauer, M. (1963) Kompassorientierung. *Ergebnisse der Biologie*, **26**, 158–181.

Linden, E. (1974) *Apes, Men and Language*. New York: Saturday Review Press.

Linsenmaier, K. E. (1987) Kin recognition in subsocial arthropods, in particular in the desert isopod *Hemilepistus reaumuri*. In *Kin Recognition in Animals*, ed. D. J. C. Fletcher and C. D. Michener. New York: John Wiley, pp. 121–208.

Lipp, H.-P., Vyssotski, A., Wolfer, D. P., *et al.* (2004) Pigeon homing along highways and exits. *Current Biology*, **14**, 1239–1249.

Loeb, J. (1900) *Comparative Physiology of the Brain and Comparative Psychology*. New York: G. P. Putnam's Sons. (Translated and extensively revised from: Loeb, J. (1899) *Einleitung in die vergleichende Gehirnphysiologie und vergleichende Psychologie*. Leipzig, Germany: Verlag von Johann Ambrosius Barth.)

Logothetis, N. K., Pauls, J., Bülthoff, H. H. and Poggio, T. (1994) View-dependent object recognition by monkeys. *Current Biology*, **4**, 401–414.

Logue, A. W. (1988) Research on self-control: an integrating framework. *Behavioral and Brain Sciences*, **11**, 665–709.

Lohman, K. J. (1992) How sea turtles navigate. *Scientific American*, **266**, 100–106.

Lopatina, N. G. (1971) *Signalling Activity in Families of the Honey Bee* (Apis mellifera). Leningrad, USSR: Nauka.

Lorenz, K. (1931) Beiträge zur Ethologie sozialer Corviden. *Journal für Ornithologie*, **79**, 67–127.

Lorenz, K. (1935) Der Kumpanin der Umvelt des Vogels: die Artgenosse als Ausloesendesmoment socialer Verhaltensweisen. *Journal für Ornithologie*, **83**, 137–213.

Lorenz, K. (1937) Über den Begriff der Instinkthandlung. *Folia Biotheoretica B*, **2**, 17–50.

Lorenz, K. (1950) The comparative method in studying innate behaviour patterns. *Symposia of the Society for Experimental Biology*, **4**, 221–268.

Lorenz, K. (1952) *King Solomon's Ring*, trans. M.K. Wilson and Th. Y. Cromwell. New York: T.Y. Crowell & Co.

Lorenz, K. (1965) *Evolution and Modification of Behavior*. Chicago, IL: University of Chicago Press.

Lorenz, K. (1966) *On Aggression*. New York: Harcourt Brace.

Lorenz, K. (1969) Innate bases of learning. In *On the Biology of Learning*, ed. K.H. Pribram. New York: Harcourt Brace, pp. 13–93.

Lorenz, K. (1979) *The Year of the Greylag Goose*. New York: Harcourt Brace Jovanovich.

Lott, D.F. (1991) *Intraspecific Variation in the Social Systems of Wild Vertebrates*. Cambridge, UK: Cambridge University Press.

Lubbock, J. (1882) *Ants, Bees, and Wasps: A Record of Observations on the Habits of the Social Hymenoptera*. London: Kegan Paul, Trench & Co.

Lubow, R.E. (1973) Latent inhibition. *Psychological Bulletin*, **79**, 398–407.

Lucas, J.R., Brodin, A., de Kort, S.R. and Clayton, N.S. (2004) Does hippocampal size correlate with the degree of caching specialisation? *Proceeding of the Royal Society of London B*, **271**, 2423–2429.

Lukas, D., Reynolds, V., Boesch, C. and Vigilant, L. (2005) To what extent does living in a group mean living with kin? *Molecular Ecology*, **14**, 2181–2196.

Luria, A.R. and Majovski, L.V. (1977) Basic approach used in American and Soviet clinical neuropsychology. *American Psychologist*, **32**, 959–971.

MacDonald, D.W. (1976) Food caching by red foxes and other carnivores. *Zoological Tierpsychology*, **42**, 170–185.

Mackintosh, N.J. (1974) *The Psychology of Animal Learning*. London: Academic Press.

Mackintosh, N.J. (1976) Overshadowing and stimulus intensity. *Animal Learning and Behavior*, **4**, 186–192.

Mackintosh, N.J. and Little, L. (1969) Intradimensional and extradimensional shift learning by pigeons. *Psychonomic Science*, **15**, 5–6.

Macphail, E.M. (1982) *Brain and Intelligence in Vertebrates*. Oxford, UK: Clarendon Press.

Macphail, E.M. (1993) *The Neuroscience of Animal Intelligence, from the Seahare to the Seahorse*. New York: Columbia University Press.

Macphail, E.M. (2002) The role of the avian hippocampus in spatial memory. *Psicológica*, **1**, 93–108.

Macphail, E.M. and Bolhuis, J.J. (2001) The evolution of intelligence: adaptive specializations versus general process. *Biological Reviews*, **76**, 341–364.

Magurran, A.E. and Higham, A. (1988) Information transfer across fish shoals under predator threat. *Ethology*, **78**, 153–158.

Maier, N.R.F. and Schneirla, T.C. (1935) *Principles of Animal Psychology*. New York: McGraw-Hill.

Mainardi, D. (1976) *Il cane e la volpe (Il racconto-verità di un libero viaggio nel dominio dell'erologia)*. Milan: Editore Rizzoli.

Makarova, E.G. (1990) *Childhood in the Beginning (Teacher's Notes)*. Moscow: Pedagogica. (In Russian.)

Makino, H. and Jitsumori, M. (2001) Category learning and prototype effect in pigeons: a study with morphed images of human faces. *Japanese Journal of Psychology*, **71**, 477–485.

Malcolm, J.R. and Marten, K. (1982) Natural selection and the communal rearing of pups in African wild dog (*Lycaon pictus*). *Behavioral Ecology and Sociobiology*, **10**, 1–13.

Mann, K.D., Turnell, E.R., Atema, J. and Gerlach, G. (2003) Kin recognition in juvenile zebrafish (*Danio rerio*) based on olfactory cues. *Biological Bulletin*, **205**, 224–225.

Mann, J. and Sargeant, B. (2003) Like mother, like calf: the ontogeny of foraging traditions in wild Indian Ocean bottlenose dolphins (*Tursiops* sp.). In *The Biology of Traditions: Models and Evidence*, ed. D.M. Fragaszy and S. Perry. Cambridge, UK: Cambridge University Press, pp. 236–266.

Mann, J., Connor, R.C., Tyack, P.L. and Whitehead, H. (eds.) (2000) *Cetacean Societies: Field Studies of Dolphins and Whales*. Chicago, IL: University of Chicago Press.

Manning, A. (1979) *An Introduction to Animal Behaviour*, 3rd edn. London: Edward Arnold.

Manser, M.R. (2001) The acoustic structure of suricates' alarm calls varies with predator type and the level of response urgency. *Proceedings of the Royal Society of London B*, **268**, 2315–2324.

Manser, M.R. and Bell, M.B. (2004) Spatial representation on shelter locations in meerkats, *Suricata suricatta*. *Animal Behaviour*, **68**, 151–157.

Marikovsky, P.I. (1969) On the biology of desert woodlouse *Hemilepistus rhinoceros*. *Zoological Journal*, **48**, 677–685. (In Russian with English summary.)

Markov, V.I. and Ostrovskaya, V.M. (1990) Organization of communication system in *Tursiops truncatus* Montagu. *In Sensory Abilities of Cetaceans*, ed. J. Thomas and R. Kastelein. New York: Plenum Press, pp. 599–622.

Markova, A. Ya. (1962) A process of elementary abstraction in monkeys. *Problems of Psychology*, **1**, 18–32. (In Russian.)

Markowitz, H., Schmidt, M., Nadal, L. and Squier, L. (1975) Do elephants ever forget? *Journal of Applied Behavioural Analysis*, **8**, 333–335.

Marler, P. (1970) A comparative approach to vocal learning: song development in white-crowned sparrows. *Journal of Comparative Psychology*, **71**, 1–25.

Marler, P. (1984) Song learning: innate species differences in the learning process. In *The Biology of Learning*, ed. P. Marler and H. S. Terrance. New York: Springer-Verlag, pp. 289–309.

Marler, P. (1997) Three models of song learning: evidence from behavior. *Journal of Neurobiology*, **33**, 501–516.

Marler, P. and Hamilton, W.J. (1966) *Mechanisms of Animal Behavior*. New York: John Wiley.

Marler, P. and Tamura, M. (1962) Song 'dialects' in three populations of white-crowned sparrows. *Condor*, **64**, 368–377.

Marler, P., and Tenaza, R. (1977) Signaling behavior of apes with special reference to vocalization. In *How Animals Communicate*, ed. T. Sebeok. Bloomington, IN: University of Indiana Press, pp. 965–1033.

Marten, K. and Psarakos, S. (1995) Evidence of self-awareness in the bottlenose dolphin (*Tursiops truncatus*). In *Self-Awareness in Animals and Humans: Developmental Perspectives*, ed. S. Taylor Parker, R.W. Mitchell and M.L. Boccia. New York: Cambridge University Press, pp. 361–379.

Maslow, A. (1936) A theory of sexual behavior of infra-human primates. *Journal of Genetic Psychology*, **48**, 310–338.

Mason, J. R. and Reidinger, R. F. (1981) Effects of social facilitation and observational learning on feeding behaviour of the red-winged blackbird (*Agelaius phoeniceus*). *Auk*, **98**, 778–784.

Massey, A. (1977) Agonistic aids and kinship in a group of pigtail macaques. *Behavioral Ecology and Sociobiology*, **2**, 31–40.

Mateo, J.M. (1996) The development of alarm-call response behaviour in free-living juvenile Belding's ground squirrels. *Animal Behaviour*, **52**, 489–505.

Mateo, J. M. (2002) Kin recognition abilities and nepotism as a function of sociality. *Proceedings of the Royal Society of London B*, **269**, 721–727.

Mateo, J. M. (2004) Recognition systems and biological organization: the perception component of social recognition. *Annales Zoologici Fennici*, **41**, 729–745.

Mateo, J. M. (2006) The nature and representation of individual recognition cues in Belding's ground squirrels. *Animal Behaviour*, **71**, 141–154.

Mathieu, M. and Bergeron, G. (1981) Piagettian assessment of cognitive development in chimpanzees (*Pan troglodytes*). In *Primate Behavior and Sociobiology*, ed. A.B. Chiarell and S. Coruccini. Berlin: Springer-Verlag, pp. 142–147.

Matsuzawa, T. (1985) Use of numbers by a chimpanzee. *Nature*, **315**, 57–59.

Matsuzawa, T. and Yamakoshi, G. (1996) Comparison of chimpanzee material culture between Bossou and Nimba, West Africa. In *Reaching into Thought: The Mind of the Great Apes*, ed. A.E. Russon, K.A. Bard and S. Parker. Cambridge, UK: Cambridge University Press, pp. 211–232.

Matthews, G.V.T. (1955) *Bird Navigation*: Cambridge, UK: Cambridge University Press.

Maynard Smith, J. (1964) Group selection and kin selection. *Nature*, **201**, 1145–1147.

Maynard Smith, J. (1974) The theory of games and animal conflicts. *Journal of Theoretical Biology*, **47**, 209–221.

Maynard Smith, J. (1982) *Evolution and the Theory of Games*. Cambridge, UK: Cambridge University Press.

Maynard Smith, J. (1989) *Evolutionary Genetics*. Oxford, UK: Oxford University Press.

Maynard Smith, J. and Michie, D. (1964) Machines that play games. *New Scientist*, **12**, 367–369.

Maynard Smith, J. and Price, G.R. (1973) The logic of animal conflict. *Nature*, **246**, 16–18.

Mayr, E. (1942) *Systematics and the Origin of Species*. New York: Columbia University Press.

Mayr, E. (2001) *What Is Evolution?* New York: Basic Books.

Mazokhin-Porshnyakov, G.A. (1969) *Insect Vision*. New York: Plenum Press.

Mazokhin-Porshnyakov, G.A. (1989) How to measure animal intelligence? *Priroda*, **4**, 18–25. (In Russian.)

Mazokhin-Porshnyakov, G.A. and Kartsev, V.M. (1979) A study of sucession of visiting feeders during flight inspections in insects (on a strategy of the visual search). *Zoological Journal*, **58**, 1281–1289. (In Russian with English summary.)

Mazokhin-Porshnyakov, G. A. and Kartsev, V. M. (1984) Peculiarities of searching behaviour of social and parasitic hymenopterans. In *Insect Behaviour*, ed. G.M. Dlussky. Moscow: Nauka, pp. 64–80. (In Russian.)

Mazokhin-Porshnyakov, G. A. and Kartsev, V. M. (2000) Learning in bees and wasps in complicated experimental tasks. In *The Cognitive Development of an Autonomous Behaving Prerational Intelligence*, vol. 1, *Adaptive Behavior and Intelligent Systems without Symbols and Logic*, ed. H. Cruse, J. Dean and H. Ritter. Berlin: Springer-Verlag, pp. 449–467.

Mazokhin-Porshnyakov, G. A. and Murzin, S. V. (1977) Feeding and nest visual cues in the ant *Cataglyphis setipes turcomanica*. *Zoological Journal*, **56**, 400–404. (In Russian with English summary.)

Mazokhin-Porshnyakov, G. A., Semyonova, S. A. and Milevskaya, I. A. (1977) Characteristic features of the identification by *Apis mellifera* of objects by their size. *Journal of General Biology*, **38**, 855–962 (In Russian with English summary.)

Mazur, J. E. (1987) An adjusting procedure for studying delayed reinforcement. In *Quantitative Analyses of Behavior*, vol. 5, *The Effect of Delay and of Intervening Events on Reinforcement Value*, ed. M. L. Commons, J. E. Mazur, J. A. Nevin, and H. Rachlin. Hillsdale, NJ: Lawrence Erlbaum, pp. 55–73.

McComb, K., Packer, C. and Pusey, A. (1994) Roaring and numerical assessment in contests between groups of female lions, *Panthera leo*. *Animal Behaviour*, **47**, 379–387.

McComb, K., Moss, C., Sayialel, S. and Baker, L. (2000) Unusually extensive networks of vocal recognition in African elephants. *Animal Behaviour*, **59**, 1103–1109.

McComb, K., Moss, C., Durant, S. M., Baker, L. and Sayialel, S. (2001) Matriarchs as repositories of social knowledge in African elephants. *Science*, **292**, 491–494.

McFarland, D. (1985) *Animal Behaviour: Psychology, Ethology and Evolution*. London: Pitman.

McGaugh, J. L. (1989) Involvement of hormonal and neuromodulatory systems in the regulation of memory storage. *Annual Review of Neuroscience*, **12**, 255–287.

McGonigle, B. and Chalmers, M. (1977) Are monkeys logical? *Nature*, **267**, 694–697.

McGonigle, B. and Chalmers, M. (1992) Monkeys are rational! *Quarterly Journal of Experimental Psychology B*, **45**, 189–228.

McGrew, W. C. (1974) Tool use by wild chimpanzees in feeding upon driver ants. *Journal of Human Evolution*, **3**, 501–508.

McGrew, W. C. (1977) Socialisation and object manipulation of wild chimpanzees. In *Primate Bio-Social Development: Biological, Social, and Ecological Determinants*, ed. S. Chevalier-Skolnikoff and E. Poirier. New York: Garland Press, pp. 159–87.

McGrew, W. C. (1992) *Chimpanzee Material Culture: Implications for Human Evolution*. Cambridge, UK: Cambridge Universtiy Press.

McGrew, W. C. (2001) The other faunivory: primate insectivory and early human diet. In *Meat-Eating and Human Evolution*, ed. C. B. Stanford and H. T. Bunn. New York: Oxford University Press, pp. 160–178.

McGrew, W. C. (2004) *The Cultured Chimpanzee: Reflections on Cultural Primatology*. Cambridge, UK: Cambridge University Press.

McGrew, W. C. and Feistner, A. T. C. (1992) Two non-human primate models for the evolution of human food sharing: chimpanzee and callitrichids. In *The Adapted Mind: Evolutionary Psychology and the Generation of Culture*, ed. J. H. Barkow, L. Cosmides and J. Tooby. New York: Oxford University Press, pp. 229–248.

McGrew, W. C. and Tutin, C. E. G. (1978) Evidence for a social custom in wild chimpanzees? *Man*, **13**, 234–51.

McGrew, W. C., Marchant, L. F. and Nishida, T. (eds.) (1996) *Great Ape Societies*. Cambridge, UK: Cambridge University Press.

Meck, W. H. and Church, R. M. (1983) A mode control model of counting and timing processes. *Journal of Experimental Psychology: Animal Behavior Processes*, **9**, 320–334.

Meinertzhagen, R. (1954) The education of young ospreys. *Ibis*, **96**, 153–155.

Meltzoff, A. N. (1988) The human infant as *Homo imitans*. In *Social Learning: Psychological and Biological Perspectives*, ed. T. Zentall and B. Galef. Hillsdale, NJ: Lawrence Erlbaum, pp. 319–341.

Melville, H. (1851) *Moby Dick; or The Whale*. New York: Harper & Brothers.

Mendes, N. and Huber, L. (2004) Object permanence in common marmosets (*Callithrix jacchus*). *Journal of Comparative Psychology*, **118**, 103–112.

Menyuk, P., Menn, L. and Silber, R. (1986) Early strategies for the perception of words and sounds. In *Language Acquisition*, ed. P. Fletcher and M. Garman. Cambridge, UK: Cambridge University Press, pp. 49–70.

Menzel, C. R. (1996) Spontaneous use of matching visual cues during foraging by long-tailed macaques (*Macaca fascicularis*). *Journal of Comparative Psychology*, **110**, 370–376.

Menzel, E. W. (1971) Communication about the environment in a group of young chimpanzees. *Folia Primatologica*, **15**, 220–232.

Menzel, E. W. (1973a) Chimpanzee spatial memory organization. *Science*, **182**, 943–945.

Menzel, E.W. (1973b) Leadership and communication in young chimpanzees. In *Precultural Primate Behavior*, ed. E.W. Menzel. Basel, Switzerland: Karger, pp. 192–225.

Menzel, E.W. (1974) A group of chimpanzees in a 1-acre field: leadership and communication. In *Behavior of Nonhuman Primates*, ed. A.M. Schrier and F. Stollintz. New York: Academic Press, pp. 83–153.

Menzel, E.W. (1975) Purposive behavior as a basis for objective communication between chimpanzees. *Science*, **189**, 652–654.

Menzel, E.W. and Draper, W.A. (1965) Primate selection of food by size: visible versus invisible rewards. *Journal of Comparative and Physiological Psychology*, **59**, 231–239.

Menzel, R. (1985) Memory traces in honey bees. In *Neurobiology and Behaviour of Honeybees*, ed. R. and A. Mercer. Berlin: Springer-Verlag, pp. 310–325.

Menzel, R. (1999) Memory dynamics in the honeybee. *Journal of Comparative Physiology*, **185**, 323–340.

Menzel, R. (2001) Searching for the memory trace in a mini-brain, the honeybee. *Learning and Memory*, **8**, 53–62.

Menzel, R. and Müller, U. (1996) Learning and memory in honeybees: from behavior to neural substrates. *Annual Review of Neuroscience*, **19**, 379–404.

Menzel, R., Geiger, K., Müller, U., Joerges, J. and Chittka, L. (1998) Bees travel novel homeward routes by integrating separately acquired vector memories. *Animal Behaviour*, **55**, 139–152.

Menzel, R., Greggers, U., Smith, A., *et al.* (2005) Honey bees navigate according to a map-like spatial memory. *Proceedings of National Academy of Sciences of the USA*, **102**, 3040–3045.

Messenger, J.B. (1973) Learning in the cuttlefish, *Sepia*. *Animal Behaviour*, **21**, 801–826.

Metzgar, L.H. (1967) An experimental comparison of screech owl predation on resident and transient white-footed mice (*Peromyscus leucopus*). *Journal of Mammalogy*, **48**, 387–391.

Michelsen, A. (1993) The transfer of information in the dance language of honeybees: progress and problems. *Journal of Comparative Physiolology A*, **173**, 135–141.

Michelsen, A. (1999) The dance language of honeybees: recent findings and problems. In *The Design of Animal Communication*, ed. M.D. Hauser and M. Konishi. Cambridge, MA: MIT Press, pp. 111–113.

Michelsen, A., Andersen, B.B., Kirchner, W. and Lindauer, M. (1990) Transfer of information during honeybee dances, studied by means of a mechanical model. In *Sensory Systems and Communication in Arthropods*, ed. F.G. Bribakin, K. Wiese and A.V. Popov. Basel, Switzerland: Birkhäuser, pp. 284–300.

Michener, C.D. (1969) Comparative social behavior of bees. *Annual Review of Entomology*, **14**, 299–342.

Michener, C.D. (1974) *The Social Behavior of the Bees: A Comparative Study*. Cambridge, MA: Harvard University Press.

Michie, D. (1998) Ants: they really can talk. *The Independent on Sunday*, 15 November.

Midford, P.E., Hailman, J.P. and Woolfeden, G.E. (2000) Social learning of a novel foraging patch in families of free-living Florida scrub-jays. *Animal Behaviour*, **59**, 1199–1207.

Miklósi, A. (1999) The ethological analysis of imitation. *Biological Reviews*, **74**, 347–374.

Miklósi, A., Polgardi, R., Topal, J. and Csanyi, V. (2000) Intentional behaviour in dog–human communication: an experimental analysis of 'showing' behaviour in the dog. *Animal Cognition*, **3**, 159–166.

Miles, H.L. (1993) Language and the orangutan: the old 'person' of the forest. In *The Great Ape Project: Equality beyond Humanity*, ed. P. Cavalieri and P. Singer. London: Fourth Estate, pp. 42–57.

Miles, H.L. (1994) Me Chantek: the development of self-awareness in a signing orangutan. In *Self-Awareness in Monkeys and Apes: Developmental Perspectives*, ed. S. Parker, R. Mitchell and M. Boccia. Cambridge, UK: Cambridge University Press, pp. 254–272.

Milius, S. (1998) Birds can remember what, where, and when. *Science News*, 19 September.

Mill, J.S. (1843) *A System of Logic*, 2 vols. London: Parker. Reprinted (2002) Honolulu, HI: University Press of the Pacific.

Miller, G.A. (1956) The magical number seven plus or minus two: some limits on our capacity for processing information. *Psychological Review*, **63**, 81–96.

Miller, R. (2004) *Imprint Training of the Newborn Foal*. Guilford, CT: Globe Pequot Press.

Millikan, G. and Bowman, R. (1967) Observation on Galapagos tool-using finches in captivity. *Living Bird*, **6**, 23–41.

Milner, B. (1963) Effects of different brain lesions on card sorting: the role of the frontal lobes. *Archives of Neurology*, **9**, 90–100.

Mineka, S. and Cook, M. (1988) Social learning and the acquisition of snake fear in monkeys. In *Comparative Social Learning*, ed. T. Zentall and B.G. Galef, Jr. Hillsdale, NJ: Lawrence Erlbaum, pp. 51–73.

Mineka, S., Keir, R. and Price, V. (1980) Fear of snakes in wild- and laboratory-reared rhesus monkeys

(*Macaca mulatta*). *Animal Learning and Behaviour*, **8**, 653–553.

Mitani, J.C. and Watts, D. (2005) Correlates of territorial boundary patrol behavior in wild chimpanzees. *Animal Behaviour*, **70**, 1079–1086.

Mitani, J.C., Hunley, K.L. and Murdoch, M.E. (1999) Geographic variation in the calls of wild chimpanzees: a reassessment. *American Journal of Primatology*, **47**, 133–151.

Mitani, J.C., Merriwether, D.A. and Zhang, C. (2000) Male affiliation, cooperation, and kinship in wild chimpanzees. *Animal Behaviour*, **59**, 885–893.

Mitani, J.C., Watts, D., Pepper, J. and Merriwether, D.A. (2002) Demographic and social constraints on male chimpanzee behaviour. *Animal Behaviour*, **63**, 727–737.

Mitchell, R.W. (1993) Mental modes of mirror self-recognition: two theories. *New Ideas in Psychology*, **11**, 295–325.

Moore, B.R. (1973) The role of directed Pavlovian reactions in simple instrumental learning in the pigeon. In *Constraints on Learning*, ed. R.A. Hinde and J. Stevenson-Hinde. London: Academic Press, pp. 159–186.

Moore, B.R. (1992) Avian movement imitation and a new form of mimicry: tracing the evolution of a complex form of learning. *Behaviour*, **122**, 231–263.

Mordvilko A.K. (1936) Ants and aphids. *Priroda*, **4**, 44–55. (In Russian.)

Morgan, C.L. (1884) Instinct. *Nature*, **29**, 370–374, 405, 451–52.

Morgan, C.L. (1896) *Habit and Instinct*. London: Edward Arnold.

Morgan, C.L. (1900) *Animal Behaviour*. London: Edward Arnold.

Morris, R.G.M. (1981) Spatial localisation does not require the presence of local cue. *Learning and Motivation*, **12**, 239–260.

Morton, J. (1971) What could possibly be innate? In *Biological and Social Factors in Psycholinguistics*, ed. J. Morton. London: Logos Press, pp. 1–12.

Mowrer, O.H. (1950) *Learning Theory and Personality Dynamics*. New York: Ronald Press.

Mowrer, O.H. (1954) A psychologist looks at language. *American Psychologist*, **9**, 660–694.

Müller, J. (1826) *Zur vergleichenden Physiologie des Gesichtssinnes des Menschen und der Thiere*. Leipzig, Germany.

Müller, P. (1834–1840) *Handbuch der Physiologie des Menschen für Vorlesungen*, 2 vols, 2nd edn. Coblenz, Germany: J. Hoelscher.

Munn, N.L. (1950) *Handbook of Psychological Research on the Rat*. Boston, MA: Houghton Mofflin.

Munz, W.R.A. (1971) Sensory processes and behavior. In *Psychobiology*, ed. J.L. McGaugh. New York: Academic Press, pp. 73–98.

Murphy, J.J. (1873) Instinct: a mechanical analogy. *Nature*, **7**, 483.

Myer, J.S. and White, R.T. (1965) Aggressive motivation in the rat. *Animal Behaviour*, **13**, 430–433.

Nadel, L. and Moscovitch, M. (1997) Memory consolidation, retrograde amnesia and the hippocampal complex. *Current Opinion in Neurobiology*, **7**, 217–227.

Nagel, T. (1974) What is it like to be a bat? *Philosophical Review*, **83**, 435–450.

Nagell, K., Olguin K. and Tomasello, M. (1993) Processes of social learning in the tool use of chimpanzees and human children. *Journal of Comparative Psychology*, **107**, 174–186.

Nakajima, S. (1997) Transfer testing after serial feature-ambiguous discrimination in Pavlovian key-peck conditioning. *Animal Learning and Behaviour*, **25**, 413–426.

Nakajima, S. (2001) Failure of hierarchical conditional rule learning in the pigeon (*Columba livia*). *Animal Cognition*, **4**, 221–226.

Neet, C.C. (1933) Visual pattern discrimination in the *Macacus rhesus* monkey. *Journal of Genetic Psychology*, **43**, 163–196.

Neiworth, J.J. and Rilling, M.E. (1987) A method for studying imagery in animals. *Journal of Experimental Psychology: Animal Behavior Processes*, **13**, 203–214.

Neiworth, J.J., Burman, M.A., Basile, B.M. and Lickteig, M.T. (2002) Use of experimenter-given cues in visual co-orienting and in an object-choice task by a New World monkey species, cotton-top tamarins (*Saguinus oedipus*). *Journal of Comparative Psychology*, **116**, 3–11.

Nelson, D.A., Hallberg, K.I. and Soha, J.A. (2004) Cultural evolution of Puget Sound white-crowned sparrow song dialects. *Ethology*, **110**, 879–908.

Nevin, J.A. and Liebold, K. (1966) Stimulus control of matching and oddity in a pigeon. *Psychonomic Science*, **5**, 351–352.

Newman, J. and Simmes, D. (1982) Inheritance and experience in the acquisition of primate acoustic behavior. In *Primate Communication*, ed. T. Snowdon, C.H. Brown and M.E. Petersen. Cambridge, UK: Cambridge University Press, pp. 259–278.

Nishida, T. (1987) Local traditions and cultural transmission. In *Primate Societies*, ed. S.S. Smuts, D.L. Cheney, R.M. Seyfarth, R.W. Wrangham and T.T. Strusaker. Chicago, IL: University of Chicago Press, pp. 462–474.

Nishida, T. (1990) A quarter century of research in the Mahale Mountains: an overview. In *The Chimpanzees of the Mahale Mountains: Sexual and Life History Strategies*, ed. T. Nishida. Tokyo: University of Tokyo Press, pp. 3–35.

Nissani, M. (1977) Gynandromorph analysis of some aspects of sexual behaviour of *Drosophila melanogaster*. *Animal Behaviour*, **25**, 555–566.

Nissani, M. (2004) Theory of mind and insight in chimpanzees, elephants, and other animals? In *Developments in Primatology: Progress and Prospects*, ed. R. H. Tuttle, vol. 4, *Comparative Vertebrate Cognition: Are Primates Superior to Non-Primates?* ed. L. J. Rogers and G. Kaplan. New York: Kluwer, pp. 227–261.

Nissani, M. (2006) Do Asian elephants (*Elephas maximus*) apply causal reasoning to tool use tasks? *Journal of Experimental Psychology: Animal Behavior Processes*, **32**, 91–96.

Nissani, M., Hoefler-Nissani, D., Lay, U T. and Htun, U W. (2005) Simultaneous visual discrimination in Asian Elephants. *Journal of the Experimental Analysis of Behavior*, **83**, 15–29.

Nissen, N. W. (1931) A field study of chimpanzee. *Comparative Psychological Monographs, Series 36*, **8**, 1–105.

Nissen, H. W. (1934) Equivalence and ambivalence of stimuli in chimpanzees. *Psychological Bulletin*, **31**, 617–618.

Nissen, H. W. (1951) Pylogenetic comparison. In *Handbook of Experimental Psychology*, ed. S. S. Stevens. New York: John Wiley, pp. 347–386.

Nottebohm, F. (1970) The ontogeny of bird song. *Science*, **167**, 950–956.

Nottebohm, F. (1993) The search for neural mechanisms that define the sensitive period for song learning in birds. *Netherlands Journal of Zoology*, **43**, 193–234.

Novgorodova, T. A. (2005) Communication system of *Formica pratensis* Retz. (Hymenoptera, Formicidae). *Abstracts 29th International Ethological Conference*, Budapest, Hungary, p. 163.

Novgorodova, T. A. (2006) Experimental investigation of foraging modes in *Formica pratensis* Retz. (Hymenoptera, Formicidae) using 'binary tree' maze. *Entomological Review*, **86**, 287–293.

O'Donnell, K. A., Rapp, P. R. and Hof, P. R. (1999) Preservation of prefrontal cortical volume in behaviorally characterized aged macaque monkeys. *Experimental Neurology*, **160**, 300–310.

Öhman, A. and Mineka, S. (2003) The malicious serpent: snakes as a prototypical stimulus for an evolved module of fear. *Current Directions in Psychological Science*, **12**, 5–9.

O'Keefe, J. and Nadel, L. (1978) *The Hippocampus as a Cognitive Map*. Oxford, UK: Clarendon Press.

Oliveira, R. F. (2004) Social modulation of androgens in vertebrates: mechanisms and function. In *Advances in the Study of Behavior*, vol. 34, ed. P. J. B. Slater, J. S. Rosenblatt, C. T. Snowdon and T. J. Roper. New York: Academic Press, pp. 165–239.

Oliveira, R. F., Carneiro, L. A. and Canario, A. V. M. (2005) Behavioural endocrinology: no hormonal response in tied fights. *Nature*, **437**, 207–208.

Olton, D. S. (1978) Characteristics of spatial memory. In *Cognitive Processes in Animal Behavior*, ed. S. H. Hulse, H. Fowler and W. K. Honig. Hillsdale, NJ: Lawrence Erlbaum, pp. 341–373.

Orians, G. S. and Pearson, N. E. (1979) On the theory of central place foraging. In *Analysis of Ecological Systems*, ed. D. J. Horn, R. D. Mitchell and C. R. Stairs. Columbus, OH: Ohio State University Press, pp. 154–177.

Orlov, T., Yakovlev, V., Amit, D., Hochstein, S. and Zohary, E. (2002) Serial memory strategies in macaque monkeys: behavioral and theoretical aspects. *Cerebral Cortex*, **12**, 306–317.

Osgood, C. E. (1953) *Method and Theory in Experimental Psychology*. New York: Oxford University Press.

Osthaus, B., Lea, S. E. G. and Slater, A. M. (2005) Dogs fail to understand means-end relationships in a string-pulling task. *Animal Cognition*, **8**, 37–47.

Ottoni, E. B. and Mannu, M. (2001) Semi-free-ranging tufted capuchin monkeys (*Cebus apella*) spontaneously use tools to crack open nuts. *International Journal of Primatology*, **22**, 347–358.

Owren, M. J. and Rendall, D. (2003) Salience of caller identity in rhesus monkeys (*Macaca mulatta*) coo and screams: perceptual experiments with human listeners. *Journal of Comparative Psychology*, **117**, 380–390.

Owren, M. J., Dieter, J. A., Seyfarth, R. M. and Cheney, D. L. (1993) Vocalization of rhesus (*Macaca mulatta*) and Japanese (*Macaca fuscata*) macaques cross-fostered between species show evidence of only limited modification. *Developmental Psychobiology*, **26**, 389–406.

Pack, A. A. and Herman, L. M. (1995) Sensory integration in the bottlenosed dolphin: immediate recognition of complex shapes across the senses of echolocation and vision. *Journal of the Acoustical Society of America*, **98**, 722–733.

Page, R. E., Robinson, G. E., Calderone, N. E. and Rothenbuhler, W. C. (1989) Genetic structure, division of labour, and the evolution of insect societies. In *The Genetics of Social Evolution*, ed. M. D. Breed and R. E. Page. Boulder, CO: Westview Press, pp. 15–30.

Panov, A. A. (1957) Development of mushroom bodies of the honeybee's brain during larva and pupae stages. *Bulletin of Moscow State University*, **2**, 47–54.

Papi, F. (1986) Pigeon navigation: solved problems and opened questions. *Monitore Zoologico Italiano (Nuova Serie)*, **20**, 471–517.

Papi, F., Ioale, P., Fiaschi, V. and Benvenuti, S. (1972) Pigeon homing: cues detected during the outward journey influence initial orientation. In *Animal Migration, Navigation, and Homing*, ed. K. Schmidt-Koenig and W. Keeton. Berlin: Springer-Verlag, pp. 65–77.

Parker, S. and Gibson, K. (1979) A developmental model for the evolution of language and intelligence in early hominids. *Behavioral and Brain Sciences*, **2**, 367–408.

Parks, E. R. (1968) The orienting reaction as a mediator of sensory preconditioning. *Psychonomic Science*, **11**, 11–12.

Parr, L. A. and de Waal, F. B. M. (1999) Visual kin recognition in chimpanzees. *Nature*, **399**, 647–648.

Parr, L. A., Winslow, J. T., Hopkins, W. D. and de Waal, F. B. M. (2000) Recognizing facial cues: individual recognition in chimpanzees (*Pan troglodytes*) and rhesus monkeys (*Macaca mulatta*). *Journal of Comparative Psychology*, **114**, 1–14.

Patricelli, G. L., Uy, J. A. C., Walsh, G. and Borgia, G. (2002) Sexual selection: male displays adjusted to female's response. *Nature*, **415**, 279–280.

Patterson, F. G. (1978) The gestures of a gorilla: language acquisition in another pongid. *Brain and Language*, **5**, 72–97.

Patterson, F. G. (1981) Can an ape create a sentence? Some affirmative evidence. *Science*, **211**, 86–87.

Patterson, F. G. and Cohn, R. H. (1994) Self-recognition and self-awareness in lowland gorillas. In *Self-Awareness in Animals and Humans: Developmental Perspectives*, ed. S. T. Parker, R. W. Mitchell, and M. L. Boccia. New York: Cambridge University Press, pp. 273–290.

Patterson, F. G. and Linden, E. (1981) *The Education of Koko*. New York: Holt Rinehart & Winston.

Patterson, F. G., Tanner, J. and Mayer, N. (1988) Pragmatic analysis of gorilla utterances: early communicative development in the gorilla Koko. *Journal of Pragmatics*, **12**, 35–54.

Pauly, P. J. (1987) *Controlling Life: Jacques Loeb and the Engineering Ideal in Biology*. New York: Oxford University Press.

Pavlov, I. P. (1903) The experimental psychology and psychopathology of animals. *Proceedings 14th Annual International Medical Congress, Madrid*.

Pavlov, I. P. (1927) *Conditioned Reflexes: An Investigation of the Physiological Activity of the Cerebral Cortex*. London: Oxford University Press.

Pavlov, I. P. (1928) *Lectures on Conditioned Reflexes: Twenty-Five Years of Objective Study of the Higher Nervous Activity (Behaviour) of Animals*, vol. 1, trans. and ed. W. H. Gantt. London: Lawrence & Wishart.

Pearce, J. M. (1987) A model for stimulus generalization for Pavlovian conditioning. *Psychological Review*, **94**, 61–73.

Pearce, J. M. (1994) Similarity and discrimination: a selective review and a connectionist model. *Psychological Review*, **101**, 587–607.

Pearce, J. M. (2000) *Animal Learning and Cognition: An Introduction*, 2nd edn. Hove, UK: Psychology Press.

Penfield, W. and Jasper, H. H. (1954) *Epilepsy and the Functional Anatomy of the Human Brain*. Boston, MA: Little, Brown.

Penfield, W. and Rasmussen, T. (1950) *The Cerebral Cortex of Man*. New York: Macmillan.

Pepperberg, I. M. (1981) Functional vocalizations by an African grey parrot (*Psittacus erithacus*). *Zoological Tierpsychology*, **55**, 139–160.

Pepperberg, I. M. (1983) Cognition in the African grey parrot: preliminary evidence for auditory vocal comprehension of the class concept. *Animal Learning and Behaviour*, **11**, 179–185.

Pepperberg, I. M. (1987) Acquisition of the same/different concept by an African grey parrot (*Psittacus erithacus*): learning with respect to color, shape, and material. *Animal Learning and Behaviour*, **15**, 423–432.

Pepperberg, I. M. (1990) Cognition in an African grey parrot (*Psittacus erithacus*): further evidence for comprehension of categories and labels. *Journal of Comparative Psychology*, **104**, 41–52.

Pepperberg, I. M. (1999) *The Alex Studies*. Cambridge, MA: Harvard University Press.

Pepperberg, I. M. (2002) Cognitive and communicative abilities of grey parrots. *Current Directions in Psychological Science*, **11**, 83–87.

Pepperberg, I. M. (2004) 'Insightful' string-pulling in grey parrots (*Psittacus erithacus*) is affected by vocal competence. *Animal Cognition*, **7**, 263–266.

Pepperberg, I. M. and Gordon, J. D. (2005) Numerical comprehension by a grey parrot (*Psittacus erithacus*), including a zero-like concept. *Journal of Comparative Psychology*, **119**, 197–209.

Pepperberg, I. M. and Funk, M. S. (1990) Object permanence in four species of psittacine birds. *Animal Learnimg and Behaviour*, **18**, 97–108.

Pepperberg, I. M. and Kozak, F. A. (1986) Object permanence in African grey parrot (*Psittacus errithacus*). *Animal Learning and Behaviour*, **14**, 322–330.

Pepperberg, I. M. and Lynn, S. K. (2000) Perceptual consciousness in grey parrots. *American Zoologist*, **40**, 893–901.

Pepperberg, I. M., Garcia, S. E., Jackson, E. C. and Marconi, S. (1995) Mirror use by African grey parrots (*Psittacus erithacus*). *Journal of Comparative Psychology*, **109**, 182–195.

Pepperberg, I. M., Willner, M. R. and Gravitz, L. B. (1997) Development of Piagetian object permanence in a grey parrot (*Psittacus erithacus*). *Journal of Comparative Psychology*, **111**, 63–75.

Perry, S. (1997) Male–female social relationships in wild white-faced capuchin monkeys, *Cebus capucinus*. *Behaviour*, **134**, 477–510.

Perry, S. and Manson, J. H. (2003) Traditions in monkeys. *Evolutionary Anthropology*, **12**, 71–81.

Petty, R. E., Cacioppo, J. T. and Schumann, D. (1983) Central and peripheral routes to advertising effectiveness: the moderating role of involvement. *Journal of Consumer Research*, **10**, 134–148.

Pfeffer, K., Fritz, J. and Kotrschal, K. (2002) Hormonal correlates of being an innovative greylag goose. *Animal Behaviour*, **63**, 687–695.

Pfennig, D. W. (1992) Polyphenism in spadefoot toad tadpoles as a locally adjusted evolutionarily stable strategy. *Evolution*, **46**, 1408–1420.

Pfennig, D. W., Collins, J. P. and Ziemba, R. E. (1999) A test of alternative hypotheses for kin recognition in cannibalistic tiger salamanders. *Behavioral Ecology*, **10**, 436–443.

Pfungst, O. (1908) *Clever Hans: The Horse of Mr Van Osten*. Reissued (1965) New York: Holt.

Piaget, J. (1932) *The Origins of Intelligence in Children*. Reissued (1952) New York: International Universities Press, Inc.

Piaget, J. (1937) *The Construction of Reality in the Child*. Reissued (1954) New York: Basic Books.

Piaget, J. (1942) *The Child's Conception of Number*. London: Routledge & Kegan Paul.

Piaget, J. and Inhelder, B. (1956) *The Child's Conception of Space*. London: Routledge & Kegan Paul.

Piersma, T. (1994) Close to the edge: energetic bottlenecks and the evolution of migratory pathways in knots. Ph.D. thesis, University of Groningen, the Netherlands.

Plenge, M., Curio, E. and Witte, K. (2000) Sexual imprinting supports the evolution of novel male traits by transference of a preference for the colour red. *Behaviour*, **137**, 741–759.

Plowright, C. M. S. (1997) Function and mechanism of mirror image ambiguity in bumble bees. *Animal Behaviour*, **53**, 1295–1303.

Plowright, C. M. S., Landry, F., Church, D., *et al.* (2001) A change in orientation: recognition of rotated patterns by bumble bees. *Journal of Insect Behavior*, **14**, 113–127.

Pollok, B., Prior, H. and Güntürkün, O. (2000) Development of object permanence in food-storing magpies *Pica pica*. *Journal of Comparative Psychology*, **114**, 148–157.

Pongrácz, P., Miklósi, A., Kubinyi, E., *et al.* (2001) Social learning in dogs: the effect of a human demonstrator on the performance of dogs in a detour task. *Animal Behaviour*, **62**, 1109–1117.

Poole, J. H. and Moss, C. J. (1983) Relationships and social structure in African elephants. In *Primate Social Relationships: An Integrated Approach*, ed. R. A. Hinde. Oxford, UK: Blackwell Scientific Publications, pp. 315–325.

Poole, J. H., Tyack, P. L., Stoeger-Horwath, A. S. and Watwood, S. (2005) Elephants are capable of vocal learning. *Nature*, **434**, 455–456.

Porter, J. (1904) A preliminary study of the psychology of the English sparrow. *American Jornal of Psychology*, **15**, 313–346.

Povinelli, D. J. (1989) Failure to find self-recognition in Asian elephants (*Elephas maximus*) in contrast to their use of mirror cues to discover hidden food. *Journal of Comparative Psycology*, **103**, 122–131.

Povinelli, D. J. (1999) Social understanding in chimpanzees: new evidence from a longitudinal approach. In *Developing Theories of Intention: Social Understanding and Self-Control*, ed. P. Zelazo, J. Astington and D. Olson. Hillsdale, NJ: Lawrence Erlbaum, pp. 195–225.

Povinelli, D. J. (2000) *Folk Physics for Apes: Chimpanzees, Tool Use, and Causal Understanding*. Oxford, UK: Oxford University Press.

Povinelli, D. J. and DeBlois, S. (1992) Young children's (*Homo sapiens*) understanding of knowledge formation in themselves and others. *Journal of Comparative Psychology*, **106**, 228–238.

Povinelli, D. J. and Eddy, T. J. (1996) Chimpanzees: joint visual attention. *Psychological Science*, **7**, 129–135.

Povinelli, D. J. and Prince, C. G. (1998) When self met other. In *Self-Awareness: Its Nature and Development*, ed. M. Ferrari and P. J. Sternberg. New York: Guilford Press, pp. 37–107.

Povinelli, D. J., Nelson, K., and Boysen, S. (1992a) Comprehension of role reversal in chimpanzees: evidence of empathy? *Animal Behaviour*, **43**, 633–640.

Povinelli, D.J., Parks, K.A. and Novak, M.A. (1992b) Role reversal by rhesus monkeys, but no evidence of empathy. *Animal Behaviour*, **44**, 269–283.

Poyarkov, A.D. (1986) Historical (biographical) method of description of social organisation and behaviour in stray dogs. In *Research Methods in Ecology and Ethology*, ed. E.N. Panov. Moscow: Pushino, pp. 172–203. (In Russian.)

Pravosudov, V.V. (2003) Long-term moderate elevation of corticosterone facilitates avian food-caching behaviour and enhances spatial memory. *Proceedings of the Royal Society of London B*, **270**, 2599–2604.

Premack, D. (1959) Toward empirical behavior laws. I. Positive reinforcement. *Psychological Review*, **66**, 219–233.

Premack, D. (1962) Reversibility of the reinforcement relation. *Science*, **136**, 235–237.

Premack, D. (1965) Reinforcement theory. In *Nebraska Symposium of Motivation*, ed. D. Levine. Lincoln, NE: University of Nebraska Press, pp. 123–180.

Premack, D. (1971) Language in chimpanzee? *Science*, **172**, 808–822.

Premack, D. (1976) On the study of intelligence in chimpanzees. *Current Anthropology*, **17**, 516–21.

Premack, D. (1983) Animal cognition. *Annual Review of Psychology*, **34**, 351–362.

Premack, D. (2004) Psychology: is language the key to human intelligence? *Science*, **303**, 318–320.

Premack, D. and Premack, A.J. (1996) Why animals lack pedagogy and some cultures have more of it than others. In *The Handbook of Education and Human Development*, ed. D.R. Olson and N. Torrance. Oxford, UK: Blackwell Scientific Publications, pp. 302–323.

Premack, D. and Woodruff, G. (1978) Does the chimpanzee have a theory of mind? *Behavior and Brain Sciences*, **1**, 515–526.

Preston, G.C., Dickinson, A. and Mackintosh, N.J. (1986) Contextual conditional discriminations. *Quarterly Journal of Experimental Psychology B*, **38**, 217–237.

Preston, S.D. and Jacobs, L.F. (2001) Conspecific pilferage but not presence affects Merriam's kangaroo rat cache strategy. *Behavioral Ecology*, **12**, 517–523.

Promptov, A.N. (1940) Development of species specific behavioural patterns in wild birds. *Dokladi Akademii Nauk USSR*, **27**, 240–244. (In Russian with English summary.)

Protopopov, V.P. (1950) Processes of distraction and integration (abstraction) in animals and humans. In *Studying High Nervous Activity by Means of Natural Experiments*, ed. V.P. Protopopov. Kiev, USSR: Gosmedizdat, pp. 3–24. (In Russian.)

Provine, R.R. (1996) Laughter. *American Scientist*, **84**, 38–45.

Pryor, K. (1969) The porpoise caper. *Psychology Today*, **3**, 7, 46–49.

Pryor, K. (1975) *Lads before the Wind: Diary of a Dolphin Trainer*. Waltham, MA: Sunshine Books.

Pryor, K. (1985) *Don't Shoot the Dog! The New Art of Teaching and Training*. New York: Bantam Books.

Pryor, K., Haag, R. and O'Reilly, J. (1969) The creative porpoise: training for novel behavior. *Journal of the Experimental Analysis of Behavior*, **12**, 653–661.

Pulliam, H.R. (1974) On the theory of optimal diet. *American Naturalist*, **105**, 575–587.

Queller, D.C., Peretes, J.M., Solis, C.R. and Strassmann, J.E. (1997) Control of reproduction in social insect colonies: individual and collective relatedness preferences in the paper wasp *Polistes annularis*. *Behavioral Ecology and Sociotiology*, **40**, 3–16.

Rachlin, H. and Green, L. (1972) Commitment, choice and self-control. *Journal of the Experimental Analysis of Behavior*, **17**, 15–22.

Rainer, J., Asaad, W.F. and Miller, E.K. (1998) Selective representation of relevant information by neurons in the primate prefrontal cortex. *Nature*, **393**, 577–579.

Randall, J., Collins, K., McCowan, B., Hooper, S.L. and Rogovin, K. (2005) Alarm signals of the great gerbil: acoustic variation by predator context, sex, age, individual and family group. *Journal of the Acoustical Society of America*, **118**, 1–9.

Rashotte, M.E., Griffin, R.W. and Sisk, C.L. (1977) Second-order conditioning of the pigeon's key peck. *Animal Learning and Behaviour*, **5**, 25–38.

Ratnieks, F.L.W. (1988) Reproductive harmony via mutual policing by workers in eusocial Hymenoptera. *American Naturalist*, **132**, 217–236.

Reaux, J.E., Theall, L.A. and Povinelli, D.J. (1999) A longitudinal investigation of chimpanzees' understanding of visual perception. *Child Development*, **70**, 275–290.

Redshaw, M. (1978) Cognitive development in human and gorilla infants. *Journal of Human Evolution*, **7**, 133–141.

Reeve, H.K. (1992) Queen activation of lazy workers in colonies of the eusocial naked mole-rat. *Nature*, **358**, 147–149.

Regolin, L., Vallortigara, G. and Zanforlin, M. (1995) Detour behaviour in the domestic chick: searching for a disappearing prey or a disappearing social partner. *Animal Behaviour*, **50**, 203–211.

Reichmuth Kastak, C. and Schusterman, R. J. (2002) Sea lions and equivalence: extending classes by exclusion. *Journal of the Experimental Analysis of Behavior*, **78**, 449–465.

Reiss, D. and Marino, L. (2001) Mirror self-recognition in the bottlenose dolphin: a case of cognitive convergence. *Proceedings of National Academy of Sciences of the USA*, **98**, 5937–5942.

Rendall, D., Rodman, P. S. and Emond, R. E. (1996) Vocal recognition of individuals and kin in free-ranging rhesus monkeys. *Animal Behaviour*, **51**, 1007–1015.

Rendell, L. and Whitehead, H. (2001) Culture in whales and dolphins. *Behavioral and Brain Sciences*, **24**, 309–382.

Rensch, B. and Dücker, G. (1959) Versuche über visuelle Generalisation bei einer Schleichkatze. *Zoological Tierpsychology*, **16**, 671–692.

Rescorla, R. and Wagner, A. R. (1972) A theory of Pavlovian conditioning: variations in the effectiveness of reinforcement and nonreinforcement. In *Classical Conditioning II: Current Research and Theory*, ed. A. H. Black and W. F. Prokasy. New York: Appleton-Century-Crofts, pp. 64–99.

Révész, G. (1924) Experiments on animal perception. *British Journal of Psychology*, **14**, 387–414.

Reznikova, Zh. I. (1974) Mechanisms of territoriality in ants *Formica pratensis* Retz. *Zoological Journal*, **53**, 212–223. (In Russian with English summary.)

Reznikova, Zh. I. (1975) Non-antagonistic relations in ant species occupying similar ecological niches. *Zoological Journal*, **54**, 1020–1031. (In Russian with English summary.)

Reznikova, Zh. I. (1979) Different forms of territoriality in ants *Formica pratensis* Retz. *Zoological Journal*, **58**, 1490–1499. (In Russian with English summary.)

Reznikova, Zh. I. (1980) Interspecies hierarchy in ants. *Zoological Journal*, **59**, 1168–1176. (In Russian with English summary.)

Reznikova, Zh. I. (1981) Behaviour of ants *Camponotus japonicus aterrimus* Em. on their feeding territory. *Bulletin of the Moscow Society of Explorers of Nature*, **86**, 36–41.

Reznikova, Zh. I. (1982) Interspecific communication among ants. *Behaviour*, **80**, 84–95.

Reznikova, Zh. I. (1983) *Interspecies Relations in Ants*. Novosibirsk, USSR: Nauka. (In Russian.)

Reznikova, Zh. I. (1994) The original pattern of manegement in ant communities: interspecies social control. *Progress to Meet the Challenge of Environmental Change* (International Ecological Congress abstracts), Manchester, UK, p. 27.

Reznikova, Zh. I. (1996) The role of social experience in early development of ant behaviour. *Proceedings 20th International Congress of Entomology*, Florence, Italy, p. 406.

Reznikova, Zh. I. (1999) Ethological mechanisms of population density control in coadaptive complexes of ants. *Russian Journal of Ecology*, **30**, 187–192.

Reznikova, Zh. I. (2001) Interspecific and interaspecific social learning in ants. *Contributions 27th International Ethological Conference*, Tubingen, Germany, p. 108.

Reznikova, Zh. I. (2003) Goverment and nepotism in social insects: new dimension provided by experimental approach. *Euroasian Entomological Journal*, **2**, 3–14.

Reznikova, Zh. I. (2004) Social learning in animals: comparative analysis of different forms and levels. *Journal of General Biology*, **65**, 136–152. (In Russian with English summary.)

Reznikova, Zh. I. (2005) Different forms of social learning in ants. *Abstracts of International Conference on Animal Social Learning*, St Andrews, UK, p. 17.

Reznikova, Zh. I. (2006) The study of tool use as the way for general estimation of cognitive abilities in animals. *Journal of General Biology*, **67**, 3–22. (In Russian with English summary.)

Reznikova, Zh. I. and Dorosheva, H. (2004) Impacts of red wood ants *Formica polyctena* on the spatial distribution and behavioural patterns of ground beetles (Carabidae). *Pedobiologia*, **48**, 15–21.

Reznikova, Zh. I and Dorosheva, H. (2005) Adaptation learning in insects: facilitation of manipulation with innate behavioural patterns. *Abstracts of 29th International Ethological Conference*, Budapest, Hungany, p. 184.

Reznikova, Zh. I. and Novgorodova, T. A. (1998) The importance of individual and social experience for interaction between ants and symbiotic aphids. *Doklady Biological Sciences*, **359**, 173–175.

Reznikova, Zh. I. and Panteleeva, S. N. (2001) Interaction of the ant *Myrmica rubra* L. as a predator with springtails (Collembola) as a mass prey. *Doklady Biological Sciences*, **380**, 475–477.

Reznikova, Zh. I. and Panteleeva, S. N. (2005) The ontogeny of hunting behaviour in ants: experimental study. *Doklady Biological Sciences*, **401**, 139–141.

Reznikova, Zh. I. and Ryabko, B. Ya. (1990) Information theory approach to communication in ants. In *Sensory Systems and Communication in Arthropods*, ed. F. G. Gribakin, K. Wiese and A. V. Popov. Basel, Switzerland: Birkhäuser, pp. 305–307.

Reznikova, Zh. I. and Ryabko, B. Ya. (1993) Ants' aptitude for the transmission of information on the number of objects. In *Sensory Systems of Arthropods*, ed. K. Wiese, F. G. Gribakin, A. V. Popov and G. Renninger. Basel, Switzerland: Birkhäuser, pp. 634–639.

Reznikova, Zh. I. and Ryabko, B. Ya. (1994) Experimental study of the ant communication system with the application of the information theory approach. *Memorabilia Zoologica*, **48**, 219–236.

Reznikova, Zh. I. and Ryabko, B. Ya. (1996) Transmission of information regarding the quantitative characteristics of objects in ants. *Neuroscience and Behavioural Psychology*, **36**, 396–405.

Reznikova, Zh. I. and Ryabko, B. Ya. (2000) Using information theory approach to study the communication system and numerical competence in ants. In *From Animals to Animats*, vol. 6, *Proceedings 6th International Conference on Simulation of Adaptive Behaviour*, ed. J.-A. Meyer, A. Berthoz, D. Floreano, H. Roitblat and S. W. Wilson. Cambridge, MA: MIT Press, pp. 501–506.

Reznikova, Zh. I. and Ryabko, B. Ya. (2001) A study of ants' numerical competence. *Electronic Transactions on Artificial Intelligence B*, **5**, 111–126.

Reznikova, Zh. I. and Ryabko, B. Ya. (2003) In the shadow of the binary tree: of ants and bits. In *Proceedings 2nd International Workshop on the Mathematics and Algorithms of Social Insects*, ed. C. Anderson and T. Balch. Atlanta, GA: Georgian Institute of Technology, pp. 139–145.

Rice, S. H. (2004) *Evolutionary Theory: Mathematical and Conceptual Foundations*. Sunderland, MA: Sinauer Associates.

Richard, P. B. (1983) Mechanisms and adaptation in the constructive behavior of the beaver *C. fiber L. Acta Zoologica Fennica*, **174**, 105–108.

Richardson, M. W. (1938) Multidimensional psychophysics. *Psychological Bulletin*, **35**, 659–660.

Riebel, K., Hall, M. L. and Langmore, N. E. (2005) Female songbirds still struggling to be heard. *Trends in Ecology and Evolution*, **20**, 419–420.

Riley, D. A., Cook, R. G. and Lamb, M. R. (1981) A classification and analysis of short-term retention codes. In *The Psychology of Learning and Motivation: Advances in Research and Theory*, vol. 15, ed. G. H. Bower. New York: Academic Press, pp. 51–79.

Rilling, M. (2000) How the challenge of explaining learning influenced the origins and development of John B. Watson's behaviorism. *American Journal of Psychology*, **113**, 275–301.

Rinderer, T. E. and Beaman, L. D. (1995) Genic control of honey bee dance language dialect. *Theoretical and Applied Genetics*, **91**, 727–732.

Ristau, C. A. (ed.) (1991) *Cognitive Ethology: The Minds of Other Animals*. Hillsdale, NJ: Lawrence Erlbaum.

Ristau, C. A. and Robbins, D. (1982) Language in the great apes: a critical review. *Advances in the Study of Behavior*, **12**, 141–255.

Ritter, C. (1988) *Nobel Prizes*. Press release, Associated Press, 24 December.

Rizley, R. C. and Rescorla, R. A. (1972) Associations in second-order conditioning and sensory preconditioning. *Journal of Comparative and Physiological Psychology*, **81**, 1–11.

Robbins, R. L. (2000) Vocal communication in free-ranging African wild dogs (*Lycaon pictus*). *Behaviour*, **137**, 1271–1298.

Roberts, R. L., Williams, J. R., Wang, A. K. and Carter, C. S. (1998) Cooperative breeding and monogamy in prairie voles: influence of the sire and geographical variation. *Animal Behaviour*, **55**, 1131–1140.

Roberts, W. A. (1979) Spatial memory in the rat on a hierarchical maze. *Learning and Motivation*, **10**, 117–140.

Roberts, W. A. and Mazmanian, D. S. (1988) Concept learning at different levels of abstraction by pigeons, monkeys, and people. *Journal of Experimental Psychology: Animal Behavioral Processes*, **14**, 247–260.

Robson, S. K. and Traniello, J. F. A. (1998) Resource assessment, recruitment behavior and the organisation of cooperative foraging in the ant *Formica schaufussi*. *Journal of Insect Behaviour*, **11**, 1–22.

Robson, S. K. and Traniello, J. F. A. (2002) Transient division of labor and behavioral specialisation in the ant *Formica schaufussi*. *Naturwissenschaften*, **89**, 128–131.

Roche, B. and Barnes, D. (1997) A transformation of respondently conditioned stimulus function in accordance with arbitrarily applicable relations. *Journal of the Experimental Analysis of Behavior*, **67**, 275–301.

Rodríguez, F., Durán, E., Vargas, J. P., Torres, B. and Salas, C. (1994) Performance of goldfish trained in allocentric and egocentric mazes procedures suggest the presence of cognitive mapping system in fishes. *Animal Learning and Behaviour*, **22**, 409–420.

Rogers, L. J. (2002) Lateralization in vertebrates: its early evolution, general pattern and development. In *Advances in the Study of Behavior*, vol. 31, ed. P. J. B. Slater, J. Rosenblatt, C. Snowdon and T. Roper. New York: Academic Press, pp. 107–162.

Rogers, L. J. and Anson, J. M. (1979) Lateralisation of function in the chicken fore-brain. *Pharmacology, Biochemistry and Behavior*, **10**, 679–686.

Roginsky, G. Z. (1948) *Acquirement and Germs of Intellectual Actions in Anthropoids (Chimpanzees)*. Leningrad, USSR: Leningrad State University Press.

Rogovin, K.A. (1992) Habitat use by two species of Mongolian marmots (*Marmota sibirica* and *M. baibacina*) in a zone of sympatry. *Acta Theriologica*, **37**, 345–350.

Roitblat, H.L. (1980) Codes and coding processes in pigeon short-term memory. *Animal Learning and Behaviour*, **8**, 341–351.

Romanes, G.J. (1881) *Animal Intelligence*. London: Kegan Paul.

Romanes, G.J. (1883) *Mental Evolution in Animals*. London: Kegan Paul.

Romanes, G.L. (1885) Homing faculty of Hymenoptera. *Nature*, **32**, 630.

Ronacher, B. and Wehner, R. (1995) Desert ants, *Cataglyphis fortis*, use self-induced optic flow to measure distances travelled. *Journal of Comparative Physiology A*, **177**, 21–27.

Roper, T.J. (1986) Cultural evolution of feeding behaviour in animals. *Science Progress*, **70**, 571–583.

Rosch, E. (1973) Natural categories. *Cognitive Psychology*, **4**, 328–350.

Rose, F.C. and Bynum, W.F. (1982) *Historical Aspects of the Neurosciences: A Festschrift for Macdonald Critchely*. New York: Raven Press.

Rose, S.P.R. (1993) *The Making of Memory*. New York: Bantam Books.

Rose, S.P.R. (2000) God's organism? The chick as a model system for memory studies. *Learning and Memory*, **7**, 1–17.

Rose, S.P.R. (2005) Memory beyond the synapse. *Neuron Glia Biology*, **1**, 1–7.

Rose, S.P.R. and Stewart, M.G. (1999) Cellular correlates of memory formation in the chick following passive avoidance training. *Behavioral and Brain Research*, **98**, 237–243.

Rosengren, R. (1971) Route fidelity, visual memory and recruitment behaviour in foraging wood ants of the genus *Formica* (Hymenoptera, Formicidae). *Acta Zoologica Fennica*, **143**, 1–102.

Rosengren, R. and Fortelius, W. (1987) Trail communication and directional recruitment to food in red wood ants (*Formica*). *Annales Zoologici Fennici*, **24**, 137–146.

Ross, R.T. and Holland, P.C. (1981) Conditioning of simultaneous and serial feature-positive discriminations. *Animal Learning and Behaviour*, **9**, 293–303.

Roush, R.S. and Snowdon, C.T. (1994) Ontogeny of food related calls in cotton-top tamarins. *Animal Behaviour*, **47**, 263–273.

Rowell, T.E. (1966) Hierarchy in the organization of a captive baboon group. *Animal Behaviour*, **14**, 430–443.

Rumbaugh, D.M. (1968) The learning and sensory capacities of the squirrel monkey in phylogenetic perspective. In *The Squirrel Monkey*, ed. L.A. Rosenblum and R.W. Cooper. New York: Academic Press, pp. 255–317.

Rumbaugh, D.M. (1977) *Language Learning by a Chimpanzee: The Lana Project*. New York: Academic Press.

Rumbaugh, D.M. and Gill, T.V. (1977) Lana's acquisition of language skills. In *Language Learning by a Chimpanzee: The Lana Project*, ed. D.M. Rumbaugh. New York: Academic Press, pp. 165–192.

Rumbaugh, D.M. and Pate, J.L. (1984) Primates learning by levels. In *Behavioral Evolution and Integrative Levels*, ed. G. Greenberg and E. Tobach. Hillsdale, NJ: Lawrence Erlbaum, pp. 221–240.

Rumbaugh, D.M. and Savage-Rumbaugh, E.S. (1994) Language in comparative perspective. In *Animal Learning and Cognition*, ed. N.J. Mackintosh. New York: Academic Press, pp. 307–333.

Ryabko, B. Ya. (1993) Methods of analysis of animal communication systems based on information theory. In *Sensory Systems of Arthropods*, ed. K. Wiese, F.G. Gribakin, A.V. Popov and G. Renninger. Basel, Switzerland: Birkhäuser, pp. 627–634.

Ryabko, B. Ya. and Reznikova, Zh. I. (1996) Using Shannon entropy and Kolmogorov complexity to study the communicative system and cognitive capacities in ants. *Complexity*, **2**, 37–42.

Sacchi, R., Saino, N. and Galeotti, P. (2002) Features of begging calls reveal general condition and need of food of barn swallow (*Hirundo rustica*) nestlings. *Behavioral Ecology*, **13**, 268–273.

Saino, N., Ambrosini, R., Martinelli, R., *et al.* (2003) Gape coloration reliably reflects immunocompetence of barn swallow (*Hirundo rustica*) nestlings. *Behavioral Ecology*, **14**, 16–22.

Santi, A. (1978) The role of physical identity of the sample and correct comparison stimulus in matching-to-sample paradigms. *Journal of the Experimental Analysis of Behavior*, **29**, 511–516.

Santos, L.R., Ericson, B., and Hauser, M.D. (1999) Constraints on problem-solving and inhibition: object retrieval in cotton-top tamarins. *Journal of Comparative Psychology*, **113**, 1–8.

Santos, L.R., Miller, C.T. and Hauser, M.D. (2003) Representing tools: how two non-human primate species distinguish between the functionally relevant and irrelevant features of a tool. *Animal Cognition*, **6**, 269–281.

Sanz, C., Morgan, D. and Gulick, S. (2004) New insights into chimpanzees, tools, and termites from the Congo Basin. *American Naturalist*, **164**, 567–581.

Sapolsky, R.M. and Share, L.J. (2004) A pacific culture among wild baboons: its emergence and transmission. *PloS Biology*, **2**, 4, e106.

Sappington, B. F. and Goldman, L. (1994) Discrimination learning and concept formation in the Arabian horse. *Journal of Animal Science*, **72**, 3080–3087.

Sarnat, H. B. and Netsky, M. G. (1981) *Evolution of the Nervous System*, 2nd edn. Oxford, UK: Oxford University Press.

Savage-Rumbaugh, E. S. (1986) *Ape Language: From Conditioned Response to Symbol*. New York: Columbia University Press.

Savage-Rumbaugh, E. S. and Brakke, K. E. (1996) Animal language: methodological and interpretive issues. In *Readings in Animal Cognition*, ed. M. Bekoff and D. Jamieson. Cambridge, MA: MIT Press, pp. 269–288.

Savage-Rumbaugh, E. S. and Lewin, R. (1994) *Kanzi: The Ape at the Brink of the Human Mind*. New York: John Wiley.

Savage-Rumbaugh, E. S., Rumbaugh, D. M., Smith, S. T. and Lawson, J. (1980) Reference: the linguistic essential. *Science*, **210**, 922–925.

Savage-Rumbaugh, E. S., Macdonald, K., Sevcik, R. A., Hopkins, R. and Rubert, E. (1986) Spontaneous symbol acquisition and communicative use by pygmy chimpanzees (*Pan paniscus*). *Journal of Experimental Psychology: General*, **115**, 211–235.

Savage-Rumbaugh, E. S., Murphy, J., Sevcik, R. A., *et al.* (1993) Language comprehension in ape and child. *Monograph of the Society for Research in Child Development*, **58**, 1–252.

Savage-Rumbaugh, E. S., Shanker, S. G. and Taylor, T. J. (1998) *Apes, Language and the Human Mind*. Oxford, UK: Oxford University Press.

Scaife, M. and Bruner, J. (1975) The capacity for joint visual attention in the infant. *Nature*, **253**, 265–266.

Schaadt, C. P. and Rymon, L. M. (1982) Innate fishing behaviour of ospreys. *Raptor Research*, **16**, 61–62.

Schäfer, M. and Wehner, R. (1993) Loading does not affect measurement of walking distance in desert ants, *Cataglyphis fortis*. *Deutsche zoologische Gesellschaft*, **86**, 270–271.

Schaik, C. P. van (2004) *Among Apes: Red Apes and the Rise of Human Culture*. Cambridge, MA: Harvard University Press.

Schaik, C. P. van, Deaner, R. O. and Merrill, M. Y. (1999) The conditions for tool use in primates: implications for the evolution of material culture. *Journal of Human Evolution*, **36**, 719–741.

Schaik, C. P. van, Ancrenaz, M., Borgen, G., *et al.* (2003) Orangutan cultures and the evolution of material culture. *Science*, **299**, 102–105.

Schaik, C. P. van, Preuschoft, S. and Watts, D. P. (2004) Great ape social systems. In *The Evolution of Thought: Evolutionary Origins of Great Ape Intelligence*, ed. A. E. Russon and D. R. Begun. Cambridge, UK: Cambridge University Press, pp. 190–209.

Schastnyi, A. I. and Firsov, L. A. (1961) Physiological analysis of ape's communication in group experiments. *Doklady Biological Sciences*, **141**, 1264–1266. (In Russian with English summary.)

Scheiner, R., Weiss, A., Malun, D. and Erber, J. (2001) Learning in honey bees with brain lesions: how partial mushroom-body ablations affect sucrose responsiveness and tactile antennal learning. *Animal Cognition*, **4**, 227–235.

Scheumann, M. and Call, J. (2004) The use of experimenter-given cues by South African fur seals (*Arctocephalus pusillus*). *Animal Cognition*, **7**, 224–230.

Schjelderup-Ebbe, T. (1922) Beiträge zur Sozialpsycholgie des Haushuhns. *Zeitschrift für Psychologie*, **88**, 225–252.

Schjelderup-Ebbe, T. (1935) *Social Behaviour of Birds*. Reissued (1967) in *Handbook of Social Psychology*, ed. C. Murchison. Worcester, MA: Clark University Press.

Schmidt-Koenig, K. (1960) The sun's azimuth compass: one factor in the orientation of homing pigeons. *Science*, **131**, 826.

Schmidt-Koenig, K. (1979) *Avian Orientation and Navigation*. London: Academic Press.

Schneider, A. M. and Tarshis, B. (1975) *An Introduction to Physiological Psychology*. New York: Random House.

Schneirla, T. C. (1946) Ant learning as a problem in comparative psychology. In *Twentieth Century Psychology*, ed. P. L. Harriman. New York: Philosophical Library, pp. 276–305.

Schradin, C. and Pillay, N. (2004) The striped mouse from the succulent Karoo, South Africa: a territorial group living solitary forager with communal breeding and helpers at the nest. *Journal of Comparative Psychology*, **118**, 37–47.

Schtodin, M. P. (1947) The facts concerning high nervous activity of anthropoids (chimpanzees). *Annales of Pavlov's Institite of Evolutionary Physiology and Pathology of High Nervous Activity, Leningrad Academy of Science*, **2**, 171–183. (In Russian.)

Schusterman, R. J. and Gisiner, R. (1988) Artificial language comprehension in dolphins and sea lions: the essential cognitive skills. *Psychological Record*, **39**, 311–348.

Schusterman, R. J. and Kastak, D. (1993) A California sea lion (*Zalophus californianus*) is capable of forming equivalence relations. *Psychological Record*, **43**, 823–839.

Schusterman, R. J. and Kastak, D. (1998) Functional equivalence in a California sea lion: relevance to social and communicative interactions. *Animal Behaviour*, **55**, 1087–1095.

Schusterman, R. J. and Krieger, K. (1986) Artificial language comprehension and size transposition by a California sea lion (*Zalophus californianus*). *Journal of Comparative Psychology*, **100**, 348–355.

Schusterman, R. J., Reichmuth Kastak, C., and Kastak, D. (2002) The cognitive sea lion: meaning and memory in the lab and in nature. In *The Cognitive Animal: Empirical and Theoretical Perspectives on Animal Cognition*, ed. M. Bekoff, C. Allen and G. Burghardt. Cambridge, MA: MIT Press, pp. 217–228.

Schutz, F. (1965) Sexuelle Praegung bei Anatiden. *Zoological Tierpsychology*, **22**, 50–103.

Scott, J. P. (1992) The phenomemon of attachment in human/non-human relationships. In *The Inevitable Bond: Examining Scientist–Animal Interactions*, ed. H. Davis and D. A. Balfour. New York: Cambridge University Press, pp. 72–92.

Scott, J. P. and Fuller, J. L. (1965) *Genetics and the Social Behavior of the Dog*. Chicago, IL: University of Chicago Press.

Sechenov, I. M. (1863) *Reflex of the Brain*. Trans. S. Belsky (1965) Cambridge, MA: MIT Press.

Seeley, T. D. (1995) *The Wisdom of the Hive: The Social Psychology of Honey Bee Colonies*. Cambridge, MA: Harvard University Press.

Seeley, T. D., Kühnholz, S. and Seeley, R. H. (2002) An early chapter in behavioral physiology and sociobiology: the science of Martin Linaduer. *Journal of Comparative Physiology A*, **188**, 439–453.

Seger, J. (1981) Kinship and covariance. *Journal of Theoretical Biology*, **91**, 191–213.

Seidel, R. J. (1959) A review of sensory preconditioning. *Psychological Bulletin*, **46**, 58–73.

Seitz, A. (1940) Die Paarbildung bei einigen Cichliden. I. Die Paarbildung bei *Astatotilapia strigigena* (Pfeffer). *Zoological Tierpsychology*, **4**, 40–84.

Seligman, M. E. P. (1970) On the generality of the laws of learning. *Psychological Review*, **77**, 406–418.

Seligman, M. E. P. (1971) Phobias and preparedness. *Behavior Therapy*, **2**, 307–320.

Sevenster, P. (1968) Motivation and learning in sticklebacks. In *The Central Nervous System and Fish Behaviour*, ed. D. Ingle. Chicago, IL: University of Chicago Press, pp. 233–245.

Seyfarth, E.-A., Hergernröder, R., Ebbes, H. and Barth, F. G. (1982) Idiothetic orientation of a wandering spider: compensations of detours and estimates of goal distance. *Behavioral Ecology and Sociobiology*, **11**, 139–148.

Seyfarth, R. M. and Cheney, D. L. (1990) The assessment by vervet monkeys of their own and another species' alarm calls. *Animal Behaviour*, **40**, 754–764.

Seyfarth, R. M. and Cheney, D. L. (1997) Communication and the minds of monkeys. In *Origin and Evolution of Intelligence*, ed. A. B. Schiebel and J. W. Schopf. Boston, MA: Jones & Bartlett, pp. 27–42.

Seyfarth, R. M. and Cheney, D. L. (2002) The structure of social knowledge in monkeys. In *The Cognitive Animal*, ed. M. Bekoff and C. Allen. Cambridge, MA: MIT Press, pp. 379–384.

Seyfarth, R. M. and Cheney, D. L. (2003) Signalers and receivers in animal communication. *Annual Review of Psychology*, **54**, 145–173.

Seyfarth, R. M. and Cheney, D. L. (2004) Meaning and emotion in animal vocalizations. In *Emotions Inside Out: 130 Years after Darwin's* The Expression of the Emotions in Man and Animals, ed. P. Ekman, J. J. Campos, R. J. Davidson and F. de Waal. *Annals of the New York Academy of Sciences*, **1000**, 32–55.

Seyfarth, R. M., Cheney, D. L. and Marler, P. (1980) Vervet monkey alarm calls semantic communication in a free-ranging primate, *Animal Behaviour*, **28**, 1070–1094.

Shannon, C. E. (1948) A mathematical theory of communication. *Bell System Technical Journal*, **27**, 379–423, 623–656.

Shapiro, K. L., Jacobs, W. J. and LoLordo, V. M. (1980) Stimulus–reinforcer interactions in Pavlovian conditioning of pigeons: implications for selective associations. *Animal Learning and Behaviour*, **8**, 586–594.

Shepard, R. N. and Cooper, L. (1982) *Mental Images and their Transformations*. Cambridge, MA: MIT Press.

Shepard, R. N. and Metzler, R. (1971) Mental rotation in three-dimensional objects. *Science*, **171**, 701–703.

Shepherd, W. T. (1915) Tests on adaptive intelligence in dogs and cats, as compared with adaptive intelligence in rhesus monkeys. *American Journal of Psychology*, **26**, 211–216.

Sherman, P. W. (1977) Nepotism and the evolution of alarm calls. *Science*, **197**, 1246–1253.

Sherman, P. W., Jarvis, J. U. M. and Alexander, R. D. (eds.) (1991) *The Biology of the Naked Mole-Rat*. Princeton, NJ: Princeton University Press.

Sherrington, C. S. (1904) *The Integrative Action of the Nervous System*. Cambridge, UK: Cambridge University Press.

Sherry, D. F. (1989) Food storing in the Paridae. *Wilson Bulletin*, **101**, 289–304.

Sherry, D. F. and Duff, S. J. (1996) Behavioural and neural bases of orientation in food-storing birds. *Journal of Experimental Biology*, **199**, 165–172.

Sherry, D. F. and Galef, B. G., Jr (1984) Cultural transmission without imitation: milk bottle opening by birds. *Animal Behaviour*, **32**, 937–938.

Sherry, D. F. and Galef, B. G., Jr (1990) Social learning without imitation: more about milk bottle opening by birds. *Animal Behaviour*, **40**, 987–989.

Sherry, D. F., Vaccarino, A. L., Buckenham, K. and Herz, R. S. (1989) The hippocampal complex of food-storing birds. *Brain, Behaviour and Evolution*, **34**, 308–317.

Sherwin, C. M., Heyes, C. M. and Nicol, C. J. (2002) Social learning influences the preferences of domestic hens for novel food. *Animal Behaviour*, **63**, 933–942.

Shettleworth, S. J. (1972) Constrains of learning. In *Advances in the Study of Behaviour*, vol. 4, ed. D. S. Lehrman, R. Hinde and E. Shaw. New York: Academic Press, pp. 1–68.

Shettleworth, S. J. (1975) Reinforcement and the organization of behaviour in golden hamsters: hunger, environment and food reinforcement. *Journal of Experimental Psychology: Animal Behavior Processes*, **105**, 56–87.

Shettleworth, S. J. (1994) Biological approaches to learning. In *Handbook of Perception and Cognition*, vol. 9, ed. N. J. Mackintosh. London: Academic Press, pp. 185–219.

Shettleworth, S. J. (1998) *Cognition, Evolution, and Behavior*. New York: Oxford University Press.

Shettleworth, S. J. (2001) Animal cognition and animal behaviour. *Animal Behaviour*, **61**, 277–286.

Shettleworth, S. J. and Krebs, J. R. (1982) How marsh tits find their hoards: the role of site preference and spatial memory. *Journal of Experimental Psychology: Animal Behaviour Processes*, **8**, 354–375.

Shillito, D. J., Gallup, G. G., Jr and Beck, B. B. (1999) Factors affecting mirror behaviour in western lowland gorillas, *Gorilla gorilla*. *Animal Behaviour*, **57**, 999–1004.

Shishimi, A. (1985) Latent inhibition experiments with goldfish (*Carasius auratus*). *Journal of Comparative Psychology*, **99**, 316–327.

Siegel, E. and Rachlin, H. (1995) Soft commitment: self-control as achieved by response persistence. *Journal of the Experimental Analysis of Behavior*, **64**, 117–128.

Silk, J. B. (1999) Male bonnet macaques use information about third-party rank relationships to recruit allies. *Animal Behaviour*, **58**, 45–51.

Sinervo, B. and Clobert, J. (2003) Morphs, dispersal, genetic similarity and the evolution of cooperation. *Science*, **300**, 1949–1951.

Sinervo, B. and Lively, C. M. (1996) The rock-paper-scissors game and the evolution of alternative male reproductive strategies. *Nature*, **380**, 240–243.

Sinn, D. L., Perrin, N. A., Mather, J. A. and Anderson, R. C. (2001) Early temperamental traits in an octopus (*Octopus bimaculoides*). *Journal of Comparative Psychology*, **115**, 351–364.

Skinner, B. F. (1938) *The Behavior of Organisms: An Experimental Analysis*. New York: Appleton-Century.

Skinner, B. F. (1945) Baby in a box: introducing the mechanical baby tender. *Ladies' Home Journal*, **62**, 30–31, 135–136, 138.

Skinner, B. F. (1948a) *Walden Two*. New York: Macmillan.

Skinner, B. F. (1948b) Superstition in the pigeon. *Journal of Experimental Psychology*, **38**, 168–172.

Skinner, B. F. (1956) A case history in scientific method. *American Psychologist*, **11**, 221–233.

Skinner, B. F. (1966) Some responses to the stimulus 'Pavlov'. *Conditional Reflex*, **1**, 74–78.

Skinner, B. F. (1967) B. F. Skinner. In *A History of Psychology in Autobiography*, vol. 5, ed. E. G. Boring and G. Lindzey. New York: Appleton-Century-Crofts, pp. 385–413.

Skinner, B. F. (1971) *Beyond Freedom and Dignity*. New York: Knopf.

Skinner, B. F. (1978) *Reflections on Behaviorism and Society*. Englewood Cliffs, NJ: Prentice-Hall.

Skinner, B. F. (1979) *The Shaping of a Behaviorist: Part Two of an Autobiography*. New York: Knopf.

Skinner, B. F. (1985) *Particulars of My Life: The Shaping of a Behaviorist, a Matter of Consequences*. New York: Oxford University Press.

Skinner, B. F. (1987) *Upon Further Reflection*. Englewood Cliffs, NJ: Prentice-Hall.

Skinner, B. F. (1989) The origins of cognitive thought. *American Psychologist*, **44**, 13–18.

Skutch, A. F. (1935) Helpers at the nest. *Auk*, **52**, 257–273.

Slater, P. J. B. (1983) Bird song learning: theme and variations. In *Perspectives in Ornithology*, ed. A. H. Brush and G. A. Clark, Jr. Cambridge, UK: Cambridge University Press, 21–70.

Slater, P. J. B. (1989) Bird song learning: causes and consequences. *Ethology, Ecology and Evolution*, **1**, 19–46.

Slater, P. J. B. (1999) *Essentials of Animal Behaviour*. Cambridge, UK: Cambridge University Press.

Slater, P. J. B. (2003) Fifty years of bird song research: a case study in animal behaviour. *Animal Behaviour*, **65**, 957–969.

Slobodchikoff, C. N., Kiriazis, J., Fischer, C. and Creef, E. (1991) Semantic information distinguishing individual predators in the alarm calls of Gunnison's prairie dogs. *Animal Behaviour*, **42**, 713–719.

Sluckin, W. (1965) *Imprinting and Early Learning*. Chicago, IL: Aldine.

Smale, L., Nunes, S. and Holekamp, K.E. (1997) Sexually dimorphic dispersal in mammals: patterns, causes and consequences. *Advances in the Study of Behavior*, **26**, 181–250.

Small, W.S. (1900) Experimental study of the mental processes of the rat. I. *American Journal of Psychology*, **11**, 133–165;

Small, W.S. (1901) Experimental study of the mental processes of the rat. II. *American Journal of Psychology*, **12**, 206–239.

Smirnova, A.A., Lazareva, O.F. and Zorina, Z.A. (2000) Use of number by crows: investigation by matching and oddity learning. *Journal of the Experimental Analysis of Behavior*, **73**, 163–176.

Smolker, R.A., Richards, A.F., Connor, R.C., Mann, J. and Berggren, P. (1997) Sponge-carrying by Indian Ocean bottlenose dolphins: possible tool-use by a delphinid. *Ethology*, **103**, 454–465.

Smulders, T.V., Sasson, A.D. and DeVoogd, T.J. (1995) Seasonal variation in hippocampal volume in a food-storing bird, the black-capped chickadee. *Journal of Neurobiology*, **27**, 15–25.

Snowdon, C.T. (1986) Vocal communication. In *Comparative Primate Biology*, vol. 2A, *Behaviour, Conservation and Ecology*, ed. G. Mitchell and J. Erwin. New York: Alan R. Liss, pp. 495–530.

Snowdon, C.T. (1999) An empiricist view of language evolution and development. In *The Origins of Language: What Nonhuman Primates Can Tell Us*, ed. B.J. King. Santa Fe, NM: School of American Research Press, pp. 79–114.

Snowdon, C.T. and Soini, P. (1988) The tamarins, genus *Saguinus*. In *Ecology and Behavior of Neotropical Primates*, vol. 2, ed. R.A. Mittermeier, A.F. Coimbra-Filho and G.A.B. da Fonseca. Washington, DC: World Wildlife Fund, pp. 223–298.

Snowdon, C.T., Brown, C.H. and Peterson, M.R. (eds.) (1982) *Primate Communication*. Cambridge, UK: Cambridge University Press.

Sober, E.R. and Wilson, D.S. (1998) *Unto Others: The Evolution and Psychology of Unselfish Behavior*. Cambridge, MA: Harvard University Press.

Sokal, R.R., Oden, N.L. and Thomson, B.A. (1988) Genetic change across language boundaries in Europe. *American Journal of Physical Anthropology*, **70**, 489–502.

Solomon, N.G. and French, J.A. (eds.) (1996) *Co-operative Breeding in Mammals*. Cambridge, UK: Cambridge University Press.

Sophian, C. (1985) Understanding the movement of objects: early developments in spatial cognition. *British Journal of Developmental Psychology*, **3**, 321–333.

Soproni, K., Miklósi, A., Topál, J. and Csányi, V. (2001) Comprehension of human communicative signs in pet dogs. *Journal of Comparative Psychology*, **115**, 122–126.

Soproni, K., Miklósi, A., Topál, J. and Csányi, V. (2002) Dogs (*Canis familiaris*) responsiveness to human pointing gestures. *Journal of Comparative Psychology*, **116**, 27–34.

Sousa, C. and Matsuzawa, T. (2001) The use of tokens as rewards and tools by chimpanzees. *Animal Cognition*, **4**, 213–221.

Spalding, D.A. (1872) On instinct. *Nature*, **6**, 485–486.

Spalding, D.A. (1873) Instinct, with original observations on young animals. *Macmillan's Magazine*, **27**, 282–293.

Spence, K.W. (1936) The nature of discrimination learning in animals. *Psychological Review*, **43**, 427–449.

Spence, K.W. (1937) The differential response in animals to stimuli varying within a single dimension. *Psychological Review*, **44**, 430–444.

Spence, K.W. (1960) *Behavior Theory and Learning*. Englewood Cliffs, NJ: Prentice-Hall.

Spencer, H. (1855) *The Principles of Psychology*. London: Longman, Brown, Green and Longman.

Spencer, H. (1852a) A theory of population, deduced from the general law of animal fertility. *Westminster Review*, **57**, 468–501.

Spencer, H. (1852b). *The Development Hypothesis*. Reissued (1892) New York: Appleton.

Spetch, M.L. (1995) Overshadowing in landmark learning: touch-screen studies with pigeons and humans. *Journal of Experimental Psychology: Animal Behavior Processes*, **21**, 166–181.

Spinozzi, G. (1989) Early sensorimotor development in *Cebus* (*Cebus apella*). In *Cognitive Structure and Development in Nonhuman Primates*, ed. F. Antinucci. Hillsdale, NJ: Lawrence Erlbaum, pp. 55–66.

Spinozzi, G. and Potì, P. (1989) Causality. I. The support problem. In *Cognitive Structure and Development in Nonhuman Primates*, ed. F. Antinucci. Hillsdale, NJ: Lawrence Erlbaum, pp. 113–119.

Spinozzi, G. and Potì, P. (1993) Piagetian stage V in two chimpanzee infants (*Pan troglodytes*): the development of permanence of objects and the spatialization of causality. *International Journal of Primatology*, **14**, 905–917.

Spitz, R. (1945) Hospitalism: an inquiry into the genesis of psychiatric condition in early childhood. *Psychoanalytic Study of the Child*, **1**, 53–74.

Srinivasan, M.V. (1998) Ants match as they march. *Nature*, **392**, 660.

Srinivasan, M.V., Lehrer, M., Zhang, S.W. and Horridge, G.A. (1989) How honey bees measure their distance from objects of unknown sizes. *Journal of Comparative Physiology A*, **165**, 605–613.

Stacey, P.B. and Koenig, W.D. (1984) Cooperative breeding in the acorn woodpecker. *Scientific American*, **251**, 114–121.

Stacey, P.B. and Koenig, W.D. (1990) *Cooperative Breeding in Birds: Long-Term Studies of Ecology and Behavior*. Cambridge, UK: Cambridge University Press.

Staddon, J.E.R. (1983) *Adaptive Behaviour and Learning*. Cambridge, UK: Cambridge University Press.

Stander, P.E. (1992) Cooperative hunting in lions: the role of the individual. *Behavioral Ecology and Sociobiology*, **29**, 445–454.

Starkey, P. (1992) The early development of numerical reasoning. *Cognition*, **43**, 93–126.

Stebaev, I.V. and Reznikova, Zh. I. (1972) Two interaction types of ants living in steppe ecosystems in South Siberia. *Ecologia Polska*, **20**, 103–109.

Steche, W. (1957) Beiträge zur Analyse der Bienentänze. *Insectes Sociaux*, **4**, 305–318.

Stephens, D.W. (2002) Discrimination, discounting and impulsivity: a role for an informational constraint. *Philosophical Transactions of the Royal Society of London B*, **357**, 1527–1537.

Stephens, D.W. and Anderson, D. (2001) The adaptive value of preference for immediacy: when short-sighted rules have farsighted consequences. *Behavioral Ecology*, **12**, 330–339.

Stephens, D.W. and Krebs, J.R. (1986) *Foraging Theory*. Princeton, NJ: Princeton University Press.

Stephens, D.W. and McLinn, C.M. (2003) Choice and context: testing a simple short-term choice rule. *Animal Behaviour*, **66**, 59–70.

Stephens, D.W., Lynch, J.F., Sorensen, A.E. and Gordon, C. (1986) Preference and profitability: theory and experiment. *American Naturalist*, **127**, 533–553.

Stern, D.L. and Foster, W.A. (1996) The evolution of sterile soldiers in aphids. *Biological Reviews*, **71**, 27–79.

Strassmann, J.E. and Queller, D.C. (2001) Selfish responses by clone invaders. *Proceedings of the National Academy of Sciences of the USA*, **98**, 11839–11841.

Straub, R.O. and Terrace, H.S. (1981) Generalisation of serial learning in the pigeon. *Animal Learning and Behaviour*, **9**, 454–468.

Straub, R.O., Seidenberg, M.S., Bever, T.G. and Terrace, H.S. (1979) Serial learning in the pigeon. *Journal of the Experimental Analysis of Behavior*, **32**, 137–148.

Stresemann, E. (1947) Baron von Pernau, pioneer student of bird behaviour. *Auk*, **64**, 35–52.

Struhsaker T. (1967) *Behavior of Vervet Monkeys* (Cercopithecus aethiops). Berkeley, CA: University of California Press.

Studdert-Kennedy, M. (1983) On learning to speak. *Human Neurobiology*, **2**, 191–195.

Suddendorf, T. and Corballis, M.C. (1997) Mental time travel and the evolution of the human mind. *Genetic, Social, and General Psychology Monographs*, **123**, 133–167.

Sutherland, W.J., Ens, B.J., Goss-Custard, J.D. and Hulscher, J.B. (1996) Specialisation. In *The Oystercatcher: From Individuals to Populations*, ed. J.D. Goss-Custard. Oxford, UK: Oxford University Press, pp. 56–76.

Swaddle, J.P., Cathey, M.G., Correll, M. and Hodkinson, B.P. (2005) Socially transmitted mate preferences in a monogamous bird: a non-genetic mechanism of sexual selection. *Proceedings of the Royal Society of London B*, **272**, 1053–1058.

Swartz, K.B., Chen, S. and Terrace, H.S. (1991) Serial learning by rhesus monkeys. I. Acquisition and retention of multiple four-item lists. *Journal of Experimental Psychology: Animal Behavior Processes*, **17**, 396–410.

Swartz, K.B., Sarauw, D. and Evans, S. (1999) Comparative aspects of mirror self-recognition in great apes. In *The Mentalities of Gorillas and Orangutans in Comparative Perspective*, ed. S.T. Parker, R.W. Mitchell and H.L. Miles. Cambridge, UK: Cambridge University Press, pp. 283–294.

Swenson, R. (1991) Order, evolution, and natural law: fundamental relations in complex systems theory. In *Handbook of Systems and Cybernetics*, ed. C. Negoita. New York: Marcel Dekker, pp. 125–148.

Takeda, K. (1961) Classical conditioning in the honey bee. *Journal of Insect Physiology*, **6**, 168–179.

Tarasyan, K.K., Filatova, O.A., Burdin, A.M., Hoyt, E. and Sato, H. (2005) Keys for the status of killer whales in Eastern Kamchatka, Russia: foraging ecology and acoustic behaviour. *Biosphere Conservation*, **6**, 73–83.

Tautz, J. (1996) Honeybee waggle dance: recruitment success depends on the dance floor. *Journal of Experimental Biology*, **199**, 1375–1381.

Tautz, J., Casas, J. and Sandeman, D. (2001) Phase reversal of vibratory signals in honeycomb may assist dancing honeybees to attract their audience. *Journal of Experimental Biology*, **204**, 3737–3746.

Tebbich, S. and Bshary, R. (2004) Cognitive abilities related to tool use in the woodpecker finch, *Cactospiza pallida*. *Animal Behaviour*, **57**, 689–697.

Tebbich, S., Taborsky, M., Fessl, B. and Blomqvist, D. (2001) Do woodpecker finches acquire tools use by social learning? *Proceedings of the Royal Society of London B*, **268**, 2189–2193.

Tebbich, S., Taborsky, M. and Fessl, B. (2002) The ecology of tool use in the woodpecker finch *Cactospiza pallida*. *Ecology Letters*, **5**, 656–664.

Temeles, E. J. (1994) The role of neighbours in territorial systems: when are they 'dear enemies'? *Animal Behaviour*, **47**, 339–350.

Temerlin, M. K. (1975) *Lucy: Growing Up Human*. Palo Alto, CA: Science and Behaviour Books.

Templeton, J. J. and Giraldeau, L.-A. (1995) Public information cues affect the scrounging decisions of starlings. *Animal Behaviour*, **49**, 1617–1626.

Terkel, J. (1996) Cultural transmission of feeding behavior in the black rat (*Rattus rattus*). In *Social Learning in Animals: The Roots of Culture*, ed. C. M. Heyes and B. G. Galef, Jr. San Diego, CA: Academic Press, pp. 17–47.

Terrace, H. S. (1979) *Nim: A Chimpanzee Who Learned Sign Language*. New York: Knopf.

Terrace, H. S. (1987) Chunking by a pigeon in a serial learning task. *Nature*, **325**, 149–151.

Terrace, H. S. (2001) Comparative psychology of chunking. In *Animal Cognition and Sequential Behavior*, ed. S. Fountain. Dordrecht, the Netherlands: Kluwer, pp. 23–56.

Terrace, H. S., Son, L. K. and Brannon, E. M. (2003) Serial expertise of rhesus macaques. *Psychological Science*, **14**, 66–73.

Theberge, J. B. and Falls, J. B. (1967) Howling as a means of communication in timber wolves. *American Zoologist*, **7**, 331–338.

Theberge, J. B. and Pimlott, D. H. (1969) Observations of wolves at a rendezvous site in Algonquin Park. *Canadian Field Naturalist*, **83**, 122–128.

Thierry, B. (1985) Patterns of agonistic interactions in three species of macaques (*M. mulatta, M. fascicularis, M. tonkeana*). *Aggressive Behaviour*, **11**, 223–233.

Thomas, R. K. (1986) Vertebrate intelligence: a review of the laboratory research. In *Animal Intelligence: Insights into the Animal Mind*, ed. B. J. Hoage and L. Goldman. Washington, DC: Smithsonian Institution Press, pp. 37–56.

Thomas, R. K. and Chase, L. (1980) Relative numerousness judgements by squirrel monkeys. *Bulletin of the Psychonomic Society*, **16**, 79–82.

Thomas, R. K., Phillips, J. A. and Young, C. D. (1999) Comparative cognition: human numerousness judgements. *American Journal of Psychology*, **112**, 215–233.

Thompson, R. K. R. (1995) Natural and relational concepts in animals. In *Comparative Approaches to Cognitive Science*, ed. H. L. Roitblat and J. A. Meyer. Cambridge, MA: MIT Press, pp. 175–224.

Thompson, R. K. R. and Oden, D. L. (2000) Categorical perception and conceptual judgements by nonhuman primates: the paleological monkey and the analogical ape. *Cognitive Science*, **24**, 363–396.

Thompson, R. K. R., Oden, D. L. and Boysen, S. T. (1997) Language-naive chimpanzees (*Pan troglodytes*) judge relations between relations in a conceptual matching-to-sample task. *Journal of Experimental Psychology: Animal Behavior Processes*, **23**, 30–43.

Thompson Seton, E. (1898) *Wild Animals I Have Known*. New York: Charles Scribner's Sons.

Thorndike, E. L. (1898) Animal intelligence: an experimental study of the associative process in animals. *Psychological Review, Monograph Supplements*, **2**, 1–109.

Thorndike, E. L. (1911) *Animal Intelligence: Experimental Studies*. New York: Macmillan.

Thorndike, E. L. (1932) *The Fundamentals of Learning*. New York: Teachers' College Press.

Thorndike, E. L. (1949) *Selected Writings from a Connectionist's Psychology*. New York: Greenwood Press.

Thorpe, W. H. (1950) A note on detour experiments with *Ammophyla pubescens* Curt. (Hymenoptera: Sphecidae). *Behaviour*, **2**, 257–263.

Thorpe, W. H. (1956) *Learning and Instinct in Animals*. London: Methuen.

Thorpe, W. H. (1958) The learning of song patterns in birds, with especial reference to the song of the chaffinch, *Fringilla coelebs*. *Ibis*, **100**, 535–570.

Thorpe, W. H. (1961) *Bird Song*. New York: Cambridge University Press.

Thorpe, W. H. (1963) *Learning and Instinct in Animals*, 2nd edn. Cambridge, MA: Harvard University Press.

Thouless, C. R., Fanshawe, J. H. and Bertram, B. C. R. (1989) Egyptian vultures *Neophron pernopterus* and ostrich *Struthio camelus* eggs: the origins of stone-throwing behavior. *Ibis*, **131**, 9–15.

Tibbetts, E. A. (2002) Visual signals of individual identity in the wasp *Polistes fuscatus*. *Proceedings of the Royal Society of London B*, **269**, 1423–1428.

Tibbetts, E. A. and Dale, J. (2004) A socially enforced signal of quality in a paper wasp. *Nature*, **432**, 218–222.

Tikh, N. A. (1950) Group living and communication in monkeys from the standpoint of anthropogenesis. Ph.D. thesis, Leningrad Academy of Sciences, USSR. (In Russian.)

Timberlake, W. (1994) Behavior systems associationism, and Pavlovian conditioning. *Psychonomic Bulletin and Review*, **1**, 405–420.

Timberlake, W. and Hoffman, C. (2002) How does the ecological foraging behavior of desert kangaroo rats (*Dipodomys deserti*) relate to their behavior on radial mazes? *Learning and Behavior*, **30**, 342–354.

Tinbergen, N. (1942) An objective study of the innate behaviour of animals. *Bibliotheca Biotheoretica*, **1**, 39–98.

Tinbergen, N. (1951) *The Study of Instinct*. Oxford, UK: Clarendon Press.

Tinbergen, N. (1963) On aims and methods of ethology. *Zeitschrift für Tierpsychologie*, **20**, 410–433.

Tinbergen, N. (1973) *Autobiography*. Available online: www.lnobel.se/laureates/medicine-1973-3-autobio.html/

Tinbergen, N. (1976) *The Herring Gull's World*. London: Collins.

Tinbergen, N. and Kruyt, W. (1938) Über die Orientierung des Bienenwolfes *Philanthus triangulum* Fabr. III. Die Bevorzugung bestimmter Wegmarken. *Zeitschrift für vergleichende Physiologie*, **25**, 292–234.

Tinbergen, N. and Perdeck, A. C. (1950) On the stimulus situation releasing the begging response in the newly hatched herring gull chick (*Larus argentatus argentatus* Pont). *Behaviour*, **3**, 1–39.

Tinklepaugh, O. L. (1928) An experimental study of representative factors in monkeys. *Journal of Comparative Psychology*, **8**, 197–236.

Tinklepaugh, O. L. (1932) Multiple delayed reaction with chimpanzees and monkeys. *Journal of Comparative Psychology*, **13**, 207–243.

Tiunova, A., Anokhin, K. V. and Rose, S. P. R. (1998) Two critical periods of protein and glycoprotein synthesis in memory consolidation for visual categorisation learning in chicks. *Learning and Memory*, **4**, 401–410.

Tobin, H. and Logue, A. W. (1994) Self-control across species (*Columba livia, Homo sapiens*, and *Rattus norvegicus*). *Journal of Comparative Psychology*, **108**, 126–133.

Tobin, H., Logue, A. W., Chelonis, J. J., Ackerman, K. T. and May, J. G. (1996) Self-control in the monkey *Macaca fascicularis*. *Animal Learning and Behaviour*, **24**, 168–174.

Todd, P. M. and Miller, G. F. (1991) On the sympatric origin of species: mercurial mating in the quicksilver model. In *Proceedings 4th International Conference on Genetic Algorithms*, ed. R. K. Belew and L. B. Booker. San Mateo, CA: Morgan Kaufmann, pp. 547–554.

Todt, D. (1975) Social learning of vocal patterns and modes of their application in grey parrots. *Zeitschrift für Tierpsychologie*, **39**, 178–188.

Tolman, E. C. (1922) A new formula for behaviorism. *Psychological Review*, **29**, 44–53.

Tolman, E. C. (1932) *Purposive Behavior in Animals and Men*. New York: Appleton-Century-Crofts.

Tolman, E. C. (1937) The acquisition of string-pulling by rats: conditioned response or sign-gestalt? *Psychological Review*, **44**, 195–211.

Tolman, E. C. (1938) The determiners of behavior at a choice point. *Psychological Review*, **45**, 1–41.

Tolman, E. C. (1948) Cognitive maps in rats and men. *Psychological Review*, **55**, 189–208.

Tolman, E. C. and Honzik, C. H. (1930a) 'Insight' in rats. *University of California Publications in Psychology*, **4**, 215–232.

Tolman, E. C. and Honzik, C. H. (1930b) Introduction and removal of reward and maze performance in rats. *University of California Publications in Psychology*, **4**, 257–275.

Tolman, E. C., Ritchie, B. F. and Kalish, D. (1946a) Studies in spatial learning. I. Orientation and the short cut. *Journal of Experimental Psychology*, **36**, 13–24.

Tolman, E. C., Ritchie, B. F. and Kalish, D. (1946b) Studies in spatial learning. II. Place learning versus response learning. *Journal of Experimental Psychology*, **36**, 221–229.

Tomasello, M. (1996) Do apes ape? In *Social Learning in Animals: The Roots of Culture*, ed. C. M. Heyes and B. G. Galef, Jr. San Diego, CA: Academic Press, pp. 319–346.

Tomasello, M. (1999) *The Cultural Origins of Human Cognition*. Cambridge, MA: Harvard University Press.

Tomasello, M. (2000) Primate cognition: introduction to the issue. *Cognitive Science*, **24**, 351–362.

Tomasello, M. and Call, J. (1997) *Primate Cognition*. Oxford, UK: Oxford University Press.

Tomasello, M. and Farrar, J. M. (1986) Joint attention and early language. *Child Development*, **57**, 1454–1463.

Tomasello, M., Davis-Dasilva, M., Camak, L. and Bard, K. (1987) Observational learning of tool-use by young chimpanzees. *Journal of Human Evolution*, **2**, 175–183.

Tomback, D. F. (1980) How nutcrackers find their seed stores. *Condor*, **82**, 10–19.

Tomonaga, M. and Matsuzawa, T. (2002) Enumeration of briefly presented items by the chimpanzee (*Pan troglodytes*) and humans (*Homo sapiens*). *Animal Learning and Behaviour*, **30**, 143–157.

Tonooka, R. (2001) Leaf-folding behavior for drinking water by wild chimpanzees (*Pan troglodytes verus*) at Bossou, Guinea. *Animal Cognition*, **4**, 325–334.

Torgerson, W. (1958) *Theory and Methods of Scaling*. New York: John Wiley.

Tóth, E. and Duffy, J.E. (2005) Coordinated group response to nest intruders in social shrimp. *Proceedings of the Royal Society of London B, Biology Letters*, **1**, 49–52.

Trivers, R.L. (1971) The evolution of reciprocal altruism. *Quarterly Review of Biology*, **46**, 35–57.

Trivers, R.L. (1974) Parent–offspring conflict. *American Zoologist*, **14**, 249–264.

Tryon, R.C. (1940) Genetic differences in maze learning in rats. *In 39th Yearbook, National Society for the Study of Education*. Bloomington, IN: Public School Publishing, pp. 111–119.

Tulving, E. (1972) Episodic and semantic memory. In *Organisation of Memory*, ed. E. Tulving and W. Donaldson. New York: Academic Press, pp. 381–403.

Tulving, E. (1983) *Elements of Episodic Memory*. Oxford, UK: Clarendon Press.

Tulving, E. and Markowitsch, H.J. (1998) Episodic and declarative memory: role of the hippocampus. *Hippocampus*, **8**, 198–204.

Turner, C.H. (1910) Experiments on color-vision of the honey bee. *Biological Bulletin*, **19**, 257–279.

Turner, C.H. (1911) Experiments on pattern vision of the honey bee. *Biological Bulletin*, **21**, 249–264.

Turner, E.R.A. (1965) Social feeding in birds. *Behaviour*, **24**, 1–65.

Uexküll, J. von (1909), *Umwelt und Innenwelt der Tiere*. Berlin: Jena. Reprinted (1989) Environment (*Umwelt*) and the inner world of animals. In *The Foundations of Comparative Ethology*, trans. C.J. Mellor and D. Gove, ed. G.M. Burghardt. New York: Van Nostrand Reinhold, pp. 222–245.

Ukhtomsky, A.A. (1950) Dominanta as a factor of behaviour. In *Collected Works*, vol. 1. Leningrad, USSR: Nauka, pp. 293–315. (In Russian.)

Ulanova, L.I. (1950) Shaping notations expressing need for food in monkeys. In *Studying High Nervous Activity by Means of Natural Experiments*, ed. V.P. Protopopov. Kiev, USSR: Gosmedizdat, pp. 103–114. (In Russian.)

Uller, C., Hauser, M.D. and Carey, S. (2001) Spontaneous representation of number in cotton-top tamarins (*Saguinus oedipus*). *Journal of Comparative Psychology*, **115**, 248–257.

Uzgiris, I. and Hunt, M. (1975) *Assessment in Infancy: Ordinal Scales of Psychological Development*. Urbana, IL: University of Illinois Press.

Vallortigara, G. (2000) Comparative neuropsychology of the dual brain: a stroll through animals' left and right perceptual worlds. *Brain and Language*, **73**, 189–219.

Vallortigara, G. (2004) Visual cognition and representation in birds and primates. In *Comparative Vertebrate Cognition: Are Primates Superior to Nonprimates?* ed. L.J. Rogers and G. Kaplan. New York: Kluwer, pp. 57–94.

Vallortigara, G., Zanforlin, M. and Pasti, G. (1990) Geometric modules in animals spatial representations: a test with chicks (*Gallus gallus domesticus*). *Journal of Comparative Psychology*, **104**, 248–254.

Valone, T.J. (1989) Group foraging, public information, and patch estimation. *Oikos*, **56**, 357–363.

Vander Wall, S.B. (1990) *Food Hoarding in Animals*. Chicago, IL: University of Chicago Press.

Vander Wall, S.B. and Jenkins, S.H. (2003) Reciprocal pilferage and the evolution of food-hoarding behavior. *Behavioral Ecology*, **14**, 656–667.

Vančatová, M. (1984) The influence of imitation on tool using in capuchin monkeys (*Cebus apella*). *Anthropologie*, **22**, 1–2.

Vatzuro, E.G. (1941) Physiological analysis of some of Köller's experiments. In *9th Conference on Physiological Problems*. Leningrad, USSR: Academy of Sciences Publishing House, pp. 20–22. (In Russian.)

Vatzuro, E.G. (1948). *A Study of High Nervous Activity of a Chimpanzee*. Moscow: Academy of Medicine. (In Russian.)

Vauclair, J. (1996) *Animal Cognition: Recent Developments in Modern Comparative Psychology*. Cambridge, MA: Harvard University Press.

Vauclair, J. (2002) Categorisation and conceptual behavior in nonhuman primates. In *The Cognitive Animal: Empirical and Theoretical Perspectives on Animal Cognition*, ed. M. Bekoff, C. Allen and G. Burghardt. Cambridge, MA: MIT Press, pp. 239–245.

Vauclair, J. and Fagot, J. (1996) Categorization of alphanumeric characters by guinea baboons: within- and between-class stimulus discrimination. *Current Psychology of Cognition*, **15**, 449–452.

Veen, T., Richardson, D.S., Blaakmeer, K. and Komdeur, J. (2000) Experimental evidence for innate predator recognition in the Seychelles warbler. *Proceedings of Royal Society of London B*, **267**, 2253–2258.

Verbeek, M.E.M., Drent, P.J. and Wiekema, P.R. (1994) Consistent individual differences in early exploratory behavior of male great tits. *Animal Behaviour*, **48**, 1113–1121.

Verplanck, W.S. (1955) Since learned behavior is innate, and vice versa, what now? *Psychological Review*, **62**, 139–144.

Verron, M.D. (1952) *A Further Study of Visual Perception*. Cambridge, UK: Cambridge University Press.

Vetter, T. and Troje, N. (1997) Separation of texture and shape in images of faces for image coding and synthesis. *Journal of the Optical Society of America A*, **14**, 2152–2161.

Viguier, C. (1882) Le sens de l'orientation et ses organes chez les animaux et chez l'homme. *Revue Philosophique de la France et de l'Etranger*, **14**, 1–36.

Vince, M.A. (1956) 'String pulling' in birds. I. Individual differences in wild adult great tits. *British Journal of Animal Behaviour*, **4**, 111–116.

Vince, M.A. (1958) 'String-pulling' in birds. II. Differences related to age in greenfinches, chaffinches, and canaries. *Behaviour*, **6**, 53–59.

Vince, M.A. (1961) 'String-pulling' in birds. III. The successful response in greenfinches and canaries. *Behaviour*, **17**, 103–129.

Visalberghi, E. (2002) Insights from capuchin monkeys' studies: ingredients of, recipes for, and flaws in capuchins' success. In *The Cognitive Animal: Empirical and Theoretical Perspectives on Animal Cognition*, ed. M. Bekoff, C. Allen and G. Burghardt. Cambridge, MA: MIT Press, pp. 405–411.

Visalberghi, E. and Addessi, E. (2001) Acceptance of novel food in capuchin monkeys: do specific social facilitation and visual stimulus enhancement play a role? *Animal Behaviour*, **62**, 567–576.

Visalberghi, E. and Fragaszy, D. (1990) Do monkeys ape? In *'Language' and Intelligence in Monkeys and Apes*, ed. S. Parker and K. Gibson. Cambridge, UK: Cambridge University Press, pp. 247–273.

Visalberghi, E. and Limongelli, L. (1996) Action and understanding: tool use revisited through the mind of capuchin monkeys. In *Reaching into Thought: The Minds of the Great Apes*, ed. A. Russon, K. Bard and S. Parker. Cambridge, UK: Cambridge University Press, pp. 57–79.

Visalberghi, E. and Trinca, L. (1989) Tool use in the capuchin monkeys: distinguishing between performing and understanding. *Primates*, **30**, 511–521.

Visscher, P.K. and Seeley, T.D. (1982) Foraging strategy of honeybee colonies in a temperate deciduous forest. *Ecology*, **63**, 1790–1801.

Vlasak, A.N. (2006) Global and local spatial landmarks: their role during foraging by Columbian ground squirrels *Spermophilus columbianus*. *Animal Cognition*, **9**, 71–80.

Voelkl, B. and Huber, L. (2000) True imitation in marmosets. *Animal Behaviour*, **60**, 195–202.

Vonk, J. and MacDonald, E. (2002) Natural concepts in a juvenile gorilla (*Gorilla gorilla gorilla*) at three levels of abstraction. *Journal of the Experimental Analysis of Behavior*, **78**, 315–332.

Voronin, L.G. and Firsov, L.A. (1967) Investigations of high nervous activity of anthropoids in the USSR. *Journal of High Nervous Activity*, **17**, 834–841. (In Russian with English summary.)

Voskresenskaya, A.K. (1957) On the role played by mushroom bodies (*corpora pedunculata*) of the supraoesophageal ganglion in the conditioned reflexes of the honey bee. *Doklady Biological Sciences*, **112**, 964–967. (In Russian with English summary.)

Vygotsky, L.S. (1962) *Thought and Language*. Cambridge, MA: MIT Press.

Waal, F.B.M. de (1982) *Chimpanzee Politics: Power and Sex among Apes*. New York: Harper and Row. Reissued (2000) Baltimore, MD: Johns Hopkins University Press.

Waal, F.B.M. de. (1989) Behavioral contrasts between bonobo and chimpanzee. In *Understanding Chimpanzees*, ed. P. Heltne and L. Marquardt. Cambridge, MA: Harvard University Press, pp. 154–175.

Waal, F.B.M. de. (1997) *Good Natured: The Origins of Right and Wrong in Humans and Other Animals*. Cambridge, MA: Harvard University Press.

Waal, F.B.M. de. (1999) Cultural primatology comes of age. *Nature*, **399**, 635–636.

Waal, F.B.M. de. (2001) *The Ape and the Sushi Master: Cultural Reflections by a Primatologist*. New York: Basic Books.

Waal, F.B.M. de (2005) A century of getting to know the chimpanzee. *Nature*, **437**, 56–59.

Waal, F.B.M. de and Berger, M.L. (2000) Payment for labour in monkeys. *Nature*, **404**, 563.

Waal, F.B.M. de and Brosnan, S.F. (2006) Simple and complex reciprocity in primates. In *Cooperation in Primates and Humans: Mechanisms and Evolution*, ed. P.M. Kappeler and C.P. van Schaik. Berlin: Springer-Velag, pp. 85–105.

Waal, F.B.M. de and Seres, M. (1997) Propagation of handclasp grooming among captive chimpanzees. *American Journal of Primatology*, **43**, 339–346.

Waal, F.B.M. de and Yoshihara, D. (1983) Reconciliation and re-directed affection in rhesus monkeys. *Behaviour*, **85**, 224–241.

Waal, F.B.M. de, Dindo, M., Freeman, C.A. and Hall, M.J. (2005) The monkey in the mirror: hardly a stranger. *Proceedings of the National Academy of Sciences of the USA*, **102**, 1140–1147.

Wallace, A.R. (1855) On the law which has regulated the introduction of new species. *Annals and Magazine of Natural History* (2nd ser.) **16**, 184–196.

Wallace, A.R. (1860) The ornithology of Northern Celebes. *Ibis*, **2**, 140–147.

Wallace, D.G., Hines, D.J., Pellis, S.M. and Whishaw, I.Q. (2002) Vestibular information is required for dead reckoning in the rat. *Journal of Neuroscience*, **22**, 10009–10017.

Wallace, R.A. (1979) *Animal Behavior: Its Development, Ecology, and Evolution*. Santa Monica, CA: Goodyear.

Wallmann, J.M. (1992) *Aping Language (Themes in the Social Sciences)*. New York: Cambridge University Press.

Wallraff, H.G. (2004) Avian olfactory navigation: its empirical foundation and conceptual state. *Animal Behaviour*, **67**, 189–204.

Wallraff, H.G., Kiepenheuer, J. and Streng, A. (1994) The role of visual familiarity with the landscape in pigeon homing. *Ethology*, **97**, 1–25.

Walsh, J.F., Grunewald, J. and Grunewald, B. (1985) Green-backed heron (*Butorides striatus*) possibly using apparent bait. *Journal für Ornithologie*, **126**, 439–442.

Warren, J.M. (1965) Comparative psychology of learning. *Annual Review of Psychology*, **16**, 95–118.

Washburn, D.A. and Rumbaugh, D.M. (1989) Rhesus monkey *Macaca mulatta* complex learning skills reassessed. *International Journal of Primatology*, **12**, 377–388.

Washburn, D.A. and Rumbaugh, D.M. (1992) Comparative assessment of psychomotor performance: target prediction by humans and macaques (*Macaca mulatta*). *Journal of Experimental Psychology: General*, **121**, 305–312.

Wasmann E. (1899) Die psychischen Fähigkeiten der Ameisen. *Zoologica*, **26**, 1–133.

Wasserman, E.A. and Miller, R.R. (1997) What's elementary about associative learning? *Annual Review of Psychology*, **48**, 573–607.

Wasserman, E.A., Kiedinger, R.E. and Bhatt, R.S. (1988) Conceptual behavior in pigeons: categories, subcategories, and pseudocategories. *Journal of Experimental Psychology: Animal Behavior Processes*, **14**, 235–246.

Wasserman, E.A., Hugart, J.A. and Kirkpatrick-Steger, K. (1995) Pigeons show same–different conceptualization after training with complex visual stimuli. *Journal of Experimental Psychology: Animal Behavior Processes*, **21**, 248–252.

Wasserman, E.A., Gagliardi, J.L., Cook, B.R., *et al.* (1996) The pigeon's recognition of drawings of depth-rotated stimuli. *Journal of Experimental Psychology: Animal Behavior Processes*, **22**, 205–221.

Wasserman, E.A., Young, M.E. and Nolan, B.C. (2000) Display variability and spatial organisation as contributors to the pigeon's discrimination of complex visual stimuli. *Journal of Experimental Psychology: Animal Behavior Processes*, **26**, 133–143.

Wasserman, E.A., Young, M.E. and Fagot, J. (2001) Effects of number of items on the baboon's discrimination of same from different visual displays. *Animal Cognition*, **4**, 163–170.

Watson, J.B. (1903) *Animal Education*. Chicago, IL: University of Chicago Press.

Watson, J.B. (1913) Psychology as the behaviorist views it. *Psychological Review*, **20**, 158–177.

Watson, J.B. (1916) The place of the conditioned reflex in psychology. *Psychological Review*, **23**, 89–116.

Watson, J.B. (1919) *Psychology from the Standpoint of a Behaviorist*. Philadelphia, PA: Lippincott.

Watson, J.B. (1928) *Psychological Care of Infant and Child*. New York: W. W. Norton.

Watson, J.B. and Rayner, R. (1920) Conditioned emotional reactions. *Journal of Experimental Psychology*, **3**, 1–14.

Wehner, R. (1997) Preparational intelligence: how insects and birds find their way. In *The Origin and Evolution of Intelligence*, ed. A.B. Scheibel and J.W. Schopf. Boston, MA: Jones & Bartlett, pp. 1–26.

Wehner, R. (1998) Navigation in context: grand theories and basic mechanisms. *Journal of Avian Biology*, **29**, 370–386.

Wehner, R. (1999) Spatial representation in small-brain insect navigators: ant algorithms. In *Learning: Rule Extraction and Representation*, ed. A.D. Friederici and R. Menzel. Berlin: Walter de Gruyter, pp. 240–257.

Wehner, R. and Menzel, R. (1990) Do insects have cognitive maps? *Annual Review of Neuroscience*, **13**, 403–414.

Wehner, R. and Müller, M. (1985) Does interocular transfer occur in visual navigation by ants? *Nature*, **315**, 228–229.

Wehner, R. and Srinivasan, M.V. (1981) Searching behaviour of desert ants, genus *Cataglyphis* (Formicidae, Hymenoptera). *Journal of Comparative Physiology*, **142**, 315–338.

Wehner, R., Lehrer, L. and Harvey, W.R. (eds.) (1996a) *Journal of Experimental Biology, Special Issue: Navigation*, **199**.

Wehner, R., Michel, B. and Antonsen, P. (1996b) Visual navigation in insects: coupling of egocentric and geocentric information. *Journal of Experimental Biology*, **199**, 129–140.

Weinrich, M.T., Schilling, M.R. and Belt, C.R. (1992) Evidence for acquisition of a novel feeding behaviour: lobtail feeding in humpback whales, *Megaptera novaeangliae*. *Animal Behaviour*, **44**, 1059–1072.

Weir, A.A.S., Chappell, J. and Kacelnik, A. (2002) Shaping of hooks in New Caledonian crows. *Science*, **297**, 981–982.

Weir, J. S. (1958) Polyethism in workers of the ant *Myrmica*. *Insectes Sociaux*, **5**, 97–128.

Weiss, D., Ghazanfar, A., Miller, C. T. and Hauser, M. D. (2002) Specialised processing of primate facial and vocal expressions: evidence for cerebral asymmetries. In *Cerebral Vertebrate Lateralization*, ed. L. Rogers and R. Andrews. New York: Cambridge University Press, pp. 480–530.

Wells, M. J. (1962) *Brain and Behaviour in Cephalopods*. Stanford, CA: Stanford University Press.

Wells, M. J. (1978) *Octopus: Physiology and Behaviour of an Advanced Invertebrate*. London: Chapman & Hall.

Welzl, H. and Stork, O. (2003) Cell adhesion molecules: key players in memory consolidation? *News in Physiological Sciences*, **18**, 147–150.

Wenner, A. M. (1962) Sound production during the waggle dance of the honey bee. *Animal Behaviour*, **10**, 79–95.

Werdenich, D. and Huber, L. (2006) A case of quick problem-solving in birds: string pulling in keas, *Nestor notabilis*. *Animal Behaviour*, **71**, 855–863.

Wernicke, C. (1881–83) *Lehrtuch der Gehirnkrankheiten für Ärzte und Studirende*, 3 vols. Kassel, Germany: Fischer.

Wertheimer, M. (1912) Experimentelle Studien über das Sehen von Bewegung. *Zeitschrift für Psychologie mit Zeitschrift für angewandte Psychologie*, **61**, 161–265.

Wheatley, B. P. (1980) Feeding and ranging of East Bornean *Macaca fascicularis*. In *The Macaques: Studies in Ecology, Behavior and Evolution*, ed. D. G. Lindburg. New York: Van Nostrand Reinhold, pp. 215–246.

Wheeler, W. N. (1928) *The Social Insects: Their Origin and Evolution*. London: Kegan Paul.

Whitehead, H. (1998) Cultural selection and genetic diversity in matrilineal whales. *Science*, **282**, 1708–1711.

Whiten, A. (1989) Transmission mechanisms in primate cultural evolution. *Trends in Ecology and Evolution*, **4**, 61–62.

Whiten, A. (1993) Evolving a theory of mind: the nature of non-verbal mentalism in other primates. In *Understanding Other Minds*, ed. S. Baron-Cohen, H. Tager-Flusberg and D. Cohen. Oxford, UK: Oxford University Press, pp. 367–396.

Whiten, A. (1998) Imitation of the sequential structure of actions by chimpanzees (*Pan troglodytes*). *Journal of Comparative Psychology*, **112**, 270–281.

Whiten, A. (1999) The evolution of deep social mind in humans. In *The Descent of Mind*, ed. M. Corballis and S. Lea. Oxford, UK: Oxford University Press, pp. 155–175.

Whiten, A. (2000) Primate culture and social learning. *Cognitive Science*, **24**, 477–508.

Whiten, A., Custance, D. M., Gomez, J. C., Texidor, P. and Bard, K. A. (1996) Imitative learning of artificial fruit processing in children (*Homo sapiens*) and chimpanzees (*Pan troglodytes*). *Journal of Comparative Psychology*, **110**, 3–14.

Whiten, A., Goodall, J., McGrew, W. C., *et al.* (1999) Culture in chimpanzees. *Nature*, **399**, 682–685.

Wich, S. A., Singleton, I., Utami-Atmoko, S. S., *et al.* (2003) The status of the Sumatran orang-utan *Pongo abelii*: an update. *Oryx*, **37**, 49–54.

Wilkinson, G. S. (1984) Reciprocal food sharing in the vampire bat. *Nature*, **308**, 181–184.

Wilkinson, G. S. (1990) Food sharing in vampire bats. *Scientific American*, **2622**, 76–82.

Willatts, P. (1999) Development of means–end behavior in young infants: pulling a support to retrieve a distant object. *Developmental Psychology*, **35**, 651–667.

Williams, J. L., Friend, T. H., Toscano, M. J., *et al.* (2002) The effects of early training sessions on the reactions of foals at 1, 2, and 3 months of age. *Applied Animal Behaviour Science*, **77**, 105–114.

Wilson, E. O. (1971) *The Insect Societies*. Cambridge, MA: The Belknap Press of Harvard University Press.

Wilson, E. O. (1975) *Sociobiology: The New Synthesis*. Cambridge, MA: The Belknap Press of Harvard University Press.

Wilson, E. O. (1987) Kin recognition: an introductory synopsis. In *Kin Recognition in Animals*, ed. D. J. C. Fletcher and C. D. Michener. New York: John Wiley, pp. 7–18.

Wilson, E. O. (1998) *Consilience: The Unity of Knowledge*. New York: Knopf.

Wilson, E. O. and Hölldobler, B. (2005) Eusociality: origin and consequences. *Proceedings of the National Academy of Sciences of the USA*, **102**, 13367–13371.

Wilsson, L. (1971) Observations and experiments on the ethology of the European beaver (*Castor fiber* L.). *Viltrevy (Swedish Wildlife Research)*, **8**, 115–266.

Wiltschko, R. and Wiltschko, W. (2003) Avian navigation: from historical to modern concepts. *Animal Behaviour*, **65**, 257–272.

Wimmer, H. and Perner, J. (1983) Beliefs about beliefs: representation and constraining function of wrong beliefs in young children's understanding of deception. *Cognition*, **13**, 103–128.

Witte, K., Hirschler, U. and Curio, E. (2000) Sexual imprinting on a novel adornment influences mate preferences in the Javanese mannikin. *Ethology*, **106**, 349–363.

Witthöft, W. (1967) Absolute Anzahl und Verteilung der Zellen im Hirn der Honigbiene. *Zeitschrift für Morphologie und Ökologie der Tiere*, **61**, 160–184.

Wolf, J. B. (1936) Effectiveness of token rewards for chimpanzee. *Comparative Psychological Monographs*, **5**, 1–72.

Wong, S. (1999) Development and behaviour of hatchlings of the Australian brush-turkey *Alectura lathami*. Ph. D. thesis, Griffith University, Brisbane, Australia.

Wood, S., Moriarty, K. M., Gardner, B. T. and Gardner, R. A. (1980) Object permanence in child and chimpanzee. *Animal Learning and Behaviour*, **8**, 3–9.

Woodruff, G. and Premack, D. (1979) Intentional communication on the chimpanzee: the development of deception. *Cognition*, **7**, 333–362.

Woodruff, G. and Premack, D. (1981) Primitive mathematical concepts in the chimpanzee: proportionality and numerosity. *Nature*, **293**, 568–570.

Woolfenden, G. E. and Fitzpatrick, J. W. (1984) *The Florida Scrub Jay: Demography of a Cooperative Breeding Bird*. Princeton, NJ: Princeton University Press.

Wozniak, R. H. (1997) *Behaviourism: The Early Years*. Available online: www.brynmawr.edu/Acads/Psych/rwozniak/behaviorism.html/

Wrangham, R. W., McGrew, W. C., de Waal, F. B. M. and Heltne, P. G. (eds). (1994) *Chimpanzee Cultures*. Cambridge, MA: Harvard University Press.

Wright, J. P., Jones, C. G. and Flecker, A. S. (2002) An ecosystem engineer, the beaver, increases species richness at the landscape scale. *Oecologia*, **132**, 96–101.

Wright, T. F. and Wilkinson, G. S. (2001) Population genetic structure and vocal dialects in an Amazon parrot. *Proceedings of the Royal Society of London B*, **268**, 609–616.

Wright, T. F., Rodriguez, A. and Fleischer, R. C. (2005) Vocal dialects, sex-biased dispersal and microsatellite population structure in the parrot *Amazona auropalliata*. *Molecular Ecology*, **14**, 1197–1205.

Wyczoikowska, A. (1913) Theoretical and experimental studies in the mechanism of speech. *Psychological Review*, **20**, 448–458.

Wyles, J. S., Kunkel, J. G. and Wilson, A. C. (1983) Birds, behavior and anatomical evolution. *Proceedings of the National Academy of Sciences of the USA*, **80**, 4394–4397.

Wynn, K. (1992) Addition and subtraction by human infants. *Nature*, **358**, 749–750.

Wynn-Edwards, V. C. (1962) *Animal Dispersion in Relation to Social Behaviour*. Edinburgh, UK: Oliver & Boyd.

Xia, L., Siemann, M. and Delius, J. D. (2000) Matching of numerical symbols with number of responses by pigeons. *Animal Cognition*, **3**, 35–43.

Yaglom, A. M. and Yaglom, I. M. (1967) *Challenging Mathematical Problems with Elementary Solutions*. San Francisco, CA: Holden-Day.

Yeagley, H. L. (1947) A preliminary study of a physical basis of bird navigation. *Journal of Applied Physiology*, **22**, 746–760.

Yerkes, R. M. (1916a) Provision for the study of monkeys and apes. *Science*, **43**, 231–234.

Yerkes, R. M. (1916b) *The Mental Life of Monkeys and Apes: A Study of Ideational Behavior*. Baltimore, MD: Behavior Monographs. Reissued (1979) New York: Delmar.

Yerkes, R. M. (1943) *Chimpanzees: A Laboratory Colony*. New York: Johnson Reprint Corporation.

Yerkes, R. M. and Morgulis, S. (1909) The method of Pavlov in animal psychology. *Psychological Bulletin*, **6**, 257–273.

Yerkes, R. M. and Yerkes, A. W. (1929) *The Great Apes: A Study of Anthropoid Life*. New Haven, CT: Yale University Press.

Young, J. Z. (1960) Regularities in the retina and optic lobes of octopus in relation to form discrimination. *Nature*, **186**, 836–839.

Young, J. Z. (1961) Learning and discrimination in the octopus. *Biological Reviews*, **36**, 32–96.

Young, M. E. and Wasserman, E. A. (1997) Entropy detection by pigeons: response to mixed visual displays after same–different discrimination training. *Journal of Experimental Psychology: Animal Behavior Processes*, **23**, 157–170.

Young, R. M. (1970) *Mind, Brain, and Adaptation in the Nineteenth Century: Cerebral Localization and its Biological Context from Gall to Ferrier*. Oxford, UK: Clarendon Press. Reprinted (1990) Oxford University Press.

Yurk, H., Barrett-Lennard, L. G., Ford, J. K. B. and Matkin, C. O. (2002) Cultural transmission within maternal lineages: vocal clans in resident killer whales in Southern Alaska. *Animal Behaviour*, **63**, 1103–1119.

Zajonc, R. B. (1965) Social facilitation. *Science*, **149**, 269–274.

Zanin, A. V., Markov, V. I. and Sidorova, I. E. (1990) The ability of bottlenose dolphins, *Tursiops truncatus*, to report arbitrary information. In *Sensory Abilities of Cetaceans: Laboratory and Field Evidence*, ed. J. A. Thomas and R. A. Kastelein. New York: Plenum Press, pp. 685–697.

Zazzo, R. (1979) Des enfants, des signes et des chiens devant la mirror. *Revue de Psychologie Appliquée*, **29**, 235–246.

Zentall, T. R. (1996) An analysis of imitative learning in animals. In *Social Learning in Animals: The Roots of Culture*, ed. C. M. Heyes and B. G. Galef, Jr. San Diego, CA: Academic Press, pp. 221–243.

Zentall, T. R. (2003) Imitation in animals: how do they do it? *Current Directions in Psychological Science*, **12**, 91–95.

Zentall, T. R. and Akins, C. (2001) Imitation in animals: evidence, function and mechanisms. In *Avian Visual Cognition*, ed. R. G. Cook. Available online: www.pigeon.psy.tufts.edu/avc/zentall/

Zentall, T. R. and Levine, J. M. (1972) Observational learning and social facilitation in the rat. *Science*, **178**, 1220–1221.

Zentall, T. R. and Riley, D. A. (2000) Selective attention in animal discrimination learning. *Journal of General Psychology*, **127**, 45–66.

Zentall, T. R., Steirn, J. N. and Jackson-Smith, P. (1990) Memory strategies in pigeons' performance of a radial-arm-maze analog task. *Journal of Experimental Psychology: Animal Behavior Processes*, **16**, 358–371.

Zorina, Z. A. (1997) Reasoning in birds. *Physiology and General Biology Reviews*, **11**, 3–47.

Zuberbühler, K. (2000) Referential labelling in Diana monkeys. *Animal Behaviour*, **59**, 917–927.

Zuberbühler, K. (2002) A syntactic rule in forest monkey communication. *Animal Behaviour*, **63**, 293–299.

Zuberbühler, K., Cheney, D. L. and Seyfarth, R. M. (1999) Conceptual semantics in a nonhuman primate. *Journal of Comparative Psychology*, **113**, 33–42.

Zucca, P., Antonelli, F. and Vallortigara, G. (2005) Detour behaviour in three species of birds: quails (*Coturnix* sp.), herring gulls (*Larus cachinnans*) and canaries (*Serinus canaria*). *Animal Cognition*, **8**, 122–128.

Zuckerman, S. (1932) *The Social Life of Monkeys and Apes*. London: Kegan Paul.

Zurowski, W. (1992) Building activity of beavers. *Acta Theriologica*, **37**, 403–411.

Index